Springer Series in the Data Sciences

Springer Series in the Data Sciences focuses primarily on monographs and graduate level textbooks. The target audience includes students and researchers working in and across the fields of mathematics, theoretical computer science, and statistics. Data Analysis and Interpretation is a broad field encompassing some of the fastest-growing subjects in interdisciplinary statistics, mathematics and computer science. It encompasses a process of inspecting, cleaning, transforming, and modeling data with the goal of discovering useful information, suggesting conclusions, and supporting decision making. Data analysis has multiple facets and approaches, including diverse techniques under a variety of names, in different business, science, and social science domains. Springer Series in the Data Sciences addresses the needs of a broad spectrum of scientists and students who are utilizing quantitative methods in their daily research. The series is broad but structured, including topics within all core areas of the data sciences. The breadth of the series reflects the variation of scholarly projects currently underway in the field of machine learning.

More information about this series at http://www.springer.com/series/13852

Yoni Nazarathy • Hayden Klok

Statistics with Julia

Fundamentals for Data Science, Machine Learning and Artificial Intelligence

Springer

Yoni Nazarathy
School of Mathematics and Physics
The University of Queensland
St Lucia, QLD, Australia

Hayden Klok
UQ Business School
The University of Queensland
St Lucia, QLD, Australia

ISSN 2365-5674 ISSN 2365-5682 (electronic)
Springer Series in the Data Sciences
ISBN 978-3-030-70903-7 ISBN 978-3-030-70901-3 (eBook)
https://doi.org/10.1007/978-3-030-70901-3

Mathematics Subject Classification: 62-07, 62B15, 62C10, 62E15, 62F03, 62F05, 62F10, 62F12, 62F15, 62F25, 62G07, 62H10, 62H25, 62J05, 62J07, 62J10, 62J12, 62M10, 65C05, 65C10, 65C40, 68N15, 68T01, 68T05, 68U10, 68U15, 90C39, 90C40, 62M45

This Springer imprint is published by the registered company Springer Nature Switzerland AG
The registered company address is: Gewerbestrasse 11, 6330 Cham, Switzerland

To my mother, Julianna Forbes,

Hayden Klok.

To my parents, Lea and Moshe,

Yoni Nazarathy.

Preface

The journey of this book began at the end of 2016 when preparing a statistics course for The University of Queensland. At the time, the Julia language was already showing itself as a powerful new and applicable tool, even though it was only at version 0.5. For this reason, we chose Julia for use in the course. By exposing students to statistics with Julia early on, they would be able to employ Julia for data science, numerical computation, and machine-learning tasks later in their careers. This choice was not without some resistance from students and colleagues, since back then, as is still now in 2021, in terms of volume, the R-language dominates the world of statistics, in the same way that Python dominates the world of machine learning. So why Julia?

There were three main reasons: performance, simplicity, and flexibility. Julia is quickly becoming a major contending language in the world of data science, statistics, machine learning, artificial intelligence, and general scientific computing. It is easy to use like R, Python, and MATLAB, but due to its type system and just-in-time compilation, it performs computations much more efficiently. This enables it to be fast, not just in terms of runtime, but also in terms of development time. In addition, there are many different Julia packages. These include advanced methods for the data scientist, statistician, or machine-learning practitioner. Hence the language and ecosystem has a broad scope of application.

Our goal in writing this book was to create a resource for understanding the fundamental concepts of statistics needed for mastering machine learning, data science, and artificial intelligence. This is with a view of introducing the reader to Julia as a computational tool. The book also aims to serve as a reference for the data scientist, machine-learning practitioner, bio-statistician, finance professional, or engineer, who has either studied statistics before, or wishes to fill gaps in their understanding. In today's world, such students, professionals, or researchers often use advanced methods and techniques. However, one is often required to take a step back and explore or revisit fundamental concepts. Revisiting these concepts with the aid of a programming language such as Julia immediately makes the concepts concrete.

Now, 5 years since we embarked on this book writing journey, Julia has matured beyond v1.0, and the book has matured along with it. Julia can be easily deployed by anyone who wishes to use it. However, currently many of Julia's users are hard-core developers that contribute to the language's standard libraries, and to the extensive package ecosystem that surrounds it. Therefore, much of the Julia material available at present is aimed at other developers rather than end users. This is where our book comes in, as it has been written with the end user in mind.

This book is about statistics, probability, data science, machine learning, and artificial intelligence. By reading it you should be able to gain a basic understanding of the concepts that underpin these fields. However in contrast to books that focus on theory, this book is code example centric. Almost all of the concepts that we introduce are backed by illustrative code examples. Similarly almost all of the figures are generated via the code examples. The code examples have been deliberately written in a simple format, sometimes at the expense of efficiency and generality, but with the advantage of being easily readable. Each of the code examples aims to convey a specific statistical point, while covering Julia programming concepts in parallel. The code examples are reminiscent of examples that a lecturer may use in a lecture to illustrate concepts. The content of the book is written in a manner that does not assume any prior statistical knowledge, and in fact only assumes some basic programming experience and a basic understanding of mathematical notation.

As you read this book, you can also run the code examples yourself. You may experiment by modifying parameters in the code examples or making any other modification that you can think of. With the exception of a few introductory examples, most of the code examples rarely focus on the Julia language directly but are rather meant to illustrate statistical concepts. They are then followed by a brief and dense description dealing with specific Julia language issues. Nevertheless, if learning Julia is your focus, by using and experimenting with the examples you can learn the basics of Julia as well. Updated code examples can be downloaded from the book's GitHub repository:

`https://github.com/h-Klok/StatsWithJuliaBook`

Further, an erratum, and supporting material can be found in the book's website:

`https://statisticswithjulia.org/`

The book contains a total of 10 chapters. The content may be read sequentially, or accessed in an ad hoc manner. The structure of the individual chapters is as follows:

Chapter 1 is an introduction to Julia, including its setup, package manager, and a list of the main packages used in the book. The reader is introduced to some basic Julia syntax, and programmatic structure through code examples that aim to illustrate some of the language's basic features. As it is central to the book, basics of random number generation are also introduced. Further, examples dealing with integration with other languages including R and Python are presented.

Chapter 2 explores basic probability, with a focus on events, outcomes, independence, and conditional probability concepts. Several typical probability examples are presented, along with exploratory simulation code.

Chapter 3 explores random variables and probability distributions, with a focus on the use of Julia's `Distributions` package. Discrete, continuous, univariate, and multi-variate probability distributions are introduced and explored as an insightful and pedagogical task. This is done through both simulation and explicit analysis, along with the plotting of associated functions of distributions, such as the PMF, PDF, CDF, and quantiles.

Chapter 4 momentarily departs from probabilistic notions to focus on data processing, data summary, and data visualizations. The concept of the `DataFrame` is introduced as a mechanism for storing heterogeneous data types with the possibility of missing values. Data frames play an integral component of data science and statistics in Julia, just as they do in R and Python. A summary of classic descriptive statistics and their application in Julia is also introduced. This is augmented by the inclusion of concepts such as kernel density estimation and the empirical cumulative distribution function. The chapter closes with some basic functionality for working with files.

Chapter 5 introduces general statistical inference ideas. The sampling distributions of the sample mean and sample variance are presented through simulation and analytic examples, illustrating the central limit theorem and related results. Then general concepts of statistical estimation are explored, including basic examples of the method of moments and maximum likelihood estimation, followed by simple confidence bounds. Basic notions of statistical hypothesis testing are introduced, and finally the chapter is closed by touching basic ideas of Bayesian statistics.

Chapter 6 covers a variety of practical confidence intervals for both one and two samples. The chapter starts with standard confidence intervals for means, and then progresses to the more modern bootstrap method and prediction intervals. The chapter also serves as an entry point for investigating the effects of model assumptions on inference.

Chapter 7 focuses on hypothesis testing. The chapter begins with standard T-tests for population means, and then covers hypothesis tests for the comparison of two means. Then, Analysis of Variance (ANOVA) is covered, along with hypothesis tests for checking independence and goodness of fit. The reader is then introduced to power curves.

Chapter 8 covers least squares, statistical linear regression models, generalized linear models, and a touch of time series. It begins by covering least squares and then moves onto the linear regression statistical model, including hypothesis tests and confidence bands. Additional concepts of regression are also explored. These include assumption checking, model selection, interactions, LASSO, and more. Generalized linear models are introduced and an introduction to time-series analysis is also presented.

Chapter 9 provides a broad overview of machine-learning concepts. The concepts presented include supervised learning, unsupervised learning, reinforcement learning, and generative adversarial networks. In a sprint of presenting methods for dealing with such problems, examples illustrate multiple machine-learning methods including random forests, support vector machines, clustering, principal component analysis, deep learning, and more.

Chapter 10 moves on to dynamic stochastic models in applied probability, giving the reader an indication of the strength of stochastic modeling and Monte Carlo simulation. It focuses on dynamic systems, where Markov chains, discrete event simulation, and reliability analysis are explored, along with several aspects dealing with random number generation. It also includes examples of basic epidemic modeling.

In addition to the core material, the book also contains 3 appendices.

Appendix A contains a list of many useful items detailing "how to perform ... in Julia", where the reader is directed to specific code examples that deal directly with these items.

Appendix B lists additional language features of the Julia language that were not used by the code examples in this book.

Appendix C lists additional Julia packages related to statistics, machine learning, data science, and artificial intelligence that were not used in this book.

Whether you are an industry professional, a student, an educator, a researcher, or an enthusiast, we hope that you find this book useful. Use it to expand your knowledge in fundamentals of statistics with a view towards machine learning, artificial intelligence, and data science. We further hope that the integration of Julia code and the content that we present help you quickly apply Julia for such purposes.

We would like to thank many colleagues, family members, and friends for their feedback, comments, suggestions, and support. These include Ali Araghian, Sergio Bacelar, Milan Bouchet-Valat, Heidi Dixon, Jaco Du Plessis, Vaughan Evans, Julianna Forbes, Liam Hodgkinson, Bogumił Kamiński, Dirk Kroese, Benoit Liquet, Ruth Luscombe, Geoff McLachlan, Sarat Moka, Moshe Nazarathy, Jack Ort, Miriam Redding, Robert Salomone, Vincent Tam, James Tanton, Ziqing Yan, and others. In particular, we thank Vektor Dewanto, Alex Stenlake, Joshua McDonald, and Jason Whyte for detailed feedback, and for catching dozens of typos and errors. We also thank many students who have engaged with early versions of the book, helping us improve the content. We also thank multiple GitHub users who have opened issues dealing with the book's source code. We also thank Joe Grotowski and Matt Davis from The University of Queensland for additional help dealing with the publishing process. Yoni Nazarathy would also like to acknowledge the Australian Research Council (ARC) for supporting part of this work via Discovery Project grant DP180101602.

Brisbane, Australia Yoni Nazarathy
Brisbane, Australia Hayden Klok

Contents

Chapter 1

Introducing Julia

Programming goes hand in hand with mathematics, statistics, data science, and many other fields. Scientists, engineers, data scientists, and statisticians often need to automate computation that would otherwise take too long or be infeasible to carry out. This is for the purpose of prediction, planning, analysis, design, control, visualization, or as an aid for theoretical research. Often, general programming languages such as Fortran, C/C++, Java, Swift, C#, Go, JavaScript, or Python are used. In other cases, more mathematical/statistical programming languages such as Mathematica, MATLAB/Octave, R, or Maple are employed. The process typically involves analyzing a problem at hand, writing code, analyzing behavior and output, re-factoring, iterating, and improving the model. At the end of the day, a critical component is speed, specifically the speed it takes to reach a solution—whatever it may be.

When trying to quantify speed, the answer is not always simple. On one hand, speed can be quantified in terms of how fast a piece of computer code runs, namely, *runtime speed*. On the other hand, speed can be quantified in terms of how long it takes to code, debug, and re-factor a solution, namely, *development speed*. Within the realm of *scientific computing* and *statistical computing*, compiled low-level languages such as Fortran or C/C++ generally yield fast runtime performance, however require more care in creation of the code. Hence they are generally fast in terms of runtime, yet slow in terms of development time. On the opposite side of the spectrum are mathematically specialized languages such as Mathematica, R, MATLAB, as well as Python. These typically allow for more flexibility when creating code, and hence generally yield quicker development times. However, runtimes are typically significantly slower than what can be achieved with a low-level language. In fact, many of the efficient statistical and scientific computing packages incorporated in these languages are written in low-level languages, such as Fortran or C/C++, which allow for faster runtimes when applied as closed modules.

A practitioner wanting to use a computer for statistical and mathematical analysis often faces a trade-off between runtime and development time. While both development and runtime speed are hard to fully and fairly quantify, Figure 1.1 illustrates an indicative schematic for general speed trade-offs between languages. As is postulated in this figure, there is a type of a *Pareto-optimal frontier* ranging from the C language on one end to languages like R on the other. The location of each language on this figure is indicative only. However, few would disagree that "R is generally faster to code than C" and "C generally runs faster than R". So, what about Julia?

Y. Nazarathy and H. Klok, *Statistics with Julia*, Springer Series in the Data Sciences,
https://doi.org/10.1007/978-3-030-70901-3_1

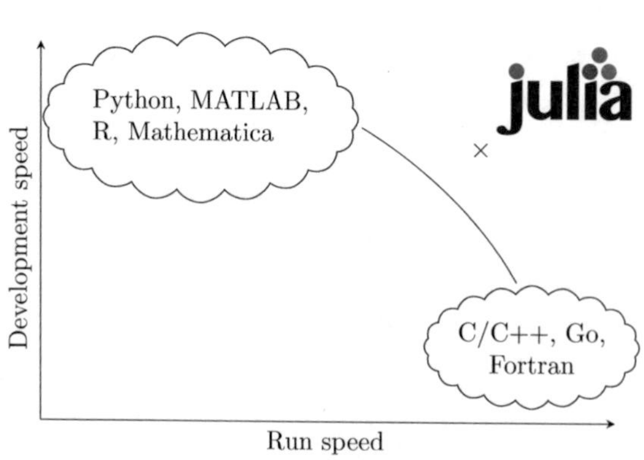

Figure 1.1: A schematic of run speed versus development speed.
Observe the Pareto-optimal frontier existing prior to Julia.

The *Julia language and framework* developed in the last decade makes use of a variety of advances in compilation, computer languages, scientific computation, and performance optimization. It is a language designed with a view of improving on the Pareto-optimal frontier depicted in Figure 1.1. With syntax, style, and feel somewhat similar to R, Python, and MATLAB/Octave, and with performance comparable to that of C/C++ and Fortran, Julia attempts to break the so-called *two-language problem*. That is, it is postulated that practitioners may quickly create code in Julia, which also runs quickly. Further, re-factoring, improving, iterating, and optimizing code can be done in Julia, and does not require the code to be ported to C/C++ or Fortran. In contrast to Python, R, and other high-level languages, the Julia standard libraries, and almost all of the Julia code base is written in Julia.

In addition to development speed and runtime speed, there is another import aspect—*learning speed*. In this context, we focus on learning how to use Julia along with the process of learning and/or strengthening knowledge of statistics, machine learning, data science, and artificial intelligence. In this respect, with the exception of some minor discussions in Section 1.1, "runtime speed and performance" is seldom mentioned in this book, as it is axiomatically obtained by using Julia. Similarly, coding and complex project development speed is not our focus. Again, the fact that Julia feels like a high-level language, very similar to Python, immediately suggests it is practical to code complex projects quickly in the language. Our focus is on learning quickly.

By following the code examples in this book (there are over 200), we allow you to learn how to use the basics of Julia quickly and efficiently. In the same go, we believe that this book will strengthen or build your understanding of probability, statistics, machine learning, data science, and artificial intelligence. In fact, the book contains a self-contained overview of these fields, taking the reader through a tour of many concepts, illustrated via Julia code examples. Even if you are a seasoned statistician, data scientist, machine learner, or probabilist, we are confident that you will find some of our discussions and examples interesting and gain further insights on problem domains that you studied previously.

Question: *Do I need to have any statistics or probability knowledge to read this book?*
Answer: Prior statistics or probability knowledge is not required. Hence, this book is a self-contained guide for the core principles of probability, statistics, machine learning, data science, and artificial intelligence. It is ideally suited for engineers, data scientists, or science professionals, wishing to

strengthen their core probability, statistics, and data-science knowledge while exploring the Julia language. However, general mathematical notation and results including basics from linear algebra, calculus, and discrete mathematics are used. So readers that have not had basic (university entry level) mathematical exposure may find some of the content and notation challenging.

Question: *What experience in programming is needed in order to use this book?*
Answer: While this book is not an introductory programming book, it does not assume that the reader is a professional software developer. Any reader that has coded in some other language at a basic level will be able to follow the code examples and their detailed descriptions.

Question: *How to read the book?*
Answer: You may either read the book sequentially, or explore the various concepts and code examples in an ad hoc manner. This book is code example centric. The code examples are the backbone of the story with each example illustrating a statistical concept together with the text, figures, and formulas that surround it. In either case, feel free to use the code repository on GitHub:

`https://github.com/h-Klok/StatsWithJuliaBook`

While exploring certain code examples you can try to modify the code to experiment with various aspects of the statistical phenomena being presented. You may often modify numerical parameters and see what effect your modification has on the output. For ad hoc Julia help, you may also use Appendix A, "How-to in Julia". It directs you to individual code listings that implicitly answer the "how to". We note that a few code examples may be updated in the GitHub repository due to library updates.

Question: *What are the unifying features of the code examples?*
Answer: With the exception of a few examples focusing on Julia basics, most code examples in this book are meant to illustrate statistical concepts. Each example is designed to run autonomously and to fit on less than a page. Hence the code examples are often not optimized for efficiency and modularity. Instead, the goal is always to "get the job done" in the clearest, cleanest, and simplest way possible. With the aid of the code, you will pick up Julia syntax, structure, and package usage. However, you should not treat the code as ideal scientific programming code but rather as illustrative code for presenting and exploring basic concepts.

The remainder of this chapter is structured as follows: In Section 1.1, we present a brief overview of the Julia language. In Section 1.2, we describe some options for setting up a Julia working environment presenting the REPL and Jupyter. Then in Section 1.3 we dive into Julia code examples designed to highlight basic powerful language features. We continue in Section 1.4 where we present code examples for plotting and graphics. Then in Section 1.5 we overview random number generation and the Monte Carlo method, used throughout the book. We close with Section 1.6 where we illustrate how other languages such as Python, R, and C can be easily integrated with your Julia code. If you are a newcomer to statistics, then it is possible that some of the examples covered in this first chapter are based on ideas that you have not previously touched. The purpose of the examples is to illustrate key aspects of the Julia language in this context. Hence, if you find the examples of this chapter overwhelming, feel free to advance to the next chapter where elementary probability is introduced starting with basic principles. The content then builds up from there gradually.

1.1 Language Overview

We now embark on a very quick tour of Julia. We start by overviewing language features in broad terms and continue with several basic code examples. This section is in no way a comprehensive description of the programming language and its features. Rather, it aims to overview a few selected language features and introduces minimal basics.

About Julia

Julia is first and foremost a *scientific programming language.* It is perfectly suited for statistics, machine learning, data science, as well as for light and heavy numerical computational tasks. It can also be integrated in user-level applications; however, one would not typically use it for front-end interfaces or game creation. It is an open-source language and platform, and the Julia community brings together contributors from the scientific computing, statistics, and data-science worlds. This puts the Julia language and package system in a good place for combining mainstream statistical methods with methods and trends of the scientific computing world. Coupled with programmatic simplicity similar to Python, and with speed similar to C, Julia is taking an active part of the data-science revolution. In fact, some believe it may overtake Python and R to become the primary language of data science in the future. Visit `https://julialang.org/` for more details.

We now discuss a few of the language's main features. If you are relatively new to programming, you may want to skip this discussion, and move to the subsection below which deals with a few basic commands. A key distinction between Julia and other high-level scientific computing languages is that Julia is *strongly typed.* This means that every variable or object has a distinct type that can either explicitly or implicitly be defined by the programmer. This allows the Julia system to work efficiently and integrate well with Julia's *just-in-time (JIT)* compiler. However, in contrast to low-level strongly typed languages, Julia alleviates the user from having to be "type-aware" whenever possible. In fact, many of the code examples in this book do not explicitly specify types. That is, Julia features *optional typing*, and when coupled with Julia's *multiple dispatch* and *type inference* system, Julia's JIT compilation system creates fast running code (compiled to *LLVM*), which is also very easy to program and understand.

The core Julia language imposes very little, and in fact the standard Julia libraries, and almost all of Julia `Base`, is written in Julia itself. Even primitive operations such as integer arithmetic are written in Julia. The language features a variety of additional packages, some of which are used in this book. All of these packages, including the language and system itself, are free and open source (MIT licensed). There are dozens of features of the language that can be mentioned. While it is possible, there is no need to vectorize code for performance. There is efficient support for *Unicode*, including but not limited to UTF-8. C can be called directly from Julia. There are even Lisp-like macros, and other meta-programming facilities.

Julia development started in 2009 by Jeff Bezanson, Stefan Karpinski, Viral Shah, and Alan Edelman. The language was launched in 2012 and has grown significantly since then, with the current version 1.5 as of the end of 2020. While the language and implementation are open source, the commercial company *Julia Computing* provides services and support for schools, universities, business, and enterprises that wish to use Julia.

A Few Basic Commands

Julia is a complete programming language supporting various programming paradigms including *procedural programming*, *object oriented programming*, *meta programming*, and *functional programming*. It is useful for *numerical computations*, *data processing*, *visualization*, *parallel computing*, *network input and output*, and much more.

As with any programming language you need to start somewhere. We start with an extended "Hello world". Look at the code listing below, and the output that follows. If you've programmed previously, you can probably figure out what each code line does. We've also added a few comments to this code example, using #. Read the code below, and look at the output that follows:

Listing 1.1: Hello world and perfect squares

```julia
println("There is more than one way to say hello:")

# This is an array consisting of three strings
helloArray = ["Hello","G'day","Shalom"]

for i in 1:3
    println("\t", helloArray[i], " World!")
end

println("\nThese squares are just perfect:")

# This construct is called a 'comprehension' (or 'list comprehension')
squares = [i^2 for i in 0:10]

# You can loop on elements of arrays without having to use indexing
for s in squares
    print(" ",s)
end

# The last line of every code snippet is also evaluated as output (in addition to
# any figures and printing output generated previously).
sqrt.(squares)
```

```
There is more than one way to say hello:
        Hello World!
        G'day World!
        Shalom World!

These squares are just perfect:
  0  1  4  9  16  25  36  49  64  81  100
11-element Array{Float64,1}:
  0.0
  1.0
  2.0
  3.0
  4.0
  5.0
  6.0
  7.0
  8.0
  9.0
 10.0
```

Most of the book contains code listings such as Listing 1.1. For brevity of future code examples, we generally omit comments. Instead most listings are followed by minor comments as seen below.

The `println()` function is used for strings such as `"There is...hello:"`. In line 4 we define an array consisting of 3 strings. The `for` *loop* in lines 6–8 executes three times, with the variable `i` incremented on each iteration. Line 7 is the body of the loop where `println()` is used to print several arguments. The first, `"\t"` is a tab spacing. The second is the `i`-th entry of `helloArray` (in Julia array indexing begins with index 1), and the third is an additional string. In line 10 the `"\n"` character is used within the string to signify printing a new line. In line 13, a *comprehension* is defined. It consists of the elements, $\{i^2 \; : \; i \in \{0, \ldots, 10\}\}$. We cover comprehensions further in Listing 1.2. Lines 16–18 illustrate that loops may be performed on all elements of an array. In this case, the loop changes the value of the variable `s` to another value of the array `squares` in each iteration. Note the use of the `print()` function to print without a newline. In Line 22, the last line of the code block applies the `sqrt()` function on each element of the array `squares` by using the "`.`" broadcast operator. The expression of the last line of every code block, unless terminated by a "`;`", is presented as output. In this case, it is an `11-element` array of the numbers $0, \ldots, 10$. The type of the output expression is also presented. It is `Array{Float64,1}`.

When exploring statistics and other forms of numerical computation, it is often useful to use a *comprehension* as a basic programming construct. As explained above, a typical form of a comprehension is:

```
[f(x) for x in A]
```

Here, `A` is some array, or more generally a collection of objects. Such a comprehension creates an array of elements, where each element `x` of `A` is transformed via `f(x)`. Comprehensions are ubiquitous in the code examples we present in this book. We often use them due to their expressiveness and simplicity. We now present a simple additional example:

Listing 1.2: Using a comprehension

```
array1 = [(2n+1)^2 for n in 1:5]
array2 = [sqrt(i) for i in array1]
println(typeof(1:5), "   ", typeof(array1), "   ", typeof(array2))
1:5, array1, array2
```

```
UnitRange{Int64}   Array{Int64,1}   Array{Float64,1}
(1:5, [9, 25, 49, 81, 121], [3.0, 5.0, 7.0, 9.0, 11.0])
```

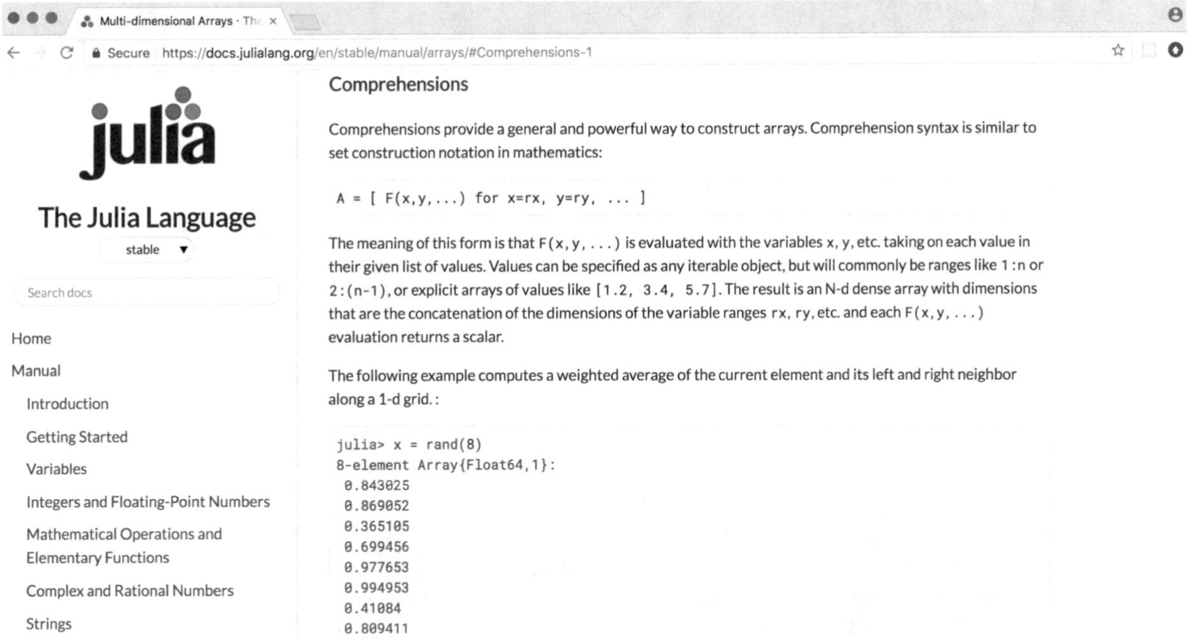

Figure 1.2: Visit `https://docs.julialang.org` for official language documentation.

The array `array1` is created in line 1 with the elements $\{(2n+1)^2 \; : \; n \in \{1,\ldots,5\}\}$ in order. Note that while mathematical sets are not ordered, comprehensions generate ordered arrays. Observe the literal 2 in the multiplication `2n`, without explicit use of the `*` symbol. In the next line, `array2` is created. An alternative would be to use `sqrt.(array1)`. In line 3, we print the `typeof()` three expressions. The type of `1:5` (used to create `array1`) is a `UnitRange` of `Int64`. It is a special type of object that encodes the integers $1,\ldots,5$ without explicitly allocating memory. Then the types of both `array1` and `array2` are `Array` types, and they contain values of types `Int64` and `Float64`, respectively. In line 4, a tuple of values is created through the use of a comma between `1:5`, `array1`, and `array2`. As it is the last line of the code, it is printed as output (when running in Jupyter as described below). Observe that in the output, the values of the second element of the tuple are printed as integers (no decimal point) while the values of the third element are printed as floating point numbers.

Getting Help

You may consult the official Julia documentation, `https://docs.julialang.org/`, for help. The documentation strikes a balance between precision and readability, see Figure 1.2.

While using Julia, help may be obtained through the use of `?`. For example try `? sqrt` and you will see output similar to Figure 1.3.

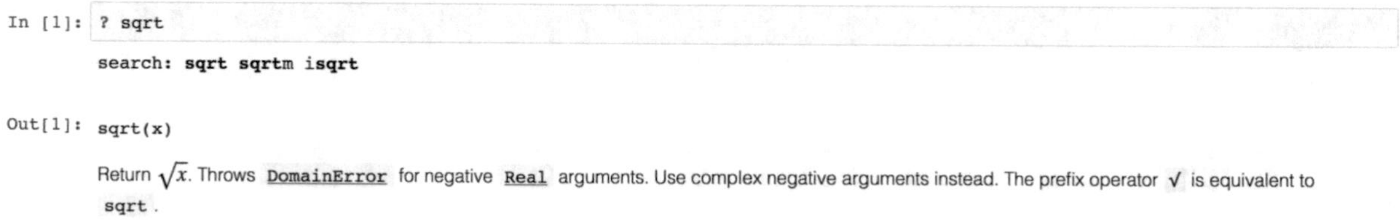
In [1]: ? sqrt
search: sqrt sqrtm isqrt
Out[1]: sqrt(x)
Return √x. Throws DomainError for negative Real arguments. Use complex negative arguments instead. The prefix operator √ is equivalent to sqrt .

Figure 1.3: Snapshot from a Julia Jupyter notebook: Keying in `? sqrt` presents help for the `sqrt()` function.

You may also find it useful to apply the `methods()` function. Try, `methods(sqrt)`. You will see output that contains lines of this sort:

```
...
sqrt(x::Float32) at math.jl:426
sqrt(x::Float64) at math.jl:425
sqrt(z::Complex{#s45} where #s45<:AbstractFloat) at complex.jl:392
sqrt(z::Complex) at complex.jl:416
sqrt(x::Real) at math.jl:434
sqrt{T<:Number}(x::AbstractArray{T,N} where N) at deprecated.jl:56
...
```

This presents different *Julia methods* that have been implemented for the function `sqrt()`. In Julia, a given function may be implemented in different ways depending on different input arguments with each different implementation being a *method*. This is called *multiple dispatch*. Here, the various methods of `sqrt()` are shown for different types of input arguments.

Runtime Speed and Performance

While Julia is fast and efficient, for most of this book we don't explicitly focus on runtime speed and performance. Rather, our aim is to help the reader learn how to use Julia while enhancing knowledge of probability, statistics, data science, and machine learning. Nevertheless, we now briefly discuss runtime speed and performance.

From a user perspective, Julia feels like an *interpreted language* as opposed to a *compiled language*. With Julia, you are not required to explicitly compile your code before it is run. However, as you use Julia, behind the scenes, the system's JIT compiler compiles every new function and code snippet as it is needed. This often means that on a first execution of a function, runtime is much slower than the second, or subsequent runs. From a user perspective, this is apparent when using other packages (as the example in Listing 1.3 illustrates, this is often done by the `using` command). On a first call (during a session) to the `using` command of a given package, you may sometimes wait a few seconds for the package to compile. However, afterwards, no such wait is needed.

For day-to-day statistics and scientific computing needs, you often don't need to give much thought to performance and run speed with Julia, since Julia is inherently fast. For instance, as we do in dozens of examples in this book, simple Monte Carlo simulations involving 10^6 random variables typically run in less than a second, and are very easy to code. However, as you progress into more complicated projects, many repetitions of the same code block may merit profiling and optimization of the code in question. Hence, you may wish to carry out basic profiling.

For basic profiling of performance the `@time` *macro* is useful. Wrapping code blocks with it (via `begin` and `end`) causes Julia to profile the performance of the block. In Listings 1.3 and 1.4, we carry out such profiling. In both listings, we populate an array, called `data`, containing 10^6 values, where each value is a mean of 500 random numbers. Hence, both listings handle half a billion numbers. However, Listing 1.3 is a much slower implementation.

Listing 1.3: Slow code example

```
using Statistics

@time begin
    data = Float64[]
    for _ in 1:10^6
        group = Float64[]
        for _ in 1:5*10^2
            push!(group,rand())
        end
        push!(data,mean(group))
    end
    println("98% of the means lie in the estimated range: ",
                (quantile(data,0.01),quantile(data,0.99)) )
end
```

```
98% of the means lie in the estimated range: (0.4699623580817418, 0.5299937027991253)
 11.587458 seconds (10.00 M allocations: 8.034 GiB, 4.69% gc time)
```

The actual output of the code gives a range, in this case approximately 0.47 to 0.53 where 98% of the sample means (averages) lie. We cover more on this type of statistical analysis in the chapters that follow.

The second line of output, generated by `@time`, states that it took about 11.6 seconds for the code to execute. There is also further information indicating how many memory allocations took place, in this case about 10 million, totaling just over 8 gigabytes (in other words, Julia writes a little bit, then clears, and repeats this process many times over). This constant read-write is what slows our processing time.

Now, look at Listing 1.4 and its output.

Listing 1.4: Fast code example

```
using Statistics

@time begin
    data = [mean(rand(5*10^2)) for _ in 1:10^6]
    println("98% of the means lie in the estimated range: ",
                    (quantile(data,0.01),quantile(data,0.99)) )
end
```

```
98% of the means lie in the estimated range: (0.469999864362845, 0.5300834606858865)
  1.705009 seconds (1.01 M allocations: 3.897 GiB, 10.76% gc time)
```

As can be seen, the output gives the same estimate for the interval containing 98% of the means. However, in terms of performance, the output of `@time` indicates that this code is clearly superior. It took about 1.7 seconds (compare with 11.6 seconds for Listing 1.3). In this case, the code is much faster because far fewer memory allocations are made. Note that "`gc time`" stands for "garbage collection" and quantifies what percentage of the running time Julia was busy with internal memory management.

Here are some comments for both code Listings 1.3 and 1.4:

In both listings we use the `Statistics` package, required for the `mean()` function. Line 4 (Listing 1.3) creates an empty array of type `Float64`, `data`. Line 6 creates an empty array, `group`. Then lines 7–9 loop 500 times, each time pushing to the array, `group`, a new random value generated from `rand()`. The `push!()` function here uses the naming convention of having an exclamation mark when the function modifies the argument. This is not part of the Julia language, but rather decorates the name of the function. In this case, it modifies `group` by appending another new element. Here is one point where the code is inefficient. The Julia compiler has no direct way of knowing how much memory to allocate for `group` initially, hence some of the calls to `push!()` imply reallocation of the array and copying. Line 10 is of a similar nature. The composition of `push!()` and `mean()` imply that the new mean (average of 500 values) is pushed into `data`. However, some of these calls to `push!()` imply a reallocation. At some point the allocated space of `data` will suddenly run out, and at this point the system will need to internally allocate new memory, and copy all values to the new location. This is a big cause of inefficiency in our example. Line 13 creates a tuple within `println()`, using `(,)`. The two elements of the tuple are return values from the `quantile()` function which computes the 0.01 and 0.99 quantiles of `data`. Quantiles are covered further in Chapter 4. The lines of Listing 1.4 are relatively simpler and in this case performance is better. All of the computation is carried out in the comprehension in Line 4, within the square brackets `[]`. Writing the code in this way allows the Julia compiler to pre-allocate 10^6 memory spaces for `data`. Then, applying `rand()` with an argument of `5*10^2`, indicating the number of desired random values, allows for faster operation. The functionality of `rand()` is covered in Section 1.5.

Julia is inherently fast, even if you don't give it much thought as a programmer. However, in order to create truly optimized code, one needs to understand the inner workings of the system a bit better. There are some general guidelines that you may follow. A key is to think about memory usage and allocation as in the examples above. Other issues involve allowing Julia to carry out type inference efficiently. Nevertheless, for simplicity, the majority of the code examples of this book ignore types as much as possible and don't focus on performance.

Types and Multiple Dispatch

Functions in Julia are invoked via *multiple dispatch.* This means the way a function is executed, i.e. its *method* is based on the *type* of its inputs, i.e. its *argument* types. Indeed functions can have multiple methods of execution, which can be checked using the `methods()` command.

Julia has a powerful type system which allows for *user-defined types.* One can check the type of a variable using the `typeof()` function, while the functions `subtypes()` and `supertype()` return the *subtypes* and *supertype* of a particular type, respectively. As an example, `Bool` is a subtype of `Integer`, while `Real` is the supertype of `Integer`. This is illustrated in Figure 1.4, which shows the type hierarchy of numbers in Julia.

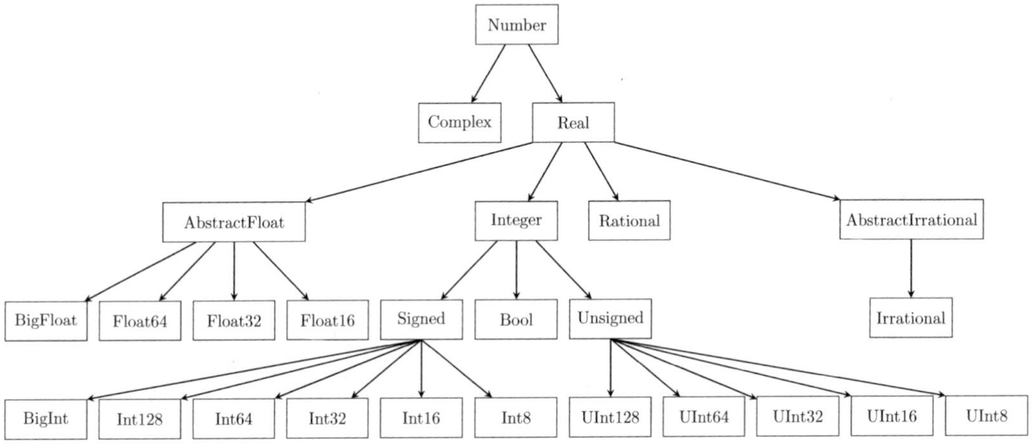

Figure 1.4: Type hierarchy for Julia numbers.

One aspect of Julia is that if the user does not specify all variable types in a given piece of code, Julia will attempt to infer what types the unspecified variables should be, and will then attempt to execute the code using these types. This is known as *type inference*, and relies on a type inference algorithm. This makes Julia somewhat forgiving when it comes to those new to coding, and also allows one to quickly mock-up fast working code. It should be noted, however, that if one wants the fastest possible code, then it is good to specify the types involved. This also helps to prevent *type instability* during code execution. We don't focus on this further in the book, but suggest you consult the documentation for the macro `@code_warntype` as an entry point for such profiling.

Variable Scope, Local Variables, and Global Variables

The *scope of a variable* is the region of code in which the variable is visible. Like almost any other programming language, Julia has rules about variable scope implying that not all variables can be accessed from everywhere within the program. Such restrictions serve several purposes including driving the programmer to create clear readable code, allowing the compiler to optimize execution, supporting concurrent operation, and reducing the possibility of name clashes.

When discussing scope, a key distinction lies between *local variables* and *global variables*. The former refers to variables defined within a `function` definition or a block of code such as a `for` loop or `while` loop. The latter refers to variables that can potentially be accessed from anywhere in the program. A detailed description of variable scope rules is in the Julia documentation, see `https://docs.julialang.org/en/v1/manual/variables-and-scoping/`. Here we only focus on a few key issues that are relevant to many of our code examples.

In general, a global variable is a variable defined outside of a function or another block of code. Since global variables have a much less restricted domain than local variables, it is typically good programming practice to minimize or even eliminate their use. This is especially true for programs that span more than a single file and/or multiple lines of code (hundreds or thousands). Some Julia mechanisms that allow you to organize variables include *structures* using the `struct` keyword and *modules* using the `module` keyword. However our code examples aim to be short self-contained scripts and don't explicitly try to encapsulate data. We don't explicitly use structures and modules and we define functions only if these are critically needed. Our aim is for simple code that fits on a minimal footprint. As a consequence our code examples often use global variables. If you wish to integrate parts of our code examples in larger projects, then it is good practice to eliminate or significantly reduce the use of globals as you carry out such integration. In less trivial coding situations, you should aim to hardly use global variables and the `global` keyword. Still, for the purposes of illustrative code snippets, our heavy use of globals is justified.

As a minimal introduction to variable scope we present Listing 1.5. One of the main goals of this listing is to show the use of the `global` keyword which is prevalent in many of the code listings that follow. You may wish to skip reading the details of this listing and then refer back to it when thinking and considering variable scope. The important point is simply to observe that the `global` keyword is sometimes needed and is thus present in quite a few of the examples that follow.

The key aspect of Listing 1.5 is the execution of a `for` loop in global scope followed by a similar loop wrapped within a function. The global variables `data`, `s`, `beta`, and `gamma` are potentially visible in all parts of the code, including in the `for` loop of lines 5–12 and the function `sumData()`. However, in certain cases, the `global` keyword needs to be used to mark the variable as "coming from" global scope. This is the case for the variable `s` that is marked as global in line 7. Interestingly, when using a Jupyter notebook for such code (Jupyter notebooks are described in the next section), such usage of the `global` keyword is not needed. The listing presents multiple other aspects of variable scope. We detail some of these aspects in the code comments that follow. A full description of variable scope rules is in the Julia documentation.

Listing 1.5: Variable scope and the `global` keyword

```
1   data = [1,2,3]
2   s = 0
3   beta, gamma = 2, 1
4
5   for i in 1:length(data)
6       print(i," ")
7       global s     #This usage of the 'global' keyword is not needed in Jupyter
8                    #But elsewhere without it:
9                    #ERROR: LoadError: UndefVarError: s not defined
10      s += beta*data[i]
11      data[i] *= -1
12  end
13  # print(i)          #Would cause ERROR: LoadError: UndefVarError: i not defined
14  println("\nSum of data in external scope: ", s)
15
16  function sumData(beta)
17      s = 0            #try adding the prefix global
18      for i in 1:length(data)
19          s += data[i] + gamma
20      end
21      return s
22  end
23  println("Sum of data in a function: ", sumData(beta/2))
24  @show s
```

```
1 2 3
Sum of data in external scope: 12
Sum of data in a function: -3
s = 12
```

In line 1 we define an array, `data`. It is a variable defined in global scope and is hence a global variable. Similarly for the variables `s`, `beta`, and `gamma` in lines 2–3. Lines 5–12 loop over the range `1:length(data)` where in each iteration the variable `i` takes the next value. The scope of the variable `i` is within the block of the for loop (lines 5–12). Note that if you were to uncomment line 13, the attempt to access `i` at that line would cause an error. Because the for loop is not inside a function, it defines a new local scope. This means that accessing the global variable `s` for modification requires an explicit declaration with the `global` keyword as is done in line 7. For code in Jupyter notebooks this can be avoided but otherwise not. Notice however that `global` declarations are not always needed. For example the global variable `beta` is used in line 10, but as it isn't modified there is no need to declare it as global with `global`. The variable (array) `data` also doesn't need to be declared even though the contents of the array is modified in line 11. In lines 16–22 we define the function `sumData()`. Here the name of the function argument is `beta` and is not the global variable `beta`. Hence when the function is called in line 23, we can pass any argument to it for the local `beta`. In this case the argument is half of the global `beta`, i.e. a value of 1. Note that we define a local variable `s` in line 17. It is a different variable from the global `s` defined in line 2. If we were to add a `global` keyword in line 17 then it would be the global variable `s`. You can try doing that and see how the `@show` macro in line 24 that displays the value of the global `s` would change. Note again that the global variables `data` and `gamma` are used inside the body of the function for read-only purposes.

1.2 Setup and Interface

There are multiple ways to run Julia including the *REPL command line interface*, *Jupyter notebooks*, the *Juno IDE* (Integrated Development Environment) on the *Atom* editor, as well as several other working environments including *VSCode* via *Julia for VSCode*, and *Pluto*, both of which are quickly gaining popularity. Here we focus on the REPL and Jupyter notebooks as these are the most mature environments to date. We also mention that in developing the code examples, we used both Jupyter notebooks and the Juno IDE. The latter is also available directly from Julia computing and is packaged as *Julia Pro*.

No matter how you run Julia, there is an instance of a Julia *kernel* running. The running kernel contains all of the currently compiled Julia functions, loaded packages, defined variables, and objects. You may even run multiple kernels, sometimes in a distributed manner. We first describe the *REPL* and *Jupyter notebooks* environments. We then describe the *package manager* which allows one to extend Julia's basic functionality by installing additional packages.

REPL Command Line Interface

The *Read Evaluate Print Loop (REPL)* command line interface is a simple and straightforward way of using Julia. It can be downloaded directly from: `https://julialang.org/downloads/`. Downloading it implies downloading the Julia kernel as well.

Once installed locally, Julia can be launched and the Julia REPL will appear, within which Julia commands can be entered and executed. For example, in Figure 1.5 the code `1+2` was entered, followed by the enter key. Note that if Julia is launched as its own stand-alone application, a new Julia instance will appear. However, if you are working in a shell/command line environment, the

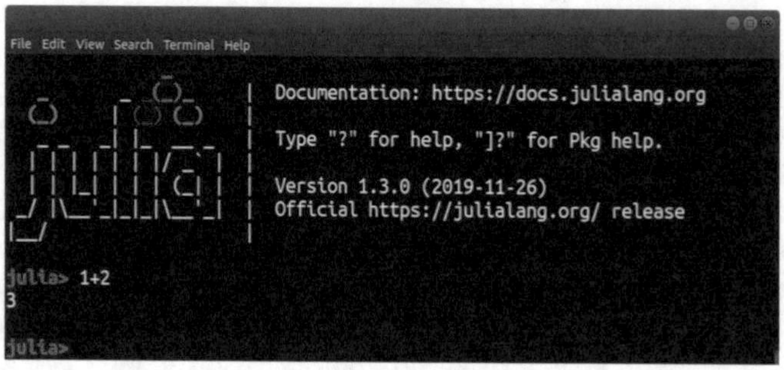

Figure 1.5: Julia's REPL interface.

REPL can also be launched from within the current environment.

When using the REPL (as well as the IDEs mentioned above), typically one will also work with Julia files which have the `.jl` extension. In fact, every code listing in this book is stored in such a file. These files are available on the book's GitHub repository.

Jupyter Notebooks

An alternative to using the REPL is to use a *Jupyter Notebook* as presented in Figure 1.6. It is a browser-based interface in which one can type and execute Julia code, as well as Python, R, and other languages. Jupyter notebooks are easy to use and allow one to combine code, output, visuals, and markdown formatting all together in one document. A Jupyter notebook is both a means of presentation and execution.

Each notebook consists of a series of cells, in which code can be typed and run. Cells can be of different type. *Code cells* allow Julia code to be entered and executed, while *markdown cells* allow for formatting of the document in *Markdown*, which is a simple formatting language that also incorporates hyperlinks, images, and *LaTeX formatting* for formulas.

Jupyter notebook files have the `.ipynb` extension. The content of notebooks can also be exported as PDF and other formats. A common way to run Jupyter for Julia is using the *Anaconda* Python distribution which installs a *Jupyter notebook server* locally. A technical note is that the `IJulia` (Julia) package is required for Julia to work within Jupyter notebooks. More on packages below. Another advantage of Jupyter notebooks is that because they are browser based, they can be configured to run over a remote connection.

The user interface for using Jupyter notebooks is easy to learn. When starting, note that there are two input modes. *Edit mode* allows code/text to be entered into a cell, while *command mode* allows keyboard-activated actions, such as toggling line numbering, copying cells, and deleting cells. Cells can be executed by first selecting the cell and then pressing `ctrl-enter` or `shift-enter`. In command mode, additional cells can be created by pressing `a` or `b` to create cells above or below, respectively.

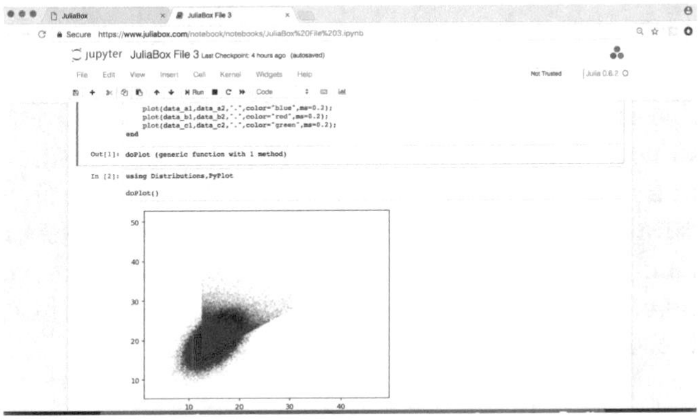

Figure 1.6: An example of a Jupyter notebook.

The Package Manager

Although Julia comes with many built-in features, the core system can be extended. This is done by installing packages, which can be added to Julia at your discretion. This allows users to customize their Julia installation depending on their needs, and at the same time offers support for developers who wish to create their own packages, enriching the Julia ecosystem. Note that packages may be either *registered*, meaning that they are part of the Julia package repository, or *unregistered*, meaning they are not. A list of currently registered packages is available at: `https://julialang.org/packages/`.

When using the REPL you can enter the *package manager mode* by typing "`]`". This mode can be exited by hitting the backspace key. In this mode, packages can be installed, updated, or removed. The following lists a few of the many useful commands available:

`] add Foo`	adds the package `Foo.jl` to the current Julia build.
`] status`	lists what packages and versions are currently installed.
`] update`	updates existing packages.
`] remove Foo`	removes package `Foo.jl` from the current Julia build.

An alternative which works both in the REPL and in Jupyter notebooks is to use functions from the `Pkg` package. The standard usage is of the form `using Pkg` followed by `Pkg.add("Foo")`. This adds the package `Foo.jl`. Similar functions exist via the `Pkg` package for other package operations.

As you study the code examples in this book, you will notice that most start with the `using` command, followed by one or more package names. This is how Julia packages are loaded into the current namespace of the kernel, so that the package's functions, objects, and types can be used. Note that the use of the `using` command does not imply installing a package. Installation of a package is a one-time operation which must be performed before the package can be used. In comparison, typing the keyword `using` is needed in every session during which package functionality is required.

Packages Used in This Book

The code in this book uses a variety of Julia packages. Some of the key packages used in the context of probability, statistics, and machine learning are `DataFrames`, `Distributions`, `Flux`, `GLM`, `Plots`, `Random`, `Statistics`, `StatsBase`, and `StatsPlots` as well as many other important packages. Some of these are built-in with the base installation, for example, `Statistics` and `Random`, while others require user installation via the package manager as described above. A short description of each of the packages that we use in the book is contained below. We have placed a "*" next to every package that is part of the basic installation.

*`Base.jl` is the basic Julia package sitting at the base of the language.

`BSON.jl` allows us to store and read data using the common Binary JSON format.

`Calculus.jl` provides tools for working with basic calculus operations including differentiation and integration both numerically and symbolically.

`CategoricalArrays.jl` provides tools for working with categorical variables.

`Clustering.jl` provides support for various clustering algorithms.

`Combinatorics.jl` allows us to enumerate combinatorics and permutations.

`CSV.jl` is a utility library for working with CSV and other delimited files.

`DataFrames.jl` is a package for working with tabular data.

`DataStructures.jl` provides support for various types of data structures.

*`Dates.jl` provides support for working with dates and times.

`DecisionTree.jl` is a package for decision trees and random forest algorithms.

`DifferentialEquations.jl` provides efficient Julia implementations of numerical solvers for various types of differential equations.

`Distributions.jl` provides support for working with probability distributions.

`Flux.jl` is a deep learning library written in pure Julia.

`GLM.jl` is a package for linear models and generalized linear models.

`HCubature.jl` is an implementation of multidimensional "h-adaptive" (numerical) integration.

`HypothesisTests.jl` implements a wide range of hypothesis tests and confidence intervals.

`HTTP.jl` provides HTTP client and server functionality.

`IJulia.jl` is required to interface Julia with Jupyter notebooks.

`Images.jl` is an image processing library.

`JSON.jl` is a package for parsing and printing JSON.

`Juno.jl` is a package needed for using the Juno development envionrment.

`KernelDensity.jl` is a kernel density estimation package.

`Lasso.jl` implements LASSO model fitting.

`LaTeXStrings.jl` makes it easier to type LaTeX equations in string literals.

`LIBSVM.jl` is a package for Support Vector Machines (SVM) using the LIBSVM library.

`LightGraphs.jl` provides support for the implementation of graphs in Julia.

*`LinearAlgebra.jl` provides linear algebra support.

`Measures.jl` allows building up and representing expressions involving differing types of units that are then evaluated, resolving them into absolute units.

`MLDatasets.jl` provides an interface for accessing common Machine Learning (ML) datasets.

`MultivariateStats.jl` is a package for multivariate statistics and data analysis, including ridge regression, PCA, dimensionality reduction and more.

`NLsolve.jl` provides methods to solve non-linear systems of equations.

`Plots.jl` is one of the main plotting packages in the Julia ecosystem. It is the main plotting package used throughout our book.

`PyCall.jl` provides the ability to directly call and fully interoperate with Python Julia.

`PyPlot.jl` provides a Julia interface to the Matplotlib plotting library from Python, and specifically to the matplotlib.pyplot module.

`QuadGK.jl` provides support for one-dimensional numerical integration using adaptive Gauss-Kronrod quadrature.

*`Random.jl` provides support for pseudo random number generation.

`RCall.jl` provides several different ways of interfacing with R from Julia.

`RDatasets.jl` provides an easy way to interface with the standard datasets that are available in the core of the R language, as well as several datasets included in R's more popular packages.

`Roots.jl` contains routines for finding roots of continuous scalar functions of a single variable.

`SpecialFunctions.jl` contains various special mathematical functions, such as Bessel, zeta, digamma, along with sine and cosine integrals, as well as others.

*`Statistics.jl` contains common statistics functions such as mean and standard deviation.

`StatsBase.jl` provides basic support for statistics including high-order moment computation, counting, ranking, covariances, sampling and cumulative distribution function estimation.

`StatsModels.jl` allows us to specify models using formulas as common in linear models.

`StatsPlots.jl` provides extensive statistical plotting recipes.

`TimeSeries.jl` provides support for working with time series data.

Many additional useful packages, not employed in our code examples, are in Appendix C.

1.3 Crash Course by Example

Almost every procedural programming language needs functions, conditional statements, loops, and arrays. Similarly, every scientific programming language needs to support plotting, matrix manipulations, and floating point calculations. Julia is no different. In this section we present several examples, and we begin to explore various basic programming elements. Each example aims to introduce another aspect of Julia. These examples are not necessarily minimal examples needed for learning the basics of Julia, nor do they build statistical foundations from the ground up. Rather, they are designed to show what can be done with Julia. Hence if you find these examples too complex from either a programming or a mathematical perspective, feel free to skip directly to Chapter 2, where basic probability is demonstrated via simple examples from the ground up.

Alternatively, if you prefer to engage with the language through more simple examples, you may wish to use other resources alongside this book. If you are a beginner to programming, we recommend the introductory book to programming with the Julia language, "Think Julia – How to Think Like a Computer Scientist" by A. Downey, B. Lauwens [DL19]. If you are a seasoned programmer and are looking for a more general-purpose text about Julia, see "Julia 1.0 Programming Cookbook" by B. Kamiński, P. Szufel [KS18]. You can also visit `https://julialang.org/learning/` for a variety of other resources.

In addition to the general Julia programming resources mentioned above, there are also several other texts that are worth considering for specific aspects of scientific computing, data science, and artificial intelligence. The book [KW19] provides an exhaustive introduction to *optimization algorithms* together with Julia code. The book [K18] focuses on *operations research* using Julia. Finally, the book [MP18] is an applied *data-science* resource, as is [V16].

We now present four selected examples which are designed to highlight a few features of Julia. The bubble sort example shows basic programming. The roots of the a polynomial example illustrates simple numerical computation. The Markov chain example shows of how to work with matrices and randomness. And finally, the text processing example shows how one can interface with the web and do basic work with strings and text.

Bubble Sort

In our first example, we construct a basic sorting algorithm using first principles. The algorithm we consider here is called *bubble sort*. This algorithm takes an input *array*, indexed $1, \dots, n$, then sorts the elements smallest to largest by allowing the larger elements, or "bubbles", to "percolate up". The algorithm is implemented in Listing 1.6. As can be seen from the code, the locations j and $j+1$ are swapped inside the two *nested loops*. This maintains an increasing (non-decreasing) order in the array. The *conditional statement* `if` is used to check if the numbers at indexes j and $j+1$ are in the wrong order, and if needed, swap them.

Listing 1.6: Bubble sort

```
1  function bubbleSort!(a)
2      n = length(a)
3      for i in 1:n-1
4          for j in 1:n-i
5              if a[j] > a[j+1]
6                  a[j], a[j+1] = a[j+1], a[j]
7              end
8          end
9      end
10     return a
11 end
12
13 data = [65, 51, 32, 12, 23, 84, 68, 1]
14 bubbleSort!(data)
```

```
8-element Array{Int64,1}:
1
12
23
32
51
65
68
84
```

In lines 1–11, we define a *function*, named `bubbleSort!()`. The input argument `a` is implicitly expected to be an array. The function sorts `a` in place, and returns a reference to the array. Note that in this case, the function name ends with "!" by convention. This exclamation mark decorates the name of the function, letting us know that the function argument, `a`, will be modified (`a` is sorted in place without memory copying). In Julia, arrays are *passed by reference.* Arrays are indexed from `1` to the length of the array, obtained by `length()`. In line 6 the elements `a[j]` and `a[j+1]` are swapped by using assignment of the form `m,n = x,y` which is syntactic shorthand for `m=x` followed by `n=y`. In line 14, the function is called on `data`. As it is the last line of the code block and is not followed by a "`;`", the expression evaluated in that line is presented as output (if running in a Jupyter notebook), in our case the sorted array. Note that it has a type `Array{Int64,1}`, meaning an array of integers. Julia inferred this type automatically. Try changing some of the values in line 13 to floating points, e.g. `[65.0, 51.0` ... (etc)`]` and see how the output changes.

Keep in mind that Julia already contains standard sorting functions such as `sort()` and `sort!()`, so you don't need to implement your own sorting function as we did. For more information on these functions use `? sort`. Also, the bubble sort algorithm is not the most efficient sorting algorithm, but is introduced here as a means of understanding Julia better. For an input array of length n, it will execute line 5 about $n^2/2$ times. For non-small n, this is much slower performance than optimal sorting algorithms where the number of comparisons can be reduced to an order of $n \log(n)$ times.

Roots of a Polynomial

We now consider a different type of programming example that comes from elementary numerical analysis. Consider the polynomial,

$$f(x) = a_n x^n + a_{n-1} x^{n-1} + \ldots + a_1 x + a_0,$$

with real-valued coefficients $a_0, \ldots, a_n$. Say we wish to find all x values that solve the equation $f(x) = 0$. We can do this numerically with Julia using the `find_zeros()` function from the `Roots` package. This general-purpose solver takes a function as input and numerically tries to find all its roots within some domain. As an example, consider the quadratic polynomial,

$$f(x) = -10x^2 + 3x + 1.$$

Ideally, we would like to supply the `find_zeros()` function with the coefficient values, −10, 3, and 1. However, `find_zeros()` is not designed for a specific polynomial, but rather for any Julia function that represents a real mathematical function. Hence one way to handle this is to define a Julia function specifically for this quadratic $f(x)$ and give it as an argument to `find_zeros()`. However, here we will take this one step further, and create a slightly more general solution. We first create a function called `polynomialGenerator` which takes a list of arguments representing the coefficients, $a_n, a_{n-1}, \ldots, a_0$ and returns the corresponding polynomial function. We then use this function as an argument to the `find_zeros()` function, which then returns the roots of the original polynomial.

Listing 1.7 shows our approach. Note that for our example it is straightforward to solve the roots analytically and verify the code. This is done using the quadratic formula as follows:

$$x = \frac{-3 \pm \sqrt{3^2 - 4(-10)}}{2(-10)} = \frac{3 \pm 7}{20} \qquad \Rightarrow \qquad x_1 = 0.5, \quad x_2 = -0.2.$$

Listing 1.7: Roots of a polynomial

```
using Roots

function polynomialGenerator(a...)
    n = length(a)-1
    poly =  function(x)
                return sum([a[i+1]*x^i for i in 0:n])
            end
    return poly
end

polynomial = polynomialGenerator(1,3,-10)
zeroVals = find_zeros(polynomial,-10,10)
println("Zeros of the function f(x): ", zeroVals)
```

```
Zeros of the function f(x): [-0.2, 0.5]
```

In line 1 we employ the `using` keyword, indicating to include elements from the package `Roots`. Note that this assumes that the package has already been added as part of the Julia configuration. Lines 3–9 define the function `polynomialGenerator()`. An argument, `a`, along with the *splat operator* `...` indicates that the function will accept comma separated parameters with an unspecified number of parameters. For our example we have three coefficients, specified in line 11. Line 4 makes use of the `length()` function to determine how many arguments were given to the function `polynomialGenerator()`. Notice that the degree of the polynomial, represented in the local variable `n`, is one less than the number of arguments. Lines 5–7 define an internal function with an input argument `x`, and then stores this function as the variable `poly`, returned from `polynomialGenerator()`. One can pass functions as arguments, and assign them to variables. The main workhorse of this function is line 6, where the `sum()` function is used to sum over an array of values. This array is implicitly defined using a *comprehension*. In this case, the comprehension is `[a[i+1]*x^i for i in 0:n]`. This creates an array of length $n+1$ where the i-th element of the array is `a[i+1]*x^i`. In line 12 the `find_zeros()` function from the `Roots` package is used to find the roots of the polynomial. The latter arguments are guesses for the roots which are used for initialization. The roots calculated are then assigned to `zeroVals` and the output printed.

Steady State of a Markov Chain

We now introduce some basic linear algebra computations and simulation through a simple *Markov chain* example. Consider a theoretical city, where the weather is described by three possible states: (1) "Fine", (2) "Cloudy", and (3) "Rain". On each day, given a certain state, there is a probability distribution for the weather state of the next day. This simplistic weather model constitutes a *discrete-time* (homogeneous) Markov chain. This Markov chain can be described by the 3×3 *transition probability matrix*, P, where the entry $P_{i,j}$ indicates the probability of transitioning to state j given that the current state is i. The transition probabilities are illustrated in Figure 1.7.

One important computable quantity for such a model is the long-term proportion of occupancy in each state. That is, in steady state, what proportion of the time is the weather in state 1, 2, or 3. Obtaining this *stationary distribution*, denoted by the vector $\pi = [\pi_1 \;\; \pi_2 \;\; \pi_3]$ (or an approximation for it), can be achieved in several ways, as shown in Listing 1.8. For pedagogical and exploratory reasons we use four methods to find the stationary distribution. Note that some of these methods involve linear algebra and/or results from the theory of Markov chains. These are not covered here, but rather discussed in Section 10.2 of Chapter 10. If you haven't been exposed to linear algebra, we suggest you only skim through this example. The four methods that we use are as follows:

1. By raising the matrix P to a high power, (repeated matrix multiplication of P with itself), the limiting distribution is obtained in any row. Mathematically,

$$\pi_i = \lim_{n\to\infty} [P^n]_{j,i} \qquad \text{for any index, } j. \tag{1.1}$$

2. We solve the (overdetermined) linear system of equations,

$$\pi P = \pi \qquad \text{and} \qquad \sum_{i=1}^{3} \pi_i = 1.$$

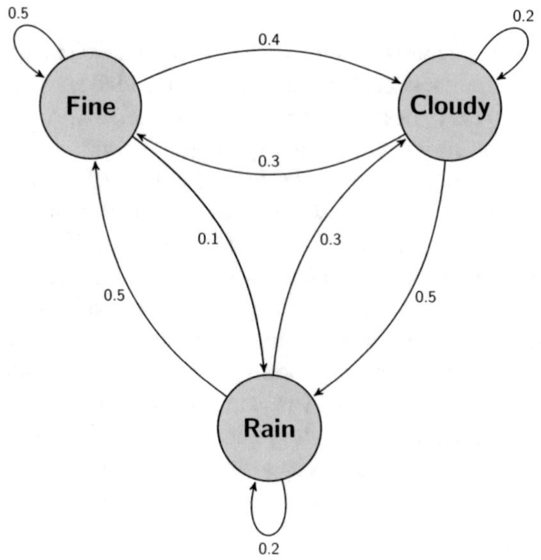

Figure 1.7: Three-state Markov chain of the weather.
Notice the sum of the arrows leaving each state is 1.

This linear system of equations can be reorganized into a system with 3 equations and 3 unknowns by realizing that one of the equations inside $\pi P = \pi$ is redundant. Written out explicitly, we have

$$\begin{bmatrix} P_{11}-1 & P_{21} & P_{31} \\ P_{12} & P_{22}-1 & P_{32} \\ 1 & 1 & 1 \end{bmatrix} \begin{bmatrix} \pi_1 \\ \pi_2 \\ \pi_3 \end{bmatrix} = \begin{bmatrix} 0 \\ 0 \\ 1 \end{bmatrix}. \tag{1.2}$$

3. By making use of the *Perron Frobenius theorem* which implies that eigenvectors corresponding to the eigenvalue of maximal magnitude which is 1 are proportional to π, we find such an eigenvector and normalize it by the sum of probabilities (L_1 norm).

4. We run a simple Monte Carlo simulation (see also Section 1.5) by generating random values of the weather according to P, and take the long-term proportions of each state. In contrast to the previous three approaches, this approach does not use any linear algebra.

The output shows that the four estimates of the vector π are very similar. Each column represents the stationary distribution obtained from methods 1 to 4, while the rows represent the stationary probability of being in each state.

Listing 1.8: Steady state of a Markov chain in several ways

```
using LinearAlgebra, StatsBase

# Transition probability matrix
P = [0.5 0.4 0.1;
     0.3 0.2 0.5;
     0.5 0.3 0.2]

# First way
piProb1 = (P^100)[1,:]

# Second way
A = vcat((P' - I)[1:2,:],ones(3)')
b = [0 0 1]'
piProb2 = A\b

# Third way
eigVecs = eigvecs(copy(P'))
highestVec = eigVecs[:,findmax(abs.(eigvals(P)))[2]]
piProb3 = Array{Float64}(highestVec)/norm(highestVec,1)

# Fourth way
numInState = zeros(Int,3)
state = 1
N = 10^6
for t in 1:N
    numInState[state] += 1
    global state = sample(1:3,weights(P[state,:]))
end
piProb4 = numInState/N

display([piProb1 piProb2 piProb3 piProb4])
```

```
3x4 Array{Float64,2}:
0.4375  0.4375  0.4375  0.437521
0.3125  0.3125  0.3125  0.312079
0.25    0.25    0.25    0.2504
```

In lines 4–6 the transition probability matrix P is defined. The notation for explicitly defining a matrix in Julia is the same as that of MATLAB. In line 9, (1.1) is implemented and n is taken as 100 (approximating ∞). The first row of the resulting matrix is obtained via `[1,:]`. Note that using `[2,:]` or `[3,:]` instead will approximately yield the same result, since the limit in Equation (1.1) is independent of j. Lines 12–14 use quite a lot of matrix operations to set up the system of Equations (1.2). The use of `vcat()` (*vertical concatenation*) creates the matrix on the left-hand side by concatenating the 2×3 matrix, `(P' - I)[1:2,:]` with a row vector of 1's, `ones(3)'`. Note the use of `I` which is the identity matrix. Finally, the solution is found by using `A\b` in the same fashion as MATLAB for solving linear equations of the form $Ax = b$. In lines 17–19 the built-in `eigvecs()` and `eigvals()` functions from `LinearAlgebra` are used to find the eigenvalues and a set of eigenvectors of P, respectively. The `findmax()` function is then used to find the index matching the eigenvalue with the largest magnitude. Note that the absolute value function `abs()` works on complex values as well. Also note that when normalizing in line 19, we use the L_1 norm which is essentially the sum of absolute values of the vector. In lines 22–29 a direct Monte Carlo simulation of the Markov chain is carried out through a million iterations and modifications of the `state` variable. We accumulate the occurrences of each state in line 26. Line 27 is the actual transition, which uses the `sample()` function from the `StatsBase` package. At each iteration the next state is randomly chosen based on the probability distribution given the current state. This is done via the use of weight vector. Note that the normalization from counts to frequency in line 29 uses the fact that Julia casts integer counts to floating point numbers upon division. That is, both the variables `numInState` and `N` are an array of integers and an integer, respectively, but the division (vector by scalar) makes `piProb4` a floating point array.

Web Interfacing, JSON, and String Processing

We now look at a different type of example which deals with text. Imagine that we wish to analyze the writings of Shakespeare. In particular, we wish to look at the occurrences of some common words in all of his known texts and present a count of a few of the most common words. One simple and crude way to do this is to pre-specify a list of words to count, and then specify how many of these words we wish to present.

To add another dimension to this problem we will use a JSON (*Java Script Object Notation*) file. This file format is widely used for storing hierarchical datasets both in data science and web development, hence the name. We use the below JSON file in the example that follows.

```
{
  "words": [ "heaven","hell","man","woman","boy","girl","king","queen",
        "prince","sir","love","hate","knife","english","england","god"],
  "numToShow": 5
}
```

The JSON format uses "`{ }`" characters to enclose a hierarchical nested structure of key value pairs. In the example above there isn't any nesting, but rather only one top level set of "`{ }`". Within this there are two keys: `words` and `numToShow`. Treating this as a JSON object means that the key `numToShow` has an associated value 5. Similarly, `words` is an array of strings, with each element a potentially interesting word to consider in Shakespeare's texts. In general, JSON files are used for much more complex descriptions of data, but here we use this simple structure for illustration.

Now with some basic understanding of JSON, we can proceed with our example. The code in Listing 1.9 retrieves Shakespeare's texts from the web and then counts the occurrences of each of the words, ignoring case. We then show a count for each of the numToShow most common words.

Listing 1.9: Web interface, JSON and string parsing

```
1  using HTTP, JSON
2
3  data = HTTP.request("GET",
4  "https://ocw.mit.edu/ans7870/6/6.006/s08/lecturenotes/files/t8.shakespeare.txt")
5  shakespeare = String(data.body)
6  shakespeareWords = split(shakespeare)
7
8  jsonWords = HTTP.request("GET",
9  "https://raw.githubusercontent.com/"*
10 "h-Klok/StatsWithJuliaBook/master/data/jsonCode.json")
11 parsedJsonDict = JSON.parse( String(jsonWords.body))
12
13 keywords = Array{String}(parsedJsonDict["words"])
14 numberToShow = parsedJsonDict["numToShow"]
15 wordCount = Dict([(x,count(w -> lowercase(w) == lowercase(x), shakespeareWords))
16                  for x in keywords])
17
18 sortedWordCount = sort(collect(wordCount),by=last,rev=true)
19 display(sortedWordCount[1:numberToShow])
```

```
5-element Array{Pair{String,Int64},1}:
"king"=>1698
"love"=>1279
"man"=>1033
"sir"=>721
"god"=>555
```

In lines 3–4 HTTP.request from the HTTP package is used to make a HTTP request. In line 5 the body of data is then parsed to a text string via the String() constructor function. In line 6 this string is then split into an array of individual words via the split() function. In lines 8–11 the JSON file is first retrieved. Then this string is parsed into a JSON object. The URL string for the JSON file doesn't fit on one line, so we use * to concatenate strings. In line 11 the parse() function from the JSON package is used to parse the body of the file and creates a dictionary. Line 13 shows the strength of using JSON as the value associated with the JSON key words is accessed. This value (i.e. array of words) is then cast to an Array{String} type. Similarly, the value associated with the key numToShow is accessed in line 14. In line 15 a Julia dictionary is created via Dict(). It is created from a comprehension of tuples, each with x (being a word) in the first element, and the count of these words in shakespeareWords as the second element. In using count we define the anonymous function as the first argument that compares an input test argument w to the given word x, only in lowercase. Finally line 18 sorts the dictionary by its values, and line 19 displays as output the first most popular numberToShow values.

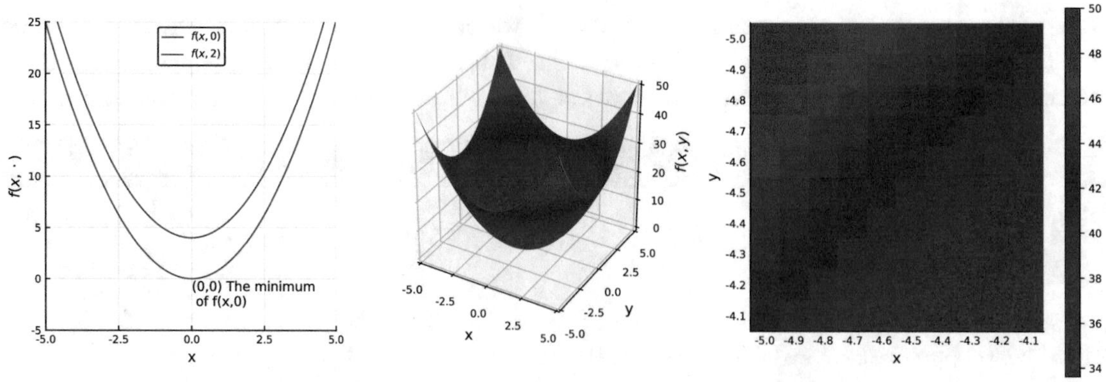

Figure 1.8: An introductory `Plots` example.

1.4 Plots, Images, and Graphics

There are many different plotting packages available in Julia, including `PyPlot`, `Gadfly`, `Makie` as well as several others. Arguably, as a starting point, two of the most useful plotting packages are the `Plots` package and the `StatsPlots` package. `Plots` simplifies the process of creating plots, as it brings together many different plotting packages under a single API. With `Plots`, you can learn a single syntax, and then use the backend of your choice to create plots. Almost all of the examples throughout this book use the `Plots` package, and in almost all of the examples the code presented directly generates the figures. That is, if you want examples of how to create certain plots, one way of doing this is to browse through the figures of the book until you find one of interest, and then look at the associated code block and use this as inspiration for your plotting needs.

In `Plots`, input data is passed positionally, while aspects of the plot can be customized by specifying keywords for specific plot attributes, such as line color or width. In general, each attribute can take on a range of values, and in addition, many attributes have aliases which empower one to write short, concise code. For example `color=:blue` can be shortened to `c=:blue`, and we make use of this alias mechanic throughout the books examples.

Since the code listings from this book can be used as direct examples, we don't present an extensive tutorial on the finer aspects of creating plots. Rather, if you are seeking detailed instructions or further references on finer points, we recommend that you visit:

`http://docs.juliaplots.org/`

As a minimal overview, the following is a brief list of some of the more commonly used `Plots` package functions for generating plots:

`plot()`—Can be used to plot data in various ways, including series data, single functions, multiple functions, as well as for presenting and merging other plots. This is the most common plotting function.

`scatter()`—Used for plotting scattered data points not connected by a line.

`bar()`—Used for plotting bar graphs.

`heatmap()`—Used to plot a matrix, or an image.

`surface()`—Used to plot surfaces (3D plots). This is the typical way in which one would plot a real-valued function of two variables.

`contour()`—Used to create a contour plot. This is an alternative way to plot a real-valued function of two variables.

`contourf()`—Similar to `contour()`, but with shading between contour lines.

`histogram()`—Used to plot histograms of data.

`stephist()`—A stepped histogram. This is a histogram with no filling.

In addition, each of these functions also has a companion function with a "`!`" suffix, e.g. `plot!()`. These functions modify the previous plot, adding additional plotting aspects to them. This is shown in many examples throughout the book. Furthermore, the `Plots` package supplies additional important functions such as `savefig()` for saving a plot, `annotate!()` for adding annotations to plots, `default()` for setting plotting default arguments, and many more. Note that in the examples throughout this book `pyplot()` is called. This activates the `PyPlot` backend for plotting.

As a basic introductory example focused solely on plotting, we present Listing 1.10. In this listing, the main object is the real-valued function of two variables, $f(x, y) = x^2 + y^2$. We use this *quadratic form* as a basic example, and also consider the cases of $y = 0$ and $y = 2$. The code generates Figure 1.8. Note the use of the `LaTeXStrings` package enabling LaTeX formatted formulas. See, for example, `http://tug.ctan.org/info/undergradmath/undergradmath.pdf`.

Listing 1.10: Basic plotting

```
using Plots, LaTeXStrings, Measures; pyplot()

f(x,y) = x^2 + y^2
f0(x) = f(x,0)
f2(x) = f(x,2)

xVals, yVals = -5:0.1:5 , -5:0.1:5
plot(xVals, [f0.(xVals), f2.(xVals)],
        c=[:blue :red], xlims=(-5,5), legend=:top,
        ylims=(-5,25), ylabel=L"f(x,\cdot)", label=[L"f(x,0)" L"f(x,2)"])
p1 = annotate!(0, -0.2, text("(0,0) The minimum\n of f(x,0)", :left, :top, 10))

z = [ f(x,y) for y in yVals, x in xVals ]
p2 = surface(xVals, yVals, z, c=cgrad([:blue, :red]),legend=:none,
        ylabel="y", zlabel=L"f(x,y)")

M = z[1:10,1:10]
p3 = heatmap(M, c=cgrad([:blue, :red]), yflip=true, ylabel="y",
        xticks=([1:10;], xVals), yticks=([1:10;], yVals))

plot(p1, p2, p3, layout=(1,3), size=(1200,400), xlabel="x", margin=5mm)
```

Line 1 includes the following packages: `Plots` for plotting; `LaTeXStrings` for displaying labels using LaTeX formatting as in line 10; and `Measures` for specifying margins such as in line 21. In line 1, as part of a second statement following "`;`", `pyplot()` is called to indicate that the PyPlot plotting backend is activated. In line 3 we define the two variable real-valued function `f()` which is the main object of this example. We then define two related single variable functions, `f0()` and `f2()`, i.e. $f(x, 0)$ and $f(x, 2)$. In line 7 we define the ranges `xVals` and `yVals`. Line 8 is the first call to `plot()` where `xVals` is the first argument indicating the horizontal coordinates, and the array `[f0.(xVals), f2.(xVals)]` represents two data series to be plotted. Then in the same function call on lines 9 and 10, we specify colors, x-limits, y-limits, location of the legend, and the labels, where `L` denotes LaTeX. In line 11 `annotate!()` modifies the current plot with an annotation. The return value is the plot object stored in `p1`. Then in lines 13–15 we create a surface plot. The "height" values are calculated via a two-way comprehension and stored in the matrix `z` on line 13. Then `surface()` is used in lines 14–15 to create the plot, which is then stored in the variable `p2`. Note the use of the `cgrad()` function to create a color gradient. In lines 17–19 a matrix of values is plotted via `heatmap()`. The argument `yflip=true` is important for orienting the matrix in the standard manner. Finally, in line 21 the three previous subplots are plotted together as a single figure via the `plot()` function.

Histogram of Hailstone Sequence Lengths

In this example we use `Plots` to create a *histogram* in the context of a well-known mathematical problem. Consider that we generate a sequence of numbers as follows: given a positive integer x, if it is even, then the next number in the sequence is $x/2$, otherwise it is $3x + 1$. That is, we start with some x_0 and then iterate $x_{n+1} = f(x_n)$ with

$$f(x) = \begin{cases} x/2 & \text{if } x \bmod 2 = 0, \\ 3x + 1 & \text{if } x \bmod 2 = 1. \end{cases}$$

The sequence of numbers arising from this function is called the *hailstone sequence.* As an example, if $x_0 = 3$, the resulting sequence is

$$3, 10, 5, 16, 8, 4, 2, 1, \ldots,$$

where the cycle $4, 2, 1$ continues forever. We call the number of steps (possibly infinite) needed to hit 1 the length of the sequence, in this case 8. Note that different values of x_0 will result in different hailstone sequences of different lengths.

It is conjectured that, regardless of the x_0 chosen, the sequence will always converge to 1. That is, the length is always finite. However, this has not been proven to date and remains an open question, known as the *Collatz conjecture.* In addition, a counter-example has not yet been computationally found. That is, there is no known x_0 for which the sequence doesn't eventually go down to 1.

Now that the context of the problem is set, we create a histogram of lengths of hailstone sequences based on different values of x_0. Our approach is shown in Listing 1.11, where we first create a function which calculates the length of a hailstone sequence based on a chosen value of x_0. We then use a comprehension to evaluate this function for each initial value, $x_0 = 2, 3, \ldots, 10^7$, and finally plot a histogram of these lengths, shown in Figure 1.9.

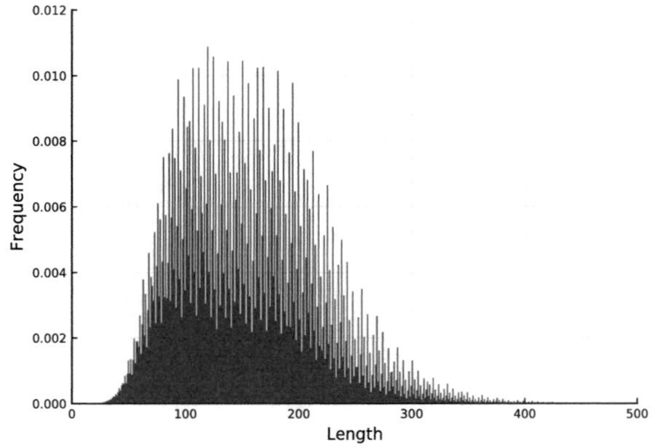

Figure 1.9: Histogram of hailstone sequence lengths.

Listing 1.11: Histogram of hailstone sequence lengths

```
1  using Plots; pyplot()
2
3  function hailLength(x::Int)
4      n = 0
5      while x != 1
6          if x % 2 == 0
7              x = Int(x/2)
8          else
9              x = 3x +1
10         end
11         n += 1
12     end
13     return n
14 end
15
16 lengths = [hailLength(x0) for x0 in 2:10^7]
17
18 histogram(lengths, bins=1000, normed=:true,
19     fill=(:blue, true), la=0, legend=:none,
20     xlims=(0, 500), ylims=(0, 0.012),
21     xlabel="Length", ylabel="Frequency")
```

In lines 3–14 the function `hailLength()` is created, which evaluates the length of a hailstone sequence, `n`, given the first number in the sequence, `x`. Note the use of `::Int`, which indicates the method implemented operates only on integer types. A `while` loop is used to sequentially and repeatedly evaluate all code contained within it, until the specified condition is `false`. In this case until we obtain a hailstone number of 1. Note the use of the *not-equals comparison operator*, `!=`. In line 6 the *modulo* operator, `%`, and *equality operator*, `==` are used in conjunction to check if the current number is even. If `true`, then we proceed to line 7, else we proceed to line 9. In line 11 our hailstone sequence length is increased by one each time we generate a new number in our sequence. In line 13 length of the sequence is returned. In line 16 a comprehension is used to evaluate our function for integer values of x_0 between 2 and 10^7. In lines 18–21 the `histogram()` function is used to plot a histogram using an arbitrary bin count of 1000.

Creating Animations

We now present an example of a live *animation* which sequentially draws the edges of a fully connected mathematical *graph*. A graph is an object that consists of *vertices*, represented by dots, and *edges*, represented by lines connecting the vertices.

In this example we construct a series of equally spaced vertices around the *unit circle*, given an integer number of vertices, n. To add another aspect to this example, we obtain the points around the unit circle by considering the complex numbers

$$z_n = e^{2\pi i \frac{k}{n}}, \qquad \text{for} \qquad k = 1, \ldots, n. \tag{1.3}$$

We then use the real and imaginary parts of z_n to obtain the horizontal and vertical coordinates for each vertex, respectively, which distributes n points evenly on the unit circle. The example in Listing 1.12 sequentially draws all possible edges connecting each vertex to all remaining vertices, and animates the process. Each time an edge is created, a frame snapshot of the figure is saved, and by quickly cycling through the frames generated, we can generate an animated *GIF*. A single frame approximately half way through the GIF animation is shown in Figure 1.10.

Listing 1.12: Animated edges of a graph

```
using Plots; pyplot()

function graphCreator(n::Int)
    vertices = 1:n
    complexPts = [exp(2*pi*im*k/n) for k in vertices]
    coords = [(real(p),imag(p)) for p in complexPts]
    xPts = first.(coords)
    yPts = last.(coords)
    edges = []
    for v in vertices, u in (v+1):n
        push!(edges,(v,u))
    end

    anim = Animation()
    scatter(xPts, yPts, c=:blue, msw=0, ratio=1,
        xlims=(-1.5,1.5), ylims=(-1.5,1.5), legend=:none)

    for i in 1:length(edges)
        u, v = edges[i][1], edges[i][2]
        xpoints = [xPts[u], xPts[v]]
        ypoints = [yPts[u], yPts[v]]
        plot!(xpoints, ypoints, line=(:red))
        frame(anim)
    end

    gif(anim, "graph.gif", fps = 60)
end

graphCreator(16)
```

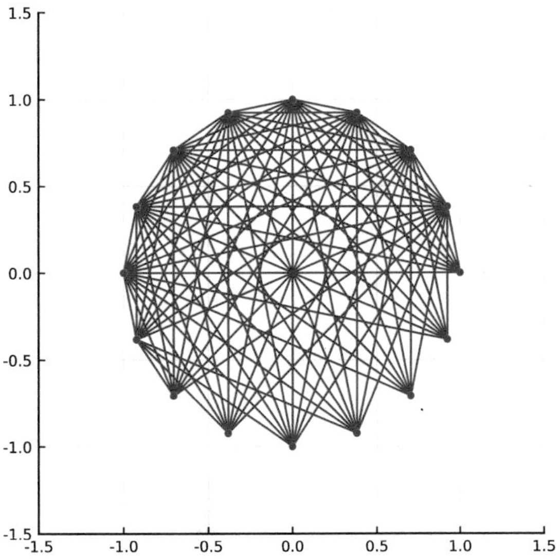

Figure 1.10: Sample frame from a graph animation.

The code defines the function `graphCreator()`, which constructs the animated GIF based on n number of vertices. In line 5 the complex points calculated via (1.3) are stored in the array `complexPoints`. In line 6 `real()` and `imag()` extract the real and imaginary parts of each complex number, respectively, and store them as paired tuples. In lines 7–8, the x- and y-coordinates are retrieved via `first()` and `last()`, respectively. Note lines 5–8 could be shortened and implemented in various other ways; however, the current implementation is useful for demonstrating several aspects of the language. Then lines 10–12 loop over `u` and `v`, and in line 11 the tuple `(u,v)` is added to `edges`. In line 14 an `Animation()` object is created. The vertices are plotted in lines 15–16 via `scatter()`. The loop in lines 18–24 plots a line for each of the edges via `plot!()`. Then `frame(anim)` adds the current figure as another frame to the animation object. The `gif()` function in line 26 saves the animation as the file `graph.gif` where `fps` defines how many frames per second are rendered.

Raster Images

We now present an example of working with *raster images*, namely, images composed of individual pixels. In Listing 1.13 we load a sample image of stars in space and locate the brightest star. Note that the image contains some amount of noise, in particular, as seen from the output, the single brightest pixel is located at $[192, 168]$ in *row major*. Therefore if we wanted to locate the brightest star by a single pixel's intensity, we would not identify the correct coordinates.

Since looking at single pixels can be deceiving, to find the highest intensity star, we use a simple method of passing a kernel over the image. This technique smoothens the image and eliminates some of the noise. The results are in Figure 1.11 where the two subplots show the original image versus the smoothed image, and the location of the brightest star for each.

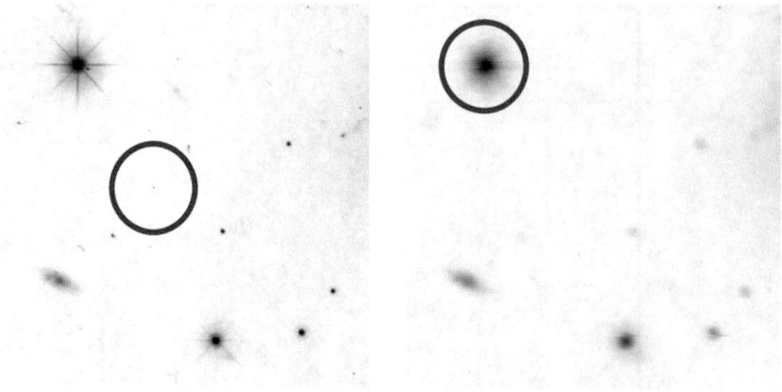

Figure 1.11: Left: Original image. Right: Smoothed image after noise removal.

Listing 1.13: Working with images

```
using Plots, Images; pyplot()

img = load("../data/stars.png")
gImg = red.(img)*0.299 + green.(img)*0.587 + blue.(img)*0.114
rows, cols = size(img)

println("Highest intensity pixel: ", findmax(gImg))

function boxBlur(image,x,y,d)
    if x<=d || y<=d || x>=cols-d || y>=rows-d
        return image[x,y]
    else
        total = 0.0
        for xi = x-d:x+d
            for yi = y-d:y+d
                total += image[xi,yi]
            end
        end
        return total/((2d+1)^2)
    end
end

blurImg = [boxBlur(gImg,x,y,5) for x in 1:cols, y in 1:rows]

yOriginal, xOriginal = argmax(gImg).I
yBoxBlur, xBoxBlur   = argmax(blurImg).I

p1 = heatmap(gImg, c=:Greys, yflip=true)
p1 = scatter!((xOriginal, yOriginal), ms=60, ma=0, msw=4, msc=:red)
p2 = heatmap(blurImg, c=:Greys, yflip=true)
p2 = scatter!((xBoxBlur, yBoxBlur), ms=60, ma=0, msw=4, msc=:red)

plot(p1, p2, size=(800, 400), ratio=:equal, xlims=(0,cols), ylims=(0,rows),
        colorbar_entry=false, border=:none, legend=:none)
```

```
Highest intensity pixel: (0.9999999999999999, CartesianIndex(192, 168))
```

In line 3 the image is read into memory via the `load()` function and stored as `img`. Since the image is 400 × 400 pixels, it is stored as a 400 × 400 array of RGBA tuples of length 4. Each element of these tuples represents one of the color layers in the following order: red, green, blue, and luminosity. In line 4 we create a grayscale image from the original image data via a linear combination of its RGB layers. This choice of coefficients is a common "Grayscale algorithm". The gray image is stored as the matrix `gImg`. In line 5 the `size()` function is used to determine then number of rows and columns of `gImg`, which are then stored as `rows` and `cols`, respectively. In line 7 `findmax()` is used to find the highest intensity element (pixel) in `gImg`. It returns a tuple of value and index, where in this case the index is of type `CartesianIndex` because `gImg` is a two-dimensional array (matrix). In lines 9–21 the function `boxBlur` is created. This function takes an array of values as input, representing an image, and then passes a kernel over the image data, taking a linear average in the process. This is known as "box blur". In other words, at each pixel, the function returns a single pixel with a brightness weighting based on the average of the surrounding pixels (or array values) in a given neighborhood within a box of dimensions $2d + 1$. Note that the edges of the image are not smoothed, as a border of un-smoothed pixels of "depth" d exists around the images edges. Visually, this kernel smoothing method has the effect of blurring the image. In line 23, the function `boxBlur()` is parsed over the image for a value of $d = 5$, i.e. a 10 × 10 kernel. The smoothed data is then stored as `blurImg`. In lines 25–26 we use the `argmax()` function which is similar to `findmax()`, but only returns the index. We use it to find the index of the pixel with the largest value, for both the non-smoothed and smoothed image data. Note the use of the trailing `.I` at the end of each `argmax()`, which extracts the `Tuple` of values of the coordinates from the `CartesianIndex` type. As the Cartesian index of matrices is row major, we reverse the row and column order for the plotting that follows. The remaining lines create Figure 1.11.

1.5 Random Numbers and Monte Carlo Simulation

Many of the code examples in this book make use of *pseudorandom number generation*, often coupled with the so-called *Monte Carlo simulation method* for obtaining numerical estimates. The phrase "Monte Carlo" associated with random number generation comes from the European province in Monaco famous for its many casinos. We now overview the core ideas and principles of random number generation and Monte Carlo simulation.

The main player in this discussion is the `rand()` function. When used without input arguments, `rand()` generates a "random" number in the interval $[0, 1]$. Several questions can be asked. How is it random? What does random within the interval $[0, 1]$ really mean? How can it be used as an aid for statistical and scientific computation? For this we discuss pseudorandom numbers in a bit more generality.

The "random" numbers we generate using Julia, as well as most "random" numbers used in any other scientific computing platform, are actually pseudorandom. That is, they aren't really random but rather appear random. For their generation, there is some deterministic (non-random and well defined) sequence, $\{x_n\}$, specified by

$$x_{n+1} = f(x_n, x_{n-1}, \ldots), \tag{1.4}$$

originating from some specified *seed*, x_0. The mathematical function, $f(\cdot)$, is often (but not always) quite a complicated function, designed to yield desirable properties for the sequence $\{x_n\}$ that make it appear random. Among other properties we wish for the following to hold:

(i) Elements x_i and x_j for $i \neq j$ should appear statistically independent. That is, knowing the value of x_i should not yield information about the value of x_j.

(ii) The distribution of $\{x_n\}$ should appear uniform. That is, there shouldn't be values (or ranges of values) where elements of $\{x_n\}$ occur more frequently than others.

(iii) The range covered by $\{x_n\}$ should be well defined.

(iv) The sequence should repeat itself as rarely as possible.

Typically, a mathematical function such as $f(\cdot)$ is designed to produce integers in the range $\{0, \dots, 2^\ell - 1\}$ where ℓ is typically 16, 32, 64, or 128 (depending on the number of bits used to represent an integer). Hence $\{x_n\}$ is a sequence of pseudorandom integers. Then if we wish to have a pseudorandom number in the range $[0, 1]$ (represented via a floating point number), we normalize via

$$U_n = \frac{x_n}{2^\ell - 1}.$$

When calling `rand()` in Julia (as well as in many other programming languages), what we are doing is effectively requesting the system to present us with U_n. Then, in the next call, U_{n+1}, and in the call after this U_{n+2}, etc. As a user, we don't care about the actual value of n, we simply trust the computing system that the next pseudorandom number will differ and adhere to the properties (i) - (iv) mentioned above.

One may ask, where does the sequence start? For this we have a special name that we call x_0. It is known as the *seed* of the pseudorandom sequence. Typically, as a scientific computing system starts up, it sets x_0 to be the current time. This implies that on different system startups, $x_0, x_1, x_2, \dots$ will be different sequences of pseudorandom numbers. However, we may also set the seed ourselves. There are several uses for this and it is often useful for reproducibility of results. Listing 1.14 illustrates setting the seed using Julia's `Random.seed!()` function.

Listing 1.14: Pseudorandom number generation

```
using Random

Random.seed!(1974)
println("Seed 1974: ",rand(),"\t", rand(), "\t", rand())
Random.seed!(1975)
println("Seed 1975: ",rand(),"\t", rand(), "\t", rand())
Random.seed!(1974)
println("Seed 1974: ",rand(),"\t", rand(), "\t", rand())
```

```
Seed 1974: 0.21334106865797864  0.12757925830167505     0.5047074487066832
Seed 1975: 0.7672833719737708   0.8664265778687816      0.5807364110163316
Seed 1974: 0.21334106865797864  0.12757925830167505     0.5047074487066832
```

As can be seen from the output, setting the seed to 1974 produces the same sequence. However, setting the seed to 1975 produces a completely different sequence.

One may ask why use random or pseudorandom numbers? Sometimes having arbitrary numbers alleviates programming tasks or helps randomize behavior. For example, when designing computer

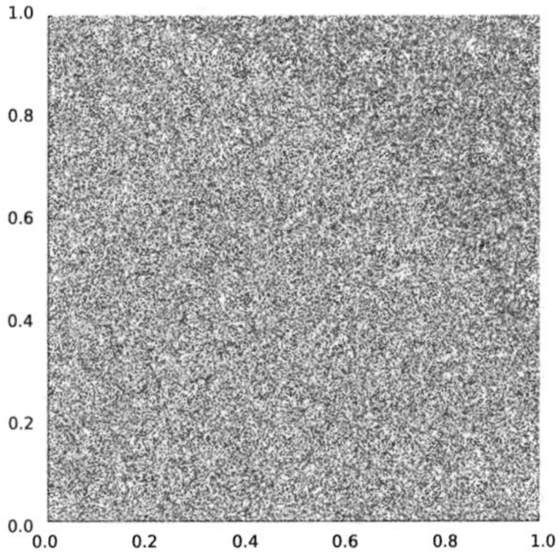

Figure 1.12: Estimating π via Monte Carlo.

video games, having enemies appear at random spots on the screen yields for a simple implementation. In the context of scientific computing and statistics, the answer lies in the Monte Carlo simulation method. Here the idea is that computations can be aided by repeated sampling and averaging out the result. Many of the code examples in our book do this and we illustrate one such simple example below.

Monte Carlo Simulation

As an example of Monte Carlo, say we wish to estimate the value of π. There are hundreds of known numerical methods to do this and here we explore one. Observe that the area of one quarter section of the unit circle is $\pi/4$. Now if we generate random points, (x, y), within a unit box, $[0, 1] \times [0, 1]$, and calculate the proportion of total points that fall within the quarter circle, we can approximate π via

$$\hat{\pi} = 4\,\frac{\text{Number of points with } x^2 + y^2 \le 1}{\text{Total number of points}}.$$

This is performed in Listing 1.15 for 10^5 points. The listing also creates Figure 1.12.

Listing 1.15: Estimating π

```
using Random, LinearAlgebra, Plots; pyplot()
Random.seed!()

N = 10^5
data     = [[rand(),rand()] for _ in 1:N]
indata   = filter((x)-> (norm(x) <= 1), data)
outdata  = filter((x)-> (norm(x) > 1), data)
piApprox = 4*length(indata)/N
println("Pi Estimate: ", piApprox)

scatter(first.(indata),last.(indata), c=:blue, ms=1, msw=0)
scatter!(first.(outdata),last.(outdata), c=:red, ms=1, msw=0,
        xlims=(0,1), ylims=(0,1), legend=:none, ratio=:equal)
```

```
Pi Estimate: 3.14068
```

In Line 2 the seed of the random number generator is set with `Random.seed!()`. This is done to ensure that each time the code is run the estimate obtained is the same. In Line 4, the number of repetitions, `N`, is set. Most code examples in this book use `N` as the number of repetitions in a Monte Carlo simulation. Line 5 generates an array of arrays. That is, the pair, `[rand(),rand()]` is an array of random coordinates in $[0,1] \times [0,1]$. Line 6 filters those points to use for the numerator of $\hat{\pi}$. It uses the `filter()` function, where the first argument is an anonymous function, `(x) -> (norm(x) <= 1)`. Here, `norm()` defaults to the L_2 norm, i.e. $\sqrt{x^2+y^2}$. The resulting `indata` array only contains the points that fall within the unit circle (with each represented as an array of length 2). Line 7 creates the analogous `outdata` array. It is not used for the estimation, but is used in plotting. Line 8 calculates the approximation, with `length()` used for the numerator of $\hat{\pi}$ and `N` for the denominator. Lines 11–13 are used to create Figure 1.12.

Inside a Simple Pseudorandom Number Generator

Number theory and related fields play a central role in the mathematical study of pseudorandom number generation, the internals of which are determined by the specific $f(\cdot)$ of (1.4). However, typically this is not of direct interest to statisticians. Nevertheless, for exploratory purposes we illustrate how one can make a simple pseudorandom number generator.

A simple to implement class of pseudorandom number generators is the class of *Linear Congruential Generators* (LCG). These types of LCGs are common in older systems. Here the function $f(\cdot)$ is nothing but an affine (linear) transformation modulo m,

$$x_{n+1} = (a\, x_n + c) \bmod m. \tag{1.5}$$

The integer parameters a, c, and m are fixed and specify the details of the LCG. Some number theory research has determined "good" values of a and c for specific values of m. For example, for $m = 2^{32}$, setting $a = 69069$ and $c = 1$ yields sensible performance (other possibilities work well, but not all). In Listing 1.16 we generate values based on this LCG, see also Figure 1.13.

Listing 1.16: A linear congruential generator

```
using Plots, LaTeXStrings, Measures; pyplot()

a, c, m = 69069, 1, 2^32
next(z) = (a*z + c) % m

N = 10^6
data = Array{Float64,1}(undef, N)

x = 808
for i in 1:N
    data[i] = x/m
    global x = next(x)
end

p1 = scatter(1:1000, data[1:1000],
    c=:blue, m=4, msw=0, xlabel=L"n", ylabel=L"x_n")
p2 = histogram(data, bins=50, normed=:true,
    ylims=(0,1.1), xlabel="Support", ylabel="Density")
plot(p1, p2, size=(800, 400), legend=:none, margin = 5mm)
```

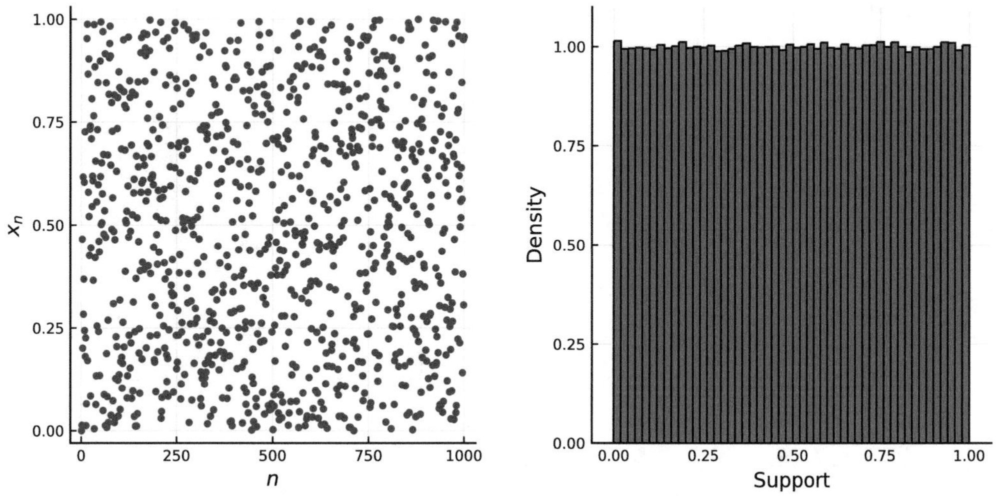

Figure 1.13: Left: The first 1,000 values generated by a linear congruential generator, plotted sequentially. Right: A histogram of 10^6 random values.

In line 4 (1.5) is implemented as the function `next()`. In line 7 an array of `Float64` of length `N` is pre-allocated. In line 9 the seed is arbitrarily set as the value `808`. In lines 10–13 a loop is used `N` times. In line 11 the current value of `x` is divided by m to obtain a number in the range $[0, 1]$. Note that in Julia division of two integers results in a floating point number. In line 12 (1.5) is applied recursively via `next()` to set a new value for `x`. In lines 15–16 a scatterplot of the first 1000 values of `data` is created, while lines 17–18 create a histogram of all values of `data` with 50 bins. As expected by the theory of LCG, a uniform distribution is obtained.

More About Julia's `rand()`

Having covered the basics, we now describe a few more aspects of Julia's random number generation. The key function at play is `rand()`. However, as you already know, a Julia function can have many methods. The `rand()` function is no different. To see this, run `methods(rand)` and you'll see dozens of different methods of `rand()`. Furthermore, if you do this after loading the `Distributions` package into the namespace (by running `using Distributions`) that number will grow substantially. Hence in short, there are many ways to use the `rand()` function in Julia. Throughout the rest of this book we use it in various ways, including in conjunction with probability distributions. However we now focus on functionality from the `Base` package.

There are other functions related to `rand()`, such as `randn()` for generating normally distributed random variables. Also after invoking `using Random`, the following functions are available: `Random.seed!()`, `randsubseq()`, `randstring()`, `randcycle()`, `bitrand()`, as well as `randperm()` and `shuffle()` for permutations. There is also the `MersenneTwister()` constructor among others. These are discussed in the Julia documentation. You may also use the built-in help to enquire about them. We now focus on the `MersenneTwister()` constructor and explain how it can be used in conjunction with `rand()` and variants.

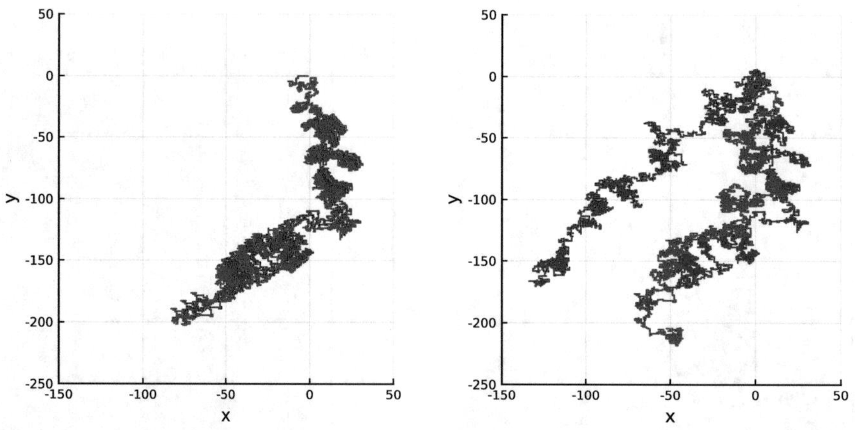

Figure 1.14: Random walks with slightly different parameters.
Left: Trajectories with same seed. Right: Different seed per trajectory.

The term *Mersenne Twister* refers to a type of pseudorandom number generator. It is an algorithm that is considerably more complicated than the LCG described above. Generally, its statistical properties are much better than those of LCG. Due to this it has made its way into most scientific programming environments in the past two decades. Julia has adopted it as the default as well.

Our interest in mentioning the Mersenne Twister is due to the fact that in Julia we can create an object representing a random number generator implemented via this algorithm. To create such an object we write, for example, `rng = MersenneTwister(seed)`, where `seed` is some initial seed value. Then the object `rng` acts as a random number generator, and may serve as an additional input to `rand()` and related functions. For example, calling `rand(rng)` uses the specific random number generator object passed to it. In addition to `MersenneTwister()`, there are also other ways to create similar objects, such as, for example, `RandomDevice()`. However we leave it to the reader to investigate these via the online help.

By creating random number generator objects, you may have more than one random sequence in your application, essentially operating simultaneously. In Chapter 10, we investigate scenarios where this is advantageous from a Monte Carlo simulation perspective. For now we show how a random number generator may be passed into a function as an argument, allowing the function to generate random values using that specific generator.

Listing 1.17 creates random paths in the plane. Each path starts at $(x, y) = (0, 0)$ and moves up, right, down, or left at each step. The movements up (`x+=1`) and right (`y+=1`) are with steps of size 1. However the movements down and left are with steps that are uniformly distributed in the range $[0, 2 + \alpha]$. Hence if $\alpha > 0$, on average the path drifts in the down-left direction. The virtue of this initial example is that by using *common random numbers* and simulating paths for varying α, we get very different behavior than if we use a different set of random numbers for each path, see Figure 1.14. We discuss more advanced applications of using multiple random number generators in Chapter 10, however we implicitly use this Monte Carlo technique throughout the book, often by setting the seed to a specific value in the code examples.

Listing 1.17: Random walks and seeds

```
1  using Plots, Random, Measures; pyplot()
2
3  function path(rng, alpha, n=5000)
4      x, y = 0.0, 0.0
5      xDat, yDat = [], []
6      for _ in 1:n
7          flip = rand(rng,1:4)
8          if flip == 1
9              x += 1
10         elseif flip == 2
11             y += 1
12         elseif flip == 3
13             x -= (2+alpha)*rand(rng)
14         elseif flip == 4
15             y -= (2+alpha)*rand(rng)
16         end
17         push!(xDat,x)
18         push!(yDat,y)
19     end
20     return xDat, yDat
21 end
22
23 alphaRange = [0.2, 0.21, 0.22]
24
25 default(xlabel = "x", ylabel = "y", xlims=(-150,50), ylims=(-250,50))
26 p1 = plot(path(MersenneTwister(27), alphaRange[1]), c=:blue)
27 p1 = plot!(path(MersenneTwister(27), alphaRange[2]), c=:red)
28 p1 = plot!(path(MersenneTwister(27), alphaRange[3]), c=:green)
29
30 rng = MersenneTwister(27)
31 p2 = plot(path(rng, alphaRange[1]), c=:blue)
32 p2 = plot!(path(rng, alphaRange[2]), c=:red)
33 p2 = plot!(path(rng, alphaRange[3]), c=:green)
34
35 plot(p1, p2, size=(800, 400), legend=:none, margin=5mm)
```

Lines 3–21 define the function `path()`. As a first argument it takes a random number generator, `rng`. That is, the function is designed to receive an object such as `MersenneTwister` as an argument. The second argument is `alpha` and the third argument is the number of steps in the path with a default value of `5000`. In lines 6–19 we loop `n` times, each time updating the current coordinate (`x` and `y`) and then pushing the values into the arrays, `xDat` and `yDat`. Line 7 generates a random value in the range `1:4`. Observe the use of `rng` as a first argument to `rand()`. In lines 13 and 15 we multiply `rand(rng)` by `(2+alpha)`. This creates uniform random variables in the range $[0, 2+\alpha]$. Line 20 returns a tuple of two arrays `xDat,yDat`. After setting `alphaRange` in line 23 and setting default plotting arguments in line 25, we create and plot paths with common random numbers in lines 26–28. This is because in each call to `path()` we use the same seed to a newly created `MersenneTwister()` object. Here `27` is just an arbitrary starting seed. In contrast, lines 31–33 have repeated calls to `path()` using a single stream, `rng`, created in line 30. Hence here, we don't have common random numbers because each subsequent call to `path()` starts at a fresh point in the stream of `rng`.

1.6 Integration with Other Languages

We now briefly overview how Julia can interface with the R-language, Python, and C. Note that there are several other packages that enable integration with other languages as well.

Using and Calling R Packages

See [LDL13] for an introduction to statistical computing with the R-language. R-code, functions, and libraries can be called in Julia via the `RCall` package which provides several different ways of interfacing with R from Julia. When working with the REPL, one may use `$` to switch between a Julia REPL and an R REPL. However, in this case, variables are not carried over between the two environments. The second way is via the `@rput` and `@rget` macros, which can be used to transfer variables from Julia to the R-environment. Finally, the `R""` (or `@R_str`) macro can also be used to parse R-code contained within the string. This macro returns an `RObject` as output, which is a Julia wrapper type around an R object.

We provide a brief example in Listing 1.18. It is related to Chapter 7 and focuses on the statistical method of ANOVA (Analysis of Variance) covered in Section 7.3. The purpose here is to demonstrate R-interoperability, and not so much on ANOVA. This example calculates the ANOVA F-statistic and *p*-value, complementing Listing 7.10. It makes use of the R `aov()` function and yields the same numerical results.

Listing 1.18: Using R from Julia

```julia
using CSV, DataFrames, RCall

data1 = CSV.read("../data/machine1.csv", header=false)[:,1]
data2 = CSV.read("../data/machine2.csv", header=false)[:,1]
data3 = CSV.read("../data/machine3.csv", header=false)[:,1]

function R_ANOVA(allData)
    data = vcat([ [x fill(i, length(x))] for (i, x) in
                    enumerate(allData) ]...)
    df = DataFrame(data, [:Diameter, :MachNo])
    @rput df

    R"""
    df$MachNo <- as.factor(df$MachNo)
    anova <- summary(aov( Diameter ~ MachNo, data=df))
    fVal <- anova[[1]]["F value"][[1]][1]
    pVal <- anova[[1]]["Pr(>F)"][[1]][1]
    """
    println("R ANOVA f-value: ", @rget fVal)
    println("R ANOVA p-value: ", @rget pVal)
end

R_ANOVA([data1, data2, data3])
```

```
R ANOVA f-value: 10.516968568709089
R ANOVA p-value: 0.00014236168817139574
```

In line 1 we specify usage of the required packages, including `RCall`. In lines 3–5 the data is loaded. In lines 7–21 we create the Julia function `R_ANOVA`, which takes a Julia array of arrays as input, `allData`. It outputs the summary results of an ANOVA test carried out in R via the `aov()` function. In lines 8–9 the array of arrays `allData` is re-arranged into a two-dimensional array, where the first column contains the observations from each of the arrays, and the second column contains the array index from which each observation has come. The data is re-arranged like this due to the format that the R `aov()` function requires. This re-arrangement is performed via the `enumerate()` function, along with the `vcat()` function and splat "`...`" operator. In line 10, the two-dimensional array `data` is converted to a `DataFrame`. Data frames are covered in Section 4.1. In line 11 the `@rput` macro is used to transfer the data frame `df` to the R-workspace. In lines 13–18 a multi-line R-code block is executed inside the `R"""` macro. In line 14, the `MachNo` column of the R-data frame `df` is defined as a factor, i.e. a categorical column via the R-code `as.factor()` and `<-`. In line 15 an ANOVA test of the `Diameter` column of the R-data frame `df` is conducted via `aov()` and passed to the `summary()` function, with the result stored as `anova`. In lines 16–17, the F-value and p-value are extracted from `anova`. Lines 19 and 20 are back to Julia where the output is printed. Note the use of `@rget` which is used to copy the variables from R back to Julia using the same name.

In addition to various R-functions, users of R will most likely also be familiar with *R-Datasets*. This is a collection of datasets commonly used in teaching and exploring statistics. You can read more about R-Datasets at

`https://vincentarelbundock.github.io/Rdatasets/datasets.html.`

Access to this collection of datasets from Julia is possible via the `RDatasets` package. Once installed in Julia, datasets can be loaded by using the `datasets()` function and specifying an "R datasets package name" followed by a "dataset name". For example, `datasets("datasets", "mtcars")` will load `mtcars`. Several code listings in this book use R-datasets.

Using and Calling Python Packages

See [MG16] for an introduction to machine learning with Python. It is possible to import Python modules and call Python functions directly in Julia via the `PyCall` package. It automatically converts types and allows data structures to be shared between Python and Julia. By default, `add PyCall` uses the `Conda` package to install a minimal Python distribution that is private to Julia. Further Python packages can then be installed from within Julia via the Julia `Conda` package.

Alternatively, one can use a pre-existing Python installation on the system. In order to do this, one must first set the Python environment variable to the path of the executable, and then re-build the `PyCall` package. For example, on a windows system with Anaconda installed, one would issue commands similar to the below from within the Julia REPL:

```
] add PyCall

ENV["PYTHON"] = "C:\\ProgramFiles\\Anaconda3\\python.exe"

] build PyCall
```

We now provide a brief example which makes use of the TextBlob Python library, which provides a simple API for conducting *Natural Language Processing* (NLP) tasks, including part-of-speech tagging, noun phrase extraction, sentiment analysis, classification, translation, and more. For our example we use TextBlob to analyze the sentiment of several sentences. The sentiment analyzer of TextBlob outputs a tuple of values, with the first value being the polarity of the sentence (a rating of positive to negative), and the second value a rating of subjectivity (factual to subjective).

In order for Listing 1.19 to work, the TextBlob Python library must first be installed. The lines below do this when executed in a shell or command prompt. Note that one can swap from the Julia REPL to a shell via "`;`".

```
pip3 install -U textblob

python -m textblob.download_corpora
```

Once Python and TextBlob are configured, Listing 1.19 can be executed. This example only briefly touches on the `PyCall` package with more information available in the package documentation.

Listing 1.19: NLP via Python's TextBlob

```
1  using PyCall
2  TB = pyimport("textblob")
3
4  str =
5  """Some people think that Star Wars The Last Jedi is an excellent movie,
6  with perfect, flawless storytelling and impeccable acting. Others
7  think that it was an average movie, with a simple storyline and basic
8  acting. However, the reality is almost everyone felt anger and
9  disappointment with its forced acting and bad storytelling."""
10
11 blob = TB.TextBlob(str)
12 [ i.sentiment for i in blob.sentences ]
```

```
(0.625, 0.636)
(-0.0375, 0.221)
(-0.46, 0.293)
```

In line 2 the `pyimport()` function is used to wrap the Python library `textblob`, which is then given the Julia alias `TB`. In lines 4–9 the string `str` is created. For this example, the string is written as a first hand account, and contains many words that give the text a negative tone. Note the use of multi-line strings using `"""`. In line 11 the `TextBlob()` function from `TB` is used to parse each sentence in `str`. The output is stored as `blob`. This is where the call to Python is made. In line 12 a comprehension is used to print the sentiment field for each sentence in `blob`. Note that `sentiment` is a Python-based field name accessible via Julia. As detailed in the TextBlob documentation, the sentiment of the blob is as an ordered pair of polarity and subjectivity, with polarity measured over $[-1.0, 1.0]$ (very negative to very positive), and subjectivity over $[0.0, 1.0]$ (very objective to very subjective). The results indicate that the first sentence is the most positive but is also the most subjective, while the last sentence is the most negative but also more objective.

Other Integrations

Julia also allows C and Fortran calls to be made directly via the `ccall()` function, which is in Julia `Base`. These calls are made without adding any extra overhead than a standard library call from C code. Note that the code to be called must be available as a shared library. For example, in Windows systems, `msvcrt` can be called instead of `libc` (`msvcrt` is a module containing C library functions, and is part of the Microsoft C Runtime Library).

When using the `ccall()` function, shared libraries are referenced in the format `(:function, "library")`. The following is an example where the C function `cos()` is called:

```
ccall( (:cos, "msvcrt"), Float64, (Float64,), pi ).
```

For this example, the `cos()` function is called from the `msvcrt` library. Here, `ccall()` takes four arguments, the first is the function and library as a tuple, the second is the return type, the third is a tuple of input types (here there is just one), and the last is the input argument, π in this case. Running this in Julia on a Windows machine returns -1.

There are also several other packages that support various other languages as well, such as `Cxx.jl` and `CxxWrap.jl` for C++, `MATLAB.jl` for Matlab, and `JavaCall.jl` for Java. Note that many of these packages are available from `https://github.com/JuliaInterop`.

Chapter 2

Basic Probability

In this chapter we introduce elementary probability concepts. We describe key notions of a probability space along with independence and conditional probability. It is important to note that most of the probabilistic analysis carried out in statistics is based on distributions of random variables. These are introduced in the next chapter. In this chapter we focus solely on probability, events, and the simple mathematical setup of a random experiment embodied in a probability space.

The notion of *probability* is the chance of something happening, quantified as a number between 0 and 1 with higher values indicating a higher likelihood of occurrence. However, how do we formally describe probabilities? The standard way is to consider a *probability space* which mathematically consists of three elements: (1) A *sample space*—the set of all possible outcomes of a certain *experiment*. (2) A collection of *events*—each event is a subset of the sample space. (3) A *probability measure* also denoted here as *probability function*—which indicates the chance of each possible event occurring. Note: do not confuse the probability function with a probability mass function, which we define in the next chapter.

As a simple example, consider the case of flipping a coin twice. Since the sample space is the set of all possible outcomes, we can represent the sample space mathematically as follows:

$$\Omega = \{hh, ht, th, tt\}.$$

Now that the sample space, Ω, is defined, we can consider individual events. For example, let A be the event of getting at least one heads. Hence,

$$A = \{hh, ht, th\}.$$

Or alternately, let B be the event of getting one heads and one tails in any order,

$$B = \{ht, th\}.$$

There can also be events that consist of a single possible outcome, for example, $C = \{th\}$ is the event of getting tails first, followed by heads. Mathematically, the important point is that events are subsets of Ω and often contain more than one outcome. Possible events also include the empty set, $\emptyset$ (nothing happening) and Ω itself (something happening). In the setup of probability, we assume there is a *random experiment* where something is bound to happen.

Y. Nazarathy and H. Klok, *Statistics with Julia*, Springer Series in the Data Sciences,
https://doi.org/10.1007/978-3-030-70901-3_2

The final component of a probability space is the probability function, also sometimes called the *probability measure*. This function, $\mathbb{P}(\cdot)$, takes an event as an input argument and returns a real number in the range $[0, 1]$. It always satisfies $\mathbb{P}(\emptyset) = 0$ and $\mathbb{P}(\Omega) = 1$. It also satisfies the fact that the probability of the union of two disjoint events is the sum of their probabilities, and furthermore the probability of the complement of an event is one minus the original probability.

We now explore all these concepts via examples. This chapter is structured as follows: In Section 2.1 we explore the basic setup of random experiments with a few examples. In Section 2.2 we explore working with sets in Julia as well as probability examples dealing with unions of events. In Section 2.3 we introduce and explore the concept of independence. In Section 2.4 we move on to conditional probability. Finally, in Section 2.5 we explore Bayes' rule for conditional probability.

2.1 Random Experiments

We now explore a few examples where we set up a *probability space*. In most examples we present a Monte Carlo simulation of the random experiment, and then compare results to theoretical ones where possible.

Rolling Two Dice

Consider the *random experiment* where two independent, fair, six-sided dice are rolled, and we wish to find the probability that the sum of the outcomes of the dice is even. Here the sample space can be represented as $\Omega = \{1, \ldots, 6\}^2$, i.e. the *Cartesian product* of the set of single roll outcomes with itself. That is, elements of the sample space are *tuples* of the form (i, j) with $i, j \in \{1, \ldots, 6\}$. Say we are interested in the probability of the event,

$$A = \{(i, j) \mid i + j \text{ is even}\}.$$

In this random experiment, since the dice have no inherent bias, it is sensible to assume a *symmetric probability function*. That is, for any $B \subset \Omega$,

$$\mathbb{P}(B) = \frac{|B|}{|\Omega|},$$

	1	**2**	**3**	**4**	**5**	**6**
1	2	3	4	5	6	7
2	3	4	5	6	7	8
3	4	5	6	7	8	9
4	5	6	7	8	9	10
5	6	7	8	9	10	11
6	7	8	9	10	11	12

Table 2.1: All possible outcomes for the sum of two dice. Even sums are shaded.

where $|\cdot|$ counts the number of elements in the set. It is called symmetric because every outcome in Ω has the same probability. Hence for our event, A, we can see from Table 2.1 that

$$\mathbb{P}(A) = \frac{18}{36} = 0.5.$$

We now obtain this in Julia via both direct calculation and Monte Carlo simulation. A direct calculation counts the number of even faces. A Monte Carlo simulation repeats the experiment many times and estimates $\mathbb{P}(A)$ based on the number of times that event A occurred.

Listing 2.1: Even sum of two dice

```
1  N, faces = 10^6, 1:6
2
3  numSol  = sum([iseven(i+j) for i in faces, j in faces]) / length(faces)^2
4  mcEst   = sum([iseven(rand(faces) + rand(faces)) for i in 1:N]) / N
5
6  println("Numerical solution = $numSol \nMonte Carlo estimate = $mcEst")
```

```
Numerical solution = 0.5
Monte Carlo estimate = 0.499644
```

In line 1 we set the number of simulation runs, `N`, and the range of faces on the dice, `1:6`. In line 3, we use a comprehension to cycle through the sum of all possible combinations of the addition of the outcomes of the two dice. The outcome of the two dice is represented by `i` and `j`, respectively, both of which take on the values of `faces`. We start with `i`=1, `j`=1 and add them, and we use the `iseven()` function to return `true` if even, and `false` if not. We then repeat the process for `i`=1, `j`=2, and so on, all the way to `i`=6, `j`=6. Finally, we count the number of `true` values by summing all the elements of the comprehension via `sum()`. The result, normalized by the total number of possible outputs, is stored in `numSol`. Line 4 also uses a comprehension, but in this case we uniformly and randomly select the values which the dice take, akin to rolling them. Again `iseven()` is used to return `true` if even and `false` if not, and we repeat this process `N` times. Using similar logic to line 3, we store the proportion of outcomes which were true in `mcEst`. Line 6 prints the results using the `println()` function. Notice the use of \n for creating a newline.

Partially Matching Passwords

We now consider an alphanumeric example. Assume that a password to a secured system is exactly 8 characters in length. Each character is one of 62 possible characters: the letters "a"–"z", the letters "A"–"Z" or the digits "0"–"9".

In this example let Ω be the set of all possible passwords, i.e. $|\Omega| = 62^8$. Now, again assuming a symmetric probability function, the probability of an attacker guessing the correct (arbitrary) password is $62^{-8} \approx 4.6 \times 10^{-15}$. Hence at a first glance, the system seems very secure.

Elaborating on this example, let us also assume that as part of the system's security infrastructure, when a login is attempted with a password that matches 1 or more of the characters, an event is logged in the system's security portal (taking up hard-drive space). For example, say the original password is 3xyZu4vN, and a login is attempted using the password **3**5xyZ**4vN**. In this case 4 of the characters match (displayed in bold) and therefore an event is logged.

While the chance of guessing a password is astronomically low, in this simple (fictional and overly simplistic) system, there exists a secondary security flaw. That is, hackers may attempt to overload the event logging system via random attacks. If hackers continuously try to log into the system with random passwords, every password that matches one or more characters will log an event, thus taking up more hard-drive space.

We now ask what is the probability of logging an event with a random password? Denote the event of logging a password A. In this case, it turns out to be much more convenient to consider the *complement*, $A^c := \Omega \setminus A$, which is the event of having 0 character matches. We have that $|A^c| = 61^8$ because given any (arbitrary) correct password, there are $61 = 62 - 1$ character options for each character, in order to ensure A^c holds. Hence,

$$\mathbb{P}(A^c) = \frac{61^8}{62^8} \approx 0.87802.$$

We then have that the probability of logging an event is $\mathbb{P}(A) = 1 - \mathbb{P}(A^c) \approx 0.12198$. So if, for example, 10^7 login attempts are made, we can expect that about 1.2 million login attempts would be written to the security log. We now simulate such a scenario in Listing 2.2.

Listing 2.2: Password matching

```
using Random
Random.seed!()

passLength, numMatchesForLog = 8, 1
possibleChars = ['a':'z' ; 'A':'Z' ; '0':'9']

correctPassword = "3xyZu4vN"

numMatch(loginPassword) =
    sum([loginPassword[i] == correctPassword[i] for i in 1:passLength])

N = 10^7

passwords = [String(rand(possibleChars,passLength)) for _ in 1:N]
numLogs   = sum([numMatch(p) >= numMatchesForLog for p in passwords])
println("Number of login attempts logged: ", numLogs)
println("Proportion of login attempts logged: ", numLogs/N)
```

```
Number of login attempts logged: 1221801
Proportion of login attempts logged: 0.1221801
```

In line 2 the seed of the random number generator is set so that the same passwords are generated each time the code is run. This is done for reproducibility. In line 4 the password length is defined along with the minimum number of character matches before a security log entry is created. In line 5 an array is created, which contains all valid characters which can be used in the password. Note the use of "`;`", which performs *array concatenation* of the three ranges of characters. In line 7 we set an arbitrary correct login password. Note that the type of `correctPassword` is a `String` containing only characters from `possibleChars`. In lines 9 and 10 the function `numMatch()` is created, which takes the password of a login attempt and checks each index against that of the actual password. If the index character is correct, it evaluates `true`, else `false`. The function then returns how many characters were correct by using `sum()`. Line 14 uses the function `rand()` and the constructor `String()` along with a comprehension to randomly generate `N` passwords. Note that `String()` is used to convert from an array of single characters to a string. Line 15 checks how many times `numMatchesForLog` or more characters were guessed correctly, for each password in our array of randomly generated passwords. It then stores how many times this occurs as the variable `numLogs`.

The Birthday Problem

For our next example, consider a room full of people. We then ask what is the probability of finding a pair of people that share the same birthday. Obviously, ignoring leap years, if there are 366 people present, then it happens with certainty via the *pigeonhole principle*. However, what if there are fewer people? Interestingly, with about 50 people, a birthday match is almost certain, and with 23 people in a room, there is about a 50% chance of two people sharing a birthday. At first glance this non-intuitive result is surprising, and hence this famous probability example earned the name *the birthday paradox*. However, we just refer to it as the *birthday problem*.

To carry out the analysis, we assume birthdays are uniformly distributed in the set $\{1, \dots, 365\}$. For n people in a room, we wish to evaluate the probability that at least two people share the same birthday. Set the sample space, Ω, to be composed of ordered tuples $(x_1, \dots, x_n)$ with $x_i \in \{1, \dots, 365\}$. Hence, $|\Omega| = 365^n$. Now set the event A to be the set of all tuples $(x_1, \dots, x_j)$ where $x_i = x_j$ for some distinct i and j.

As in the previous example, we consider A^c instead. It consists of tuples where $x_i \neq x_j$ for all distinct i and j (the event of no birthday pair in the group). In this case,

$$|A^c| = 365 \cdot 364 \cdot \ldots \cdot (365 - n + 1) = \frac{365!}{(365 - n)!}.$$

Hence we have,

$$\mathbb{P}(A) = 1 - \mathbb{P}(A^c) = 1 - \frac{|A^c|}{|\Omega|} = 1 - \frac{365 \cdot 364 \cdot \ldots \cdot (365 - n + 1)}{365^n}. \tag{2.1}$$

From this we can compute that for $n = 23$, $\mathbb{P}(A) \approx 0.5073$, and for $n = 50$, $\mathbb{P}(A) \approx 0.9704$.

The code in Listing 2.3 calculates both the analytic probabilities, as well as estimates them via Monte Carlo (MC) simulation. The results are presented in Figure 2.1. For the numerical solutions, it employs two alternative implementations, `matchExists1()` and `matchExists2()`. The maximum error between the two numerical implementations is presented.

Listing 2.3: The birthday problem

```
using StatsBase, Combinatorics, Plots ; pyplot()

matchExists1(n) = 1 - prod([k/365 for k in 365:-1:365-n+1])
matchExists2(n) = 1- factorial(365,365-big(n))/365^big(n)

function bdEvent(n)
    birthdays = rand(1:365,n)
    dayCounts = counts(birthdays, 1:365)
    return maximum(dayCounts) > 1
end

probEst(n) = sum([bdEvent(n) for _ in 1:N])/N

xGrid = 1:50
analyticSolution1 = [matchExists1(n) for n in xGrid]
analyticSolution2 = [matchExists2(n) for n in xGrid]
println("Maximum error: $(maximum(abs.(analyticSolution1 - analyticSolution2)))")

N = 10^3
mcEstimates = [probEst(n) for n in xGrid]

plot(xGrid, analyticSolution1, c=:blue, label="Analytic solution")
scatter!(xGrid, mcEstimates, c=:red, ms=6, msw=0, shape=:xcross,
        label="MC estimate", xlims=(0,50), ylims=(0, 1),
        xlabel="Number of people in room",
        ylabel="Probability of birthday match",
        legend=:topleft)
```

```
Maximum error: 2.4611723650627278208929385e-16
```

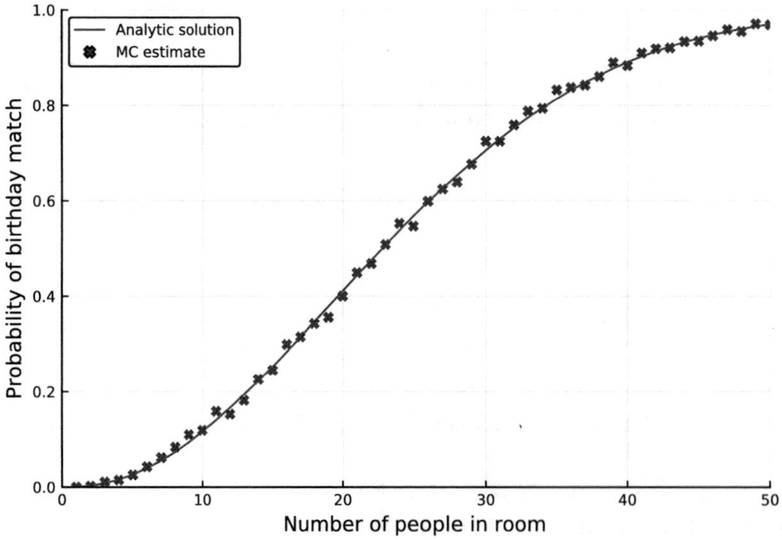

Figure 2.1: Probability that in a room of n people, at least two people share a birthday.

In lines 3 and 4, two alternative functions for calculating the probability in (2.1) are defined, `matchExists1()` and `matchExists2()`, respectively. The first uses the `prod()` function to apply a product over a comprehension. This is in fact a numerically stable way of evaluating the probability. The second implementation evaluates (2.1) in a much more explicit manner. It uses the `factorial()` function from the `Combinatorics` package. Note that the basic `factorial()` function is included in Julia `Base`; however, the method with two arguments comes from the `Combinatorics` package. Also, the use of `big()` ensures the input argument is a `BigInt` type. This is needed to avoid overflow for non-small values of `n`. Lines 6–10 define the function `bdEvent()`, which simulates a room full of n people, and if at least two people share a birthday, returns `true`, otherwise returns `false`. We now explain how it works. Line 7 creates the array `birthdays` of length n, and uniformly and randomly assigns an integer in the range $[1, 365]$ to each index. The values of this array can be thought of as the birth dates of individual people. Line 8 uses the function `counts()` from the `StatsBase` package to count how many times each birth date occurs in `birthdays`, and assigns these counts to the new array `dayCounts`. The logic can be thought of as follows: if two indices have the same value, then this represents two people having the same birthday. Line 9 checks the array `dayCounts`, and if the maximum value of the array is greater than one (i.e. if at least two people share the same birth date) then returns `true`, else `false`. Line 12 defines the function `probEst()`, which, when given n number of people, uses a comprehension to simulate `N` rooms, each containing `n` people. For each element of the comprehension, i.e. room, the `bdEvent()` function is used to check if at least one birthday pair exists. Then, for each room, the total number of at least one birthday pair is summed up and divided by the total number of rooms `N`. For large `N`, the function `probEst()` will be a good estimate for the analytic solution of finding at least one birthday pair in a room of `n` people. Lines 14–17 evaluate the analytic solutions over the grid, `xGrid`, and prints the maximal absolute error between the solutions. The output shows that the numerical error is negligible. Line 20 evaluates the Monte Carlo estimates. Lines 22–27 plot the analytic and numerical estimates of these probabilities on the same graph.

Sampling With and Without Replacement

Consider a small pond with a small population of 7 fish, 3 of which are gold and 4 of which are silver. Now say we fish from the pond until we catch 3 fish, either gold or silver. Let G_n denote the event of catching n gold fish. It is clear that unless $n = 0, 1, 2$, or 3, $\mathbb{P}(G_n) = 0$. However, what is $\mathbb{P}(G_n)$ for $n = 0, 1, 2, 3$? Before continuing, let us make a distinction between two sampling policies:

Catch and keep —We sample from the population *without replacement.* That is, whenever we catch a fish, we remove it from the population.

Catch and release —We sample from the population *with replacement.* That is, whenever we catch a fish, we return it to the population (pond) before continuing to fish.

The computation of the probabilities $\mathbb{P}(G_n)$ for these two cases of catch and keep, and catch and release, may be obtained via the *Hypergeometric distribution* and *Binomial distribution*, respectively. These are both covered in more detail in Section 3.5. We now estimate these probabilities using Monte Carlo simulation. Listing 2.4 simulates each policy N times; counts how many times zero, one, two, and three gold fish are sampled in total; and finally presents these as proportions of the total number of simulations. Note that the total probability in both cases sum to one. The probabilities are plotted in Figure 2.2.

Listing 2.4: Fishing with and without replacement

```
using StatsBase, Plots ; pyplot()

function proportionFished(gF,sF,n,N,withReplacement = false)
    function fishing()
        fishInPond = [ones(Int64,gF); zeros(Int64,sF)]
        fishCaught = Int64[]

        for fish in 1:n
            fished = rand(fishInPond)
            push!(fishCaught,fished)
            if withReplacement == false
                deleteat!(fishInPond, findfirst(x->x==fished, fishInPond))
            end
        end
        sum(fishCaught)
    end

    simulations = [fishing() for _ in 1:N]
    proportions = counts(simulations,0:n)/N

    if withReplacement
        plot!(0:n, proportions,
             line=:stem, marker=:circle, c=:blue, ms=6, msw=0,
             label="With replacement",
             xlabel="n",
             ylims=(0, 0.6), ylabel="Probability")
    else
        plot!(0:n, proportions,
            line=:stem, marker=:xcross, c=:red, ms=6, msw=0,
            label="Without replacement")
    end
end

N = 10^6
goldFish, silverFish, n = 3, 4, 3
plot()
proportionFished(goldFish, silverFish, n, N)
proportionFished(goldFish, silverFish, n, N, true)
```

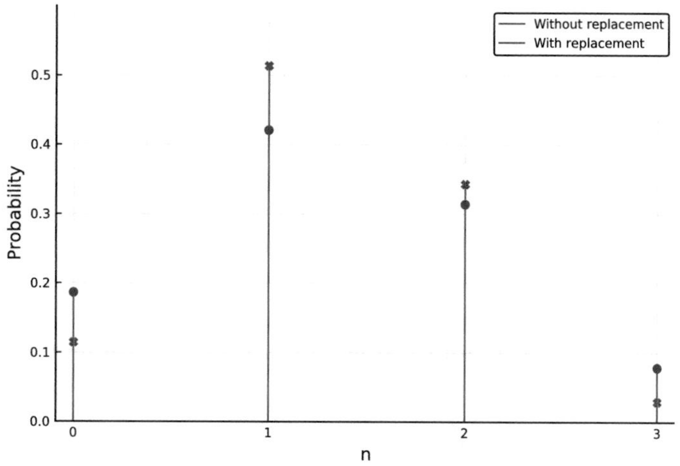

Figure 2.2: Estimated probabilities of catching n of gold fish, with and without replacement.

Lines 3–32 define the function `proportionFished()`, which takes five arguments: the number of gold fish in the pond `gF`, the number of silver fish in the pond `sF`, the number of times we catch a fish `n`, the total number of simulation runs `N`, and a policy of whether we throw back (i.e. replace) each caught fish, `withReplacement`, which is set to `false` by default. In lines 4–16 we create an inner function `fishing()` that generates one random instance of a fishing day, returning the number of gold fish caught. Line 5 generates an array, where the values in the array represent fish in the pond, with 0's and 1's representing silver and gold fish, respectively. Notice the use of the `zeros()` and `ones()` functions, each with a first argument, `Int64` indicating the Julia type. Line 6 initializes an empty array, which represents the fish to be caught. Lines 8–14 perform the act of fishing `n` times via the use of a `for` loop. Lines 9–10 randomly sample a "fish" from our "pond", and then stores this in value in our `fishCaught` array. Line 12 is only run if `false` is used, in which case we "remove" the caught "fish" from the pond via the function `deleteat!()`. Note that technically we don't remove the exact caught fish, but rather a fish with the same value (0 or 1) via `findfirst()`. Our use of this function returns the first index in `fishInPond` with a value equaling `fished`. Line 15 is the (implicit) return statement for the function `fishing()` and is the sum of how many gold fish were caught (since gold fish are stored as 1's and silver fish as 0's). Line 18 implements our chosen policy N times total, with the total number of gold fish each time stored in the array `simulations`. Line 19 uses the `counts()` function to return the proportion of times $0, \dots, n$ gold fish were caught. Lines 21–31 then use `plot!()` to overlay the existing plot with the probabilities. The `proportionFished()` function is then called twice in lines 37 and 38 to generate the resulting plot.

Lattice Paths

We now consider a square grid on which an ant walks from the southwest corner to the northeast corner, taking either a step north or a step east at each grid intersection. This is illustrated in Figure 2.3 where it is clear that there are many possible paths the ant could take. Let us set the sample space to be

$$\Omega = \text{All possible lattice paths},$$

where the term *lattice path* describes a trajectory of the ant going from the southwest point, $(0,0)$ to the northeast point, (n,n). Since Ω is finite, we can consider the number of elements in it, denoted $|\Omega|$. For a general $n \times n$ grid,

$$|\Omega| = \binom{2n}{n} = \frac{(2n)!}{(n!)^2}.$$

For example if $n = 5$ then $|\Omega| = 252$. The use of the *binomial coefficient* here is because out of the $2n$ steps that the ant needs to take, n steps need to be "north" and n need to be "east".

Within this context of lattice paths, there are a variety of questions. One common question has to do with the event (or set):

$$A = \text{Lattice paths that stay above the diagonal the whole way from } (0,0) \text{ to } (n,n).$$

The set A then describes all lattice paths where at any point, the ant has not taken more easterly steps than northerly steps. The question of the size of A, namely $|A|$, has interested many people in combinatorics, and it turns out that

$$|A| = \frac{\binom{2n}{n}}{n+1}.$$

For each counting value of n, the above is called the n-th *Catalan Number*. For example, if $n = 1$ then $|A| = 1$, if $n = 2$, $|A| = 2$, and if $n = 3$ then $|A| = 5$. You can try to sketch all possible paths in A for $n = 3$ (there are 5 in total).

So far we have discussed the sample space Ω, and a potential event A. One interesting question to ask deals with the probability of A. That is: *What is the chance that the ant stays on or above the diagonal as it journey's from* $(0,0)$ *to* (n,n)*?*

The answer to this question depends on the probability function/measure that we specify for this experiment (sometimes called a *probability model*). There are infinitely many choices for the model and the choice of the right model depends on the context. Here we consider two examples:

Model I—As in the previous examples, assume a symmetric probability space, i.e. each lattice path is equally likely. For this model, obtaining probabilities is a question of counting and the result just follows the combinatorial expressions above:

$$\mathbb{P}_{\text{I}}(A) = \frac{|A|}{|\Omega|} = \frac{1}{n+1}. \tag{2.2}$$

Model II—We assume that at each grid intersection where the ant has an option of where to go ("east" or "north"), it chooses either east or north, both with equal probability $1/2$. In the case where there is no option for the ant (i.e. it hits the east or north border) then it simply continues along the border to the final destination (n, n). For this model, it isn't as simple to obtain an expression for $\mathbb{P}(A)$. One way to do it is by considering a *recurrence relation* for the probabilities (sometimes known as *first step analysis*). We omit the details and present the result:

$$\mathbb{P}_{\text{II}}(A) = \frac{\binom{2n-1}{n}}{2^{2n-1}}.$$

Hence we see that the probability of the event depends on the probability model used—and this choice is not always a straightforward nor obvious one. For example, for $n = 5$ we have

$$\mathbb{P}_{\text{I}}(A) = \frac{1}{6} \approx 0.166, \qquad \mathbb{P}_{\text{II}}(A) = \frac{126}{512} \approx 0.246.$$

We now verify these values for $\mathbb{P}_{\text{I}}(A)$ and $\mathbb{P}_{\text{II}}(A)$ by simulating both Model I and Model II in Listing 2.5, which also creates Figure 2.3.

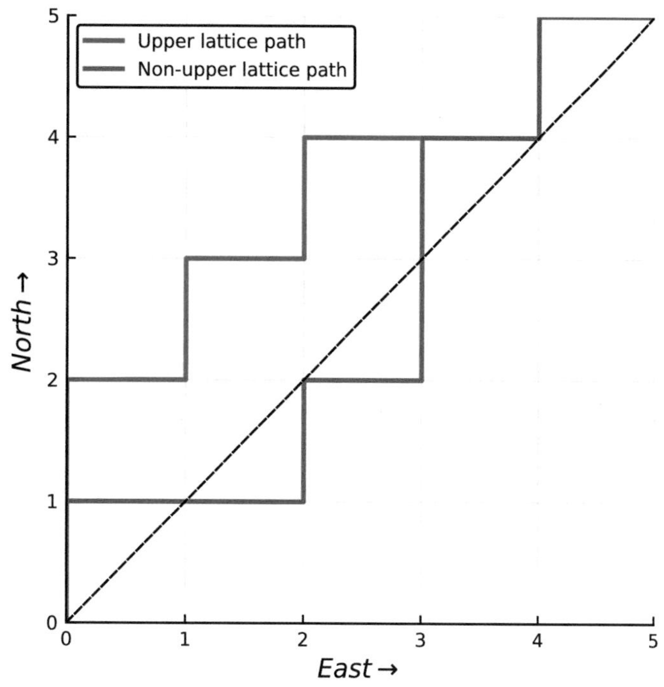

Figure 2.3: Example of two different lattice paths.

Listing 2.5: Lattice paths

```
using Random, Combinatorics, Plots, LaTeXStrings ; pyplot()
Random.seed!(12)

n, N = 5, 10^5

function isUpperLattice(v)
    for i in 1:Int(length(v)/2)
        sum(v[1:2*i-1]) >= i ? continue : return false
    end
    return true
end

omega = unique(permutations([zeros(Int,n);ones(Int,n)]))
A = omega[isUpperLattice.(omega)]
pA_modelI = length(A)/length(omega)

function randomWalkPath(n)
    x, y = 0, 0
    path = []
    while x<n && y<n
       if rand()<0.5
           x += 1
           push!(path,0)
        else
           y += 1
           push!(path,1)
        end
    end
    append!(path, x<n ? zeros(Int64,n-x) : ones(Int64,n-y))
    return path
end

pA_modelIIest = sum([isUpperLattice(randomWalkPath(n)) for _ in 1:N])/N
println("Model I: ",pA_modelI, "\t Model II: ", pA_modelIIest)

function plotPath(v,l,c)
    x,y = 0,0
    graphX, graphY = [x], [y]
    for i in v
        if i == 0
            x += 1
        else
            y += 1
        end
        push!(graphX,x), push!(graphY,y)
    end
    plot!(graphX, graphY,
            la=0.8, lw=2, label=l, c=c, ratio=:equal, legend=:topleft,
            xlims=(0,n), ylims=(0,n),
            xlabel=L"East\rightarrow", ylabel=L"North\rightarrow")
end
plot()
plotPath(rand(A), "Upper lattice path", :blue)
plotPath(rand(setdiff(omega,A)), "Non-upper lattice path", :red)
plot!([0, n], [0,n], ls=:dash, c=:black, label="")
```

```
Model I: 0.16666666666666666       Model II: 0.24696
```

In the code, a path is encoded by a sequence of 0 and 1 values, indicating "move east" or "move north", respectively. The function `isUpperLattice()` defined in lines 5–10 checks if a path is an upper lattice path by summing all the odd partial sums, and returning false if any sum ends up at a coordinate below the diagonal. Note the use of the `? :` operator in line 8. Also note that in line 7, `Int()` is used to convert the division `length(v)/2` to an integer type. In line 12, a collection of all possible lattice paths is created by applying the `permutations()` function from the `Combinatorics` package to an initial array of `n` zeros and `n` ones. The `unique()` function is then used to remove all duplicates. In line 14 the `isUpperLattice()` function is applied to each element of `omega` via the "`.`" operator just after the function name. The result is a Boolean array. Then `omega[]` selects the indices of `omega` where the value is `true` and in the next line `pA_modelI` is calculated. In lines 17–31 the function `randomWalkPath()` is implemented, which creates a random path according to Model II. Note that the code in line 29 appends either zeros or ones to the path, depending on if it hit the north boundary or east boundary first. Then in line 33, the Monte Carlo estimate, `pA_modelIIest`, is determined. The function `plotPath()` defined in lines 36–51 plots a path with a specified label and color. It is then invoked in line 53 for an upper lattice path selected via `rand(A)` and again in the next line for a non-upper path by using `setdiff(omega,A)` to determine the collection of non-upper lattice paths. Functions dealing with sets are covered in more detail in the next section.

2.2 Working with Sets

As evident from the examples in Section 2.1, mathematical *sets* play an integral part in the evaluation of probability models. Subsets of the sample space Ω are also called *events*. By carrying out *intersections*, *unions*, and *differences* of sets, we may often express more complicated events based on smaller ones.

A set is an unordered collection of unique *elements*. A set A is a *subset* of the set B if every element that is in A is also an element of B. The *union* of two sets, A and B, denoted $A \cup B$ is the set of all elements that are either in A or B, or both. The *intersection* of the two sets, denoted $A \cap B$, is the set of all elements that are in both A and B. The *difference*, denoted $A \setminus B$, is the set of all elements that are in A but not in B.

In the context of probability, the sample space Ω is often considered as the *universal set*. This allows us to then consider the *complement* of a set A, denoted A^c, which can be constructed via all elements of Ω that are not in A. Note that $A^c = \Omega \setminus A$. Also observe that in the presence of a universal set: $A \setminus B = A \cap B^c$.

Representing Sets in Julia

Julia includes built-in capability for working with sets. Unlike an `Array`, a `Set` is an unordered collection of unique objects. Listing 2.6 illustrates how to construct a `Set` in Juila, and illustrates the use of the `union()`, `intersect()`, `setdiff()`, `issubset()`, and `in()` functions. There are also other functions related to sets that you may explore independently. These include `issetequal()`, `symdiff()`, `union!()`, `setdiff!()`, `symdiff!()`, and `intersect!()`. See the online Julia documentation under "Collections and Data Structures".

Listing 2.6: Basic set operations

```
1  A = Set([2,7,2,3])
2  B = Set(1:6)
3  omega = Set(1:10)
4
5  AunionB = union(A, B)
6  AintersectionB = intersect(A, B)
7  BdifferenceA = setdiff(B,A)
8  Bcomplement = setdiff(omega,B)
9  AsymDifferenceB = union(setdiff(A,B),setdiff(B,A))
10 println("A = $A, B = $B")
11 println("A union B = $AunionB")
12 println("A intersection B = $AintersectionB")
13 println("B diff A = $BdifferenceA")
14 println("B complement = $Bcomplement")
15 println("A symDifference B = $AsymDifferenceB")
16 println("The element '6' is an element of A: $(in(6,A))")
17 println("Symmetric difference and intersection are subsets of the union: ",
18         issubset(AsymDifferenceB,AunionB),", ", issubset(AintersectionB,AunionB))
```

```
A = Set([7, 2, 3]), B = Set([4, 2, 3, 5, 6, 1])
A union B = Set([7, 4, 2, 3, 5, 6, 1])
A intersection B = Set([2, 3])
B diff A = Set([4, 5, 6, 1])
B complement = Set([7, 9, 10, 8])
A symDifference B = Set([7, 4, 5, 6, 1])
The element '6' is an element of A: false
Symmetric difference and intersection are subsets of the union: true, true
```

In lines 1–3 three different sets are created via the `Set()` function (a constructor). Note that `A` contains only three elements, since sets are meant to be a collection of unique elements. Also note that unlike arrays order is not preserved. Lines 5–9 perform various operations using the sets created. Lines 10–18 create the listing output. Note the use of the functions `in()` and `issubset()` in lines 16–18.

The Probability of a Union

Consider now two events (sets) A and B. If $A \cap B = \emptyset$, then $\mathbb{P}(A \cup B) = \mathbb{P}(A) + \mathbb{P}(B)$. However more generally, when A and B are not *disjoint*, the probability of the *intersection*, $A \cap B$ plays a role. For such cases the *inclusion exclusion formula* is useful:

$$\mathbb{P}(A \cup B) = \mathbb{P}(A) + \mathbb{P}(B) - \mathbb{P}(A \cap B). \tag{2.3}$$

To help illustrate this, consider the simple example of choosing a random lowercase letter, "a"–"z". Let A be the event that the letter is a vowel (one of "a", "e", "i", "o", "u"). Let B be the event that the letter is one of the first three letters (one of "a", "b", "c"). Now since $A \cap B = \{\text{'a'}\}$, a set with one element, we have

$$\mathbb{P}(A \cup B) = \frac{5}{26} + \frac{3}{26} - \frac{1}{26} = \frac{7}{26}.$$

For another similar example, consider the case where A is the set of vowels as before, but $B = \{\text{"x"}, \text{"y"}, \text{"z"}\}$. In this case, since the intersection of A and B is empty, we immediately know that

$\mathbb{P}(A \cup B) = (5+3)/26 \approx 0.3077$. While this example is elementary, we now use it to illustrate a type of conceptual error that one may make when using Monte Carlo simulation.

Consider code Listing 2.7, and compare `mcEst1` and `mcEst2` from lines 12 and 13, respectively. Both variables are designed to be estimators of $\mathbb{P}(A \cup B)$. However, one of them is a correct estimator and the other is faulty. In the following we look at the output given from of both, and explore the fault in the underlying logic.

Listing 2.7: An innocent mistake with Monte Carlo

```
1   using Random, StatsBase
2   Random.seed!(1)
3
4   A = Set(['a','e','i','o','u'])
5   B = Set(['x','y','z'])
6   omega = 'a':'z'
7
8   N = 10^6
9
10  println("mcEst1 \t \tmcEst2")
11  for _ in 1:5
12      mcEst1 = sum([in(sample(omega),A) || in(sample(omega),B) for _ in 1:N])/N
13      mcEst2 = sum([in(sample(omega),union(A,B)) for _ in 1:N])/N
14      println(mcEst1,"\t",mcEst2)
15  end
```

First observe line 12. In Julia, `||` means "or", so at first glance the estimator `mcEst1` looks sensible, since:

$$A \cup B = \text{the set of all elements that are in } A \text{ or } B.$$

Hence we are generating a random element via `sample(omega)` and checking if it is an element of A or an element of B. However there is a subtle error. Each of the N random experiments involves two separate calls to `sample(omega)`. Hence the code in line 12 simulates a situation where conceptually, the sample space, Ω, is composed of pairs of letters (2-tuples), not single letters!

Hence the code computes probabilities of the event, $A_1 \cup B_2$, where

$$\begin{aligned} A_1 &= \text{First element of the tuple is a vowel,} \\ B_2 &= \text{Second element of the tuple is an "x", "y", or "z" letter.} \end{aligned}$$

Now observe that A_1 and B_2 are not disjoint events, hence,

$$\mathbb{P}(A_1 \cup B_2) = \mathbb{P}(A_1) + \mathbb{P}(B_2) - \mathbb{P}(A_1 \cap B_2).$$

Further it holds that $\mathbb{P}(A_1 \cap B_2) = \mathbb{P}(A_1)\mathbb{P}(B_2)$. This follows from independence (further explored in Section 2.3). Now that we have identified the error, we can predict the resulting output.

$$\mathbb{P}(A_1 \cup B_2) = \mathbb{P}(A_1) + \mathbb{P}(B_2) - \mathbb{P}(A_1)\mathbb{P}(B_2) = \frac{5}{26} + \frac{3}{26} - \frac{5}{26}\frac{3}{26} \approx 0.2855.$$

It can be seen from the code output, which repeats the comparison 5 times, that `mcEst1` consistently underestimates the desired probability, yielding estimates near 0.2855 instead.

```
mcEst1      mcEst2
0.285158    0.307668
0.285686    0.307815
0.285022    0.308132
0.285357    0.307261
0.285175    0.306606
```

In lines 11–15 a `for` loop is implemented, which generates 5 Monte Carlo predictions. Note that lines 12 and 13 contain the main logic of this example. Line 12 is our incorrect simulation, and yields incorrect estimates. See the text above for a detailed explanation as to why the use of two separate calls to `sample()` are incorrect in this case. Line 13 is our correct simulation, and for large `N` yields results close to the expected result. Note that the `union()` function is used on `A` and `B`, instead of the "or" operator, `||`, used in line 12. The important point is that only a single sample is generated for each iteration of the composition.

Secretary with Envelopes

Now consider a more general form of the *inclusion exclusion principle* applied to a collection of sets, $C_1, \ldots, C_n$. It is presented below, written in two slightly different forms:

$$\mathbb{P}\Big(\bigcup_{i=1}^{n} C_i\Big) = \sum_{i=1}^{n} \mathbb{P}(C_i) - \sum_{\text{pairs}} \mathbb{P}(C_i \cap C_j) + \sum_{\text{triplets}} \mathbb{P}(C_i \cap C_j \cap C_k) - \ \ldots \ + (-1)^{n-1}\mathbb{P}(C_1 \cap \ldots \cap C_n)$$

$$= \sum_{i=1}^{n} \mathbb{P}(C_i) - \sum_{i<j} \mathbb{P}(C_i \cap C_j) + \sum_{i<j<k} \mathbb{P}(C_i \cap C_j \cap C_k) - \ \ldots \ + (-1)^{n-1}\, \mathbb{P}\Big(\bigcap_{i=1}^{n} C_i\Big).$$

Notice that there are n major terms. The first term deals with probabilities of individual events; the second term deals with pairs; the third with triplets; and the sequence continues until a single final term involving a single intersection is reached. The ℓ-th term has $\binom{n}{\ell}$ summands. For example, there are $\binom{n}{2}$ pairs, $\binom{n}{3}$ triplets, etc. Notice also the alternating signs via $(-1)^{\ell-1}$. It is possible to conceptually see the validity of this formula for the case of $n = 3$ by drawing a *Venn diagram* and seeing the role of all summands. In this case,

$$\mathbb{P}(C_1 \cup C_2 \cup C_3) = \mathbb{P}(C_1) + \mathbb{P}(C_2) + \mathbb{P}(C_3) - \mathbb{P}(C_1 \cap C_2) - \mathbb{P}(C_1 \cap C_3) - \mathbb{P}(C_2 \cap C_3) + \mathbb{P}(C_1 \cap C_2 \cap C_3).$$

Let us now consider a classic example that uses this inclusion exclusion principle. Assume that a secretary has an equal number of pre-labeled envelopes and business cards, n. Suppose that at the end of the day, he is in such a rush to go home that he puts each business card in an envelope at random without any thought of matching the business card to its intended recipient on the envelope. The probability that each of the business cards will go to the correct envelope is easy to obtain. It is $1/n!$, which goes to zero very quickly as n grows. However, what is the probability that each of the business cards will go to a wrong envelope?

As an aid, let A_i be the event that the i-th business card is put in the correct envelope. We have a handle on events involving intersections of distinct A_i values. For example, if $n = 10$, then $\mathbb{P}(A_1 \cap A_4 \cap A_6) = 7!/10!$, or more generally, the probability of an intersection of k such events is $p_k := (n-k)!/n!$.

The event we are seeking to evaluate is $B = A_1^c \cap A_2^c \cap \ldots \cap A_n^c$. Hence by *De Morgan's laws*, $B^c = A_1 \cup \ldots \cup A_n$. Hence using the inclusion exclusion formula together with p_k, we can simplify factorials and binomial coefficients to obtain

$$\mathbb{P}(B) = 1 - \mathbb{P}(A_1 \cup \ldots \cup A_n) = 1 - \sum_{k=1}^{n} (-1)^{k+1} \binom{n}{k} p_k = 1 - \sum_{k=1}^{n} \frac{(-1)^{k+1}}{k!} = \sum_{k=0}^{n} \frac{(-1)^k}{k!}. \quad (2.4)$$

Observe that as $n \to \infty$ this probability converges to $1/e \approx 0.3679$, yielding a simple *asymptotic approximation*. Listing 2.8 evaluates $\mathbb{P}(B)$ in several alternative ways for $n = 1, 2, \ldots, 8$. The function `bruteSetsProbabilityAllMiss()` works by creating all possibilities and counting. Although a highly inefficient way of evaluating $\mathbb{P}(B)$, it is presented here as it is instructive. The function `formulaCalcAllMiss()` evaluates the analytic solution from (2.4). Finally, the function `mcAllMiss()` estimates the probability via Monte Carlo simulation.

Listing 2.8: Secretary with envelopes

```
using Random, StatsBase, Combinatorics
Random.seed!(1)

function bruteSetsProbabilityAllMiss(n)
    omega = collect(permutations(1:n))
    matchEvents = []
    for i in 1:n
        event = []
        for p in omega
            if p[i] == i
                push!(event,p)
            end
        end
        push!(matchEvents,event)
    end
    noMatch = setdiff(omega,union(matchEvents...))
    return length(noMatch)/length(omega)
end

formulaCalcAllMiss(n) = sum([(-1)^k/factorial(k) for k in 0:n])

function mcAllMiss(n,N)
    function envelopeStuffer()
        envelopes = Random.shuffle!(collect(1:n))
        return sum([envelopes[i] == i for i in 1:n]) == 0
    end
    data = [envelopeStuffer() for _ in 1:N]
    return sum(data)/N
end

N = 10^6

println("n\tBrute Force\tFormula\t\tMonte Carlo\tAsymptotic",)
for n in 1:6
    bruteForce = bruteSetsProbabilityAllMiss(n)
    fromFormula = formulaCalcAllMiss(n)
    fromMC = mcAllMiss(n,N)
    println(n,"\t",round(bruteForce,digits=4),"\t\t",round(fromFormula,digits=4),
        "\t\t",round(fromMC,digits=4),"\t\t",round(1/MathConstants.e,digits=4))
end
```

n	Brute Force	Formula	Monte Carlo	Asymptotic
1	0.0	0.0	0.0	0.3679
2	0.5	0.5	0.4994	0.3679
3	0.3333	0.3333	0.3337	0.3679
4	0.375	0.375	0.3747	0.3679
5	0.3667	0.3667	0.3665	0.3679
6	0.3681	0.3681	0.3678	0.3679

Lines 4–18 define the function `bruteSetsProbabilityAllMiss()`, which uses a brute force approach to calculate $\mathbb{P}(B)$. The nested loops in lines 7–15 populate the array `matchEvents` with elements of `omega` that have a match. The inner loop in lines 9–13, puts elements from `omega` in `event` if they satisfy an i-th match. In line 16, notice the use of the 3 dots *splat operator*, `...`. Here `union()` is applied to all the elements of `matchEvents`. The return value in line 17 is a direct implementation via counting the elements of `noMatch`. The function on line 20 implements (2.4) in straightforward manner. Lines 22–29 implement the function `mcAllMiss()` that estimates the probability via Monte Carlo. The inner function, `envelopeStuffer()`, returns a result from a single experiment. Note that `shuffle!()` is used to create a random permutation in line 24. The remainder of the code prints the output, and compares the results to the asymptotic formula obtained via `1/MathConstants.e`.

An Occupancy Problem

We now consider a problem related to the previous example. Imagine now the secretary placing r identical business cards randomly into n envelopes, with $r \geq n$ and no limit on the number of business cards that can fit in an envelope. We now ask what is the probability that all envelopes are non-empty (i.e. occupied)?

To begin, denote A_i as the event that the i-th envelope is empty, and hence A_i^c is the event that the i-th envelope is occupied. Hence as before, we are seeking the probability of the event $B = A_1^c \cap A_2^c \cap \ldots \cap A_n^c$. Using the same logic as in the previous example,

$$\begin{aligned}\mathbb{P}(B) &= 1 - \mathbb{P}(A_1 \cup \ldots \cup A_n)\\ &= 1 - \sum_{k=1}^{n} (-1)^{k+1} \binom{n}{k} \tilde{p}_k,\end{aligned}$$

where $\tilde{p}_k$ is the probability of at least k envelopes being empty. Now from basic counting considerations,

$$\tilde{p}_k = \frac{(n-k)^r}{n^r} = \left(1 - \frac{k}{n}\right)^r.$$

Thus we arrive at

$$\mathbb{P}(B) = 1 - \sum_{k=1}^{n} (-1)^{k+1} \binom{n}{k} \left(1 - \frac{k}{n}\right)^r = \sum_{k=0}^{n} (-1)^{k} \binom{n}{k} \left(1 - \frac{k}{n}\right)^r. \tag{2.5}$$

We now calculate $\mathbb{P}(B)$ in Listing 2.9 and compare the results to Monte Carlo simulation estimates. In the code we consider several situations by varying the number of envelopes in the range $n = 1, \ldots, 100$, and for every n, consider the number of business cards $r = Kn$ for $K = 2, 3, 4$. The results are displayed in Figure 2.4.

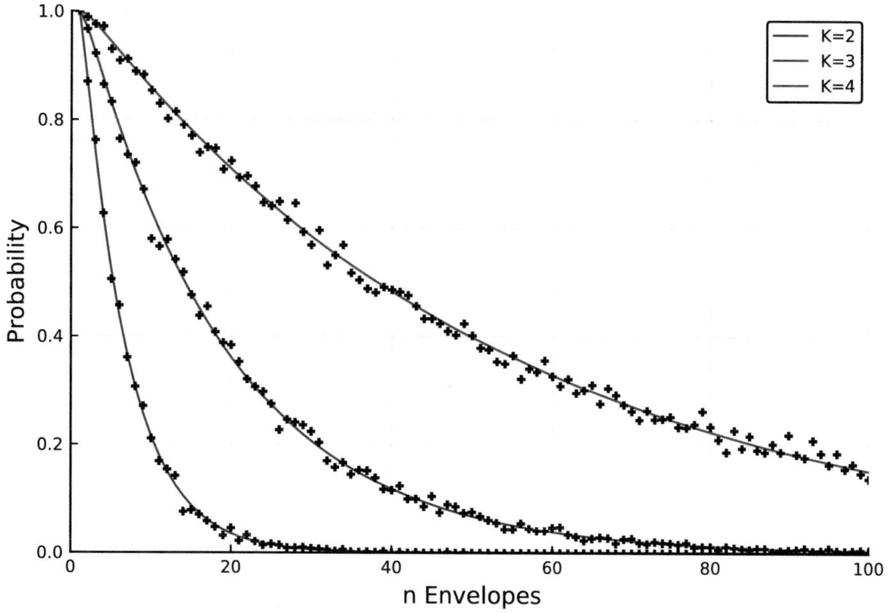

Figure 2.4: Analytic and estimated probabilities that no envelopes are empty, for various cases of n envelopes, and Kn business cards.

Listing 2.9: An occupancy problem

```
using Plots ; pyplot()

occupancyAnalytic(n,r) =  sum([(-1)^k*binomial(n,k)*(1 - k/n)^r for k in 0:n])

function occupancyMC(n,r,N)
    fullCount = 0
    for _ in 1:N
        envelopes = zeros(Int,n)
        for k in 1:r
            target = rand(1:n)
            envelopes[target] += 1
        end
        numFilled = sum(envelopes .> 0)
        if numFilled == n
            fullCount += 1
        end
    end
    return fullCount/N
end

max_n, N, Kvals = 100, 10^3, [2,3,4]

analytic = [[occupancyAnalytic(big(n),big(k*n)) for n in 1:max_n] for k in Kvals]
monteCarlo = [[occupancyMC(n,k*n,N) for n in 1:max_n] for k in Kvals]

plot(1:max_n, analytic, c=[:blue :red :green],
        label=["K=2" "K=3" "K=4"])
scatter!(1:max_n, monteCarlo, mc=:black, shape=:+,
        label="", xlims=(0,max_n),ylims=(0,1),
        xlabel="n Envelopes", ylabel="Probability", legend=:topright)
```

In line 3 we create the function `occupancyAnalytic()`, which evaluates (2.5). Note the use of the `binomial()` function. Lines 5–19 define the function `occupancyMC()`, which approximates $\mathbb{P}(B)$ for specific inputs via Monte Carlo simulation. Note the additional argument `N`, which is the total number of simulation runs. Line 5 defines the variable `fullcount`, which represents the total number of times all envelopes are full. Lines 7–17 contain the core logic of this function, and represent the act of the secretary assigning all business cards randomly to the envelopes, and repeating this process `N` times total. Observe that in this `for` loop, there is no need to keep a count of the loop iteration number, hence for clarity we use an underscore in line 7. Line 13 checks each element of `envelopes` to see if they are empty (i.e 0), and evaluates the total number of envelopes which are not empty. Note the use of element-wise comparison `.>`, resulting in an array of Boolean values that can be summed. Lines 14–16 check if all envelopes have been filled, and if so increments `fullCount` by 1. In lines 23 and 24 we create `analytic` and `monteCarlo`, respectively. Each of these is an array of arrays, with an internal array for `k=2`, `k=3`, and `k=4`. The results are then plotted.

2.3 Independence

We now consider *independence* and *independent events*. Two events, A and B, are said to be independent if the probability of their *intersection* is the product of their probabilities:

$$\mathbb{P}(A \cap B) = \mathbb{P}(A)\mathbb{P}(B).$$

A classic example is a situation where a random experiment involves physical components that are assumed to not interact, for example, flipping two coins. Independence is often a modeling assumption and plays a key role in many models presented in the remainder of the book.

Note that "independent events" should not be confused with "disjoint events" as these concepts are completely different. Take disjoint events A and B, with $\mathbb{P}(A) > 0$ and $\mathbb{P}(B) > 0$. This means that $\mathbb{P}(A)\mathbb{P}(B) > 0$. It is easy to see that the events are not independent. Since they are disjoint, $A \cap B = \emptyset$ and $\mathbb{P}(\emptyset) = 0$, however,

$$0 = \mathbb{P}(\emptyset) = \mathbb{P}(A \cap B) \neq \mathbb{P}(A)\mathbb{P}(B).$$

To explore independence, it is easiest to consider a situation where it does not hold. Consider drawing a number uniformly from the range $10, 11, \dots, 25$. What is the probability of getting the number 13? Clearly there are $25 - 10 + 1 = 16$ options, and hence the probability is $1/16 = 0.0625$. However, the event of obtaining 13 could be described as the intersection of the events $A := \{\text{first digit is } 1\}$ and $B := \{\text{second digit is } 3\}$. The probabilities of which are $10/16 = 0.625$ and $2/16 = 0.125$, respectively. Notice that the product of these probabilities is not 0.0625, but rather $20/256 = 0.078125$. Hence we see that, $\mathbb{P}(AB) \neq \mathbb{P}(A)\mathbb{P}(B)$ and the events are not independent.

One way of viewing this lack of independence is as follows. Witnessing the event A gives us some information about the likelihood of B. Since if A occurs, we know that the number is in the range $10, \dots, 19$ and hence there is a $1/10$ chance for B to occur. However, if A does not occur then we lie in the range $20, \dots, 25$ and there is a $1/6$ chance for B to occur.

If however we change the range of random digits to be $10, \dots, 29$ then the two events are independent. This can be demonstrated by running Listing 2.10, and then modifying line 4.

Listing 2.10: Independent events

```
1  using Random
2  Random.seed!(1)
3
4  numbers = 10:25
5  N = 10^7
6
7  firstDigit(x) = Int(floor(x/10))
8  secondDigit(x) = x%10
9
10 numThirteen, numFirstIsOne, numSecondIsThree = 0, 0, 0
11
12 for _ in 1:N
13     X = rand(numbers)
14     global numThirteen += X == 13
15     global numFirstIsOne += firstDigit(X) == 1
16     global numSecondIsThree += secondDigit(X) == 3
17 end
18
19 probThirteen, probFirstIsOne, probSecondIsThree =
20     (numThirteen,numFirstIsOne,numSecondIsThree)./N
21
22 println("P(13) = ", round(probThirteen, digits=4),
23         "\nP(1_) = ",round(probFirstIsOne, digits=4),
24         "\nP(_3) = ", round(probSecondIsThree, digits=4),
25         "\nP(1_)*P(_3) = ",round(probFirstIsOne*probSecondIsThree, digits=4))
```

```
P(13) = 0.0626
P(1_) = 0.6249
P(_3) = 0.1252
P(1_)*P(_3) = 0.0783
```

Lines 4 and 5 set the range of numbers considered and the number of simulation runs, respectively. Line 7 defines a function that returns the first digit of our number through the use of the `floor()` function, and converts the resulting value to an integer type. Line 8 defines a function that uses the *modulus* operator `%` to return the second digit of our number. In line 10 we initialize three placeholder variables, which represent the number chosen, and its first and second digits, respectively. Lines 12–17 contain the core logic of this example, where `N` random digits are generated. For each random digit, `X` that is generated, lines 14, 15, and 16 increment the count by 1 if the specified condition is met. Lines 19–20 evaluate the total proportions.

2.4 Conditional Probability

It is often the case that knowing an event has occurred, say B, modifies our belief about the chances of another event occurring, say A. This concept is captured via the *conditional probability* of A given B, denoted by $\mathbb{P}(A \mid B)$ and defined for B where $\mathbb{P}(B) > 0$. In practice, given a probability model, $\mathbb{P}(\cdot)$, we construct the conditional probability, $\mathbb{P}(\cdot \mid B)$ via

$$\mathbb{P}(A \mid B) := \frac{\mathbb{P}(A \cap B)}{\mathbb{P}(B)}. \tag{2.6}$$

This immediately shows that if events A and B are independent then $P(A \mid B) = P(A)$.

As an elementary example, refer back to Table 2.1 depicting the outcome of rolling two dice. Set B as the event of the sum being greater than or equal to 10. In other words,

$$B = \{(i,j) \mid i + j \geq 10\}.$$

To help illustrate this further, consider a game player who rolls the dice without showing us the result, and then poses to us the following: "The sum is greater or equal to 10. Is it even or odd?". Let A be the event of the sum being even. We then evaluate

$$\mathbb{P}(A \mid B) = \mathbb{P}(\text{Sum is even} \mid B) = \frac{\mathbb{P}(A \cap B)}{\mathbb{P}(B)} = \frac{\mathbb{P}(\text{Sum is 10 or 12})}{\mathbb{P}(\text{Sum is } \geq 10)} = \frac{4/36}{6/36} = \frac{2}{3},$$
$$\mathbb{P}(A^c \mid B) = \mathbb{P}(\text{Sum is odd} \mid B) = \frac{\mathbb{P}(A^c \cap B)}{\mathbb{P}(B)} = \frac{\mathbb{P}(\text{Sum is 11})}{\mathbb{P}(\text{Sum is } \geq 10)} = \frac{2/36}{6/36} = \frac{1}{3}.$$

It can be seen that given B, it is more likely that A occurs (even) as opposed to A^c (odd), hence we are better off answering "even".

The Law of Total Probability

Often our probability model is comprised of conditional probabilities as elementary building blocks. In such cases, (2.6) is better viewed as

$$\mathbb{P}(A \cap B) = \mathbb{P}(B)\, \mathbb{P}(A \mid B).$$

This is particularly useful when there exists some *partition* of Ω, namely, $\{B_1, B_2, \ldots\}$. A partition of a set U is a collection of non-empty sets that are mutually disjoint and whose union is U. Such a partition allows us to represent A as a disjoint union of the sets $A \cap B_k$, and treat $\mathbb{P}(A \mid B_k)$ as model data. In such a case, we have the *law of total probability*

$$\mathbb{P}(A) = \sum_{k=0}^{\infty} \mathbb{P}(A \cap B_k) = \sum_{k=0}^{\infty} \mathbb{P}(A \mid B_k)\, \mathbb{P}(B_k).$$

As an exotic fictional example, consider the world of semi-conductor manufacturing. Room cleanliness in the manufacturing process is critical, and dust particles are kept to a minimum. Let A be the event of a manufacturing failure, and assume that it depends on the number of dust particles via

$$\mathbb{P}(A \mid B_k) = 1 - \frac{1}{k+1},$$

where B_k is the event of having k dust particles in the room ($k = 0, 1, 2, \ldots$). Clearly the larger k, the higher the chance of manufacturing failure. Furthermore assume that

$$\mathbb{P}(B_k) = \frac{6}{\pi^2 (k+1)^2} \qquad \text{for } k = 0, 1, \ldots.$$

From the well-known *Basel Problem*, we have $\sum_{k=1}^{\infty} k^{-2} = \pi^2/6$. This implies that $\sum_k \mathbb{P}(B_k) = 1$.

Now we ask, what is the probability of manufacturing failure? The analytic solution is given by

$$\mathbb{P}(A) = \sum_{k=0}^{\infty} \mathbb{P}(A \mid B_k)\, \mathbb{P}(B_k) = \sum_{k=0}^{\infty} \Big(1 - \frac{1}{k+1}\Big) \frac{6}{\pi^2 (k+1)^2}.$$

With some calculus, the infinite series can be explicitly evaluated to

$$\mathbb{P}(A) = 1 - \frac{6\,\zeta(3)}{\pi^2} \approx 0.2692,$$

where $\zeta(\cdot)$ is the *Riemann Zeta Function,*

$$\zeta(s) = \sum_{n=1}^{\infty} \frac{1}{n^s},$$

and $\zeta(3) \approx 1.2021$. Note that the appearance of $\zeta(\cdot)$ in this example is by design due to the fact that we chose $\mathbb{P}(A \mid B_k)$ and $\mathbb{P}(B_k)$ to have the specific structure. Listing 2.11 approximates the infinite series numerically (truncating at `n = 2000`) and compares the result to the analytic solution.

Listing 2.11: Defects in manufacturing

```
1  using SpecialFunctions
2
3  n = 2000
4
5  probAgivenB(k) = 1- 1/(k+1)
6  probB(k) = 6/(pi*(k+1))^2
7
8  numerical= sum([probAgivenB(k)*probB(k) for k in 0:n])
9  analytic = 1 - 6*zeta(3)/pi^2
10
11 println("Analytic: ", analytic, "\tNumerical: ", numerical)
```

```
Analytic: 0.26923703059856086    Numerical: 0.26893337073278945
```

This listing is self-explanatory, however note the use of the Julia function `zeta()` from the `SpecialFunctions` package in line 9. Note also that `pi` is a defined constant.

2.5 Bayes' Rule

Bayes' rule, also known as *Bayes' theorem*, is nothing but a simple manipulation of (2.6) yielding

$$\mathbb{P}(A \mid B) = \frac{\mathbb{P}(B \mid A)\mathbb{P}(A)}{\mathbb{P}(B)}. \tag{2.7}$$

However, the consequences are far reaching. Often we observe a *posterior outcome* or measurement, say the event B, and wish to evaluate the probability of a *prior condition*, say the event A. That is, given some measurement or knowledge we wish to evaluate how likely is it that a prior condition occurred. Equation (2.7) allows us to do just that.

Was it a 0 or a 1?

As an example, consider a communication channel involving a stream of transmitted bits (0's and 1's), where 70% of the bits are 1, and the rest 0. A typical snippet from the channel ...0101101011101111101....

The channel is imperfect due to physical disturbances such as interfering radio signals, and furthermore the bits received are sometimes distorted. Hence there is a chance (ε_0) of interpreting a bit as 1 when it is actually 0, and similarly there is a chance (ε_1) of interpreting a bit as 0 when it is actually 1.

Now say that we received (Rx) a bit, and interpreted it as 1. This is the posterior outcome. What is the chance that it was in fact transmitted (Tx) as a 1? Applying Bayes' rule:

$$\mathbb{P}(\text{Tx } 1 \mid \text{Rx } 1) = \frac{\mathbb{P}(\text{Rx } 1 \mid \text{Tx } 1)\mathbb{P}(\text{Tx } 1)}{\mathbb{P}(\text{Rx } 1)} = \frac{(1-\varepsilon_1)0.7}{0.7(1-\varepsilon_1)+0.3\varepsilon_0}. \tag{2.8}$$

For example, if $\varepsilon_0 = 0.1$ and $\varepsilon_1 = 0.05$ we have that $\mathbb{P}(\text{Tx } 1 \mid \text{Rx } 1) = 0.9568$. Listing 2.12 illustrates this via simulation.

Listing 2.12: Tx Rx Bayes

```
using Random
Random.seed!(1)

N = 10^5
prob1 = 0.7
eps0, eps1 = 0.1, 0.05

flipWithProb(bit,prob) = rand() < prob ? xor(bit,1) : bit

TxData = rand(N) .< prob1
RxData = [x == 0 ? flipWithProb(x,eps0) : flipWithProb(x,eps1) for x in TxData]

numTx1 = 0
totalRx1 = 0
for i in 1:N
    if RxData[i] == 1
        global totalRx1 += 1
        global numTx1 += TxData[i]
    end
end

monteCarlo = numTx1/totalRx1
analytic = ((1-eps1)*0.7)/((1-eps1)*0.7+0.3*eps0)

println("Monte Carlo: ", monteCarlo, "\t\tAnalytic: ", analytic)
```

```
Monte Carlo: 0.9576048007598325		Analytic: 0.9568345323741007
```

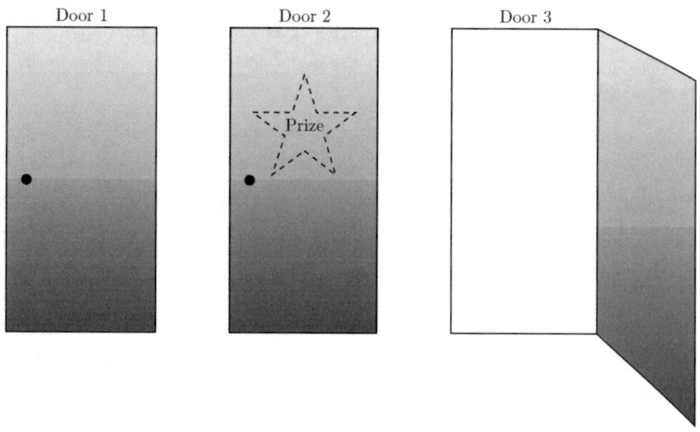

Figure 2.5: Monty Hall: If the prize is behind Door 2 and Door 1 is chosen, the game show host must reveal Door 3.

In lines 8 the function `flipWithProb()` is defined. It uses the `xor()` function to randomly flip the input argument `bit`, according to the rate given by the argument `prob`. Line 10 generates the array `TxData`, which contains true and false values representing our transmitted bits of 1's and 0's, respectively. It does this by uniformly and randomly generating numbers on the range $[0, 1]$, and then evaluating element-wise if they are less than the specified probability of receiving a 1, `prob1`. Line 11 generates the array `RxData`, which represents our simulated received data. First the type of received bit is checked, and the `flipWithProb()` function is used to flip received bits at the rates specified in line 6 if the received bit is a 0 or 1. Lines 13–20 are used to check the nature of all bits. If the bit received is 1, then it increments the counter `totalRx1` by 1. It also increments the counter `numTx1` by the value of the transmitted bit (which may be 1, but could also be 0). The remaining lines then calculate the Monte Carlo-based estimate and compare to the analytic solution from (2.8).

The Monty Hall Problem

The *Monty Hall problem* is a famous problem which was first posed and solved in 1975 by the mathematician Steve Selvin [SBK75]. It is a famous example illustrating how probabilistic reasoning may sometimes yield to surprising results.

Consider a contestant on a television game show, with three doors in front of her. One of the doors contains a prize, while the other two are empty. The contestant is then asked to guess which door contains the prize, and she makes a random guess. Following this, the game show host (GSH) reveals an empty (losing) door from one of the two remaining doors not chosen. The contestant is then asked if she wishes to stay with their original choice, or if she wishes to switch to the remaining closed door. Following the choice of the contestant to stay or switch, the door with the prize is revealed. The question is: should the contestant stay with their original choice, or switch? Alternatively, perhaps it doesn't matter.

For example, in Figure 2.5 we see the situation where the hidden prize is behind door 2. Say the contestant has chosen door 1. In this case, the GSH has no choice but to reveal door 3. Alternatively, if the contestant has chosen door 2, then the GSH will reveal either door 1 or door 3.

The two possible policies (or strategies of play) for the contestant are as follows:

Policy I —Stay with their original choice after the door is revealed.

Policy II —Switch after the door is revealed.

Let us consider the probability of winning for the two different policies. If the player adopts Policy I then she always stays with her initial guess regardless of the GSH action. In this case, her chance of success is 1/3, that is, she wins if her initial choice is correct.

However if she adopts Policy II then she always switches after the GSH reveals an empty door. In this case we can show that her chance of success is 2/3, that is, she actually wins if her initial guess is incorrect. This is because the GSH must always reveal a losing door. If she originally chose a losing door, then the GSH must reveal the second losing door every time (otherwise he would reveal the prize). That is, if the player chooses an incorrect door at the start, the non-revealed door will always be the winning door. The chance of such an event is 2/3.

As a further aid for understanding imagine a case of 100 doors and a single prize behind one of them. In this case assume that the player chooses a door, for example, door 1, and following this the GSH reveals 98 losing doors. There are now only two doors remaining, her choice door 1, and (say for example), door 38. The intuition of the problem suddenly becomes obvious. The player's original guess was random and hence door 1 had a 1/100 chance of containing the prize; however, the GSH's actions were constrained. He had to reveal only losing doors, and hence there is a 99/100 chance that door 38 contains the prize. Hence, Policy II is clearly superior.

We now analyze the case of 3 doors by applying Bayes' theorem. Let A_i be the event that the prize is behind door i. Let B_i be the event that door i is revealed by the GSH. Then, for example, if the player initially chooses door 1 and then the GSH reveals door 2, we have the following:

$$\mathbb{P}(A_1 \mid B_2) = \frac{\mathbb{P}(B_2 \mid A_1)\mathbb{P}(A_1)}{\mathbb{P}(B_2)} = \frac{\frac{1}{2} \times \frac{1}{3}}{\frac{1}{2}} = \frac{1}{3}, \qquad \textbf{(Policy I)}$$

$$\mathbb{P}(A_3 \mid B_2) = \frac{\mathbb{P}(B_2 \mid A_3)\mathbb{P}(A_3)}{\mathbb{P}(B_2)} = \frac{1 \times \frac{1}{3}}{\frac{1}{2}} = \frac{2}{3}. \qquad \textbf{(Policy II)}$$

In the second case note that $\mathbb{P}(B_2 \mid A_3) = 1$ because the GSH must reveal door 2 if the prize is behind door 3 since door 1 was already picked. Hence, we see that while neither policy guarantees a win, Policy II clearly dominates Policy I.

Now that we have shown this analytically, we perform a Monte Carlo simulation of the Monty Hall problem in Listing 2.13.

Listing 2.13: The Monty Hall problem

```
1   using Random
2   Random.seed!(1)
3
4   function montyHall(switchPolicy)
5       prize, choice = rand(1:3), rand(1:3)
6       if prize == choice
7           revealed = rand(setdiff(1:3,choice))
8       else
9           revealed = rand(setdiff(1:3,[prize,choice]))
10      end
11
12      if switchPolicy
13          choice = setdiff(1:3,[revealed,choice])[1]
14      end
15      return choice == prize
16  end
17
18  N = 10^6
19  println("Success probability with policy I (stay): ",
20          sum([montyHall(false) for _ in 1:N])/N)
21  println("Success probability with policy II (switch): ",
22          sum([montyHall(true) for _ in 1:N])/N)
```

```
Success probability with policy I (stay): 0.332913
Success probability with policy II (switch): 0.667027
```

In lines 4–16 the function `montyHall()` is defined, which performs one simulation run of the problem given a policy, with `false` indicating policy I and `true` indicating policy II (switching). At the start of the game, the location of the `prize` and the player's door `choice` are uniformly and randomly initialized. Lines 6–10 contain the logic and action of the GSH. Since he knows the location of both the prize and the chosen door, he first mentally checks if they are the same. If they are, he reveals a door according to line 7. If not, then he proceeds to reveal a door according to the logic in line 9. In either case, the revealed door is stored in the variable `revealed`. Line 7 represents his action if the initial `choice` door is the same as the `prize` door. In this case, he is free to reveal either of the remaining two doors, i.e. the set difference between all doors and the player's `choice` door. In this case the set difference has 2 elements. Line 9 represents the GSH action if the `choice` door is different to the `prize` door. In this case, his hand is forced. As he cannot reveal the player's chosen door or the prize door, he is forced to reveal the one remaining door, which can be thought of as the set difference between `1:3` (all doors) and `[prize, choice]`. In this case the set difference has a single element. Line 13 represents the contestant's action, after the GSH revelation, based on either a switch (`true`) or stay (`false`) policy. If the contestant chooses to stay with her initial guess (`false`), then we skip to Line 15. However, if she chooses to swap (`true`), then we reassign our initial `choice` to the one remaining door in line 13. Note the use of `[1]`, which is used to assign the value from the array to `choice`, rather than the array itself. Line 15 checks if the player's choice is the same as the prize, and returns `true` if she wins, or `false` if she loses. Lines 19–22 repeat this experiment `N` times for each of the policies and print the Monte Carlo estimates.

Chapter 3

Probability Distributions

In this chapter, we introduce random variables, different types of distributions, and related concepts. In the previous chapter, we explored probability spaces without much emphasis on numerical random values. However, when carrying out random experiments, there are almost always numerical values involved. In the context of probability, these values are often called *random variables.* Mathematically, a random variable X is a function of the sample space, Ω, and takes on integer, real, complex, or even a vector of values. That is, for every possible outcome $\omega \in \Omega$, there is some possible numerical value, $X(\omega)$.

The study of random variables and their probability distributions deals with quantification of different random numerical values of $X(\omega)$. Such a quantification is summarized via the probability distribution of X and can be described by different means including probability density functions, cumulative distribution functions, and the like. Here numerical summaries such as mean, variance, and related measures also appear. A good part of the study of probability distributions involves dealing with specific parametric forms of distributions. Examples include the normal distribution, the binomial distribution, beta distributions, and many other forms.

The chapter is organized as follows: In Section 3.1, we introduce the concept of a random variable and its probability distribution. In Section 3.2, we introduce the mean, variance, and other numerical descriptors of probability distributions. In Section 3.3, we explore several alternative functions for describing probability distributions. In Section 3.4, we focus on Julia's `Distributions` package which is useful when working with probability distributions. Then Section 3.5 explores a variety of discrete distributions. This is followed by Section 3.6 where we explore some continuous distributions together with additional concepts such as hazard rates and more. We close with Section 3.7, where we explore multidimensional probability distributions.

3.1 Random Variables

As an example, consider a sample space Ω which consists of six names. Assume that the probability function (or probability measure), $\mathbb{P}(\cdot)$, assigns uniform probabilities to each of the names. Let

Y. Nazarathy and H. Klok, *Statistics with Julia*, Springer Series in the Data Sciences,
https://doi.org/10.1007/978-3-030-70901-3_3

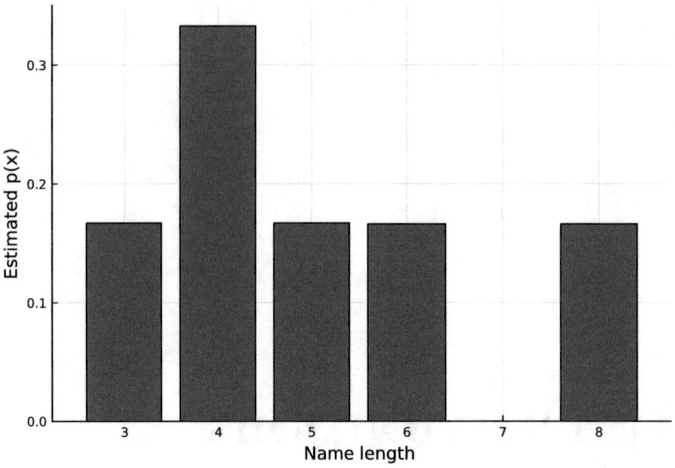

Figure 3.1: A discrete probability distribution taking values on $\{3, 4, 5, 6, 8\}$.

now, $X : \Omega \to \mathbb{Z}$, be the function (i.e. random variable) that counts the number of letters in each name. The question is then finding:

$$p(x) := \mathbb{P}\big(X = x\big), \qquad \text{for} \qquad x \in \mathbb{Z}.$$

The function $p(x)$ represents the *probability distribution* of the random variable X. In this case, since X measures name lengths, X is a discrete random variable, and its probability distribution may be represented by a *Probability Mass Function* (*PMF*), such as $p(x)$.

To illustrate this, we carry out a simulation of many such random experiments, yielding many replications of the random variable X, which we then use to estimate $p(x)$. This is performed in Listing 3.1.

Listing 3.1: A simple random variable

```
using StatsBase, Plots; pyplot()

names = ["Mary","Mel","David","John","Kayley","Anderson"]
randomName() = rand(names)
X = 3:8
N = 10^6
sampleLengths = [length(randomName()) for _ in 1:N]

bar(X,counts(sampleLengths)/N, ylims=(0,0.35),
    xlabel="Name length", ylabel="Estimated p(x)", legend=:none)
```

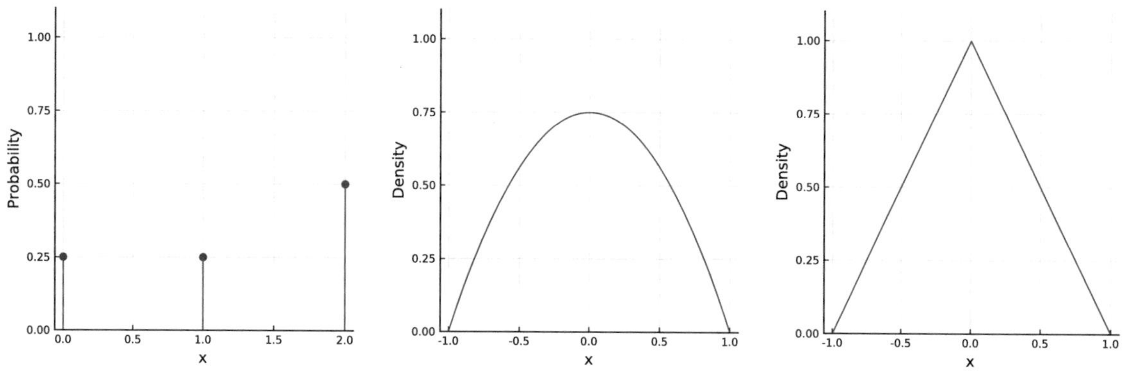

Figure 3.2: Three different examples of probability distributions.

In line 3, we create the array `names`, which contains names with different character lengths. Note that two names have four characters, namely, "Mary" and "John", while there is no name with seven characters. In line 4, we define the function `randomName()` which randomly selects, with equal probability, an element from the array `names`. In line 5, we specify that we will count names of three to eight characters in length. Line 6 specifies how many random experiments of choosing a name we will perform. Line 7 uses a comprehension and the function `length()` to count the length of each random name, and stores the results in the array `sampleLengths`. Here the Julia function `length()` is the analog of the random variable. That is, it is a function of the sample space, Ω, yielding a numerical value. Line 9 uses the function `counts()` to count how many words are of length 3, 4, up to 8. The `bar()` function is then used to plot a barchart of the proportion of counts for each word length. Two key observations can be made. It can be seen that words of length 4 occurred twice as much as words of lengths 3, 5, 6, and 8. In addition, no words of length 7 were selected, as no name in our original array had a length of 7.

Types of Random Variables

In the previous example, the random variable X took on discrete values and is thus called a *discrete random variable.* However, quantities measured in nature are often continuous, in which case a *continuous random variable* better describes the situation. For example, the weights of people randomly selected from a big population.

In describing the probability distribution of a continuous random variable, the probability mass function, $p(x)$, is no longer applicable. This is because for a continuous random variable X, $\mathbb{P}(X = x)$ for any particular value of x is 0. Hence, in this case, the *Probability Density Function (PDF)*, $f(x)$ is used, where

$$f(x)\Delta \;\approx\; \mathbb{P}\big(x \le X \le x + \Delta\big).$$

Here the approximation becomes exact as $\Delta \to 0$. Figure 3.2 illustrates three examples of probability distributions. The one on the left is discrete and the other two are continuous.

The discrete probability distribution appearing on the left in Figure 3.2 can be represented mathematically by the probability mass function

$$p(x) = \begin{cases} 0.25 & \text{for } x = 0, \\ 0.25 & \text{for } x = 1, \\ 0.5 & \text{for } x = 2. \end{cases} \tag{3.1}$$

The smooth continuous probability distribution is defined by the probability density function,

$$f_1(x) = \frac{3}{4}(1 - x^2) \qquad \text{for } -1 \leqslant x \leqslant 1.$$

Finally, the triangular probability distribution is defined by the probability density function,

$$f_2(x) = \begin{cases} x + 1 & \text{for } x \in [-1, 0], \\ 1 - x & \text{for } x \in (0, 1]. \end{cases}$$

Note that for both the probability mass function and the probability density function, it is implicitly assumed that $p(x)$ and $f(x)$ are zero for x values not specified in the equation.

It can be verified that for the discrete distribution,

$$\sum_x p(x) = 1,$$

and for the continuous distributions,

$$\int_{-\infty}^{\infty} f_i(x)\, dx = 1 \qquad \text{for} \qquad i = 1, 2.$$

There are additional descriptors of probability distributions other than the PMF and PDF, and these are further discussed in Section 3.3. Note that Figure 3.2 was generated by Listing 3.2.

Listing 3.2: Plotting discrete and continuous distributions

```
1  using Plots, Measures; pyplot()
2
3  pDiscrete = [0.25, 0.25, 0.5]
4  xGridD = 0:2
5
6  pContinuous(x) = 3/4*(1 - x^2)
7  xGridC = -1:0.01:1
8
9  pContinuous2(x) = x < 0 ? x+1 : 1-x
10
11 p1 = plot(xGridD, line=:stem, pDiscrete, marker=:circle, c=:blue, ms=6, msw=0)
12 p2 = plot(xGridC, pContinuous.(xGridC), c=:blue)
13 p3 = plot(xGridC, pContinuous2.(xGridC), c=:blue)
14
15 plot(p1, p2, p3, layout=(1,3), legend=false, ylims=(0,1.1), xlabel="x",
16         ylabel=["Probability" "Density" "Density"], size=(1200, 400), margin=5mm)
```

In line 3, we define an array specifying the PMF of our discrete distribution, and in lines 6 and 9 we define functions specifying the PDFs of our continuous distributions. In lines 11–16, we create plots of each of our distributions. Note that in the discrete case we use the `line=:stem` argument together with `marker=:circle`.

3.2 Moment-Based Descriptors

The probability distribution of a random variable fully describes the probabilities of the events, $\{\omega \in \Omega \,:\, X(\omega) \in A\}$, for all sensible $A \subset \mathbb{R}$. However, it is often useful to describe the nature of a random variable via a single number or a few numbers. The most common example of this is the *mean* which describes the center of mass of the probability distribution. Other examples include the *variance* and *moments* of the probability distribution. We expand on these now.

Mean

The mean, also known as the *expected value* of a random variable X, is a measure of the central tendency of the distribution of X. It is represented by $\mathbb{E}[X]$, and is the value we expect to obtain "on average" if we continue to take observations of X and average out the results. The mean of a discrete distribution with PMF $p(x)$ is

$$\mathbb{E}[X] = \sum_x x\, p(x).$$

In the example of the discrete distribution given by (3.1), it is

$$\mathbb{E}[X] = 0 \times 0.25 \;+\; 1 \times 0.25 \;+\; 2 \times 0.5 = 1.25.$$

The mean of a continuous random variable, with PDF $f(x)$ is

$$\mathbb{E}[X] = \int_{-\infty}^{\infty} x\ f(x)\ dx,$$

which in the examples of $f_1(\cdot)$ and $f_2(\cdot)$ from Section 3.1 yield

$$\int_{-1}^{1} x\ \frac{3}{4}(1 - x^2) = 0,$$

and

$$\int_{-1}^{0} x + 1\ \ dx\ +\ \int_{0}^{1} 1 - x\ \ dx = 0,$$

respectively. As can be seen, both continuous distributions have the same mean even though their shapes are different. For illustration purposes, we now carry out this integration numerically in Listing 3.3.

Listing 3.3: Expectation via numerical integration

```
using QuadGK

sup = (-1,1)
f1(x) = 3/4*(1-x^2)
f2(x) = x < 0 ? x+1 : 1-x

expect(f,support) = quadgk((x) -> x*f(x),support...)[1]

println("Mean 1: ", expect(f1,sup))
println("Mean 2: ", expect(f2,sup))
```

```
Mean 1: 0.0
Mean 2: -2.0816681711721685e-17
```

In line 1, we specify usage of the QuadGK package, which contains functions that support one-dimensional numerical integration via a method called *adaptive Gauss-Kronrod quadrature.* In lines 4 and 5, we define the PDFs of the distributions via `f1()` and `f2()`. In line 7, we define the function `expect()` which takes two arguments, a function to integrate `f`, and a domain over which to integrate the function `support`. It uses the `quadgk()` function to evaluate the one-dimensional integral given above. For this, an anonymous function `(x) -> x*f(x)` is created. Note that the start and end points of the integral are `support[1]` and `support[2]`, respectively. These are "splatted" into the second and third arguments of `quadgk()` via the "`...`" operator. Note also that the function `quadgk()` returns two arguments, the evaluated integral and an estimated upper bound on the absolute error. Hence, `[1]` is included at the end of the function, so that only the integral is returned. Lines 9–10 then evaluate the numerical integrals of the functions `f1` and `f2` over the interval `sup` and display the output. As can be seen, both integrals are effectively evaluated to zero.

General Expectation and Moments

In general, for a function $h : \mathbb{R} \to \mathbb{R}$ and a random variable X, we can consider the random variable $Y := h(X)$. The distribution of Y will typically be different from the distribution of X. As for the mean of Y, we have

$$\mathbb{E}[Y] = \mathbb{E}[h(X)] = \begin{cases} \displaystyle\sum_x h(x)\, p(x) & \text{for discrete,} \\ \displaystyle\int_{-\infty}^{\infty} h(x)\, f(x)\, dx & \text{for continuous.} \end{cases} \tag{3.2}$$

Note that the above expression does not require explicit knowledge of the distribution of Y but rather uses the distribution (PMF or PDF) of X.

A common case is $h(x) = x^\ell$, in which case we call $\mathbb{E}[X^\ell]$, the ℓ-th moment of X. Then, for a random variable X with PDF $f(x)$, the ℓth *moment* of X is

$$\mathbb{E}[X^\ell] = \int_{-\infty}^{\infty} x^\ell \; f(x) \; dx.$$

Note that the first moment is the mean and the zeroth moment is always 1. The second moment is related to the variance as we explain below.

Variance

The *variance* of a random variable X, often denoted as $\mathrm{Var}(X)$ or σ^2, is a measure of the spread, or *dispersion*, of the distribution of X. It is defined by

$$\mathrm{Var}(X) := \mathbb{E}[(X - \mathbb{E}[X])^2] = \mathbb{E}[X^2] - \Big(\mathbb{E}[X]\Big)^2. \tag{3.3}$$

Here we apply (3.2) by considering $h(x) = (x - \mathbb{E}[X])^2$. The second expression of (3.3) illustrates the role of the first and second moments in the variance. It follows from the first expression by expansion.

For the discrete distribution, (3.1), we have

$$\text{Var}(X) = (0 - 1.25)^2 \times 0.25 \;+\; (1 - 1.25)^2 \times 0.25 \;+\; (2 - 1.25)^2 \times 0.5 = 0.6875.$$

For the continuous distributions from Section 3.1, $f_1(\cdot)$ and $f_2(\cdot)$, with respective random variables X_1 and X_2, we have

$$\begin{aligned}
\text{Var}(X_1) &= \int_{-1}^{1} x^2 \, \frac{3}{4}(1 - x^2)\, dx - \big(\mathbb{E}[X_1]\big)^2 = \frac{3}{4}\left[\frac{x^3}{3} - \frac{x^5}{5}\right]_{-1}^{1} - 0 \;=\; 0.2,\\
\text{Var}(X_2) &= \int_{-1}^{0} x^2(x+1)\, dx + \int_{0}^{1} x^2(1-x)\, dx - \big(\mathbb{E}[X_2]\big)^2 \;=\; \frac{1}{6}.
\end{aligned}$$

The variance of X can also be considered as the expectation of a new random variable, $Y := (X - \mathbb{E}[X])^2$. However, when considering variance, the distribution of Y is seldom mentioned. Nevertheless, as an exercise we explore this now. Consider a random variable X, with density,

$$f(x) = \begin{cases} x - 4 & \text{for } x \in [4, 5],\\ 6 - x & \text{for } x \in (5, 6]. \end{cases}$$

This density is similar to $f_2(\cdot)$ previously covered, but with support $[4, 6]$. In Listing 3.4, we generate random observations from X, and calculate data points for Y based on these observations. We then plot both the distribution of X and Y, and show that the sample mean of Y is the sample variance of X. Note that our code uses some elements from the `Distributions` package, which is covered in more detail in Section 3.4.

Listing 3.4: Variance of X as the mean of Y

```
using Distributions, Plots; pyplot()

dist = TriangularDist(4,6,5)
N = 10^6
data = rand(dist,N)
yData=(data .- 5).^2

println("Mean: ", mean(yData), " Variance: ", var(data))

p1 = histogram(data, xlabel="x", bins=80, normed=true, ylims=(0,1.1))
p2 = histogram(yData, xlabel="y", bins=80, normed=true, ylims=(0,15))
plot(p1, p2, ylabel="Proportion", size=(800, 400), legend=:none)
```

```
Mean(Y) = 0.16671191478072614   Variance(X) = 0.1667120530661165
```

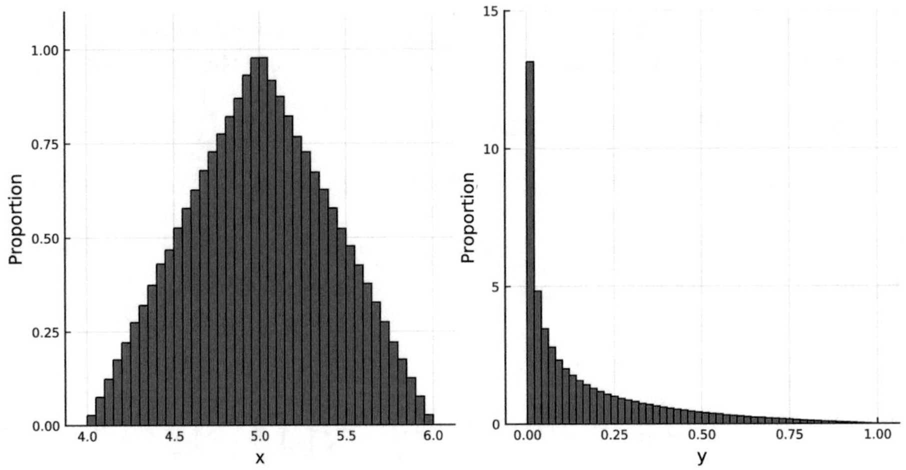

Figure 3.3: Histograms for samples of the random variables X and Y.

Line 1 calls the `Distributions` package. This package supports a variety of distribution types through the many functions it contains. We expand further on the use of the `Distributions` package in Section 3.4. Line 2 uses the `Triangular()` function from the `Distributions` package to create a triangular distribution-type object with a mean of 5 and a symmetric shape over the bound $[4, 6]$. We assign this as the variable `dist`. In line 5, we generate an array of `N` observations from the distribution by applying the `rand()` function on the distribution `dist`. Line 6 takes the observations in `data` and from them generates observations for the new random variable Y. The values are stored in the array `yData`. Line 8 uses the functions `mean()` and `var()` on the arrays `yData` and `data`, respectively. It can be seen from the output that the mean of the distribution Y is the same as the variance of X. Lines 10–12 are used to plot histograms of the data in the arrays `data` and `yData`. It can be observed that the histogram on the left approximates the PDF of our triangular distribution, while the histogram on the right approximates the distribution of the new variable Y. The distribution of Y is seldom considered when evaluating the variance of X.

Higher Order Descriptors: Skewness and Kurtosis

As described previously, the second moment plays a role defining the dispersion of a distribution via the variance. What about higher order moments? We now briefly define the skewness and kurtosis of a distribution utilizing the first three moments and first four moments, respectively.

Take a random variable X with $\mathbb{E}[X] = \mu$ and $\text{Var}(X) = \sigma^2$, then the *skewness* is defined as

$$\gamma_3 = \mathbb{E}\Big[\Big(\frac{X-\mu}{\sigma}\Big)^3\Big] = \frac{\mathbb{E}[X^3] - 3\mu\sigma^2 - \mu^3}{\sigma^3},$$

and the *kurtosis* is defined as

$$\gamma_4 = \mathbb{E}\Big[\Big(\frac{X-\mu}{\sigma}\Big)^4\Big] = \frac{\mathbb{E}[(X-\mu)^4]}{\sigma^4}.$$

Note that, γ_3 and γ_4 are invariant to changes in location and scale of the distribution.

The skewness is a measure of the asymmetry of the distribution. For a distribution having a symmetric density function about the mean, we have $\gamma_3 = 0$. Otherwise, it is either positive or negative depending on the distribution being *skewed to the right* or *skewed to the left*, respectively.

The kurtosis is a measure of the tails of the distribution. As a benchmark, any normal probability distribution (covered in detail in Section 3.6) has $\gamma_4 = 3$. Then, a probability distribution with a higher value of γ_4 can be interpreted as having "heavier tails" (than a normal distribution), while a probability distribution with a lower value is said to have "lighter tails" (than a normal distribution). This benchmark even yields a term called *excess kurtosis* defined as $\gamma_4 - 3$. Hence, a positive excess kurtosis implies "heavy tails" and a negative value implies "light tails".

Laws of Large Numbers

Throughout this book, our Monte Carlo experiments rely on *laws of large numbers.* This suite of mathematical statements claim that empirical averages converge to expected values. Stated as mathematical theorems, these laws come in different forms including the *weak law of large numbers* and the *strong law of large numbers.* In both cases, a sequence of independent and identically distributed random variables, $X_1, X_2, \ldots$, is considered. Then for each n, we compute the sample mean

$$\overline{X}_n = \frac{1}{n}\sum_{k=1}^{n} X_k,$$

and consider the sequence of sample means,

$$\overline{X}_1, \overline{X}_2, \ldots .$$

If the mean of each of the random variables X_i is μ, then a law of large numbers is a claim that the sequence $\{\overline{X}_n\}_{n=1}^{\infty}$ converges to μ. The distinction between "weak" and "strong" lies with the *mode of convergence.* For example, the weak law of large numbers claims that the sequence of probabilities

$$w_n = \mathbb{P}(|\overline{X}_n - \mu| > \varepsilon)$$

converges to 0 for any positive ε. That is, as n grows, the likelihood of the sample mean $\overline{X}_n$ to be farther away than ϵ from the mean μ vanishes. This is a statement about the sequence of probabilities, $w_1, w_2, \ldots$. In contrast, the strong law of large numbers states that

$$\mathbb{P}\Big(\lim_{n\to\infty} \overline{X}_n = \mu\Big) = 1. \tag{3.4}$$

This means that with certainty, every sequence of sample means converges to the expectation. From a practical perspective, the implication is similar to the weak law of large numbers; however, mathematically, the statement is different. In fact, the strong law of large numbers condition (3.4) implies the weak law of large numbers.

It turns out that proving the weak law of large numbers is much easier than proving the strong law of large numbers. Also, for the strong law of large numbers, if we are willing to assume that $\mathbb{E}[X_i^4] < \infty$ then a proof isn't too difficult; however, the minimal conditions are that $\mathbb{E}[X_i]$ is finite, and under these conditions a proof is more involved. See [R06] for an introduction to such aspects of *rigorous probability theory,* including proofs. Also related is the example presented later in Listing 3.30. It deals with the Cauchy distribution and illustrates a scenario where the law of large numbers breaks because $\mathbb{E}[X_i]$ does not exist.

Keep in mind that in many cases, we convert the sequence $X_1, X_2, \ldots$ into the sequence $I_1, I_2, \ldots$ via,

$$I_i = \begin{cases} 1 & \text{if } X_i \text{ satisfies some condition,} \\ 0 & \text{if } X_i \text{ does not satisfy the condition.} \end{cases}$$

In such a case,

$$\mathbb{E}[I_i] = \mathbb{P}(X_i \text{ satisfies the condition}),$$

and the average,

$$\overline{I}_n = \frac{1}{n} \sum_{i=1}^{n} I_i,$$

is the proportion of samples over $1, \ldots, n$ that satisfy the condition. Here laws of large numbers (weak or strong) imply that empirical proportions converge to probabilities.

3.3 Functions Describing Distributions

As alluded to in Section 3.2, a probability distribution can be described by a Probability Mass Function (PMF) in the discrete case, or a Probability Density Function (PDF) in the continuous case. However, there are other popular descriptors of probability distributions, such as the *Cumulative Distribution Function* (CDF), the *Complementary Cumulative Distribution Function* (CCDF), and the *Inverse Cumulative Distribution Function* (ICDF) . There are also transform-based descriptors including the *Moment-Generating Function* (MGF), *Probability-Generating Function* (PGF), as well as related functions such as the *Characteristic Function* (CF), or alternative names, including the *Laplace transform, Fourier transform,* or *z-transform.* Then, for non-negative random variables there is also the *hazard function* which we explore along with the Weibull distribution in Section 3.6. The main point to take away here is that a probability distribution can be described in many alternative ways. We now explore a few of these descriptors.

Cumulative Probabilities

Consider first the CDF of a random variable X, defined as

$$F(x) := \mathbb{P}(X \leq x),$$

where X can be discrete, continuous, or a more general random variable. The CDF is a very popular descriptor because unlike the PMF or PDF, it is not restricted to just the discrete or just the continuous case. A closely related function is the CCDF, $\bar{F}(x) := 1 - F(x) = \mathbb{P}(X > x)$.

From the definition of the CDF, $F(\cdot)$,

$$\lim_{x \to -\infty} F(x) = 0 \qquad \text{and} \qquad \lim_{x \to \infty} F(x) = 1.$$

Furthermore, $F(\cdot)$ is a non-decreasing function. In fact, any function with these properties constitutes a valid CDF and hence a probability distribution of a random variable.

In the case of a continuous random variable, the PDF $f(\cdot)$ and the CDF $F(\cdot)$ are related via

$$f(x) = \frac{d}{dx}F(x) \qquad \text{and} \qquad F(x) = \int_{-\infty}^{x} f(u)\, du.$$

Also, as a consequence of the CDF properties,

$$f(x) \geq 0, \qquad \text{and} \qquad \int_{-\infty}^{\infty} f(x)\, dx = 1. \tag{3.5}$$

Analogously, while less appealing than the continuous counterpart, in the case of discrete random variable, the PMF $p(\cdot)$ is related to the CDF via

$$p(x) = F(x) - \lim_{t \to x^-} F(t) \qquad \text{and} \qquad F(x) = \sum_{k \leq x} p(k). \tag{3.6}$$

Note that here we consider $p(x)$ to be 0 for x not in the support of the random variable. The important point in presenting (3.5) and (3.6) is to show that $F(\cdot)$ is a valid description of the probability distribution.

In Listing 3.5, we look at an elementary example, where we consider the PDF $f_2(\cdot)$ of Section 3.1 and integrate it via a crude *Riemann sum* to obtain the CDF:

$$F(x) = \mathbb{P}(X \leq x) = \int_{-\infty}^{x} f_2(u)\, du \approx \sum_{u=-\infty}^{x} f_2(u)\, \Delta u. \tag{3.7}$$

Listing 3.5: CDF from the Riemann sum of a PDF

```
1  using Plots, LaTeXStrings; pyplot()
2
3  f2(x) = (x<0 ? x+1 : 1-x)*(abs(x)<1 ? 1 : 0)
4  a, b = -1.5, 1.5
5  delta = 0.01
6
7  F(x) = sum([f2(u)*delta for u in a:delta:x])
8
9  xGrid = a:delta:b
10 y = [F(u) for u in xGrid]
11 plot(xGrid, y, c=:blue, xlims=(a,b), ylims=(0,1),
12          xlabel=L"x", ylabel=L"F(x)", legend=:none)
```

In line 3, we define the function `f2()`. The second set of brackets in the equation are used to ensure that the PDF is zero outside of the region $[-1, 1]$, as it acts like an*indicator function*, and evaluates to 0 everywhere else. In lines 4 and 5, we set the limits of our integral, and the stepwise `delta` used. In line 7, we create a function that approximates the value of the CDF through a crude Riemann sum by evaluating the PDF at each point `u`, multiplying this by `delta`, and repeating this process for each progressively larger interval up to the specified value `x`. The total area is then approximated via the `sum()` function, see (3.7). In line 9, we specify the grid of values over which we will plot our approximated CDF. Line 10 uses the function `F()` to create the array `y`, which contains the actual approximation of the CDF over the grid of value specified. Lines 11–12 are plotted in Figure 3.4.

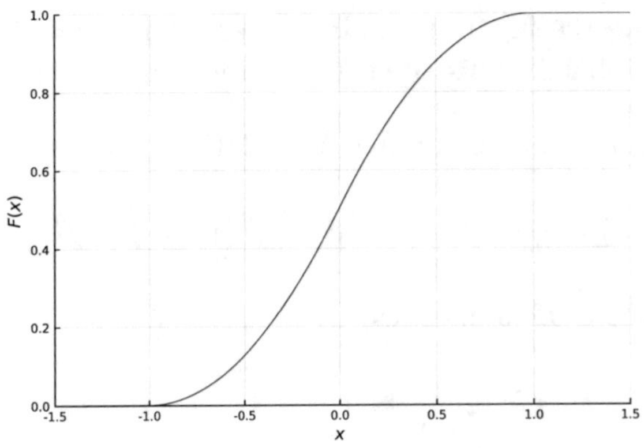

Figure 3.4: The CDF associated with the PDF $f_2(x)$.

Inverse and Quantiles

Where the CDF answers the question "what is the probability of being less than or equal to x", a dual question often asked is "what value of x corresponds to a probability of the random variable being less than or equal to u". Mathematically, we are looking for the *inverse function* of $F(x)$. In cases where the CDF is continuous and strictly increasing over all values, the inverse, $F^{-1}(\cdot)$ is well defined, and can be found via the equation,

$$F\big(F^{-1}(u)\big) = u, \qquad \text{for } u \in [0,1]. \tag{3.8}$$

For example, take the *sigmoid function* as the CDF, which is as a type of *logistic function*

$$F(x) = \frac{1}{1+e^{-x}}.$$

Solving for $F^{-1}(u)$ in (3.8) yields

$$F^{-1}(u) = \log \frac{u}{1-u}.$$

Observe that as $u \to 0^+$ we get $F^{-1}(u) \to -\infty$ and as $u \to 1^-$ we get $F^{-1}(u) \to \infty$. This is the *inverse CDF* for the distribution. Schematically, given a specified probability u, it allows us to find x values such that

$$\mathbb{P}(X \le x) = u. \tag{3.9}$$

The value x satisfying (3.9) is also called the u-th *quantile* of the distribution. If u is given as a percent, then it is called a *percentile*. The *median* is another related term, and is also known as the 0.5-th quantile. Other related terms are the *quartiles*, with the *first quartile* at $u = 0.25$, the *third quartile* at $u = 0.75$, and the *inter-quartile range*, which is defined as $F^{-1}(0.75) - F^{-1}(0.25)$. These same terms used again in respect to summarizing datasets in Section 4.2.

In more general cases, where the CDF is not necessarily strictly increasing and continuous, we may still define the inverse CDF via

$$F^{-1}(u) := \inf\{x \ : \ F(x) \ge u\}.$$

As an example of such a case, consider an arbitrary customer arriving to a queue where the server is utilized 80% of the time, and an average service takes 1 minute. How long does such a customer

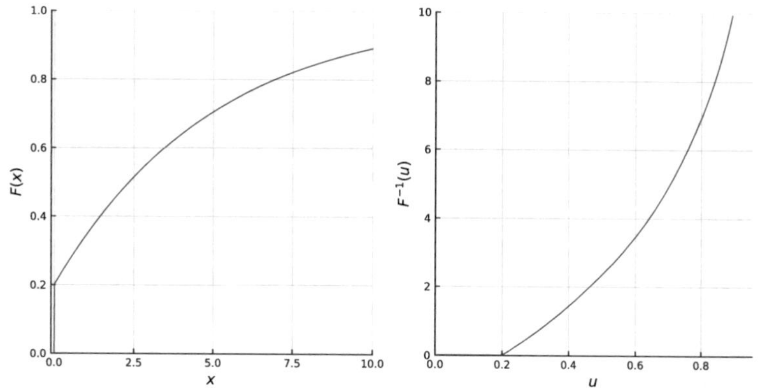

Figure 3.5: The CDF $F(x)$ and its inverse $F^{-1}(u)$.

wait in the queue until service starts? Some customers won't wait at all (20% of the customers), whereas others will need to wait until those that arrived before them are serviced. Results from the field of *queuing theory* (some of which are partially touched in Chapter 10) give rise to the following distribution function for the waiting time:

$$F(x) = 1 - 0.8e^{-(1-0.8)x} \qquad \text{for} \quad x \geq 0. \tag{3.10}$$

Notice that at $x = 0$, $F(0) = 0.2$, indicating the fact that there is a 0.2 chance for zero wait. Such a distribution is an example of a *mixed discrete and continuous distribution.* Notice that this distribution function only holds for a specific case of assumptions known as the stationary stable M/M/1 queue, explored further in Section 10.3. We now plot both $F(x)$ and $F^{-1}(u)$ in Listing 3.6 where we construct $F^{-1}(\cdot)$ programmatically. Observe Figure 3.5 where the CDF $F(x)$ exhibits a jump at 0 indicating the "probability mass". The inverse CDF then evaluates to 0 for all values of $u \in [0, 0.2]$.

Listing 3.6: The inverse CDF

```
using Plots, LaTeXStrings; pyplot()

xGrid = 0:0.01:10
uGrid = 0:0.01:1
busy = 0.8

F(t)= t<=0 ? 0 : 1 - busy*exp(-(1-busy)t)

infimum(B) = isempty(B) ? Inf : minimum(B)
invF(u) = infimum(filter((x) -> (F(x) >= u),xGrid))

p1 = plot(xGrid,F.(xGrid), c=:blue, xlims=(-0.1,10), ylims=(0,1),
        xlabel=L"x", ylabel=L"F(x)")

p2 = plot(uGrid,invF.(uGrid), c=:blue, xlims=(0,0.95), ylims=(0,maximum(xGrid)),
        xlabel=L"u", ylabel=L"F^{-1}(u)")

plot(p1, p2, legend=:none, size=(800, 400))
```

Line 3 defines the grid over which we will evaluate the CDF. Line 4 defines the grid over which we will evaluate the inverse CDF. In line 5, we define the time proportion during which the server is busy. In line 7, we define the function `F()` as in (3.10). Note that for values less than zero, the CDF evaluates to 0. In line 9, we define the function `infimum()`, which implements similar logic to the mathematical operation $\inf\{\}$. It takes an input and checks if it is empty via the `isempty()` function, and if it is returns `Inf`, else returns the minimum value of the input. This agrees with the typical mathematical notation where the infimum of the empty set is ∞. In line10, we define the function `invF()`. It first creates an array (representing a set) $\{x : F(x) \geq u\}$ directly via the Julia `filter()` function. Note that as a first argument, we use an anonymous Julia function, `(x) -> (F(x) >= u)`. We then use this function as a filter over `xGrid`. Finally, we apply the infimum over this mathematical set (represented by a vector of coordinates on the x-axis). Lines 12–18 are used to plot both the original CDF, via the `F()` function, and the inverse CDF, via the `invF()` functions, respectively.

Integral Transforms

In general terms, an *integral transform* of a probability distribution is a representation of the distribution on a different domain. Here we focus on the moment generating function (MGF). Other examples include the Characteristic Function (CF), Probability-Generating Function (PGF), and similar transforms.

For a random variable X and a real or complex fixed value s, consider the expectation, $\mathbb{E}[e^{sX}]$. When viewed as a function of s, this is the moment generating function. We present this here for a continuous random variable with PDF $f(\cdot)$:

$$M(s) = \mathbb{E}[e^{sX}] = \int_{-\infty}^{\infty} f(x)\, e^{sx}\, dx. \tag{3.11}$$

This is also known as the bi-lateral *Laplace transform* of the PDF (with argument $-s$). Many useful Laplace transform properties carry over from the theory of Laplace transforms to the MGF. A full exposition of such properties is beyond the scope of this book, however we illustrate a few via an example.

Consider two distributions with densities,

$$\begin{aligned} f_1(x) &= 2x && \text{for} && x \in [0,1], \\ f_2(x) &= 2-2x && \text{for} && x \in [0,1], \end{aligned}$$

where the respective random variables are denoted as X_1 and X_2. Computing the MGF of these distributions, we obtain

$$M_1(s) = \int_0^1 2x\, e^{sx}\, dx = 2\frac{1+e^s(s-1)}{s^2},$$

$$M_2(s) = \int_0^1 (2-2x)\, e^{sx}\, dx = 2\frac{e^s-1-s}{s^2}.$$

Define now a random variable, $Z = X_1 + X_2$ where X_1 and X_2 are assumed independent. In this case, it is known that the MGF of Z is the product of the MGFs of X_1 and X_2. That is

$$M_Z(s) = M_1(s)M_2(s) = 4\frac{\big(1+e^s(s-1)\big)\big(e^s-1-s\big)}{s^4}. \tag{3.12}$$

The new MGF $M_Z(\cdot)$ fully specifies the distribution of Z. It also yields a rather straightforward computation of moments, hence the name MGF. A key property of any MGF $M(s)$ of a random variable X is that

$$\frac{d^n}{ds^n} M(s)\Big|_{s=0} = \mathbb{E}[X^n]. \tag{3.13}$$

This can be easily verified from (3.11). Hence to calculate the n-th moment, one can simply evaluate the derivative of the MGF at $s = 0$. Note that in certain cases, evaluating the limit of $s \to 0$ is required.

In Listing 3.7, we estimate both the PDF and MGF of Z and compare the estimated MGF to $M_Z(s)$ above. The listing also creates Figure 3.6 where on the right-hand side plot it can be seen that the slope of the tangent line to the MGF at $s = 0$ is 1.0, in agreement with the mean.

Listing 3.7: A sum of two triangular random variables

```
1   using Distributions, Statistics, Plots; pyplot()
2
3   dist1 = TriangularDist(0,1,1)
4   dist2 = TriangularDist(0,1,0)
5   N=10^6
6
7   data1, data2 = rand(dist1,N), rand(dist2,N)
8   dataSum = data1 + data2
9
10  mgf(s) = 4(1+(s-1)*MathConstants.e^s)*(MathConstants.e^s-1-s)/s^4
11
12  mgfPointEst(s) = mean([MathConstants.e^(s*z) for z in
13                          rand(dist1,20) + rand(dist2,20)])
14
15  p1 = histogram(dataSum, bins=80, normed=:true,
16          ylims=(0,1.4), xlabel="z", ylabel="PDF")
17
18  sGrid = -1:0.01:1
19  p2 = plot(sGrid, mgfPointEst.(sGrid), c=:blue, ylims=(0,3.5))
20  p2 = plot!(sGrid, mgf.(sGrid), c=:red)
21  p2 = plot!( [minimum(sGrid),maximum(sGrid)],
22          [minimum(sGrid),maximum(sGrid)].+1,
23          c=:black, xlabel="s", ylabel="MGF")
24
25  plot(p1, p2, legend=:none, size=(800, 400))
```

In lines 3 and 4, we create two separate triangular distribution-type objects `dist1` and `dist2`, matching the densities $f_1(x)$ and $f_2(x)$, respectively. Note that the third argument of the `TriangularDist()` function is the location of the "peak" of the triangle (or the mode of the distribution). Distribution objects are covered further in Section 3.4. In line 7, we generate random observations from `dist1` and `dist2`, and store these observations separately in the two arrays `data1` and `data2`, respectively. In line 8, we generate observations for Z by performing element-wise summation of the values in our arrays `data1` and `data2`. In line 10, we implement the MGF function as in (3.12). In lines 12–13, we define the function `mgfPointEst()`, which crudely estimates the MGF at the point `s`. We purposefully only use 20 observations, each time estimating the sample mean of e^{sZ} for a specified s. The remainder of the code uses the data and the defined functions to generate Figure 3.6. Lines 21–23 plot the black line.

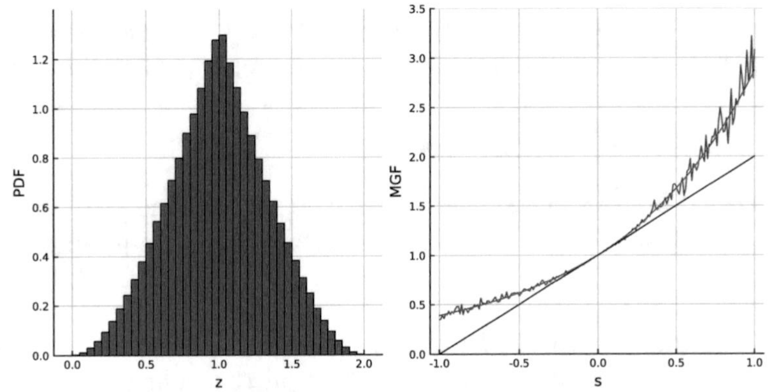

Figure 3.6: Left: The estimate of the PDF of Z via a histogram.
Right: The theoretical MGF in red vs. a Monte Carlo estimate in blue.
The slope of the black line is the mean.

3.4 `Distributions` and Related Packages

As touched on previously in Listing 3.4 and Listing 3.7, Julia has a well-developed package for distributions. The `Distributions` package allows us to create distribution-type objects based on what family they belong to (more on families of distributions in Sections 3.5 and 3.6). These distribution objects can then be used as arguments for other functions, for example, `mean()` and `var()`. Of key importance is the ability to randomly sample from a distribution using `rand()`. We can also use distributions with other functions including `pdf()`, `cdf()`, and `quantile()` to name a few. In addition, the built-in `Statistics` package as well as the `StatsBase` package contain many functions which have methods for distribution-type objects, extended by the `Distributions` package. A useful paper describing the distributions package is [BAABLPP19].

Weighted Vectors

In the case of discrete distributions of finite support, the `StatsBase` package provides the "weight vector" object via `Weights()`, which allows for an array of values, or outcomes, to be given probabilistic weights. This is also known as a *probability vector*. In order to generate observations, we use the `sample()` function (from `StatsBase`) on a vector given its weights, instead of the `rand()` function. Note that an alternative is to use the `Categorical` distribution supplied via the `Distributions` package. Listing 3.8 provides a brief example of the use of weight vectors.

Listing 3.8: Sampling from a weight vector

```
using StatsBase, Random
Random.seed!(1)

grade = ["A","B","C","D","E"]
weightVect = Weights([0.1,0.2,0.1,0.2,0.4])

N = 10^6
data = sample(grade,weightVect,N)
[count(i->(i==g),data) for g in grade]/N
```

```
5-element Array{Float64,1}:
 0.099901
 0.200248
 0.099704
 0.20068
 0.399467
```

In line 4, we define an array of strings "A" to "E", which represent possible outcomes. In line 5, we define their weights. Note the fact that `Weights()` is capitalized, signifying the fact that the function creates a new object. This type of function is known as a *Constructor*. Line 8 uses the function `sample()` to sample `N` observations from our array `grade`, according to the weights given by the weight vector `weightVect`. Line 9 uses the `count()` function to count how many times each entry `g` in `grade` occurs in `data`, and then evaluates the proportion of times total each grade occurs. It can be observed that the grades have been sampled according to the probabilities specified in the array `weightVect`. Note that you can also use the `Categorical()` object in the `Distributions` package as alternative.

Using Distribution Type Objects

We now introduce some important functionality of the `Distributions` package and distribution-type objects through an example. Consider a distribution from the "Triangular" family, with the following density:

$$f(x) = \begin{cases} x & \text{for } x \in [0,1], \\ 2-x & \text{for } x \in (1,2]. \end{cases}$$

In Listing 3.9, rather than creating the density manually as in the previous sections, we use the `TriangularDist()` constructor to create a distribution-type object, and then use this to create plots of the PDF, CDF, and inverse CDF as shown in Figure 3.7.

Listing 3.9: Using the `pdf()`, `cdf()`, and `quantile()` functions with `Distributions`

```
using Distributions, Plots, LaTeXStrings; pyplot()

dist = TriangularDist(0,2,1)
xGrid = 0:0.01:2
uGrid = 0:0.01:1

p1 = plot( xGrid, pdf.(dist,xGrid), c=:blue,
                xlims=(0,2), ylims=(0,1.1),
                xlabel="x", ylabel="f(x)")
p2 = plot( xGrid, cdf.(dist,xGrid), c=:blue,
                xlims=(0,2), ylims=(0,1),
                xlabel="x", ylabel="F(x)")
p3 = plot( uGrid,quantile.(dist,uGrid), c=:blue,
                xlims=(0,1), ylims=(0,2),
                xlabel="u", ylabel=(L"F^{-1}(u)"))

plot(p1, p2, p3, legend=false, layout=(1,3), size=(1200, 400))
```

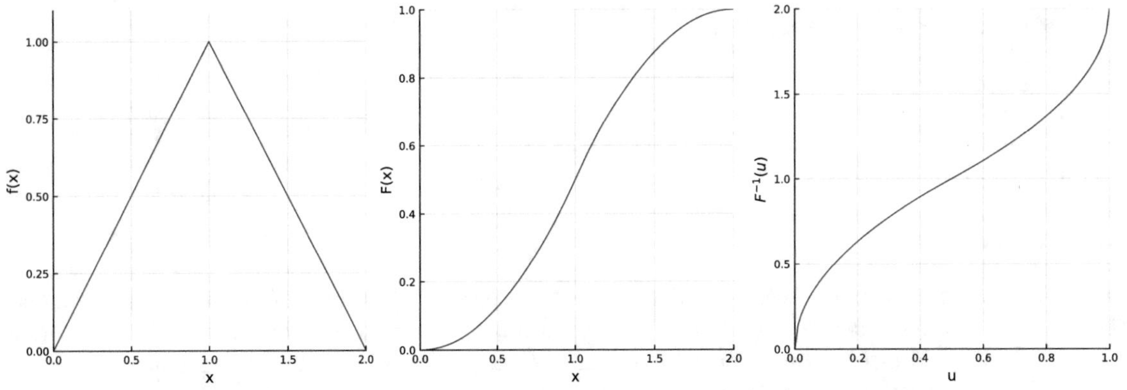

Figure 3.7: The PDF, CDF, and inverse CDF a triangular distribution.

In line 3, we use the `TriangularDist()` function to create a distribution-type object. The first two arguments are the start and end points of the support, and the third argument is the location of the "peak" (or mode). The essence of this example is in lines 7, 10, and 13 where we use the `pdf()`, `cdf()`, and `quantile()` functions, respectively. In each case, we use `dist` as the first argument and broadcast over the second argument via the "`.`" broadcast operator.

In addition to evaluating functions associated with the distribution, we can also query a distribution object for a variety of properties and parameters. Given a distribution object, you may apply `params()` on it to retrieve the distributional parameters. You may query for the `mean()`, `median()`, `var()` (variance), `std`, (standard deviation), `skewness()`, and `kurtosis()`. You can also query for the minimal and maximal value in the support of the distribution via `minimum()` and `maximum()`, respectively. You may also apply `mode()` or `modes()` to either get a single mode (value of x where the PMF or PDF is maximized) or an array of modes where applicable. Listing 3.10 illustrates some of these for our `TriangularDist`.

Listing 3.10: Descriptors of `Distribution` objects

```
using Distributions
dist = TriangularDist(0,2,1)

println("Parameters: \t\t\t",params(dist))
println("Central descriptors: \t\t",mean(dist),"\t",median(dist))
println("Dispersion descriptors: \t", var(dist),"\t",std(dist))
println("Higher moment shape descriptors: ",skewness(dist),"\t",kurtosis(dist))
println("Range: \t\t\t\t", minimum(dist),"\t",maximum(dist))
println("Mode: \t\t\t\t", mode(dist), "\tModes: ",modes(dist))
```

```
Parameters:                         (0.0, 2.0, 1.0)
Central descriptors:                1.0     1.0
Dispersion descriptors:             0.16666666666666666     0.408248290463863
Higher moment shape descriptors: 0.0     -0.6
Range:                              0.0     2.0
Mode:                               1.0     Modes: [1.0]
```

In Listing 3.11, we look at another example, where we generate random observations from a distribution-type object via the `rand()` function, and compare the sample mean against the specified mean. Note that two different types of distributions are created here, a continuous distribution and a discrete distribution. These are discussed further in Sections 3.5 and 3.6, respectively.

Listing 3.11: Using `rand()` with `Distributions`

```
1  using Distributions, StatsBase, Random
2  Random.seed!(1)
3
4  dist1 = TriangularDist(0,10,5)
5  dist2 = DiscreteUniform(1,5)
6  theorMean1, theorMean2 = mean(dist1), mean(dist2)
7
8  N = 10^6
9  data1 = rand(dist1,N)
10 data2 = rand(dist2,N)
11 estMean1, estMean2 = mean(data1), mean(data2)
12
13 println("Symmetric Triangular Distiribution on [0,10] has mean $theorMean1
14         (estimated: $estMean1)")
15 println("Discrete Uniform Distiribution on {1,2,3,4,5} has mean $theorMean2
16         (estimated: $estMean2)")
```

```
Symmetric Triangular Distiribution on [0,10] has mean 5.0
        (estimated: 4.999164797766807)
Discrete Uniform Distiribution on {1,2,3,4,5} has mean 3.0
        (estimated: 3.001862)
```

In line 4, we use the `TriangularDist()` function to create a symmetrical triangular distribution about 5, and store this as `dist1`. In line 5, we use the `DiscreteUniform()` function to create a discrete uniform distribution, and store this as `dist2`. Note that observations from this distribution can take on values from $\{1, 2, 3, 4, 5\}$, each with equal probability. In line 6, we evaluate the mean of the two distribution objects created above by applying the function `mean()` to both of them. These methods of `mean()` only use the parameters of the distribution to evaluate the mean. No data manipulation is taking place. In lines 8–11, we estimate the means of the two distributions by randomly sampling from our distributions `dist1` and `dist2`. In lines 9–10, the `Distribution` object is given as a first argument to `rand()`. Lines 13–16 print the results. It can be seen that the estimated means are a good approximation of the actual means.

The Inverse Probability Transform

One may ask how does Julia (or any software package) generate random values from a given distribution? There are a variety of techniques for transforming pseudorandom numbers from a uniform distribution into numbers from a given distribution. An extensive treatment is in [KTB11]. One basic method which stands above the rest is *inverse transform sampling.*

Let X be a random variable distributed with CDF $F(\cdot)$ and inverse CDF $F^{-1}(\cdot)$. Now take U to be a uniform random variable over $[0, 1]$, and let $Y = F^{-1}(U)$. It holds that Y is distributed like X. This useful property is called the *inverse probability transform* and constitutes a generic method for generating random variables from an underlying distribution.

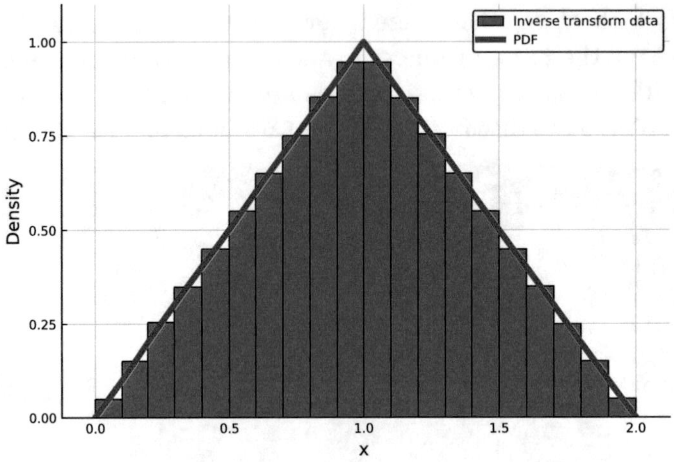

Figure 3.8: A histogram generated using the inverse probability transform compared to the PDF of a triangular distribution.

To see why the method works, consider a uniform random variable U and apply to it the inverse probability transform $F^{-1}(\cdot)$. In such a case, consider the CDF of $Y = F^{-1}(U)$ and see that it is $F(\cdot)$:

$$F_Y(y) = \mathbb{P}\big(Y \le y\big) = \mathbb{P}\big(F^{-1}(U) \le y\big) = \mathbb{P}\big(U \le F(y)\big) = F_U\big(F(y)\big) = F(y).$$

The third equality follows because $F(\cdot)$ is a monotonic function and can be applied to both sides of the inequality. The last step follows because the CDF of uniform $(0, 1)$ random variable is

$$F_U(z) = \begin{cases} 0 & \text{for } z < 0, \\ z & \text{for } 0 \le z \le 1, \\ 1 & \text{for } 1 < z. \end{cases}$$

Keep in mind that when using the `Distributions` package, we would typically generate random variables using the `rand()` function on a distribution-type object, as performed in Listing 3.11. The implementation of `rand()` may use the inverse probability transform or alternatively may use a different type of method depending on the distribution at hand. However, in Listing 3.12, we illustrate how to use the inverse probability transform with the results presented in Figure 3.8. Observe that we can implement $F^{-1}(\cdot)$ via the `quantile()` function.

Listing 3.12: Inverse transform sampling

```
using Distributions, Plots; pyplot()

triangDist = TriangularDist(0,2,1)
xGrid = 0:0.1:2
N = 10^6
inverseSampledData = quantile.(triangDist,rand(N))

histogram( inverseSampledData, bins=30, normed=true,
        ylims=(0,1.1), label="Inverse transform data")
plot!( xGrid, pdf.(triangDist,xGrid), c=:red, lw=4,
        xlabel="x", label="PDF", ylabel = "Density", legend=:topright)
```

In line 3, we create our triangular distribution `triangDist`. In lines 4 and 5, we define the support over which we plot our data, as well as how many data points we simulate. In line 6, we generate `N` random observations from a continuous uniform distribution over the domain $[0, 1]$ via the `rand()` function. Then the `quantile()` function, along with the `dot` operator (`.`) to calculate each corresponding quantile of `triangDist`. Lines 8 and 9 plot a histogram of this `inverseSampledData`, using 30 bins. For large `N`, the histogram generated is a close approximation of the PDF of the underlying distribution. Lines 10–11 then plot the analytic PDF of the underlying distribution.

3.5 Families of Discrete Distributions

A *family of probability distributions* is a collection of probability distributions having some functional form that is parameterized by a well-defined set of parameters. In the discrete case, the PMF, $p(x\,;\,\theta) = \mathbb{P}(X = x)$, is parameterized by the *parameter* $\theta \in \Theta$ where Θ is called the *parameter space*. The (scalar or vector) parameter θ then affects the actual form of the PMF, including possibly the support of the random variable. Hence, technically a family of distributions is the collection of PMFs $p(\cdot\,;\,\theta)$ for all $\theta \in \Theta$.

In this section, we present some of the most common families of *discrete distributions*. We consider the following: *discrete uniform distribution*, *binomial distribution*, *geometric distribution*, *negative binomial distribution*, *hypergeometric distribution*, and *Poisson distribution*. Each of these is implemented in the Julia `Distributions` package. The approach that we take in the code examples of this section is to generate random variables from each distribution using first principles, as opposed to applying `rand()` on a distribution object, as was demonstrated in Listing 3.11. Understanding how to generate a random variable from a given distribution using first principles helps strengthen understanding of the associated probability models and processes.

In Listing 3.13, we illustrate how to create a distribution object for each of the discrete distributions that we investigate in this section. As output we print the parameters and the support of each distribution.

Listing 3.13: Families of discrete distributions

```
using Distributions
dists = [
    DiscreteUniform(10,20),
    Binomial(10,0.5),
    Geometric(0.5),
    NegativeBinomial(10,0.5),
    Hypergeometric(30, 40, 10),
    Poisson(5.5)]

println("Distribution \t\t\t\t\t\t Parameters \t Support")
reshape([dists ;  params.(dists) ;
                ((d)->(minimum(d),maximum(d))).(dists) ],
                length(dists),3)
```

```
Distribution                                    Parameters      Support
6×3 Array{Any,2}:
 DiscreteUniform(a=10, b=20)                     (10, 20)        (10, 20)
 Binomial{Float64}(n=10, p=0.5)                  (10, 0.5)       (0, 10)
 Geometric{Float64}(p=0.5)                       (0.5,)          (0, Inf)
 NegativeBinomial{Float64}(r=10.0, p=0.5)        (10.0, 0.5)     (0, Inf)
 Hypergeometric(ns=30, nf=40, n=10)              (30, 40, 10)    (0, 10)
 Poisson{Float64}(λ=5.5)                         (5.5,)          (0, Inf)
```

Lines 2–8 are used to define an array of distribution objects. The help provided by the distributions package is useful. Use `?` <<Name>> where <<Name>> may be `DiscreteUniform`, `Binomial`, etc. Lines 10–13 result in output that is a 6×3 array of type `Any`. The first column is the actual distributions object, the second column has the distributional parameters, and the third column represents the support. The parameters and the support for each distribution are presented in more detail later in this section. Note the use of an anonymous function `(d)->(minimum(d),maximum(d))` applied via "`.`" to each element of `dists`. This function returns a tuple. The use of `reshape()` transforms the array of arrays into a matrix of the desired dimensions.

Discrete Uniform Distribution

The *discrete uniform distribution* is simply a probability distribution that places equal probabilities for all equal outcomes. One example is given by the probability of the outcomes of a die toss. The probability of each possible outcome for a fair, six-sided die is given by

$$\mathbb{P}(X = x) = \frac{1}{6} \qquad \text{for } x = 1, \ldots, 6.$$

Listing 3.14 simulates N tosses of a die, and then calculates and plots the proportion of times each possible outcome occurs, along with the PMF. The plot is shown in Figure 3.9. For large values of N, the proportion of counts for each outcome converges to $1/6$.

Listing 3.14: Discrete uniform die toss

```
using StatsBase, Plots; pyplot()

faces, N = 1:6, 10^6
mcEstimate = counts(rand(faces,N), faces)/N

plot(faces, mcEstimate,
        line=:stem, marker=:circle,
        c=:blue, ms=10, msw=0, lw=4, label="MC estimate")
plot!([i for i in faces], [1/6 for _ in faces],
        line=:stem, marker=:xcross, c=:red,
        ms=6, msw=0, lw=2, label="PMF",
        xlabel="Face number", ylabel="Probability", ylims=(0,0.22))
```

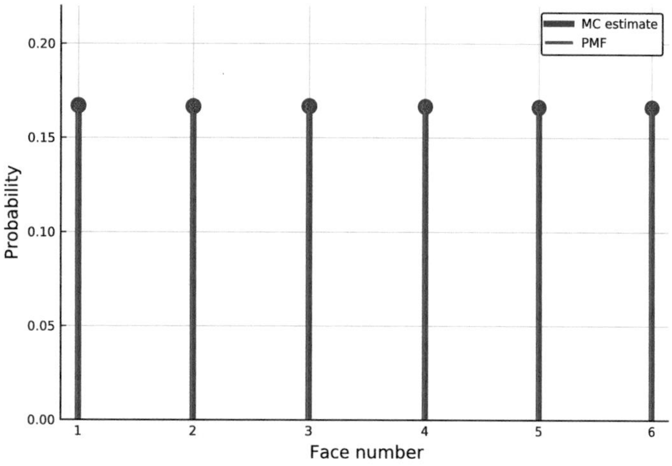

Figure 3.9: A discrete uniform PMF.

In line 3, we define all possible outcomes of our six-sided die, along with how many die tosses we will simulate. Line 4 uniformly and randomly generates `N` observations from our die, and then uses the `counts()` function to calculate proportion of times each outcome occurs. Note that applying `rand(DiscreteUniform(1,6),N)` would yield a statistically identical result to `rand(faces,N)`. Line 5 uses the `stem` function to create a stem plot of the proportion of times each outcome occurs, while line 6 plots the analytic PMF of our six-sided die.

Binomial Distribution

The *binomial distribution* is a discrete distribution which arises where multiple identical and independent yes/no, true/false, success/failure trials (also known as *Bernoulli trials*) are performed. For each trial, there can only be two outcomes, and the probability weightings of each unique trial must be the same.

As an example, consider a two-sided coin, which is flipped n times in a row. If the probability of obtaining a head in a single flip is p, then the probability of obtaining x heads total is given by the PMF,

$$\mathbb{P}(X = x) = \binom{n}{x} p^x (1-p)^{n-x} \qquad \text{for } x = 0, 1, \ldots, n.$$

Listing 3.15 simulates $n = 10$ tosses of a fair coin ($p = 1/2$), `N` times total, with success probability p, and calculates the proportion of times each possible outcome occurs. Observe that in the `Distributions` package, `pdf()` applied to a discrete distribution yields the PMF. In fact, the PMF is often loosely called a PDF (density) in statistics. The results are presented in Figure 3.10.

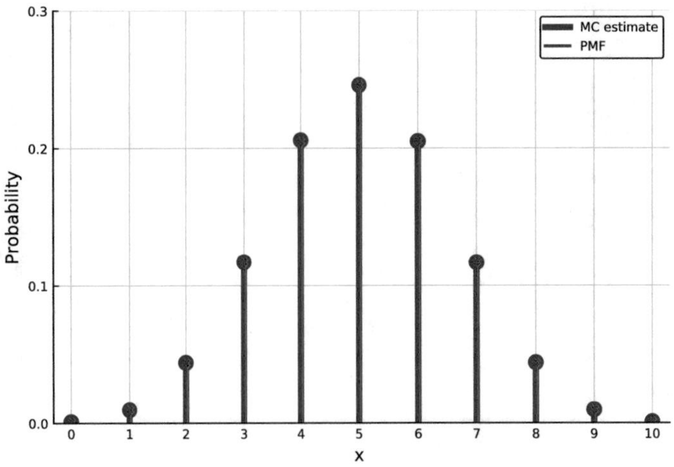

Figure 3.10: Binomial PMF for number of heads in 10 flips each with $p = 0.5$.

Listing 3.15: Coin flipping and the binomial distribution

```
1  using StatsBase, Distributions, Plots; pyplot()
2
3  binomialRV(n,p) = sum(rand(n) .< p)
4
5  p, n, N = 0.5, 10, 10^6
6
7  bDist = Binomial(n,p)
8  xGrid = 0:n
9  bPmf = [pdf(bDist,i) for i in xGrid]
10 data = [binomialRV(n,p) for _ in 1:N]
11 pmfEst = counts(data,0:n)/N
12
13 plot( xGrid, pmfEst,
14         line=:stem, marker=:circle,
15         c=:blue, ms=10, msw=0, lw=4, label="MC estimate")
16 plot!( xGrid, bPmf,
17         line=:stem, marker=:xcross, c=:red,
18         ms=6, msw=0, lw=2, label="PMF", xticks=(0:1:10),
19         ylims=(0,0.3), xlabel="x", ylabel="Probability")
```

In line 3, we define the function `binomialRV()`. It generates a binomial random variable from first principles by creating an array of uniform $[0, 1]$ values of length `n` with `rand(n)`. We then use `.<` to compare each value (element-wise) to `p`. The result is a vector of Booleans, with each one set to `true` with probability `p`. Summing up this vector creates the binomial random variable. In line 9, we create a vector incorporating the values of the binomial PMF. Note that in the Julia distributions package, PMFs are created via `pdf()`. Line 10 is where we generate `N` random values. In line 11, we use `counts()` from the `StatsBase` package to count how may times each outcome occurred, for `0:n` heads. We then normalize via division by `N`. The remainder of the code creates the plot.

Note that the binomial distribution describes part of the fishing example in Section 2.1, where we sample with replacement. This is because the probability of success (i.e. fishing a gold fish) remains unchanged regardless of how many times we have sampled from the pond.

Geometric Distribution

Another distribution associated with Bernoulli trials is the *geometric distribution*. In this case, consider an infinite sequence of independent trials, each with success probability p, and let X be the first trial that is successful. Using first principles it is easy to see that the PMF is

$$\mathbb{P}(X = x) = p\,(1-p)^{x-1} \qquad \text{for } x = 1, 2, \ldots. \tag{3.14}$$

An alternative version of the geometric distribution is the distribution of the random variable $\tilde{X}$, counting the number of failures until success. Observe that for every sequence of trials, $\tilde{X} = X - 1$. From this, it is easy to relate the PMFs of the random variables and see that

$$\mathbb{P}(\tilde{X} = x) = p(1-p)^x \qquad \text{for } x = 0, 1, 2, \ldots.$$

In the Julia `Distributions` package, `Geometric` stands for the distribution of $\tilde{X}$, not X.

We now look at an example involving the popular casino game of roulette. Roulette is a game of chance, where a ball is spun on the inside edge of a horizontal wheel. As the ball loses momentum, it eventually falls vertically down, and lands on one of 37 spaces, numbered 0 to 36. There are 18 black spaces, 18 red, and a single space ("zero") is green. Each spin of the wheel is independent, and each of the possible 37 outcomes is equally likely. Now let us assume that a gambler goes to the casino and plays a series of roulette spins. There are various ways to bet on the outcome of roulette, but in this case he always bets on black (if the ball lands on black he wins, otherwise he loses). Say that the gambler plays until his first win. In this case, the number of plays is a geometric random variable with support $x = 1, 2, \ldots$. Listing 3.16 simulates this scenario and creates Figure 3.11.

Listing 3.16: The geometric distribution

```
using StatsBase, Distributions, Plots; pyplot()

function rouletteSpins(p)
    x = 0
    while true
        x += 1
        if rand() < p
            return x
        end
    end
end

p, xGrid, N = 18/37, 1:7, 10^6
mcEstimate = counts([rouletteSpins(p) for _ in 1:N],xGrid)/N
gDist = Geometric(p)
gPmf = [pdf(gDist,x-1) for x in xGrid]
plot(xGrid, mcEstimate, line=:stem, marker=:circle,
    c=:blue, ms=10, msw=0, lw=4, label="MC estimate")
plot!( xGrid, gPmf, line=:stem, marker=:xcross,
    c=:red, ms=6, msw=0, lw=2, label="PMF",
    ylims=(0,0.5), xlabel="x", ylabel="Probability")
```

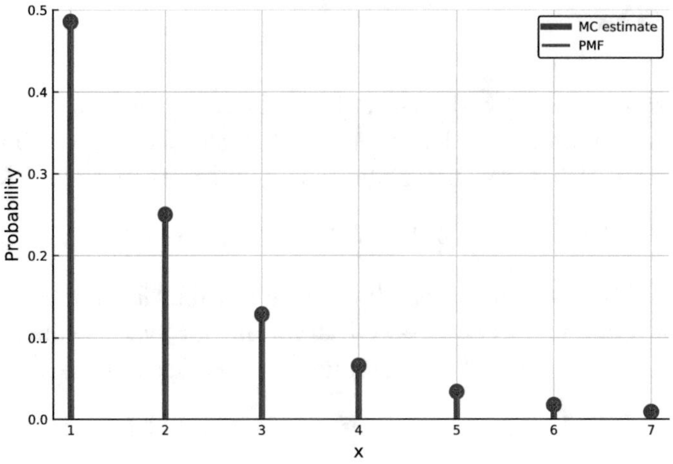

Figure 3.11: A geometric PMF.

The function `rouletteSpins()` defined in lines 3–11 is a straightforward way to generate a geometric random variable with support $1, 2, \ldots$ as X above. Lines 5–10 loop until a value is returned from the function. In each iteration, we increment `x` and check if we have a success (an event happening with probability p) via, `rand()` $<$ `p`. The remainder of the code is similar to the previous listing. Consider the second argument to `pdf()` in line 18. Here `x-1` is used because the built-in geometric distribution is for the random variable $\tilde{X}$ above, which starts at 0, while we are interested in the geometric random variable starting at 1.

Negative Binomial Distribution

Recall the previous example above of a roulette gambler. Assume now that the gambler plays until he wins for the r-th time (in the previous example $r = 1$). The *negative binomial distribution* describes this situation. That is, a random variable X follows this distribution, if it describes the number of trials until the r-th success. The PMF is given by

$$\mathbb{P}(X = x) = \binom{x-1}{r-1} p^r (1-p)^{x-r} \quad \text{for} \quad x = r, r+1, r+2, \ldots.$$

Notice that with $r = 1$ the expression reduces to the geometric PMF (3.14). Similarly to the geometric case, there is an alternative version of the negative binomial distribution. Let $\tilde{X}$ denote the number of failures until the r-th success. Here, like in the geometric case, when both random variables are coupled on the same sequence of trials, we have $\tilde{X} = X - r$. As a result

$$\mathbb{P}(\tilde{X} = x) = \binom{x+r-1}{x} p^r (1-p)^x \quad \text{for } x = 0, 1, 2, \ldots.$$

To help reinforce this, in Listing 3.17, we simulate a gambler who bets consistently on black much like in the previous example, and determine the PMF for $r = 5$. That is, we determine the probabilities that x plays will occur up to the 5-th success (or win).

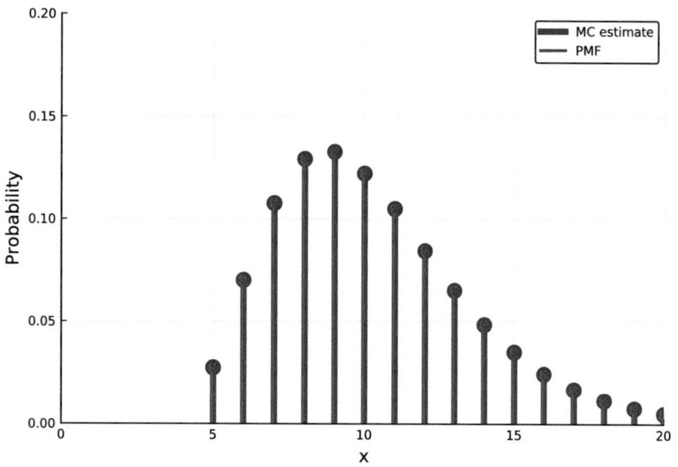

Figure 3.12: The PMF of negative binomial with $r = 5$ and $p = 18/37$.

Listing 3.17: The negative binomial distribution

```
using StatsBase, Distributions, Plots

function rouletteSpins(r,p)
    x = 0
    wins = 0
    while true
        x += 1
        if rand() < p
            wins += 1
            if wins == r
                return x
            end
        end
    end
end

r, p, N = 5, 18/37,10^6
xGrid = r:r+15

mcEstimate = counts([rouletteSpins(r,p) for _ in 1:N],xGrid)/N

nbDist = NegativeBinomial(r,p)
nbPmf = [pdf(nbDist,x-r) for x in xGrid]

plot( xGrid, mcEstimate,
        line=:stem, marker=:circle, c=:blue,
        ms=10, msw=0, lw=4, label="MC estimate")
plot!( xGrid, nbPmf, line=:stem,
         marker=:xcross, c=:red, ms=6, msw=0, lw=2, label="PMF",
         xlims=(0,maximum(xGrid)), ylims=(0,0.2),
         xlabel="x", ylabel="Probability")
```

This code is similar to the previous listing. The main difference is in the function `rouletteSpins()`, which now accepts both `r` and `p` as arguments. It is a straightforward implementation of the negative binomial story. A value is returned in line 11 only once the number of wins equals `r`. In a similar manner to the geometric example notice that in line 23, we use `x-r` for the argument of the `pdf()` function. This is because `NegativeBinomial` in the `Distributions` package stands for a distribution with support, $x = 0, 1, 2, \ldots$ and not $x = r, r+1, r+2, \ldots$ as we desire.

Hypergeometric Distribution

Moving on from Bernoulli trials, we now consider the *hypergeometric distribution*. To put it in context, consider the fishing problem discussed in Section 2.1, specifically the case where we fish without replacement. In this scenario, each time we sample from the population it decreases, and hence the probability of success changes for each subsequent sample. The hypergeometric distribution describes this situation. The PMF is given by

$$p(x) = \frac{\binom{K}{x}\binom{L-K}{n-x}}{\binom{L}{n}} \qquad \text{for } x = \max(0, n+K-L), \ldots, \min(n, K).$$

Here the parameter L is the population size, and K is the number of successes present in the population (this implies that $L-K$ is the number of failures present in the population). The parameter n is the number of samples taken from the population, and the input argument x is the number of successful samples observed. Hence, a hypergeometric random variable X with $\mathbb{P}(X = x) = p(x)$ describes the number of successful samples when *sampling without replacement*. Note that the expression for $p(x)$ can be deduced directly via combinatorial counting arguments.

To understand the support of the distribution, first consider the least possible value, $\max(0, n+K-L)$. It is either 0 or $n+K-L$ if $n > L-K$. The latter case stems from a situation where the number of samples n is greater than the number of failures present in the population. That is, in such a case, the least possible number of successes that can be sampled is

$$\text{number of samples } (n) \quad - \quad \text{number of failures in the population } (L-K).$$

As for the upper value of the support, it is $\min(n, K)$ because if $K < n$ then it isn't possible to sample only successes. Note that, in general, if the sample size n is not "too big" then the support reduces to $x = 0, \ldots, n$.

To help illustrate this distribution, we look at an example where we compare several hypergeometric distributions simultaneously. As before, let us consider a pond which contains a combination of gold and silver fish. In this example, there are $N = 500$ fish total, and we will define the catch of a gold fish a success, and a silver fish a failure. Now say that we sample $n = 30$ fish without replacement. We consider several of these cases, where the only difference between each is the number of successes, K, (gold fish) in the population.

Listing 3.18 plots the PMFs of five different hypergeometric distributions based on the number of successes in the population. The results are shown in Figure 3.13. It can be observed that as

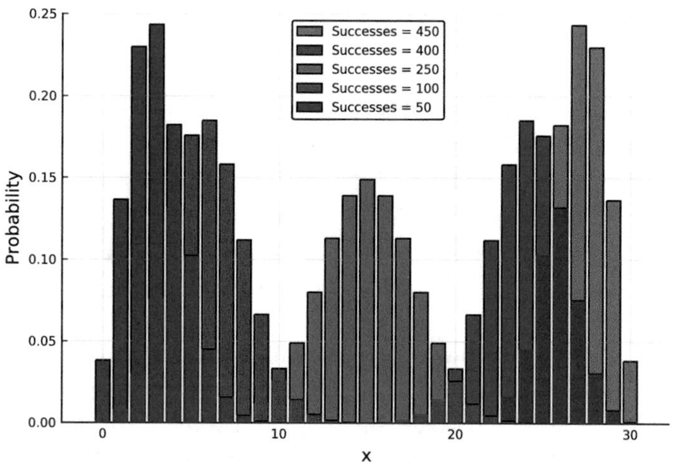

Figure 3.13: A comparison of several hypergeometric distributions for different proportions of successes in a population.

the number of successes present in the population increases, the PMF shifts further towards the right. Note that in the Julia `Distributions` package, `Hypergeometric` is parameterized via the number of successes (first argument) and number of failures (second argument), with the third argument being the sample size. This is slightly different to our parameterization above, which uses N, K, and n.

Listing 3.18: Comparison of several hypergeometric distributions

```
1  using Distributions, Plots; pyplot()
2
3  L, K, n  = 500, [450, 400, 250, 100, 50], 30
4  hyperDists = [Hypergeometric(k,L-k,n) for k in K]
5  xGrid = 0:1:n
6  pmfs = [ pdf.(dist, xGrid) for dist in hyperDists ]
7  labels = "Successes = " .* string.(K)
8
9  bar( xGrid, pmfs,
10         alpha=0.8, c=[:orange :purple :green :red :blue ],
11         label=hcat(labels...), ylims=(0,0.25),
12         xlabel="x", ylabel="Probability", legend=:top)
```

In line 3, we define the population size, `L`, the sample size `n`, and the array `K`, which contains the number of successes in the population, for each of our five scenarios. In line 4, the `Hypergeometric()` constructor is used to create several hypergeometric distributions. The constructor takes three arguments, the number of successes in the population `k`, the number of failures in the population `L-k`, and the number of times we sample from the population without replacement `n`. This constructor is then wrapped in a comprehension in order to create an array of different hypergeometric distributions, `hyperDists`. We then create an array of arrays, `pmfs` in line 6, by applying the `pdf()` function on each distribution. In lines 9–12, the `bar()` function is used to plot a bar chart of the PMF for each hypergeometric distribution in `hyperDists`. Notice the use of `hcat(labels...)` to convert `labels` from `Array{String,1}` to `Array{String,2}` which is required to label the plots in `bar()`.

Poisson Distribution and Poisson Process

The *Poisson process* is a *stochastic process* (random process) which can be used to model occurrences of events over time (or more generally in space). It may be used to model the arrival of customers to a system, the emission of particles from radioactive material, or packets arriving to a communication router. The Poisson process is the canonical example of a *point process* capturing the most sensible model for completely random occurrences over time. A full description and analysis of the Poisson process is beyond our scope; however, we provide an overview of the basics.

In a Poisson process, during an infinitesimally small time interval, Δt, it is assumed that (as $\Delta t \to 0$) there is an occurrence with probability $\lambda \Delta t$, and no occurrence with probability $1 - \lambda \Delta t$. Furthermore, as $\Delta t \to 0$, it is assumed that the chance of 2 or more occurrences during an interval of length Δt tends to 0. Here $\lambda > 0$ is the *intensity* of the Poisson process, and has the property that when multiplied by an interval of length T, the mean number of occurrences during the interval is λT.

The exponential distribution, discussed in the next section, is closely related to the Poisson process since the times between occurrences in the Poisson process are exponentially distributed. Another closely related distribution is the *Poisson distribution* that we discuss now. For a Poisson process over the time interval $[0, T]$, the number of occurrences satisfies

$$\mathbb{P}(x \text{ Poisson process occurrences during interval } [0,T]) = e^{-\lambda T} \frac{(\lambda T)^x}{x!} \qquad \text{for } x = 0, 1, \ldots .$$

The PMF $p(x) = e^{-\lambda} \lambda^x / x!$ for $x = 0, 1, 2, \ldots$ describes the Poisson distribution, the mean of which is λ. Hence, the number of occurrences in a Poisson process during $[0, T]$ is Poisson distributed with parameter (and mean) λT. Note that in applied statistics, the Poisson distribution is also sometimes taken as a model for occurrences, without explicitly considering a Poisson process. For example, assume that based on previous measurements, on average 5.5 people arrive at a hair salon during rush hour, then the probability of observing x people during rush hour can be modeled by the PMF of the Poisson distribution.

The Poisson process possesses many elegant analytic properties, and these sometimes come as an aid when considering Poisson distributed random variables. One such (seemingly magical) property is to consider the random variable $N \geq 0$ such that

$$\prod_{i=1}^{N} U_i \geq e^{-\lambda} > \prod_{i=1}^{N+1} U_i, \tag{3.15}$$

where $U_1, U_2, \ldots$ is a sequence of i.i.d. uniform$(0,1)$ random variables and $\prod_{i=1}^{0} U_i \equiv 1$. It turns out that seeking such a random variable N produces an efficient recipe for generating a Poisson random variable. That is, the N defined by (3.15) is Poisson distributed with mean λ. Notice that the recipe dictated by (3.15) is to continue multiplying uniform random variables to a "running product" until the product goes below the desired level $e^{-\lambda}$.

Returning to the hair salon example mentioned above, Listing 3.19 simulates this scenario, and compares the numerically estimated result against the PMF. The results are presented in Figure 3.14.

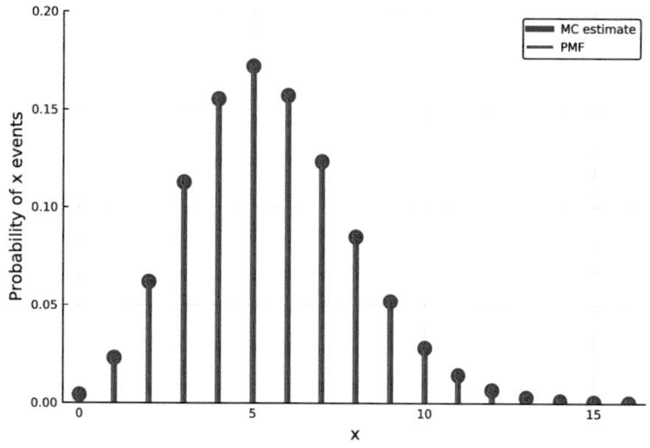

Figure 3.14: The PMF of a Poisson distribution with mean $\lambda = 5.5$.

Listing 3.19: The Poisson distribution

```
1  using StatsBase, Distributions, Plots; pyplot()
2
3  function prn(lambda)
4      k, p = 0, 1
5      while p > MathConstants.e^(-lambda)
6          k += 1
7          p *= rand()
8      end
9      return k-1
10 end
11
12 xGrid, lambda, N = 0:16, 5.5, 10^6
13
14 pDist = Poisson(lambda)
15 bPmf = pdf.(pDist,xGrid)
16 data = counts([prn(lambda) for _ in 1:N],xGrid)/N
17
18 plot( xGrid, data,
19         line=:stem, marker=:circle,
20         c=:blue, ms=10, msw=0, lw=4, label="MC estimate")
21 plot!( xGrid, bPmf, line=:stem,
22         marker=:xcross, c=:red, ms=6, msw=0, lw=2, label="PMF",
23         ylims=(0,0.2), xlabel="x", ylabel="Probability of x events")
```

In lines 3–10, the function `prn()`, standing for "Poisson random number", is defined. It implements (3.15) in a straightforward manner and takes a single argument, the expected arrival rate for our interval `lambda`. Line 16 calls `prn()` a total of `N` times, counts occurrences, and normalizes them by `N` to obtain Monte Carlo estimates of the Poisson probabilities. Lines 18–23 plot these Monte Carlo estimates as well as the PMF.

3.6 Families of Continuous Distributions

Like families of discrete distributions, families of continuous distributions are each parametrized by a well-defined set of parameters. Typically, the PDF, $f(x\,;\,\theta)$, is parameterized by the *parameter* $\theta \in \Theta$. Hence, technically a family of continuous distributions is the collection of PDFs $f(\cdot\,;\,\theta)$ for all $\theta \in \Theta$.

In this section, we present some of the most common families of *continuous distributions*. We consider the following: *continuous uniform distribution*, *exponential distribution*, *gamma distribution*, *beta distribution*, *Weibull distribution*, *Gaussian (normal) distribution*, *Rayleigh distribution*, and *Cauchy distribution*. As was done with discrete distributions, the approach taken in the code examples involves generating random variables from each distribution using first principles. We also occasionally dive into related concepts that naturally arise in the context of a given distribution. These include the squared coefficient of variation, special functions (gamma and beta), hazard rates, various transformations, and heavy tails.

In Listing 3.20, we illustrate how to create a distribution object for each of the continuous distributions we cover. The listing and its output style is similar to Listing 3.13 used for discrete distributions.

Listing 3.20: Families of continuous distributions

```
using Distributions
dists = [
    Uniform(10,20),
    Exponential(3.5),
    Gamma(0.5,7),
    Beta(10,0.5),
    Weibull(10,0.5),
    Normal(20,3.5),
    Rayleigh(2.4),
    Cauchy(20,3.5)]

println("Distribution \t\t\t Parameters \t Support")
reshape([dists ;  params.(dists) ;
            ((d)->(minimum(d),maximum(d))).(dists) ],
            length(dists),3)
```

```
Distribution                              Parameters        Support
8×3 Array{Any,2}:
 Uniform{Float64}(a=10.0, b=20.0)     (10.0, 20.0)   (10.0, 20.0)
 Exponential{Float64}(θ=3.5)          (3.5,)         (0.0, Inf)
 Gamma{Float64}(α=0.5, θ=7.0)         (0.5, 7.0)     (0.0, Inf)
 Beta{Float64}(α=10.0, β=0.5)         (10.0, 0.5)    (0.0, 1.0)
 Weibull{Float64}(α=10.0, θ=0.5)      (10.0, 0.5)    (0.0, Inf)
 Normal{Float64}(μ=20.0, σ=3.5)       (20.0, 3.5)    (-Inf, Inf)
 Rayleigh{Float64}(σ=2.4)             (2.4,)         (0.0, Inf)
 Cauchy{Float64}(μ=20.0, σ=3.5)       (20.0, 3.5)    (-Inf, Inf)
```

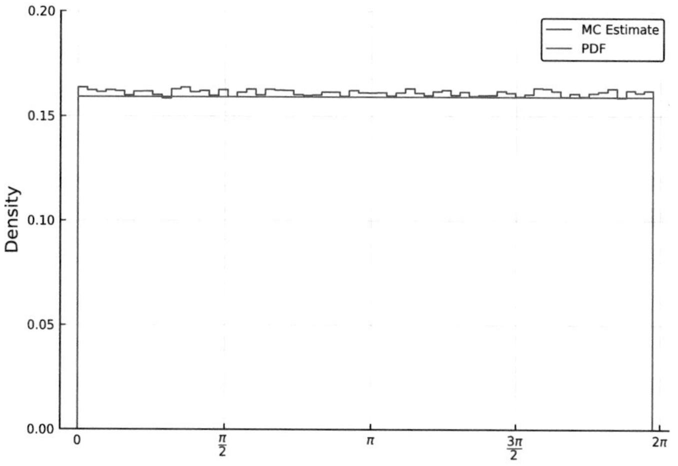

Figure 3.15: The PDF of a continuous uniform distribution over $[0, 2\pi]$.

Continuous Uniform Distribution

The *continuous uniform distribution* describes the case where the outcome of a continuous random variable X has a constant likelihood of occurring over some finite interval. Since the integral of the PDF must equal one, given an interval (a, b), the PDF is given by

$$f(x) = \begin{cases} \dfrac{1}{b-a} & \text{for } a \leqslant x \leqslant b, \\ 0 & \text{for } x < a \text{ or } x > b. \end{cases}$$

As an example, consider the case of a fast spinning circular disk, such as a hard drive. Imagine now there is a small defect on the disk, and we define X as the clockwise angle (in radians) the defect makes with the read head at an arbitrary time. In this case, X is modeled by the continuous uniform distribution over $x \in [0, 2\pi]$. Listing 3.21 creates Figure3.15 where we compare the PDF and a Monte Carlo-based estimate.

Listing 3.21: Uniformly distributed angles

```
using Distributions, Plots, LaTeXStrings; pyplot()

cUnif = Uniform(0,2π)
xGrid, N = 0:0.1:2π, 10^6

stephist( rand(N)*2π, bins=xGrid,
          normed=:true, c=:blue,
          label="MC Estimate")
plot!( xGrid, pdf.(cUnif,xGrid),
          c=:red,ylims=(0,0.2),label="PDF", ylabel="Density",xticks=([0:π/2:2π;],
          ["0", L"\dfrac{\pi}{2}", L"\pi", L"\dfrac{3\pi}{2}", L"2\pi"]))
```

In line 3, the `Uniform()` function is used to create a continuous uniform distribution over the domain $[0, 2\pi]$. In Julia you can use the Unicode character π or `pi`. In line 6, `rand(N)*2`π is used to generate `N` uniform random values on $[0, 2\pi]$. An alternative would be to use `rand(cUnif,N)`. In our case, we simulate `N` continuous uniform random variables over the domain $[0, 1]$ via the `rand()` function, and then scale each of these by a factor of 2π. A histogram of this data is then plotted using `stephist()`. Notice that the `bins` argument is set to the range `xGrid`. An alternative would be to specify an integer number of bins. Line 9 uses `pdf()` on the distribution object `cUnif` to plot the analytic PDF. Notice the use of `L` from the `LaTexStrings` package in line 11 for creating formulas.

Exponential Distribution

As alluded to in the discussion of the Poisson process above, the *exponential distribution* is often used to model random durations between occurrences. A non-negative random variable X, exponentially distributed with a rate parameter $\lambda > 0$, has PDF

$$f(x) = \lambda e^{-\lambda x}.$$

As can be verified, the mean is $1/\lambda$, the variance is $1/\lambda^2$, and the CCDF is $\bar{F}(x) = e^{-\lambda x}$. Note that in Julia, the distribution is parameterized by the mean, rather than by λ. Hence, to create an exponential distribution object with $\lambda = 0.2$ (for example), one would use `Exponential(5.0)`.

Exponential random variables possess a *lack of memory* property. It can be verified that

$$\mathbb{P}(X > t + s \mid X > t) = \mathbb{P}(X > s).$$

To show this, expand the conditional probability and use the CCDF. A similar property holds for geometric random variables. This hints at the fact that exponential random variables are the continuous analogs of geometric random variables.

To explore this further, consider a transformation of an exponential random variable X, $Y = \lfloor X \rfloor$, where $\lfloor \cdot \rfloor$ represents the mathematical *floor function*. In this case, Y is no longer a continuous random variable, but is discrete in nature, taking on values in the set $\{0, 1, 2, \ldots\}$.

We can show that the PMF of Y is

$$p_Y(y) = \mathbb{P}(\lfloor X \rfloor = y) = \int_y^{y+1} \lambda e^{-\lambda x}\, dx = (e^{-\lambda})^y (1 - e^{-\lambda}) \qquad \text{for} \qquad y = 0, 1, 2, \ldots.$$

If we set $p = 1 - e^{-\lambda}$, we observe that Y is a geometric random variable which starts at 0 and has success parameter p.

In Listing 3.22, we present a comparison between the PMF of the floor of an exponential random variable, and the PMF of the geometric distribution covered in Section 3.5. Remember that in Julia the support of `Geometric()` starts at $x = 0$. The listing creates Figure 3.16.

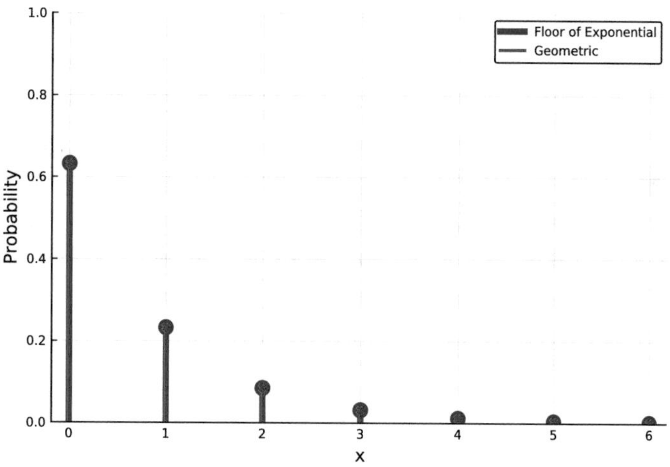

Figure 3.16: The PMF of the floor of an exponential random variable is a geometric distribution.

Listing 3.22: Flooring an exponential random variable

```
1  using StatsBase, Distributions, Plots; pyplot()
2
3  lambda, N = 1, 10^6
4  xGrid = 0:6
5
6  expDist = Exponential(1/lambda)
7  floorData = counts(convert.(Int,floor.(rand(expDist,N))), xGrid)/N
8  geomDist = Geometric(1-MathConstants.e^-lambda)
9
10 plot( xGrid, floorData,
11         line=:stem, marker=:circle,
12         c=:blue, ms=10, msw=0, lw=4,
13         label="Floor of Exponential")
14 plot!( xGrid, pdf.(geomDist,xGrid),
15         line=:stem, marker=:xcross,
16         c=:red, ms=6, msw=0, lw=2,
17         label="Geometric", ylims=(0,1),
18         xlabel="x", ylabel="Probability")
```

In line 6, the `Exponential()` function is used to create the exponential distribution object, `expDist`. Note that the function takes one argument, the inverse of the mean, hence `1/lambda` is used. In line 7, we use the `rand()` function to sample `N` times from the exponential distribution `expDist`. The `floor()` function is then used to round each observation down to the nearest integer, and the `convert()` function is used to convert the values from `Float64` to `Int` type. The function `counts()` is then used to count how many times each integer in `xGrid` occurs, and the proportions are stored in the array `floorData`. In line 8, we use the `Geometric()` function, covered previously, to create a geometric distribution object with probability of success `1-MathConstants.e^-lambda`. Lines 10–18 plot the results where `pdf()` is applied to `geomDist` in line 14.

Gamma Distribution and the Squared Coefficient of Variation

The *gamma distribution* is commonly used for modeling asymmetric non-negative data. It generalizes the exponential distribution and the chi-squared distribution (covered in Section 5.2, in the context of statistical inference). To introduce this distribution, consider the following example, where the lifetimes of light bulbs are exponentially distributed with mean λ^{-1}. Now imagine we are lighting a room continuously with a single light bulb, and that we replace the bulb with a new one when it burns out. If we start at time 0, what is the distribution of time until n bulbs are replaced?

One way to describe this time is by the random variable T, where

$$T = X_1 + X_2 + \ldots + X_n,$$

and X_i are i.i.d. exponential random variables representing the lifetimes of light bulbs. It turns out that the distribution of T is a gamma distribution. In this case, since it is a sum of i.i.d. exponential random variables it is also called an *Erlang distribution.*

We now introduce the PDF of the gamma distribution. It is a function (in x) proportional to $x^{\alpha-1}e^{-\lambda x}$, where the non-negative parameters λ and α are called the *rate parameter* and *shape parameter*, respectively. In order to normalize this function, we need to divide by

$$\int_0^\infty x^{\alpha-1}e^{-\lambda x}\,dx.$$

It turns out that this integral can be represented by $\Gamma(\alpha)/\lambda^\alpha$, where $\Gamma(\cdot)$ is a well-known mathematical special function called the *gamma function*, see (3.16). We investigate the gamma function, and the related *beta function* and *beta distribution* below. After using the gamma function for normalization, the PDF of the gamma distribution is

$$f(x) = \frac{\lambda^\alpha}{\Gamma(\alpha)} x^{\alpha-1}e^{-\lambda x}.$$

In the light bulbs case, we have that $T \sim \text{Gamma}(n, \lambda)$, with shape parameter $\alpha = n$. In general for a gamma random variable, $Y \sim \text{Gamma}(\alpha, \lambda)$, the shape parameter α does not have to be a whole number. It can analytically be evaluated that

$$\mathbb{E}[Y] = \frac{\alpha}{\lambda}, \qquad \text{and} \qquad \text{Var}(Y) = \frac{\alpha}{\lambda^2}.$$

We also take the opportunity here to introduce another general notion of variability, often used for non-negative random variables, namely, the *squared coefficient of variation*,

$$\text{SCV} = \frac{\text{Var}(Y)}{\mathbb{E}[Y]^2}.$$

The SCV is a normalized or unit-less version of the variance. The lower it is, the less variability in the random variable. It can be seen that for a gamma random variable, the SCV is $1/\alpha$ and for our light bulb example above, $\text{SCV}(T) = 1/n$. Hence for large n, i.e. more light bulbs, there is less variability.

Listing 3.23 considers the three cases of $n = 1$, $n = 10$, and $n = 50$ light bulbs (the case of $n = 1$ is exponential). For each scenario, gamma random variables are simulated by generating sums of

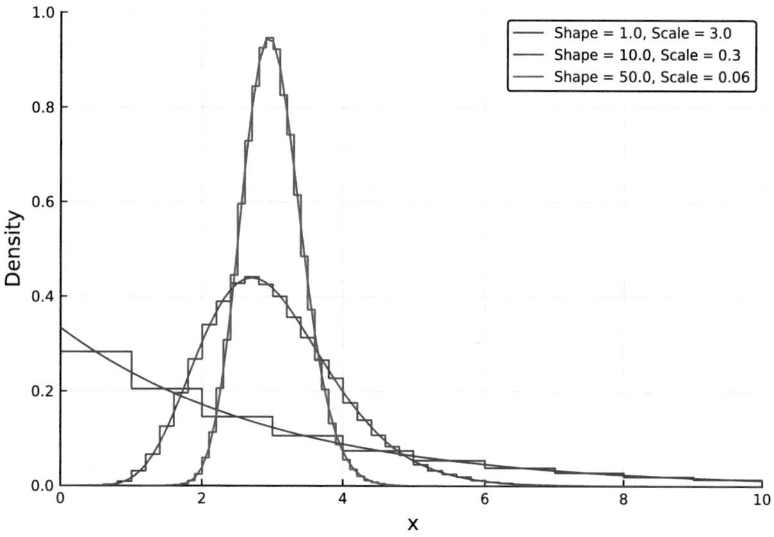

Figure 3.17: Plot of histograms of Monte Carlo-simulated gamma observations, against their analytic PDFs.

exponential random variables. In each case, we set the rate parameter for the light bulbs at λn, so that the mean time until all light bulbs run out is $1/\lambda$, independent of n. The resulting histograms are then compared to the theoretical gamma PDFs. Note that the Julia function `Gamma()` is not parametrized by λ, but by $1/\lambda$ in a similar fashion to the `Exponential()` function. This inverse of the rate parameter is called the *scale parameter*.

Listing 3.23: Gamma random variable as a sum of exponentials

```
using Distributions, Plots; pyplot()

lambda, N = 1/3, 10^5
bulbs = [1,10,50]
xGrid = 0:0.1:10
C = [:blue :red :green]
dists = [Gamma(n,1/(n*lambda)) for n in bulbs]

function normalizedData(d::Gamma)
    sh = Int64(shape(d))
    data = [sum(-(1/(sh*lambda))*log.(rand(sh))) for _ in 1:N]
end

L = [ "Shape = "*string.(shape.(i))*", Scale = "*
    string.(round.(scale.(i),digits=2)) for i in dists ]

stephist( normalizedData.(dists), bins=50,
    normed=:true, c=C, xlims=(0,maximum(xGrid)),ylims=(0,1),
    xlabel="x", ylabel="Density", label="")
plot!(xGrid, [pdf.(i,xGrid) for i in dists], c=C, label=reshape(L, 1,:))
```

In lines 3–6, we define the main variables of our problem. In line 4, we create the array `bulbs` which stores the number of bulbs for each case. In line 6, we create an array of colors which are used later for formatting the plots. In line 7, the `Gamma()` function is used along with a comprehension to create a gamma distribution for each of our cases. The three gamma distributions are stored in the array `dists`. Lines 9–12 define the function `normalizedData()` which operates on a gamma distribution as specified via `::Gamma`. The function obtains the shape parameter of the input distribution via `shape()` and converts this to an integer. Then `-log.(rand(sh))` is a raw way of generating a unit mean collection of `sh` exponential random variables using the inverse probability transform. These are then scaled by the scalar, `(1/(sh*lambda))`. Lines 14–15 generate the string array `L` used for the legend. Notice the use of the `round()` function. The remainder of the code plots the histograms and the actual PDFs.

Beta Distribution and Mathematical Special Functions

The *beta distribution* is a commonly used distribution when seeking a parameterized shape over a finite support. It is parametrized by two non-negative parameters, α and β. It has a density proportional to $x^{\alpha-1}(1-x)^{\beta-1}$ for $x \in [0,1]$. By using different positive values of α and β, a variety of shapes can be produced. You may want to try and create such plots yourself to experiment. One common example is $\alpha = 1, \beta = 1$, in which case the distribution defaults to the uniform(0,1) distribution.

As was with the gamma distribution above, in the case of beta, we are also left to seek a normalizing constant K, such that when multiplied by $x^{\alpha-1}(1-x)^{\beta-1}$, the resulting function has a unit integral over $[0,1]$. In our case,

$$K = \frac{1}{\displaystyle\int_0^1 x^{\alpha-1}(1-x)^{\beta-1}\,dx},$$

and hence the PDF is $f(x) = K\,x^{\alpha-1}(1-x)^{\beta-1}$.

We now explore the beta distribution. By focusing on the normalizing constant, we gain further insight into the mathematical gamma function $\Gamma(\cdot)$, which is a component of the gamma distribution covered previously. It turns out that

$$K = \frac{\Gamma(\alpha+\beta)}{\Gamma(\alpha)\Gamma(\beta)}.$$

Mathematically, this is called the inverse of the *beta function*, evaluated at α and β. Let us focus solely on the gamma functions, with the purpose of demystifying their use in the gamma and beta distributions. The *gamma function* is a type of *special function*, and is defined as

$$\Gamma(z) = \int_0^\infty x^{z-1}e^{-x}\,dx. \tag{3.16}$$

It is a continuous generalization of factorial. We know that for positive integer n,

$$n! = n \cdot (n-1)!, \qquad \text{with } 0! \equiv 1.$$

This is the recursive definition of factorial. The gamma function exhibits similar properties, and one can evaluate it via integration by parts,

$$\Gamma(z) = (z-1) \cdot \Gamma(z-1).$$

Note furthermore that, $\Gamma(1) = 1$. Hence, we see that for integer values of z,

$$\Gamma(z) = (z-1)!.$$

We now illustrate this in Listing 3.24 and in the process take into consideration the mathematical function beta and the beta PDF. Observe the difference in Julia between lower case `gamma()`, the special mathematical function, and `Gamma()`, the constructor for the distribution. The same applies for `beta()` and `Beta()`.

Listing 3.24: The gamma and beta special functions

```
1  using SpecialFunctions, Distributions
2
3  a,b = 0.2, 0.7
4  x = 0.75
5
6  betaAB1 = beta(a,b)
7  betaAB2 = (gamma(a)gamma(b))/gamma(a+b)
8  betaAB3 = (factorial(a-1)factorial(b-1))/factorial(a+b-1)
9  betaPDFAB1 = pdf(Beta(a,b),x)
10 betaPDFAB2 = (1/beta(a,b))*x^(a-1) * (1-x)^(b-1)
11
12 println("beta($a,$b)      = $betaAB1,\t$betaAB2,\t$betaAB3 ")
13 println("betaPDF($a,$b) at $x = $betaPDFAB1,\t$betaPDFAB2")
```

```
beta(0.2,0.7)       = 5.576463695849875,    5.576463695849875,      5.576463695849877
betaPDF(0.2,0.7) at 0.75 = 0.34214492891381176, 0.34214492891381176
```

We use the `SpecialFunctions` package for `gamma()` and `beta()`. This package also introduces a method for `factorial()` that allows us to evaluate $\Gamma(z)$ via `factorial(z-1)` even for non-integer z. In lines 6–8, the `beta()` special function at `a` and `b` is evaluated in three different ways.

Another important property of the gamma function that we encounter later on (in the context of the chi-squared distribution, which we touch on in Section 5.2) is that $\Gamma(1/2) = \sqrt{\pi}$. In Listing 3.25, we show this via numerical integration.

Listing 3.25: The gamma function at 1/2

```
using QuadGK, SpecialFunctions

g(x) = x^(0.5-1) * MathConstants.e^-x
quadgk(g,0,Inf)[1], sqrt(pi), gamma(1/2),  factorial(1/2-1)
```

```
(1.7724538355037913, 1.7724538509055159, 1.772453850905516, 1.772453850905516)
```

This example uses the `QuadGK` package, in the same manner as introduced in Listing 3.3. We can see that the numerical integration is in agreement with the analytically expected result.

Weibull Distribution and Hazard Rates

We now explore the *Weibull distribution* along with the concept of the *hazard rate function*, which is often used in *reliability analysis* and *survival analysis*. For a random variable T, representing the lifetime of an individual or a component, an interesting quantity is the instantaneous chance of failure at any time, given that the component has been operating without failure up to time x. This can be expressed as

$$h(x) = \lim_{\Delta \to 0} \frac{1}{\Delta} \mathbb{P}(T \in [x, x+\Delta) \mid T > x).$$

Alternatively, by using the conditional probability and noticing that the PDF $f(x)$ satisfies $f(x)\Delta \approx \mathbb{P}(x \le T < x + \Delta)$ for small Δ, we can express the above as

$$h(x) = \frac{f(x)}{1 - F(x)}. \tag{3.17}$$

Here the function $h(\cdot)$ is called the hazard rate, and it is a common method of viewing the distribution for lifetime random variables T. In fact, we can reconstruct the CDF $F(x)$ by

$$1 - F(x) = \exp\Big(-\int_0^x h(t)\,dt\Big). \tag{3.18}$$

Hence, every continuous non-negative random variable can be described uniquely by its hazard rate. The *Weibull distribution* is naturally defined through the hazard rate by considering hazard rate functions that have a specific simple form. It is a distribution with

$$h(x) = \lambda x^{\alpha-1}, \tag{3.19}$$

where λ is positive and α takes on any real value. Notice that the parameter α gives the Weibull distribution different modes of behavior. If $\alpha = 1$ then the hazard rate is constant, in which case the Weibull distribution is actually an exponential distribution with rate λ. If $\alpha > 1$, then the hazard rate increases over time. This depicts a situation of "aging components", i.e. the longer a component has lived, the higher the instantaneous chance of failure. This is sometimes called *Increasing Failure Rate (IFR)*. Conversely, $\alpha < 1$ depicts a situation where the longer a component has lasted, the lower the chance of it failing (as is perhaps the case with totalitarian political regimes). This is sometimes called *Decreasing Failure Rate (DFR)*.

Based on (3.19) and using (3.18), we obtain the CDF and PDF

$$F(x) = 1 - e^{-\frac{\lambda}{\alpha}x^{\alpha}}, \qquad \text{and} \qquad f(x) = \lambda x^{\alpha-1} e^{-\frac{\lambda}{\alpha}x^{\alpha}}. \tag{3.20}$$

Note that in Julia, the distribution is parameterized slightly differently via

$$f(x) = \frac{\alpha}{\theta}\Big(\frac{x}{\theta}\Big)^{\alpha-1} e^{-(x/\theta)^{\alpha}} = \alpha\theta^{-\alpha} x^{\alpha-1} e^{-\theta^{-\alpha}x^{\alpha}},$$

where the bijection from λ to θ is

$$\lambda = \alpha\theta^{-\alpha}, \qquad \text{and} \qquad \theta = \Big(\frac{\alpha}{\lambda}\Big)^{1/\alpha}. \tag{3.21}$$

In this case, θ is called the scale parameter and α is the shape parameter.

In Listing 3.26, we look at several hazard rate functions for different Weibull distributions using the parameterization (3.20), and show their differences in Figure 3.18. The example also shows how to use the `shape()` and `scale()` functions from the `Distributions` package.

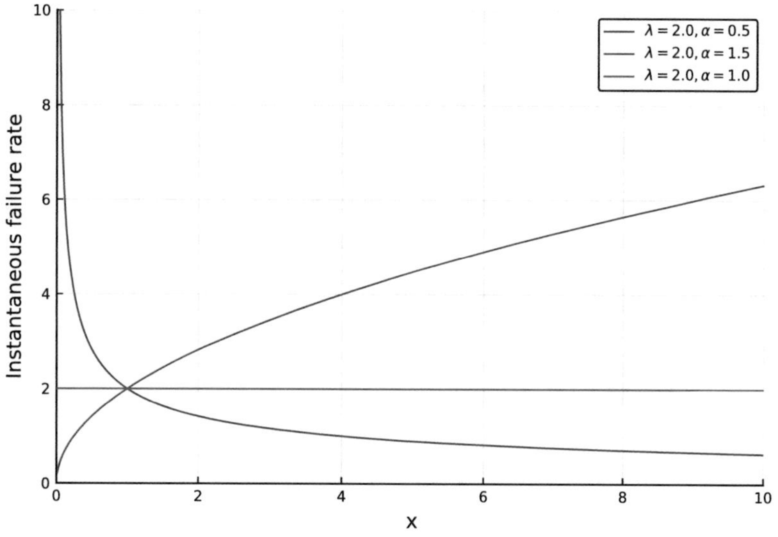

Figure 3.18: Hazard rate functions for different Weibull distributions.

Listing 3.26: Hazard rates and the Weibull distribution

```
using Distributions, Plots, LaTeXStrings; pyplot()

alphas = [0.5, 1.5, 1]
lam = 2

lambda(dist::Weibull) = shape(dist)*scale(dist)^(-shape(dist))
theta(lam,alpha) = (alpha/lam)^(1/alpha)

dists = [Weibull.(a,theta(lam,a)) for a in alphas]

hA(dist,x) = pdf(dist,x)/ccdf(dist,x)
hB(dist,x) = lambda(dist)*x^(shape(dist)-1)

xGrid = 0.01:0.01:10
hazardsA = [hA.(d,xGrid) for d in dists]
hazardsB = [hB.(d,xGrid) for d in dists]

println("Maximum difference between two implementations of hazard: ",
    maximum(maximum.(hazardsA-hazardsB)))

Cl = [:blue :red :green]
Lb = [L"\lambda=" * string(lambda(d)) * ",   " * L"\alpha =" * string(shape(d))
        for d in dists]

plot(xGrid, hazardsA, c=Cl, label=reshape(Lb, 1,:), xlabel="x",
        ylabel="Instantaneous failure rate", xlims=(0,10), ylims=(0,10))
```

```
Maximum difference between two implementations of hazard: 1.7763568394002505e-15
```

In line 6, we define the function `lambda()`, which operates on a `Weibull`-type distribution and implements the first equation in (3.21). Note the type specification `::Weibull` and the use of the `shape()` and `scale()` functions. In line 7, we define the function `theta()` which implements the second equation in (3.21). Line 9 constructs three `Weibull` objects in the array `dists`. Lines 11 and 12 implement two alternative ways of calculating the hazard rate function, `hA()` and `hB()`. The first uses (3.17) and the second uses (3.19). Then in lines 18–19, we verify that the two implementations are in agreement. The remainder of the code creates Figure 3.18.

Gaussian (Normal) Distribution

Arguably, the most well-known distribution is the *normal distribution*, also known as the *Gaussian distribution.* It is a symmetric "*bell curved*"-shaped distribution, which can be found throughout nature. Examples include the distribution of heights among adult humans and noise disturbances of electrical signals. It is commonly exhibited due to the central limit theorem, which is covered in more depth in Section 5.3.

The Gaussian distribution is defined by two parameters, μ and σ^2, which are the mean and variance respectively. The mean μ can take on any value and σ^2 is restricted to be positive. The phrase *standard normal* signifies the case of a normal distribution with $\mu = 0$ and $\sigma^2 = 1$. In the general case, the PDF is given by

$$f(x) = \frac{1}{\sigma\sqrt{2\pi}} e^{-\dfrac{(x-\mu)^2}{2\sigma^2}}.$$

The CDF of the normal distribution is not available as a simple expression. However, it is frequently needed and hence statistical tables or software are often used. The CDF of the standard normal random variable is

$$\Phi(x) = \int_{-\infty}^{x} \frac{1}{\sqrt{2\pi}} e^{-\frac{t^2}{2}}\, dt = \frac{1}{2}\Big(1 + \text{erf}\big(\frac{x}{\sqrt{2}}\big)\Big). \tag{3.22}$$

The second expression represents $\Phi(\cdot)$ in terms of the *error function* $\text{erf}(\cdot)$. It is a mathematical special function defined as

$$\text{erf}(x) = \frac{2}{\sqrt{\pi}} \int_0^x e^{-t^2}\, dt.$$

With $\Phi(\cdot)$ (or alternately $\text{erf}(\cdot)$) tabulated, one can move on to a general normal random variable with mean μ and variance σ^2. In this case, the CDF is available via

$$\Phi\Big(\frac{x-\mu}{\sigma}\Big).$$

As an illustrative example, Listing 3.27 plots the standard normal PDF, along with its first and second derivatives in Figure 3.21. The first derivative is clearly 0 at the PDF's unique maximum at $x = 0$. The second derivative is 0 at the points $x = -1$ and $x = +1$. These are exactly the *inflection points* of the normal PDF (points where the function switches between being locally convex to locally concave or vice versa). This code example also illustrates the use numerical derivatives from

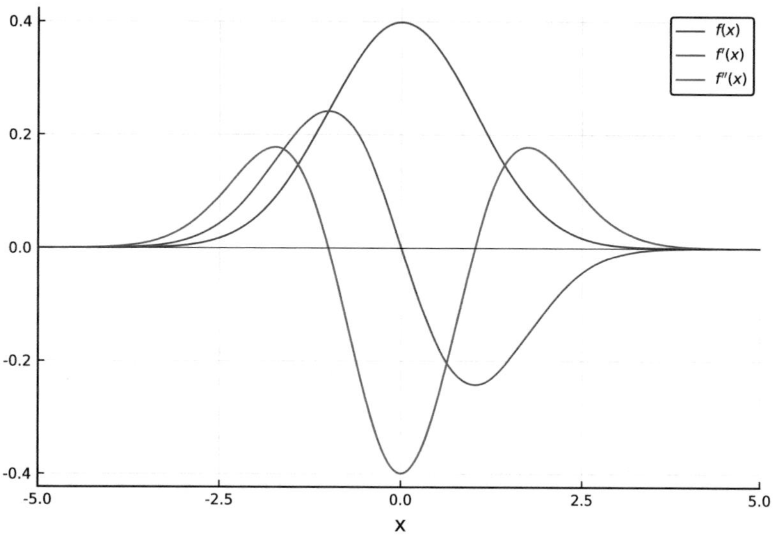

Figure 3.19: Plot of the standard normal PDF and its first and second derivatives.

the `Calculus` package. The code also presents two alternative ways of implementing $\Phi(\cdot)$ of (3.22) and shows they are equivalent. One way uses `cdf()` from the `Distributions` package and the other way uses `erf()` from the `SpecialFunctions` package.

Listing 3.27: Numerical derivatives of the normal density

```
1  using Distributions, Calculus, SpecialFunctions, Plots; pyplot()
2
3  xGrid = -5:0.01:5
4
5  PhiA(x) = 0.5*(1+erf(x/sqrt(2)))
6  PhiB(x) = cdf(Normal(),x)
7
8  println("Maximum difference between two CDF implementations: ",
9                  maximum(PhiA.(xGrid) - PhiB.(xGrid)))
10
11 normalDensity(z) = pdf(Normal(),z)
12
13 d0 = normalDensity.(xGrid)
14 d1 = derivative.(normalDensity,xGrid)
15 d2 = second_derivative.(normalDensity, xGrid)
16
17 plot(xGrid, [d0 d1 d2], c=[:blue :red :green],label=[L"f(x)" L"f'(x)" L"f''(x)"])
18 plot!([-5,5],[0,0], color=:black, lw=0.5, xlabel="x", xlims=(-5,5), label="")
```

```
Maximum difference between two CDF implementations: 1.1102230246251565e-16
```

Lines 5–9 are dedicated to showing the equivalence of the two ways of implementing $\Phi(\cdot)$. In line 11, we define the function `normalDensity()`, which takes an input z, and returns the corresponding value of the PDF of a standard normal distribution. Then in lines 14–15, the functions `derivative()` and `second_derivative()` are used to evaluate the first and second derivatives of `normalDensity`, respectively. The curves are plotted in lines 17–18.

Rayleigh Distribution and the Box-Muller Transform

We now consider an exponentially distributed random variable, X, with rate parameter $\lambda = \sigma^{-2}/2$ where $\sigma > 0$. If we set a new random variable, $R = \sqrt{X}$, what is the distribution of R? To work this out analytically, we have for $y \geq 0$

$$F_R(y) = \mathbb{P}(\sqrt{X} \leq y) = \mathbb{P}(X \leq y^2) = F_X(y^2) = 1 - \exp\Big(-\frac{y^2}{2\sigma^2}\Big),$$

and by differentiating, we get the density

$$f_R(y) = \frac{y}{\sigma^2} \exp\Big(-\frac{y^2}{2\sigma^2}\Big).$$

This is the density of the *Rayleigh Distribution* with parameter σ. We see it is related to the exponential distribution via a square root transformation. Hence, the implication is that since we know how to generate exponential random variables via $-\frac{1}{\lambda}\log(U)$ where $U \sim \text{uniform}(0,1)$, then by applying a square root we can generate Rayleigh random variables.

The Rayleigh distribution is important because of another distributional relationship. Consider two independent normally distributed random variables, N_1 and N_2, each with mean 0 and standard deviation σ. In this case, it turns out that $\tilde{R} = \sqrt{N_1^2 + N_2^2}$ is Rayleigh distributed just as R above. As we see in the next example, this property yields a method for generating normal random variables. It also yields a statistical model often used in radio communications called *Rayleigh fading*.

Listing 3.28 demonstrates three alternative ways of generating Rayleigh random variables. It generates R and $\tilde{R}$ as above, as well as by applying `rand()` to a `Rayleigh` object from the `Distributions` package. The mean of a Rayleigh random variable is $\sigma\sqrt{\pi/2}$ and is approximately 2.1306 when $\sigma = 1.7$, as in the code below.

Listing 3.28: Alternative representations of Rayleigh random variables

```
using Distributions, Random
Random.seed!(1)

N = 10^6
sig = 1.7

data1 = sqrt.(-(2* sig^2)*log.(rand(N)))

distG = Normal(0,sig)
data2 = sqrt.(rand(distG,N).^2 + rand(distG,N).^2)

distR = Rayleigh(sig)
data3 = rand(distR,N)

mean.([data1, data2, data3])
```

```
3-element Array{Float64,1}:
 2.1309969895700465
 2.1304634508886053
 2.1292020616665392
```

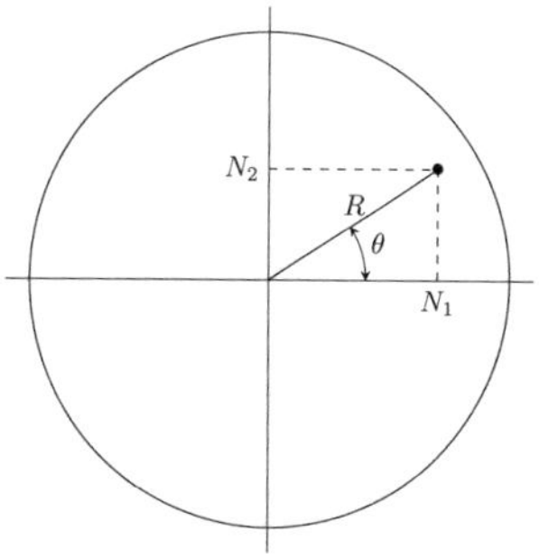

Figure 3.20: Geometry of the Box-Muller transform.

Line 7 generates `data1`, according to R above. Note the use of element-wise mapping of `sqrt()` and `log()`. Lines 9 and 10 generate `data2`, as in $\tilde{R}$ above. Here we use `rand()` applied to the normal distribution object `distG`. Lines 12 and 13 use `rand()` applied to a `Rayleigh` distribution object. Line 15 produces the output by applying `mean()` to `data1`, `data2`, and `data3` individually. Observe that the sample mean is very similar to the theoretical mean presented above.

A common way to generate normal random variables, called the *Box-Muller Transform*, is to use the relationship between the Rayleigh distribution and a pair of independent zero-mean normal random variables, as mentioned above. Consider Figure 3.20 representing the relationship between the pair (N_1, N_2) and their *polar coordinate* counterparts, R and θ. Assume now that the Cartesian coordinates of the point (N_1, N_2) are identically normally distributed, with N_1 independent of N_2 and set $\sigma = 1$. In this case, by representing N_1 and N_2 in polar coordinates (θ, R) we have that the angle θ is uniformly distributed on $[0, 2\pi]$ and that the radius R is distributed as a Rayleigh random variable.

Given this, a recipe for generating N_1 and N_2 is to first generate θ and R and then transform them via

$$N_1 = R\cos(\theta), \qquad N_2 = R\sin(\theta).$$

Often, N_2 is not needed. Hence, in practice, given two independent uniform (0,1) random variables U_1 and U_2, we set $Z = \sqrt{-2\ln U_1}\cos(2\pi U_2)$. Here Z has a standard normal distribution. Listing 3.29 uses this method to generate normal random variables and compares their histogram to the standard normal PDF. The output is presented in Figure 3.21.

Listing 3.29: The Box-Muller transform

```
using Random, Distributions, Plots; pyplot()
Random.seed!(1)

Z() = sqrt(-2*log(rand()))*cos(2*pi*rand())
xGrid = -4:0.01:4

histogram([Z() for _ in 1:10^6], bins=50,
                normed=true, label="MC estimate")
plot!(xGrid, pdf.(Normal(),xGrid),
        c=:red, lw=4, label="PDF",
        xlims=(-4,4), ylims=(0,0.5), xlabel="x", ylabel="f(x)")
```

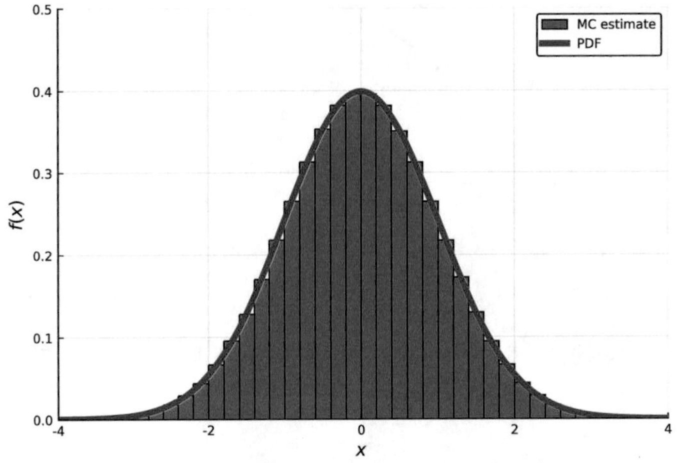

Figure 3.21: The Box-Muller transform can be used to generate a normally distributed random variable.

In line 4, we define a function `Z()`, which implements the Box-Muller transform and generates a single standard normal random variable. The remaining lines plot a histogram of 10^6 random variables from `Z()`, and compares the standard normal PDF. Notice the use of the `L` macro for latex formatting in line 11.

Cauchy Distribution

We now introduce the *Cauchy distribution*, also known as the *Lorentz distribution*. At first glance, a plot of the PDF looks very similar to the normal distribution. However, it is fundamentally different as its mean and standard deviation are undefined. The PDF of the Cauchy distribution is given by

$$f(x) = \frac{1}{\pi\gamma\left(1 + \left(\frac{x - x_0}{\gamma}\right)^2\right)}, \tag{3.23}$$

where x_0 is the location parameter at which the peak is observed and γ is the scale parameter.

In order to better understand the context of this type of distribution, we will develop a physical example of a Cauchy distributed random variable. Consider a drone hovering stationary in the sky at unit height. A pivoting laser is attached to its undercarriage, which pivots back and forth as it shoots pulses at the ground. At any point the laser fires, it makes an angle θ from the vertical $(-\pi/2 \leq \theta \leq \pi/2)$ as illustrated in Figure 3.22.

Since the laser fires at a high frequency as it is pivoting, we can assume that the angle θ is distributed uniformly on $[-\pi/2, \pi/2]$. For each shot from the laser, a point can be measured, X, horizontally on the ground from the point above which the drone is hovering. We can now consider

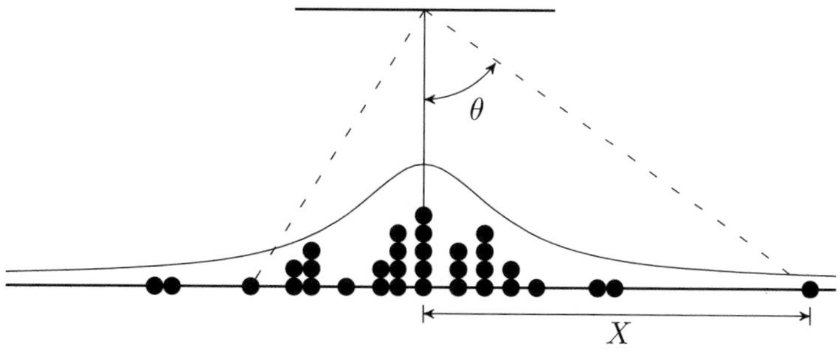

Figure 3.22: Indicative distribution of laser shots from a hovering drone.

this horizontal measurement as a new random variable, X. Hence, the CDF is

$$F_X(x) = \mathbb{P}\big(\tan(\theta) \leqslant x\big) = \mathbb{P}\big(\theta \leqslant \text{atan}(x)\big) = F_\theta\big(\text{atan}(x)\big) = \begin{cases} 0, & \text{atan}(x) \leq -\pi/2, \\ \frac{1}{\pi}\text{atan}(x), & \text{atan}(x) \in (-\pi/2, \pi/2), \\ 1, & \pi/2 \leq \text{atan}(x). \end{cases}$$

Now since it always holds that $\text{atan}(x) \in (-\pi/2, \pi/2)$ we can obtain the density by taking the derivative of $\frac{1}{\pi}\text{atan}(x)$ which evaluates to

$$f(x) = \frac{1}{\pi(1+x^2)}.$$

This is a special case of the more complicated density (3.23), with $x_0 = 0$ and $\gamma = 1$. Importantly, the expectation integral,

$$\int_{-\infty}^{\infty} x f(x)\, dx,$$

is not defined since each of the one-sided improper integrals does not converge. Hence, a Cauchy random variable is an example of a *distribution without a mean.*

You may now revisit the law of large numbers (Section 3.2) and ask what happens to sample averages of such random variables. That is, would the sequence of sample averages converge to anything? The answer is no. We illustrate this in Listing 3.30 and the associated Figure 3.24. In this example, occasional large values create huge spikes due to angles near $-\pi/2$ or $\pi/2$. There is no strong law of large numbers in this case since the mean is not defined.

Listing 3.30: The law of large numbers breaks down with very heavy tails

```
using Random, Plots; pyplot()
Random.seed!(808)

n = 10^6
data = tan.(rand(n)*pi .- pi/2)
averages = accumulate(+,data)./collect(1:n)

plot( 1:n, averages,
        c=:blue, legend=:none,
        xscale=:log10, xlims=(1,n), xlabel="n", ylabel="Running average")
```

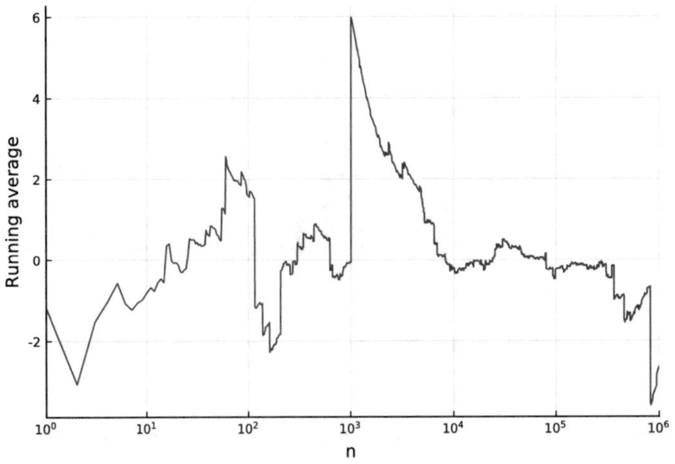

Figure 3.23: A contour plot and a three-dimensional surface plot of $f(x, y)$.

In line 2, the seed of the random number generator is set, so that the same stream of random numbers is generated each time. In line 5, we create `data`, an array of `n` Cauchy random variables constructed through the angle mechanism described and illustrated in Figure 3.22. In line 6, we use the `accumulate()` function to create a running sum, and then divide this element-wise via `./` by the array `collect(1:n)`. Notice that "+" is used as the first argument to `accumulate()`. Here the addition operator is treated as a function. The remainder of the code plots the running average.

3.7 Joint Distributions and Covariance

We now consider pairs and vectors of random variables. In general, in a probability space, we may define multiple random variables, $X_1, \ldots, X_n$, where we consider the vector or tuple, $\boldsymbol{X} = (X_1, \ldots, X_n)$, as a *random vector.* A key question deals with representing and evaluating probabilities of the form $\mathbb{P}(\boldsymbol{X} \in B)$, where B is some subset $\mathbb{R}^n$. Our focus here is on the case of a pair of random variables (X, Y), which are continuous and have a density function. The probability distribution of (X, Y) is called a *bivariate distribution*, and more generally the probability distribution of $\boldsymbol{X}$ is called a *multivariate distribution.*

The Joint PDF

A function, $f_{\boldsymbol{X}} : \mathbb{R}^n \to \mathbb{R}$, is said to be a *Joint Probability Density Function* (PDF) if for any input, $x_1, \ldots, x_n$, it holds that $f_{\boldsymbol{X}}(x_1, x_2, \ldots, x_n) \geq 0$ and

$$\int_{-\infty}^{\infty} \int_{-\infty}^{\infty} \cdots \int_{-\infty}^{\infty} f_{\boldsymbol{X}}(x_1, x_2, \ldots, x_n) \, dx_1 dx_2 \ldots dx_n = 1. \tag{3.24}$$

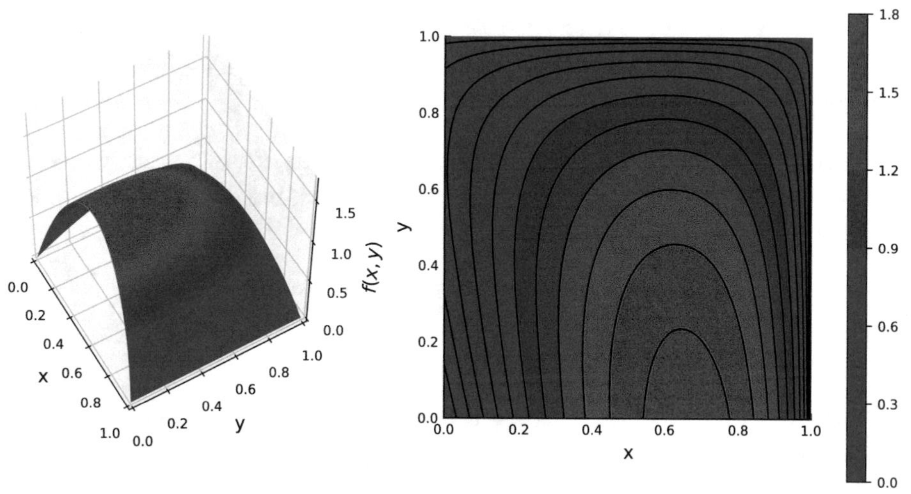

Figure 3.24: Cumulative average of Cauchy distributed random variables.

Hence, if we consider now $B \subset \mathbb{R}^n$, then the probabilities of a random vector $\boldsymbol{X}$, distributed with density $f_{\boldsymbol{X}}$, can be evaluated via

$$\mathbb{P}\Big(\boldsymbol{X} \in B\Big) = \int_B f_{\boldsymbol{X}}(\boldsymbol{x})\, d\,\boldsymbol{x}.$$

As an example let $\boldsymbol{X} = (X, Y)$ and consider the joint density, simply denoted $f(\cdot,\cdot)$,

$$f(x,y) = \begin{cases} \frac{9}{8}(4x+y)\sqrt{(1-x)(1-y)}, & x \in [0,1],\ y \in [0,1], \\ 0, & \text{otherwise.} \end{cases}$$

This PDF is plotted in Figure 3.23. We may now obtain all kinds of probabilities. For example, set $B = \{(x,y) \mid x < y\}$, then

$$\mathbb{P}\Big((x,y) \in B\Big) = \int_{x=0}^{1} \int_{y=x}^{1} f(x,y)\, dy\, dx = \frac{31}{80} = 0.3875. \tag{3.25}$$

The joint distribution of X and Y allows us to also obtain related distributions. We may obtain the *marginal densities* of X and Y, denoted $f_X(\cdot)$ and $f_Y(\cdot)$, via

$$f_X(x) = \int_{y=0}^{1} f(x,y)\, dy \qquad \text{and} \qquad f_Y(y) = \int_{x=0}^{1} f(x,y)\, dx.$$

For our example by explicitly integrating, we obtain

$$f_X(x) = \frac{3}{10}\sqrt{1-x}(1+10x) \qquad \text{and} \qquad f_Y(y) = \frac{3}{20}\sqrt{1-y}(8+5y).$$

In general, the random variables X and Y are said to be *independent* if $f(x,y) = f_X(x) f_Y(y)$. In our current example, this is not the case. Furthermore, whenever we have two densities of scalar random variables, we may multiply them to make the joint distribution of the random vector composed of independent random variables. That is, if we take our $f_X(\cdot)$ and $f_Y(\cdot)$ above, we may create $\tilde{f}(x,y)$ via

$$\tilde{f}(x,y) = f_X(x) f_Y(y) = \frac{9}{200}\sqrt{(1-x)(1-y)}(1+10x)(8+5y).$$

Observe that $\tilde{f}(x,y) \neq f(x,y)$. Hence, we see that while both bivariate distributions have the same marginal distribution, they are different bivariate distributions and hence describe different relationships between X and Y.

Of further interest is the *conditional density* of X given Y, and vice versa. It is denoted by $f_{X\,|\,Y=y}(x)$ and describes the distribution of the random variable X, given the specific value $Y = y$. It can be obtained from the joint density via

$$f_{X\,|\,Y=y}(x) = \frac{f(x,y)}{f_Y(y)} = \frac{f(x,y)}{\int_{x=0}^{1} f(x,y)\,dx}.$$

In Listing 3.31, we generate Figure 3.23, and in addition use crude Riemann sums to approximate the integral (3.25) as well as the integral over the total density.

Listing 3.31: Visualizing a bivariate density

```
1   using Plots, LaTeXStrings, Measures; pyplot()
2
3   delta = 0.01
4   grid = 0:delta:1
5   f(x,y) = 9/8*(4x+y)*sqrt((1-x)*(1-y))
6   z = [f(x,y) for y in grid, x in grid]
7
8   densityIntegral = sum(z)*delta^2
9   println("2-dimensional Riemann sum over density: ", densityIntegral)
10
11  probB = sum([sum([f(x,y)*delta for y in x:delta:1])*delta for x in grid])
12  println("2-dimensional Riemann sum to evaluate probability: ", probB)
13
14  p1 = surface(grid, grid, z,
15          c=cgrad([:blue, :red]), la=1, camera=(60,50),
16          ylabel="y", zlabel=L"f(x,y)", legend=:none)
17  p2 = contourf(grid, grid, z,
18          c=cgrad([:blue, :red]))
19  p2 = contour!(grid, grid, z,
20          c=:black, xlims=(0,1), ylims=(0,1), ylabel="y", ratio=:equal)
21
22  plot(p1, p2, size=(800, 400), xlabel="x", margin=5mm)
```

```
2-dimensional Riemann sum over density: 1.0063787264382458
2-dimensional Riemann sum to evaluate probability: 0.3932640388868346
```

In line 5, we define the bivariate density function, `f()`. In line 6, we evaluate the density over a grid of x- and y-values. This grid is then used to obtain a crude approximation of the integral in line 8 with the result printed in line 9. Similarly, the nested integral (3.25) is approximated via two Riemann sums in line 11 with the result printed in line 12. The remainder of the code creates Figure 3.23.

Covariance and Vectorized Moments

Given two random variables, X and Y, with respective means, μ_X and μ_Y, the *covariance* is defined by

$$\mathrm{Cov}(X,Y) = \mathbb{E}\big[(X-\mu_X)(Y-\mu_Y)\big] = \mathbb{E}[X\,Y] - \mu_x\mu_y.$$

The second formula follows by expansion. Notice also that $\mathrm{Cov}(X,X) = \mathrm{Var}(X)$ by comparing with (3.3). The covariance is a common measure of the relationship between the two random variables, where if $\mathrm{Cov}(X,Y) = 0$, we say the random variables are *uncorrelated.* Furthermore, if $\mathrm{Cov}(X,Y) \neq 0$, the sign of the covariance gives an indication of the direction of the relationship.

Another important concept is the *correlation coefficient,*

$$\rho_{XY} = \frac{\mathrm{Cov}(X,Y)}{\sqrt{\mathrm{Var}(X)\mathrm{Var}(Y)}}. \tag{3.26}$$

It is a normalized form of the covariance with $-1 \leq \rho_{XY} \leq 1$. Values nearing ± 1 indicate a very strong *linear relationship* between X and Y, whereas values near or at 0 indicate a lack of a linear relationship.

Note that if X and Y are independent random variables, then $\mathrm{Cov}(X,Y) = 0$ and hence $\rho_{XY} = 0$. However, the opposite case does not always hold, since, in general, $\rho_{XY} = 0$ does not imply independence. Nevertheless as described below, for jointly normal random variables it does.

Consider now a random vector $\boldsymbol{X} = (X_1, \ldots, X_n)$, taken as a column vector. It can be described by moments in an analogous manner to a scalar random variable as was detailed in Section 3.2. A key quantity is the *mean vector,*

$$\mu_X := \big[\mathbb{E}[X_1], \mathbb{E}[X_2], \ldots, \mathbb{E}[X_n]\big]^\top.$$

Furthermore, the *covariance matrix* is the matrix defined by the expectation (taken element-wise) of the (*outer product*) random matrix given by $(X-\mu_X)(X-\mu_X)^\top$, and is expressed as

$$\Sigma_X = \mathrm{Cov}(X) = \mathbb{E}[(X-\mu_x)(X-\mu_x)^\top]. \tag{3.27}$$

As can be verified, the i,j-th element of Σ_X is $\mathrm{Cov}(X_i, X_j)$ and hence the diagonal elements are the variances.

Linear Combinations and Transformations

We now consider *linear transformations* applied to random vectors. For any collection of random variables,

$$\mathbb{E}[X_1 + \ldots + X_n] = \mathbb{E}[X_1] + \ldots + \mathbb{E}[X_n].$$

For uncorrelated random variables,

$$\mathrm{Var}(X_1 + \ldots + X_n) = \mathrm{Var}(X_1) + \ldots + \mathrm{Var}(X_n).$$

More generally, if we allow the random variables to be correlated, then

$$\mathrm{Var}(X_1 + \ldots + X_n) = \mathrm{Var}(X_1) + \ldots + \mathrm{Var}(X_n) + 2\sum_{i<j} \mathrm{Cov}(X_i, X_j). \tag{3.28}$$

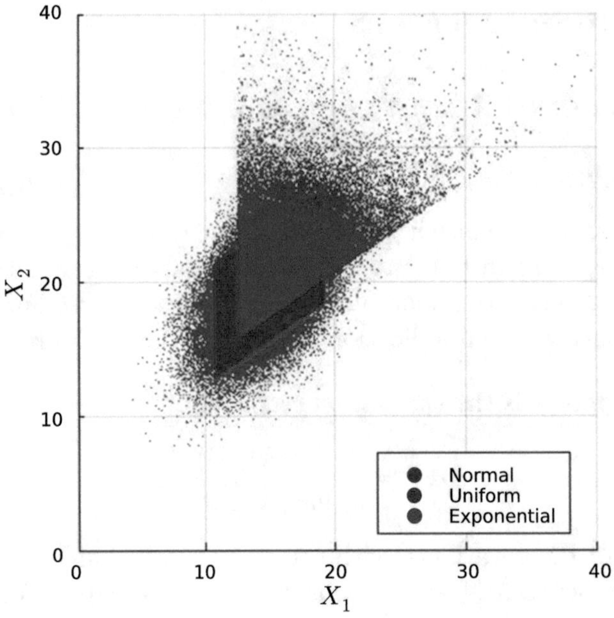

Figure 3.25: Random vectors from three different distributions, each sharing the same mean and covariance matrix.

Note that the right-hand side of (3.28) is the sum of the elements of the matrix $\text{Cov}\big((X_1, \ldots, X_n)\big)$. This is a special case of a more general *affine transformation*, where we take a random vector $\boldsymbol{X} = (X_1, \ldots, X_n)$ with covariance matrix $\Sigma_{\boldsymbol{X}}$, and an $m \times n$ matrix A and m vector $\boldsymbol{b}$. We then set

$$\boldsymbol{Y} = A\boldsymbol{X} + \boldsymbol{b}. \tag{3.29}$$

In this case, the new random vector $\boldsymbol{Y}$ exhibits mean and covariance

$$\mathbb{E}[\boldsymbol{Y}] = A\,\mathbb{E}[\boldsymbol{X}] + \boldsymbol{b} \qquad \text{and} \qquad \text{Cov}(\boldsymbol{Y}) = A\Sigma_{\boldsymbol{X}}A^\top. \tag{3.30}$$

Now to retrieve (3.28), we use the $1 \times n$ matrix $A = [1, \ldots, 1]$ and observe that $A\Sigma_{\boldsymbol{X}}A^\top$ is a sum of all of the elements of $\Sigma_{\boldsymbol{X}}$.

The Cholesky Decomposition and Generating Random Vectors

Say now that you wish to create an n-dimensional random vector $\boldsymbol{Y}$ with some specified mean vector $\mu_{\boldsymbol{Y}}$ and covariance matrix $\Sigma_{\boldsymbol{Y}}$. That is, $\mu_{\boldsymbol{Y}}$ and $\Sigma_{\boldsymbol{Y}}$ are known.

The formulas in (3.30) yield a potential recipe for such a task if we are given a random vector $\boldsymbol{X}$ with zero-mean and identity-covariance matrix ($\Sigma_{\boldsymbol{X}} = I$). For example, in the context of Monte Carlo random variable generation, creating such a random vector $\boldsymbol{X}$ is trivial—just generate a sequence of n i.i.d. normal(0,1) random variables.

Now apply the affine transformation (3.29) on $\boldsymbol{X}$ with $\boldsymbol{b} = \mu_{\boldsymbol{Y}}$ and a matrix A that satisfies

$$\Sigma_{\boldsymbol{Y}} = AA^\top. \tag{3.31}$$

Now (3.30) guarantees that $\boldsymbol{Y}$ has the desired $\mu_{\boldsymbol{Y}}$ and $\Sigma_{\boldsymbol{Y}}$.

The question is now how to find a matrix A that satisfies (3.31). For this the *Cholesky decomposition* comes as an aid and finds a matrix A. As an example assume we wish to generate a random vector $\boldsymbol{Y}$ with

$$\mu_{\boldsymbol{Y}} = \begin{bmatrix} 15 \\ 20 \end{bmatrix} \qquad \text{and} \qquad \Sigma_{\boldsymbol{Y}} = \begin{bmatrix} 6 & 4 \\ 4 & 9 \end{bmatrix}.$$

Listing 3.32 generates random vectors with these mean vector and covariance matrix using three alternative forms of zero-mean, identity-covariance matrix random variables. As you can see from Figure 3.25, such distributions can be very different in nature even though they share the same first- and second-order characteristics. The output also presents mean and variance estimates of the random variables generated, showing they agree with the specifications above.

Listing 3.32: Generating random vectors with desired mean and covariance

```
using Distributions, LinearAlgebra, LaTeXStrings, Random, Plots; pyplot()
Random.seed!(1)

N = 10^5

SigY = [ 6 4 ;
         4 9]
muY = [15 ;
       20]
A = cholesky(SigY).L

rngGens = [()->rand(Normal()),
           ()->rand(Uniform(-sqrt(3),sqrt(3))),
           ()->rand(Exponential())-1]

rv(rg) = A*[rg(),rg()] + muY

data = [[rv(r) for _ in 1:N] for r in rngGens]

stats(data) = begin
    data1, data2 = first.(data),last.(data)
    println(round(mean(data1),digits=2), "\t",round(mean(data2),digits=2),"\t",
            round(var(data1),digits=2), "\t", round(var(data2),digits=2), "\t",
            round(cov(data1,data2),digits=2))
end

println("Mean1\tMean2\tVar1\tVar2\tCov")
for d in data
    stats(d)
end

scatter(first.(data[1]), last.(data[1]), c=:blue, ms=1, msw=0, label="Normal")
scatter!(first.(data[2]), last.(data[2]), c=:red, ms=1, msw=0, label="Uniform")
scatter!(first.(data[3]),last.(data[3]),c=:green, ms=1,msw=0,label="Exponential",
        xlims=(0,40), ylims=(0,40), legend=:bottomright, ratio=:equal,
    xlabel=L"X_1", ylabel=L"X_2")
```

```
Mean1   Mean2   Var1    Var2    Cov
14.99   19.99   6.01    9.0     4.0
15.0    20.0    6.01    8.96    3.97
15.0    19.98   6.03    8.85    4.01
```

We define the covariance matrix `SigY` and the mean vector `muY` in lines 6–9. In line 10, we use `cholesky()` from `LinearAlgebra` together with `.L` to compute a lower triangular matrix A that satisfies (3.31). In lines 12–14, we define an array of functions, `rngGens`, where each element is a function that generates a scalar random variable with zero mean and unit variance. The first entry is a standard normal, the second entry is a uniform on $[-\sqrt{3}, \sqrt{3}]$, and the third entry is a unit exponential shifted by -1. The function we define in line 16, `rv()`, assumes an input argument which is a function to generate a random value and then implements the transformation (3.29). In line 18, we create an array of three arrays, with each internal array consisting of `N` two-dimensional random vectors. We then define a function `stats()` in lines 20–25 which calculates and prints first- and second-order statistics. Note the use of `begin` and `end` to define the function. The function is then used in lines 27–30 for printing output. The remainder of the code creates Figure 3.25 using `data`.

Bivariate Normal

One of the most ubiquitous families of multivariate distributions is the *multivariate normal distribution*. Similarly to the fact that a scalar (*univariate*) normal distribution is parametrized by the mean μ and the variance σ^2, a multivariate normal distribution is parametrized by the mean vector $\mu_{\boldsymbol{X}}$ and the covariance matrix $\Sigma_{\boldsymbol{X}}$.

We begin first with the *standard multivariate* having $\mu_{\boldsymbol{X}} = \mathbf{0}$ mean and $\Sigma_{\boldsymbol{X}} = I$. In this case, the PDF for the random vector $\boldsymbol{X} = (X_1, \dots, X_n)$ is

$$f(\boldsymbol{x}) = (2\pi)^{-n/2} e^{-\frac{1}{2}\boldsymbol{x}^\top \boldsymbol{x}}. \tag{3.32}$$

Listing 3.33 illustrates numerically that this is a valid PDF for increasing dimensions. The example also illustrates how to use numerical integration. The integral (3.24) is carried out. As is observed from the output, the integral is accurate for dimensions $n = 1, \dots, 8$ after which accuracy is lost for the given level of computational effort specified (up to 10^7 function evaluations).

Listing 3.33: Multidimensional integration

```
using HCubature

M = 4.5
maxD = 10

f(x) = (2*pi)^(-length(x)/2) * exp(-(1/2)*x'x)

for n in 1:maxD
    a = -M*ones(n)
    b = M*ones(n)
    I,e = hcubature(f, a, b, maxevals = 10^7)
    println("n = $(n), integral = $(I), error (estimate) = $(e)")
end
```

```
n = 1, integral = 0.9999932046537506, error (estimate) = 4.365848932375016e-10
n = 2, integral = 0.9999864091389514, error (estimate) = 1.487907641465839e-8
n = 3, integral = 0.9999796140804286, error (estimate) = 1.4899542976517278e-8
n = 4, integral = 0.9999728074508313, error (estimate) = 4.4447365681340567e-7
n = 5, integral = 0.999965936103044, error (estimate) = 2.3294669134930872e-5
n = 6, integral = 0.9999639124757695, error (estimate) = 0.0003937954462609516
```

```
n = 7, integral = 1.0001623151630603, error (estimate) = 0.0031506650163379375
n = 8, integral = 1.0074827348433588, error (estimate) = 0.023275741664597824
n = 9, integral = 1.2233043761463287, error (estimate) = 0.3731125349186617
n = 10, integral = 0.42866209316161175, error (estimate) = 0.22089760603668285
```

We use the `HCubature` package. In line 3, we define `M`. Then the integration is performed over a square of width twice of `M`, centered at the origin. In line 4, we define `maxD` as the number of dimensions up to which we wish to carry out integration. The function definition in line 6 implements (3.32). We loop over the dimensions in lines 8–13, each time computing the integral in line 11 where we specify `maxevals` as the maximum number of evaluations. The result is a tuple of the integral value and error, which are assigned to `I` and `e`, respectively.

Now, in general, using an affine transformation like (3.29), it can be shown that for arbitrary $\mu_{\boldsymbol{X}}$ and $\Sigma_{\boldsymbol{X}}$ (positive definite),

$$f(\boldsymbol{x}) = |\Sigma_{\boldsymbol{X}}|^{-1/2}(2\pi)^{-n/2}e^{-\frac{1}{2}(\boldsymbol{x}-\mu_X)^\top \Sigma_X^{-1}(\boldsymbol{x}-\mu_X)},$$

where $|\cdot|$ is the determinant. In the case of $n = 2$, this becomes the *bivariate normal distribution* with a density represented as

$$f_{XY}(x, y; \sigma_X, \sigma_Y, \mu_X, \mu_Y, \rho) = \frac{1}{2\pi\sigma_X\sigma_Y\sqrt{1-\rho^2}} \times \exp\left\{\frac{-1}{2(1-\rho^2)}\left[\frac{(x-\mu_X)^2}{\sigma_X^2} - \frac{2\rho(x-\mu_X)(y-\mu_Y)}{\sigma_X\sigma_Y} + \frac{(y-\mu_Y)^2}{\sigma_Y^2}\right]\right\}.$$

Here the elements of the mean and covariance matrix are spelled out via

$$\mu_{\boldsymbol{X}} = \begin{bmatrix}\mu_X \\ \mu_Y\end{bmatrix} \quad \text{and} \quad \Sigma_{\boldsymbol{Y}} = \begin{bmatrix}\sigma_X^2 & \sigma_X\sigma_Y\rho \\ \sigma_X\sigma_Y\rho & \sigma_Y^2\end{bmatrix}.$$

Note that $\rho \in (-1, 1)$ is the correlation coefficient as defined in (3.26).

In Section 4.2, we fit the five parameters of a bivariate normal to weather data and keep the results as assignment statements to `meanVect` and `covMat` in the file `mvParams.jl`. The example below illustrates a plot of random vectors generated from a distribution matching these parameters. Here we use the `MvNormal()` constructor from `Distributions` to create a multivariate normal distribution object. The listing generates Figure 3.26.

Listing 3.34: Bivariate normal data

```
1  using Distributions, Plots; pyplot()
2
3  include("../data/mvParams.jl")
4  biNorm = MvNormal(meanVect,covMat)
5
6  N = 10^3
7  points = rand(MvNormal(meanVect,covMat),N)
8
9  support = 15:0.5:40
10 z = [ pdf(biNorm,[x,y]) for y in support, x in support ]
11
12 p1 = scatter(points[1,:], points[2,:], ms=0.5, c=:black, legend=:none)
13 p1 = contour!(support, support, z,
14              levels=[0.001, 0.005, 0.02], c=[:blue, :red, :green],
15              xlims=(15,40), ylims=(15,40), ratio=:equal, legend=:none,
16              xlabel="x", ylabel="y")
17 p2 = surface(support, support, z, lw=0.1, c=cgrad([:blue, :red]),
18              legend=:none, xlabel="x", ylabel="y",camera=(-35,20))
19
20 plot(p1, p2, size=(800, 400))
```

In line 3, we include another Julia file defining `meanVect` and `covMat`. This file is generated in Listing 4.12 of Chapter 4. In line 4, we create an `MvNormal` distribution object representing the bivariate distribution. In line 7, we use `rand()` with a method provided via the `Distributions` package to generate random points. The rest of the code plots are shown in Figure 3.26. Notice the call to `contour()` in lines 13–16, with specified `levels`. In lines 17–18, the parameters supplied via `camera` are horizontal rotation and vertical rotation in degrees.

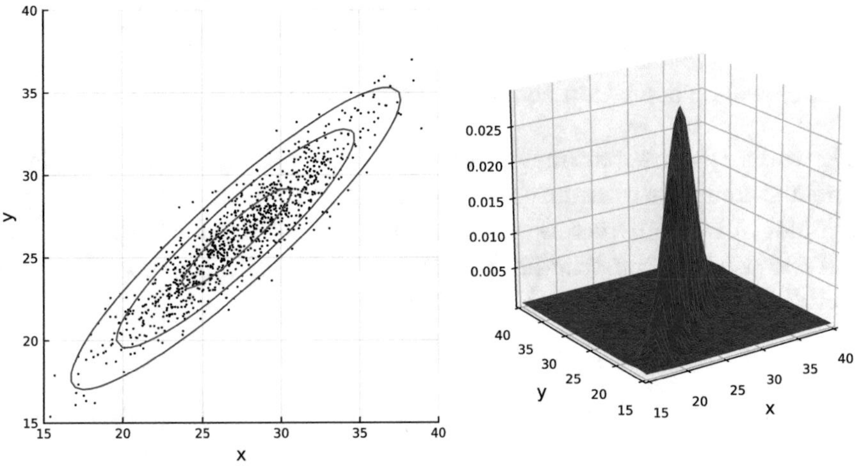

Figure 3.26: Contour lines and a surface plot for a bivariate normal distribution with randomly generated points on the contour plot.

Chapter 4

Processing and Summarizing Data

In this chapter, we introduce methods and techniques for processing and summarizing data. In statistics nomenclature, the act of summarizing data is known as *descriptive statistics*. In data science nomenclature such activities take the names of *analytics* and *dash-boarding*, while the process of manipulating and pre-processing data is sometimes called *data cleansing*, or *data cleaning*.

The statistical techniques and tools that we introduce include summary statistics and methods for data visualization, sometimes called *Exploratory Data Analysis (EDA)*. We introduce several Julia tools for this, including the `DataFrames` package which allows one to store datasets that contain non-homogeneous data and includes support for missing entries. We also use the `Statistics` and `StatsBase` packages, which contain useful functions for summarizing data.

In practice, statisticians and data scientists often collect data in various ways, including *experimental studies*, *observational studies*, *longitudinal studies*, *survey sampling*, and *data scraping*. Then to gain insight from the data, one may consider different *data configurations* such as

Single sample: A case where all observations are considered to represent items from a homogeneous population. The configuration of the data takes the form: $x_1, x_2, \ldots, x_n$.

*Single sample over time (*time-series*):* The configuration of the data takes the form: $x_{t_1}, x_{t_2}, \ldots, x_{t_n}$ with time points $t_1 < t_2 < \ldots < t_n$.

Two samples: Similar to the single sample case, only now there are two populations (x's and y's). The configuration of the data takes the form: $x_1, \ldots, x_n$ and $y_1, \ldots, y_m$.

Generalizations from two samples to k samples (each of potentially different sample size, $n_1, \ldots, n_k$).

Observations in pairs (2-tuples): In this case, although similar to the two sample case, each observation is a tuple of points, (x, y). Hence, the configuration of data is $(x_1, y_1), (x_2, y_2), \ldots, (x_n, y_n)$.

Generalizations from pairs to vectors of observations. $(x_{11}, \ldots, x_{1p}), \ldots, (x_{n1}, \ldots, x_{np})$.

Other configurations including relationship data (graphs of connections), images, and many more.

This chapter is structured as follows: In Section 4.1 we see how to manipulate tabular data via data frames in Julia. In Section 4.2, we deal with methods of summarizing data including basic

Y. Nazarathy and H. Klok, *Statistics with Julia*, Springer Series in the Data Sciences,
https://doi.org/10.1007/978-3-030-70901-3_4

elements of descriptive statistics. We then move on to plotting where, in Section 4.3, we present a variety of methods for plotting single sample data. In Section 4.4, we present plots for comparing samples. Section 4.5 presents plots for multivariate and high-dimensional data. We then present more simplistic business style plots in Section 4.6. The chapter closes with Section 4.7, where we show several ways of handling files using Julia as well as how to interact with a server side database.

For readers who wish to better understand the concepts of copies and mutability used in Section 4.1, the subsection below provides an optional overview. It can be skipped on a first reading.

Mutability, References, Shallow Copies and Deep Copies in Julia

When using any programming language, it is useful to have a basic understanding of how data is organized and referenced in memory. For this reason we now briefly overview the differences between *mutable* types, *immutable* types, *reference copying*, *shallow copies*, and *deep copies* in Julia. We also introduce the basic programming concepts of "call by value" and "call by reference". This basic understanding is important in its own right, however it may also help readers better understand certain aspects of Julia's `DataFrame` package, described in the sequel.

As a starting point, we review the difference between two mechanisms for passing variables to functions. Assume you have a variable `x`, a function `f()`, and then you then execute `f(x)`. One can envision two general mechanisms by which this can take place. The first is named *call by value* and describes a situation where the code implementing `f()` gets a copy of the variable `x`. As `f()` executes, even if its code appears to modify `x`, it is actually modifying a local copy. The second mechanism is named *call by reference* and describes a situation where `f()` obtains a *memory reference* (or *pointer*) to `x`. In such a case, as `f()` executes, if it modifies `x`, then it actually modifies values in the original memory location of `x`.

In Julia, both mechanisms exist under a unified umbrella called *pass by sharing*. This means that variables are not copied when passed to functions. However, if a value is about to be changed within a function then depending on the mutability attribute of its type, either of the mechanisms may be employed. If the variable type is *immutable* then a local copy is made and the behavior follows the "call by value" type. However, if the type is *mutable* then the called function does not create a local copy. Instead, it can modify the original variable according to the "call by reference" mechanism. Hence, the variable's property, mutable or immutable, determines which function calling mechanism is exhibited.

As a general rule, primitive types such as `Int64` or `Float32` are immutable. The same goes for composite types defined using the `struct` keyword. An exception to this is for composite types that are explicitly defined as `mutable struct`. Note that the code examples in this book seldom define types—however, many of the types we use from packages are composite types. While not often used, if you wish to programmatically check if the type of a variable is immutable or not, you can use the `isimmutable()` function.

Importantly, arrays are mutable. Listing 4.1 implements two different methods for the function `f()`. The first method is for `Int`, a primitive type (immutable), and the second is for `Array{Int}` (mutable). It then demonstrates the "call by value" behavior exhibited for the primitive type, while the "call by reference" behavior is exhibited for the array.

Listing 4.1: Call by value vs. call by reference

```
1  f(z::Int) = begin z = 0 end
2  f(z::Array{Int}) = begin z[1] = 0 end
3
4  x = 1
5  @show typeof(x)
6  @show isimmutable(x)
7  println("Before call by value: ", x)
8  f(x)
9  println("After call by value: ", x,"\n")
10
11 x = [1]
12 @show typeof(x)
13 @show isimmutable(x)
14 println("Before call by reference: ", x)
15 f(x)
16 println("After call by reference: ", x)
```

```
typeof(x) = Int64
isimmutable(x) = true
Before call by value: 1
After call by value: 1

typeof(x) = Array{Int64,1}
isimmutable(x) = false
Before call by reference: [1]
After call by reference: [0]
```

In line 1, we implement a method of `f()` for integer types. The code `z = 0` will operate on a local copy of `z`. In line 2, we implement a method of `f()`for arrays. Here, the code `z[1] = 0` will modify the first entry of the input argument `z`. Lines 4–9 use the first method, passing the variable `x` into `f()`. As can be see from the output, the operation of the function `f()` does not modify `x`. Also note the use of the `@show` macro, useful for debugging or understanding code. Lines 11–16 invoke the method of `f()` for arrays of integers (this is multiple dispatch). The key point is that `f(x)` in line 15 modifies the original `x` from global scope.

Ideally, for performance reasons, the level of actual copying of memory should be kept to a minimum. This is the underlying motivation for having a default "pass by reference" mechanism when working with arrays, as you can give functions references to huge data arrays without any memory duplication. However, this entails some level of danger because function calls may modify variables that are passed to them as arguments. For this reason, Julia offers explicit functions for creating copies of variables, namely `copy()` and `deepcopy()`. The former creates a "shallow copy" of the variable and copies all entries but does not do it recursively. The latter recursively produces a copy until a completely independent copy of the variable is created.

We demonstrate the different types of copies and their interaction with mutability in Listing 4.2. The basic example on which we apply a deep copy is a doubly nested array, e.g. `[[10]]`. In this case, using `copy()` will not be applied to the inner array `[10]`, however using `deepcopy()` recursively copies all mutable entries.

Listing 4.2: Deep copy and shallow copy

```
1  println("Immutable:")
2  a = 10
3  b = a
4  b = 20
5  @show a
6
7  println("\nNo copy:")
8  a = [10]
9  b = a
10 b[1] = 20
11 @show a
12
13 println("\nCopy:")
14 a = [10]
15 b = copy(a)
16 b[1] = 20
17 @show a
18
19 println("\nShallow copy:")
20 a = [[10]]
21 b = copy(a)
22 b[1][1] = 20
23 @show a
24
25 println("\nDeep copy:")
26 a = [[10]]
27 b = deepcopy(a)
28 b[1][1] = 20
29 @show a;
```

```
Immutable:
a = 10

No copy:
a = [20]

Copy:
a = [10]

Shallow copy:
a = Array{Int64,1}[[20]]

Deep copy:
a = Array{Int64,1}[[10]]
```

Lines 1–5 exhibit no surprise due to immutability. The `Int64` `a` is assigned to `b` and `b` is modified in line 4. At this point, Julia creates a copy because the variable is immutable. Lines 7–11 demonstrate different behavior. The array `a` is mutable and hence after `b` is assigned to `a` in line 9, the modification of `b` in line 10 also modifies `a`. Lines 13–17 show a case where a `copy()` of `a` is created. In this case, modification of `b` in line 16 does not alter `a`. Lines 19–23 are similar, however in this case, the fact that `copy()` is only a shallow copy matters. The variable `b` has a new outer array, however the inner array is still shared with `a`. Hence, the modification in line 22 modifies the inner array of `a` as well. Finally, in lines 25–29, this is resolved by creating a `deepcopy()`.

4.1 Working with Data Frames

In cases where data is homogeneous, arrays, matrices, and tensors are popular ways of organizing data. However, more commonly datasets are heterogeneous in nature or contain incomplete or missing entries. In addition, datasets are often large, and commonly require "cleaning". In such cases, more advanced data storage mechanisms are needed.

The Julia `DataFrames` package introduces a data storage structure known as a `DataFrame`, which is aimed at overcoming these challenges. It can be used to store columns of different types, and also introduces the `missing` variable type which, as the name suggests, is used in place of missing entries. The `missing` type has an important property, in that it "poisons" other types it interacts with. For example, if `x` represents a value, then `x + missing == missing`. This ensures that missing values do not 'infect' and skew results when operations are performed on data. For example, if `mean()` is used on a column with a `missing` value present, the result will evaluate as `missing`. We show ways of dealing with missing values in Listing 4.7.

Data frames are easy to work with. They can be created manually or data can be imported from a `csv` or `txt` file. Columns and rows can be referenced by their position index, name (i.e. symbol), or according to a set of user-defined rules. We now explore some of their functionally. See `http://juliadata.github.io/DataFrames.jl/stable/` for further documentation. We note that the package `DataFrames.jl` and related packages such as `CSV.jl` have undergone extensive revisions during 2019–2020 as this book already entered production. Some of the code snippets below use an earlier interface. Refer to the book's online GitHub repository for updated code snippets.

Introducing the Data Frame

We now introduce data frames through the exploration and formatting of an example dataset. The dataset has four fields; `Name`, `Date`, `Grade`, and `Price`. In addition, as is often the case with real datasets, there are missing values present. Therefore, before analysis can start, some *data cleaning* must be performed.

Any variable in a dataset can be classified as either a *numerical variable*, or *categorical variable*. A numerical variable is a variable in which the location of the measurement on the number line is meaningful. Examples include height and weight. A categorical variable communicates some information based on categories or characteristics via grouping. Categorical variables can be further split into *nominal variables*, such as blood type, or names, and *ordinal variables*, in which some order is communicated, such as grades on a test, A–E. In our example, `Price` is a numerical variable, while `Name` is a nominal categorical variable. Since, in our example, `Grade` can be thought of as a rating (A being best and E being worst) it is an ordinal categorical variable.

Having covered types of variables, we begin a step-by-step example of using data frames. In Listing 4.3, we load the data from the file `purchaseData.csv` into a data frame and inspect its contents. Often *comma separated values files* (csv files) contain a *header row* which gives details of each column. However, in other cases, no header row appears. In our case, the file's first row is a header row and it contains the names of the columns. Hence, the first row of the file is

```
Name, Date, Grade, Price
```

Listing 4.3: Creating and inspecting a DataFrame

```
1 using DataFrames, CSV
2 data = CSV.read("../data/purchaseData.csv", copycols = true)
3
4 println(size(data),"\n")
5 println(names(data),"\n")
6 println(first(data, 6),"\n")
7 println(describe(data),"\n")
```

```
(200, 4)

Symbol[:Name, :Date, :Grade, :Price]

6×4 DataFrame
| Row | Name     | Date       | Grade   | Price   |
|     | String   | String     | String  | Int64   |
-------------------------------------------------
| 1   | MARYANNA | 14/09/2008 | A       | 79700   |
| 2   | REBECCA  | 11/03/2008 | B       | missing |
| 3   | ASHELY   | 5/08/2008  | E       | 24311   |
| 4   | KHADIJAH | 2/09/2008  | missing | 38904   |
| 5   | TANJA    | 1/12/2008  | C       | 47052   |
| 6   | JUDIE    | 17/05/2008 | D       | 34365   |

4×8 DataFrame
| Row | variable | mean    | min       | median  | max       | nunique | nmissing | eltype                 |
|     | Symbol   | Union   | Any       | Union   | Any       | Union   | Int64    | Union                  |
-------------------------------------------------------------------------------------------------------
| 1   | Name     |         | ABBEY     |         | ZACHARY   | 182     | 17       | Union{Missing, String} |
| 2   | Date     |         | 1/07/2008 |         | 9/10/2008 | 141     | 4        | Union{Missing, String} |
| 3   | Grade    |         | A         |         | E         | 5       | 13       | Union{Missing, String} |
| 4   | Price    | 39702.0 | 8257      | 38045.5 | 79893     |         | 14       | Union{Missing, Int64}  |
```

In line 1, we specify use of the `DataFrames` package, which allows us to use `DataFrame` type objects. We also use the `CSV` package for reading `csv` files. In line 2, `CSV.read()` is used to create a data frame object, populated with data from the file specified. Note that our file has a header row, however in cases where there isn't a header use `header = false`. We use `copycols = true` to create a data frame with mutable columns (the default is `false`). If the default was used, each column would be of the read-only `CSV.Column` type. In line 4 the `size()` function is used to return the number of rows and columns of the data frame as a tuple. Two other useful functions not shown here are `nrow()` and `ncol()`, which return the number of rows and number of columns, respectively. In line 5, the `names()` function is used to return an array of all column names as symbols. In line 6, the `first()` function is used to display the first six lines of the data frame, as specified by the second argument. Note that `last()` can be used to display the last several rows instead. In line 7, `describe()` is used to create a data frame with a summary of the data in each column of the input data frame (`data` in our case). On inspection, by looking at the `nmissing` column, one can see there are missing values present, and we return to this problem in Listing 4.7.

Referencing Data

We now look at ways in which entries within a data frame can be referenced. Individual entries can be referenced by both row and column indexes. Columns can be referenced by their name represented as a symbol, or by their index. Multiple rows or multiple columns can be referenced via a collection of symbols or indices. We demonstrate several aspects of this in Listing 4.4 below.

Listing 4.4: Referencing data in a `DataFrame`

```
1  using DataFrames, CSV
2  data = CSV.read("../data/purchaseData.csv", copycols = true)
3
4  println("Grade of person 1: ", data[1, 3],
5          ", ", data[1,:Grade],
6          ", ", data.Grade[1], "\n")
7  println(data[[1,2,4], :], "\n")
8  println(data[13:15, :Name], "\n")
9  println(data.Name[13:15], "\n")
10 println(data[13:15, [:Name]])
```

```
Grade of person 1: A, A, A

3×4 DataFrame
| Row | Name      | Date       | Grade   | Price   |
|     | String    | String     | String  | Int64   |
---------------------------------------------------
| 1   | MARYANNA  | 14/09/2008 | A       | 79700   |
| 2   | REBECCA   | 11/03/2008 | B       | missing |
| 3   | KHADIJAH  | 2/09/2008  | missing | 38904   |

Union{Missing, String}["SAMMIE", missing, "STACEY"]

Union{Missing, String}["SAMMIE", missing, "STACEY"]

3×1 DataFrame
| Row | Name    |
|     | String  |
-----------------
| 1   | SAMMIE  |
| 2   | missing |
| 3   | STACEY  |
```

In lines 4–6, we see different ways of accessing the element from the first row and third column labeled `:Date`. In line 7, the rows and columns to be extracted are designated by the first and second arguments, `[1,2,4]`, and "`:`", respectively. Note that "`:`" can be used to select either all rows or all columns. Line 8 is somewhat similar, but here a unit range is used to select rows 13–15, while the symbol `:Name` is used so that only the `Name` column is extracted. Alternatively, the column could have been referenced by its index, i.e. `data[13:15, 1]`, or the syntax `data.Name[13:15]` could have been used instead. Note that although lines 9 and 10 look similar, there is an important difference. The code in line 9 creates an array, while that of line 10 creates a data frame object, due to the extra set of `[]`. If one wanted, additional columns could also be selected by including them in `[]`, separated by "`,`".

Modifying Data

In general, entries of a data frame can be updated like entries of a matrix, however in certain cases, care is required. Functions performed on a data frame will return a copy of that data frame. In other words, no underlying change will be made to the data frame object, but rather a shallow copy will be made and returned as output. Often one wants to change the values within a data frame. However, by default, the columns of a data frame are *immutable*, which means that the values within them cannot be changed. In order to make changes to a column, the column must first be mutable. One way to do this is by including `copycols=true` when a data frame is created from a `csv` file. This argument has the effect of making all columns mutable. Another way is by using `!` when referencing the rows of a data frame. For example, `df[!, :X]` references the underlying data in column `:X`, while `df[:, :X]` simply references a shallow copy. In Listing 4.5 below, we show how these approaches work.

Listing 4.5: Editing and copying a `DataFrame`

```
using DataFrames, CSV
data1 = CSV.read("../data/purchaseData.csv")
data2 = CSV.read("../data/purchaseData.csv", copycols=true)

try data1[1, :Name] = "YARDEN" catch; @warn "Cannot: data1 is immutable" end

data2[1, :Name] = "YARDEN"
println("\n", first(data2, 3), "\n")

data1[!, :Price] ./= 1000
rename!(data1, :Price=>Symbol("Price(000's)"))
println(first(data1, 3), "\n")

replace!(data2[!,:Grade], ["E"=>"F", "D"=>"E"]...)
println(first(data2, 3),"\n")
```

```
Warning: Cannot: data1 is immutable

| Row | Name     | Date       | Grade  | Price   |
|     | String   | String     | String | Int64   |
-------------------------------------------------
| 1   | YARDEN   | 14/09/2008 | A      | 79700   |
| 2   | REBECCA  | 11/03/2008 | B      | missing |
| 3   | ASHELY   | 5/08/2008  | E      | 24311   |

3×4 DataFrame
| Row | Name      | Date       | Grade  | Price(000's) |
|     | String    | String     | String | Float64      |
------------------------------------------------------
| 1   | MARYANNA  | 14/09/2008 | A      | 79.7         |
| 2   | REBECCA   | 11/03/2008 | B      | missing      |
| 3   | ASHELY    | 5/08/2008  | E      | 24.311       |

3×4 DataFrame
| Row | Name     | Date       | Grade  | Price   |
|     | String   | String     | String | Int64   |
-------------------------------------------------
| 1   | YARDEN   | 14/09/2008 | A      | 79700   |
| 2   | REBECCA  | 11/03/2008 | B      | missing |
| 3   | ASHELY   | 5/08/2008  | F      | 24311   |
```

In lines 2–3, two dataframes are created, the first `data1` has immutable columns, while `data2` has mutable columns due to the second argument in `CSV.read()`. In line 5, we try to change the value of the first row and first column in `data1`. This is done within the `try/catch` structure. If an error occurs within `try`, the code jumps to `catch` and continues. Since `data1` is immutable, an error is returned, and so the code after `catch` runs. Here, we use the `@warn` macro to return a warning. In line 7, we try the same change for `data2`, and since this data frame is mutable, we are able to make the change. In line 10, we perform division on every row element in the `:Price` column of `data1` by using `!` to reference all rows. By using this syntax, the underlying `:Price` column data is referenced, and the column changed to mutable, which then allows us to make make the change. The actual change is done via the combination of the broadcast "`.`" operator, which extends the in-place division via \= to each row. Note the column type changes from `Int64` to `Float64`. In line 11, `rename!()` is used to rename the `:Price` column as shown, with a pair of values, separated via `=>`, given as the second argument. Finally, in line 14, `replace!()` is used to replace all `D` and `E` entries in the `:Grade` column to `E` and `F` respectively. Note that `replace!()` operates on an iterable, hence the use of the `...` splat operator, and finally, note that the order of replacement does not matter, as the replacement does not advance one after the other sequentially. Again note that `!` was used for row referencing.

Copying a Data Frame

When copying a data frame, the same rules and principles that are relevant for other Julia types apply. These were discussed at the start of this chapter and demonstrated in Listing 4.2. We now show how `copy()` and `deepcopy()` can be used with data frames in Listing 4.6.

Listing 4.6: Using `copy()` and `deepcopy()` with a `DataFrame`

```
using DataFrames, CSV
data1 = CSV.read("../data/purchaseData.csv", copycols=true)
println("Original value: ", data1.Name[1],"\n")

data2 = data1
data2.Name[1] = "EMILY"
@show data1.Name[1]

data1 = CSV.read("../data/purchaseData.csv", copycols=true)
data2 = copy(data1)
data2.Name[1] = "EMILY"
@show data1.Name[1]
println()

data1 = DataFrame()
data1.X = [[0,1],[100,101]]
data2 = copy(data1)
data2.X[1][1] = -1
@show data1.X[1][1]

data1 = DataFrame(X = [[0,1],[100,101]])
data2 = deepcopy(data1)
data2.X[1][1] = -1
@show data1.X[1][1];
```

```
Original value: MARYANNA

data1.Name[1] = "EMILY"
data1.Name[1] = "MARYANNA"

(data1.X[1])[1] = -1
(data1.X[1])[1] = 0
```

We first create a data frame from a csv file where `data1.Name[1]` is the string `MARYANNA`. Then in lines 5–7, setting `data2 = data1` simply implies that `data2` refers to the same object as `data1`. Hence, modifying `data2` in line 6 results in a modification of `data1`. In lines 9–13, we circumvent such a situation by using the `copy()` function. In this case setting, the new name into `data2`, `EMILY`, does not affect `data1`. However, in other cases a shallow copy isn't enough for separating data frames. This is the case in lines 15–19 where we create a data frame with a column named `X` comprised of arrays. In this case, the copied data frame, `data2`, still refers to the original entries (arrays), because these are mutable and were not copied via `copy()` in line 17. The consequence is that modifying a specific entry of `data2` as in line 18 actually modifies `data1`. This is then circumvented by using `deepcopy()` as in lines 21–24.

Handling Missing Entries

We now look more closely at the case when `missing` values are present in a data frame. As discussed at the start of this section, `missing` "poisons" other types on interaction, and this property ensures that missing values do not "infect" and skew results when operations are performed on a dataset. The `DataFrames` package comes with several useful functions for dealing with missing entries. The `Missing.jl` package also provides extra functionality for dealing with missing values. In Listing 4.7, below we elaborate on some of the functions useful for dealing with missing values.

Listing 4.7: Handling missing entries

```julia
using Statistics, DataFrames, CSV
data = CSV.read("../data/purchaseData.csv", copycols=true)

println(mean(data.Price),"\n")
println(mean(skipmissing(data.Price)),"\n")
println(coalesce.(data.Grade, "QQ")[1:4],"\n")
println(first(dropmissing(data,:Price), 4),"\n")
println(sum(ismissing.(data.Name)),"\n")
println(findall(completecases(data))[1:4])
```

```
missing

39702.01075268817

["A", "B", "E", "QQ"]

4×4 DataFrame
| Row | Name     | Date       | Grade  | Price |
|     | String   | String     | String | Int64 |
-----------------------------------------------
| 1   | MARYANNA | 14/09/2008 | A      | 79700 |
```

```
| 2   | ASHELY   | 5/08/2008  | E       | 24311 |
| 3   | KHADIJAH | 2/09/2008  | missing | 38904 |
| 4   | TANJA    | 1/12/2008  | C       | 47052 |

17

[1, 3, 5, 6]
```

In line 4, we attempt to calculate the mean of the `:Price` column of `data`, however `missing` is returned as this column contains missing values. By comparison, in line 5, `skipmissing()` is first used to return a copy of the data from the `:Price` column which has no missing entries, and after this `mean()` applied. In line 6, `data.Grade` is used to obtain a reference to the `:Grade` column, and then the `coalesce()` function is used to replace all `missing` values with the string "QQ". The first four values are accessed via `[1:4]` to verify the replacement has occurred. In line 7, `dropmissing()` is used to drop all rows which have `missing` in the `:Price` column. If no second argument is given, `dropmissing()` will drop all rows that contain `missing`. In line 8, `ismissing()` is used with the broadcast operator to check if values in the `:Name` column are `missing`. If they are, `true` is returned, else `false`. Then `sum()` is used to calculate how many `missing` entries are present. The result, `17`, can be verified from the output of Listing 4.3 where we see the number of missing entries. In line 9, `completecases()` is used to check if each row contains fully completed fields, i.e. no `missing` values. If no `missing` values are present, `true` is returned, else `false`, for each row. Then `findall()` is used on this array to return an array of row indexes which have no `missing` values, and to shorten the output, the first four values of this array are printed.

Reshaping, Joining, and Manipulating Data Frames

When working with data it is not uncommon to want to perform operations such as merges or joins between several datasets, or to split or reshape the structure of a dataset. The `DataFrames` package makes this easy, as it provides many useful functions to do these types of operations. In Listing 4.8, we present brief examples of some of the more useful functions for merging and joining data frames.

Listing 4.8: Reshaping, joining and merging data frames

```julia
using DataFrames, CSV
data = CSV.read("../data/purchaseData.csv", copycols = true)

newCol = DataFrame(Validated=ones(Int, size(data,1)))
newRow = DataFrame([["JOHN", "JACK"] [123456, 909595]], [:Name, :PhoneNo])
newData = DataFrame(Name=["JOHN", "ASHELY","MARYANNA"],
                    Job=["Lawyer", "Doctor","Lawyer"])

data = hcat(data, newCol)
println(first(data, 3), "\n")

data = vcat(data, newRow, cols=:union)
println(last(data, 3), "\n")

data = join(data, newData, on=:Name)
println(data, "\n")

select!(data,[:Name,:Job])
println(data, "\n")

unique!(data,:Job)
println(data)
```

```
3×5 DataFrame
| Row | Name     | Date       | Grade  | Price   | Validated |
|     | String   | String     | String | Int64   | Int64     |
-----------------------------------------------------------------
| 1   | MARYANNA | 14/09/2008 | A      | 79700   | 1         |
| 2   | REBECCA  | 11/03/2008 | B      | missing | 1         |
| 3   | ASHELY   | 5/08/2008  | E      | 24311   | 1         |

3×6 DataFrame
| Row | Name | Date       | Grade   | Price   | Validated | PhoneNo |
|     | Any  | String     | String  | Int64   | Int64     | Any     |
-----------------------------------------------------------------------
| 1   | RIVA | 30/12/2008 | E       | 21842   | 1         | missing |
| 2   | JOHN | missing    | missing | missing | missing   | 123456  |
| 3   | JACK | missing    | missing | missing | missing   | 909595  |

3×7 DataFrame
| Row | Name     | Date       | Grade   | Price   | Validated | PhoneNo | Job    |
|     | Any      | String     | String  | Int64   | Int64     | Any     | String |
------------------------------------------------------------------------------------
| 1   | MARYANNA | 14/09/2008 | A       | 79700   | 1         | missing | Lawyer |
| 2   | ASHELY   | 5/08/2008  | E       | 24311   | 1         | missing | Doctor |
| 3   | JOHN     | missing    | missing | missing | missing   | 123456  | Lawyer |

3×2 DataFrame
| Row | Name     | Job    |
|     | Any      | String |
---------------------------
| 1   | MARYANNA | Lawyer |
| 2   | ASHELY   | Doctor |
| 3   | JOHN     | Lawyer |

2×2 DataFrame
| Row | Name     | Job    |
```

```
|     | Any       | String |
----------------------------
| 1   | MARYANNA  | Lawyer |
| 2   | ASHELY    | Doctor |
```

In line 2, we create `data` in the same manner as the previous listings. In lines 4–6, we create three separate data frames. The first, `newCol`, consists of a single column `:Validated` with the same number of rows as `data`. The second, `newRow`, consists of two rows with `:Name` and `:PhoneNumber` columns. The third, `newData`, has two rows and two columns, `:Name` and `:Job`. In line 9, `hcat()` is used to horizontally concatenate `newCol` to `data`. In line 12, `vcat()` is used to vertically concatenate `data` and `newRow`, with the new row appended to the bottom of the data frame. Note `cols=:union` is used so that all columns from both data frames are kept, and `missing` entries recorded where applicable. Alternatively, `:equal` or `:intersect` could have been used, or an array of columns to be kept instead. In line 15, `join()` is used to join `data` and `newData` together, based on the `:Name` column. Note that `join()` can be used in several different ways. The functions `select!()` and `unique!()` are demonstrated in lines 18–22. Another function not shown here is `stack()`, which can be used to stack a data frame from a wide format to a long format. We recommend the reader consult the `DataFrames` manual for further information on each of the functions listed here.

Useful Operations for Data Frames

We have already covered some of the many useful functions available in the `DataFrames` package, such as `replace!()` and `rename!()`, both introduced in Listing 4.5. We now provide insight into several more concepts, including sorting, changing a column of strings to `Date` types, how to make a column `Categorical` (useful when constructing models, as covered in Section 8.4), and finally how to split, apply, and combine data all in one via the `by()` function. Listing 4.9 demonstrates these.

Listing 4.9: Manipulating `DataFrame` objects

```
using DataFrames, CSV, Dates, Statistics
data = dropmissing(CSV.read("../data/purchaseData.csv", copycols=true))

data[!,:Date] = Date.(data[!,:Date], "d/m/y")
println(first(sort(data, :Date), 3),"\n")

println(first(filter(row -> row[:Price] > 50000, data),3 ),"\n")

categorical!(data, :Grade)
println(first(data, 3), "\n")

println(
    by(data, :Grade, :Price =>
        x -> ( NumSold=length(x), AvgPrice=mean(x)) )
    )
```

```
3×4 DataFrame
│ Row │ Name       │ Date       │ Grade  │ Price │
│     │ String     │ Date       │ String │ Int64 │
├─────┼────────────┼────────────┼────────┼───────┤
│ 1   │ STEPHEN    │ 2008-02-11 │ D      │ 33155 │
│ 2   │ JACKELINE  │ 2008-02-12 │ E      │ 8257  │
│ 3   │ ARDELL     │ 2008-03-03 │ C      │ 46911 │

3×4 DataFrame
│ Row │ Name     │ Date       │ Grade  │ Price │
│     │ String   │ Date       │ String │ Int64 │
├─────┼──────────┼────────────┼────────┼───────┤
│ 1   │ MARYANNA │ 2008-09-14 │ A      │ 79700 │
│ 2   │ NOE      │ 2008-08-15 │ A      │ 79344 │
│ 3   │ SAMMIE   │ 2008-11-05 │ B      │ 61730 │

3×4 DataFrame
│ Row │ Name     │ Date       │ Grade       │ Price │
│     │ String   │ Date       │ Categorical │ Int64 │
├─────┼──────────┼────────────┼─────────────┼───────┤
│ 1   │ MARYANNA │ 2008-09-14 │ A           │ 79700 │
│ 2   │ ASHELY   │ 2008-08-05 │ E           │ 24311 │
│ 3   │ TANJA    │ 2008-12-01 │ C           │ 47052 │

5×3 DataFrame
│ Row │ Grade          │ NumSold │ AvgPrice │
│     │ Categorical    │ Int64   │ Float64  │
├─────┼────────────────┼─────────┼──────────┤
│ 1   │ A              │ 15      │ 76606.7  │
│ 2   │ B              │ 19      │ 59873.9  │
│ 3   │ C              │ 33      │ 45285.8  │
│ 4   │ D              │ 35      │ 34656.8  │
│ 5   │ E              │ 51      │ 20492.5  │
```

In line 2 `dropmissing()` is used so that all rows with missing entries are excluded. This is done as some of the functions here require all values to be non-missing. In line 4 the `Date()` function from the `Dates` package is applied to every row from the `:Date` column, converting each entry from a string to a `Date` type, according to the string formatting given as the second argument. In line 5 `sort()` is used to sort by the `:Date` column. In line 7 `filter()` is used to return only rows which have a `:Price` greater than `50000`. In line 9 the type of the `:Grade` column is changed to categorical via `categorical!()`. In lines 13–14 the powerful `by()` function is demonstrated. Here `data` is split according to `:Grade`. The third argument is where calculations are defined. The columns to be referenced in the calculations are put to the left of "=>", in our case only `:Price` is used. The calculations are specified by the anonymous function in line 14. Note that => is used to define a `Pair` and -> is used to define an anonymous function. The anonymous function creates a `NamedTuple` defining two new columns, `:NumSold` and `:AvgPrice`. For the first, the total number of each `:Grade` is calculated based on the length, i.e. number of entries in the price column. For the second, the average price is calculated via `mean()`. Note that the `by()` function can be used in many ways, and calculations can be done over data in more than one column. There are also several other related functions not touched on here, including `mapcols()`, which can be used to transform all values in a data frame, and `aggregate()`, which has functionality similar to `by()`. Further functionality is also available via the `DataFramesMeta` package which provides a macro-based framework to interface with data frames, such as via the `@linq` macro and the |> operator. Consult the documentation for further information.

A Cleaning and Imputation Example

In practice, it is not uncommon for datasets to contain many missing values, or require a certain amount of cleaning before one can use the data. Furthermore, it may not always be practical to simply exclude every observation which has a missing value. For example, if we were to simply exclude all rows with missing values in `purchaseData.csv`, then we would lose almost 25% of the dataset.

Instead of simply deleting rows, one way of dealing with `missing` values is to use *imputation*. This involves substituting missing entries with values based on either the data observed or according to some other logic. Various methods can be used to impute missing values, and care must be taken when imputing, as it can lead to bias in the data. The exact type of imputation scheme used should ideally take into account both the nature of the data and the eventual statistical analysis. A comprehensive treatment is in [V12].

We now present an example of one way in which one might consider cleaning and imputing a dataset. In Listing 4.10 below, we clean the data and impute missing values. First, we replace all missing names with the string "QQ" and all missing dates with the string "`31/06/2008`". We then calculate the average price of each grade based on the data available and use these averages to impute missing entries in both the price and grade columns.

Listing 4.10: Cleaning and imputing data

```
using DataFrames, CSV, Statistics
data = CSV.read("../data/purchaseData.csv")

rowsKeep = .!(ismissing.(data.Grade) .& ismissing.(data.Price))
data = data[rowsKeep, :]

replace!(x -> ismissing(x) ? "QQ" : x, data.Name)
replace!(x -> ismissing(x) ? "31/06/2008" : x, data.Date)

grPr = by(dropmissing(data), :Grade, :Price=>x ->
        AvgPrice = round(mean(x), digits=-3))

d = Dict(grPr[:,1] .=> grPr[:,2])
nearIndx(v, x) = findmin(abs.(v.-x))[2]
for i in 1:nrow(data)
    if ismissing(data[i, :Price])
        data[i, :Price] = d[data[i, :Grade]]
    end
    if ismissing(data[i, :Grade])
        data[i, :Grade] = grPr[ nearIndx(grPr[:,2], data[i, :Price]), :Grade]
    end
end

println(first(data, 5), "\n")
println(describe(data))
```

```
5×4 DataFrame
| Row | Name     | Date       | Grade  | Price |
|     | String   | String     | String | Int64 |
--------------------------------------------------
| 1   | MARYANNA | 14/09/2008 | A      | 79700 |
| 2   | REBECCA  | 11/03/2008 | B      | 60000 |
| 3   | ASHELY   | 5/08/2008  | E      | 24311 |
| 4   | KHADIJAH | 2/09/2008  | D      | 38904 |
| 5   | TANJA    | 1/12/2008  | C      | 47052 |

4×8 DataFrame
| Row | variable | mean    | min       | median  | max       | nunique | nmissing | eltype                  |
|     | Symbol   | Union   | Any       | Union   | Any       | Union   | Int64    | Union                   |
------------------------------------------------------------------------------------------------------------
| 1   | Name     |         | ABBEY     |         | ZACHARY   | 183     | 0        | Union{Missing, String} |
| 2   | Date     |         | 1/07/2008 |         | 9/10/2008 | 142     | 0        | Union{Missing, String} |
| 3   | Grade    |         | A         |         | E         | 5       | 0        | Union{Missing, String} |
| 4   | Price    | 40037.9 | 8257      | 38045.5 | 79893     |         | 0        | Union{Missing, Int64}  |
```

In lines 4–5, we check if there are any rows with missing values in both the `:Grade` and `:Price` columns, and we remove them if present. First `ismissing()` is applied element-wise over all values in each column, `.&` is then used to evaluate to `true` if both columns contain `missing`, and finally the preceding `.!` is used to flip the result, evaluating to `true` if the row should be kept. In our example, there are no rows with missing values in both columns, so all rows are kept. In lines 7–8, we replace all missing names with the strings `"QQ"` and `"31/06/2008"`, respectively, via `replace!()`. In lines 10–11, `dropmissing()` and `by()` are used to calculate the mean price of each group, excluding rows with `missing` values. The results are rounded to the nearest thousand (`digits = -3`) and stored as the data frame `grPr`. In line 14, the dictionary `d` is created based on the values from `grPr`, with grade the key, and average price the value. In line 14, the `nearIndx()` function is created. It takes a value as input, `x`, and then finds the index of the nearest value from a given vector of values, `v`. In lines 15–22, we loop over each row in the data frame and impute missing values in the price and grade columns. In lines 16–18, if the price entry is missing, then the grade is used to return the corresponding value stored in the dictionary `d`. Similarly, in lines 19–21, if the grade entry is missing, then the `nearIndx()` function is used to find the index of the closest value in `grPr` based on the price in `data`, and then `missing` is replaced by the corresponding grade. In lines 24–25, the first several rows of the data frame are printed, along with a summary of the cleaned data frame. At this point, no `missing` values are present.

4.2 Summarizing Data

Now that we have introduced data frames and methods of data processing, we can explore basic methods of *descriptive statistics* to obtain data summaries. We focus on numerical data in a single sample, observations in pairs, and observations in vectors.

Single Sample

Given a set of observations, $x_1, \ldots, x_n$, we can compute a variety of descriptive statistics. The *sample mean* is often the most basic and informative *measure of centrality*. It is denoted by $\overline{x}$ and is given by

$$\bar{x} = \frac{\sum_{i=1}^{n} x_i}{n}.$$

It is the *arithmetic mean* of the observations. However, the term "arithmetic mean", is not often used, unless we want to disambiguate it with the *geometric mean* or the *harmonic mean*, each, respectively, calculated via

$$\overline{x}_g = \sqrt[n]{\prod_{i=1}^{n} x_i}, \qquad \text{and} \qquad \overline{x}_h = \frac{n}{\sum_{i=1}^{n} \frac{1}{x_i}}.$$

These two other *Pythagorean means* are not as popular in statistics as the arithmetic mean, however they are occasionally useful.

The geometric mean is useful for averaging growths factors. For example, if $x_1 = 1.03$, $x_2 = 1.05$, and $x_3 = 1.07$ are growth factors, the geometric mean, $\overline{x}_g = 1.049714$ is a good summary statistic of the "average growth factor". This is because the growth factor obtained by $\overline{x}_g^3$ equals the growth factor $x_1 \cdot x_2 \cdot x_3$. Hence, we if we start with an original base level of say 100 units (e.g. dollars) and exhibit growths of x_1, x_2, and x_3 above in three consecutive periods, then after three periods, we have

$$\text{Value after three periods} = 100 \cdot x_1 \cdot x_2 \cdot x_3 = 100 \cdot \overline{x}_g^3 = 115.7205.$$

Here, the average growth factor is $\overline{x}_g$. Using the arithmetic mean, $\overline{x} = 1.05$ in such a case would yield 115.7625, which is slightly off from the correct value. Hence, in such scenarios, using the arithmetic mean to describe "average growth" is not adequate.

The harmonic mean is useful for averaging rates or speeds. For example assume that you are on a brisk hike, walking 5 km up a mountain and then 5 km back down. Say your speed going up is $x_1 = 5$ km/h and your speed going down is $x_2 = 10$ km/h. What is your "average speed" for the whole journey? You travel up for 1 h and down for 0.5 h and hence your total travel time is 1.5 h. Hence the average speed is $10/1.5 = 6.6\overline{6}$ km/h. This is not the arithmetic mean which is 7.5 km/h but rather exactly equals the harmonic mean.

Also note that for any dataset $\overline{x}_h \le \overline{x}_g \le \overline{x}$. Here, the inequalities become equalities only if all observations are equal. For $n = 2$, the second inequality can be obtained by manipulating the basic inequality $0 \le (x_1 - x_2)^2$. Then for higher n, it can be obtained by induction. The first inequality can then be obtained from the second inequality since the harmonic mean is the reciprocal of the arithmetic mean of reciprocals.

A different breed of descriptive statistics is based on *order statistics*. This term is used to describe the *sorted sample*, and is sometimes denoted by

$$x_{(1)} \le x_{(2)} \le \ldots \le x_{(n)}.$$

Based on the order statistics, we can define a variety of statistics such as the *minimum*, $x_{(1)}$, the *maximum*, $x_{(n)}$, and the *median*, which in the case of n being odd is $x_{((n+1)/2)}$ and in case of n being even is the arithmetic mean of $x_{(n/2)}$ and $x_{(n/2+1)}$. Like the sample mean, the median is a measure of centrality. It is often preferable due to the fact that it isn't influenced by very high or very low measurements.

Related statistics are the *α-quantile*, for $\alpha \in [0,1]$ which is effectively $x_{(\widetilde{\alpha n})}$, where $\widetilde{\alpha n}$ denotes a rounding of αn to the nearest element of $\{1,\dots,n\}$, or alternatively an interpolation similar to the case of the median with n even. For $\alpha = 0.25$ and $\alpha = 0.75$, these values are known as the *first quartile* and *third quartile* respectively. Finally, the *inter quartile range (IQR)* is the difference between these two quartiles and the *range* is $x_{(n)} - x_{(1)}$.

The range and the IQR are *measures of dispersion*, meaning that the greater their magnitude, the more spread in the data. When dealing with measures of dispersion, the most popular and useful measure is the *sample variance*,

$$s^2 = \frac{\sum_{i=1}^{n}(x_i - \overline{x})^2}{n-1} = \frac{\sum_{i=1}^{n} x_i^2 - n\,\overline{x}^2}{n-1}.$$

The sample variance is approximately the arithmetic mean of squared deviations from the sample mean, but isn't exactly because we divide by $n-1$ and not n (the latter is sometimes called *population variance*). If all observations are constant then $s^2 = 0$, otherwise, $s^2 > 0$, and the bigger it is, the more dispersion we have in the data. A related quantity is the *sample standard deviation* s, where $s := \sqrt{s^2}$. Also of interest is the *standard error* $s/\sqrt{n}$. Variances, standard deviations, and standard errors play a major role in the chapters that follow.

In Julia, functions for these descriptive statistics are implemented in the built-in `Statistics` package, with some additional functionality in the `StatsBase` package. Listing 4.11 illustrates their usage.

Listing 4.11: Summary statistics

```
using CSV, Statistics, StatsBase
data = CSV.read("../data/temperatures.csv")[:,4]

println("Sample Mean: ", mean(data))
println("Harmonic <= Geometric <= Arithmetic ",
        (harmmean(data), geomean(data), mean(data)))
println("Sample Variance: ",var(data))
println("Sample Standard Deviation: ",std(data))
println("Minimum: ", minimum(data))
println("Maximum: ", maximum(data))
println("Median: ", median(data))
println("95th percentile: ", percentile(data, 95))
println("0.95 quantile: ", quantile(data, 0.95))
println("Interquartile range: ", iqr(data),"\n")

summarystats(data)
```

```
Sample Mean: 27.1554054054054
Harmonic <= Geometric <= Arithmetic (26.52, 26.84, 27.155)
Sample Variance: 16.12538955837281
Sample Standard Deviation: 4.015643106449178
Minimum: 16.1
Maximum: 37.6
```

```
Median: 27.7
95th percentile: 33.0
0.95 quantile: 33.0
Interquartile range: 6.100000000000001

Summary Stats:
Length:         777
Missing Count:  0
Mean:           27.155405
Minimum:        16.100000
1st Quartile:   24.000000
Median:         27.700000
3rd Quartile:   30.100000
Maximum:        37.600000
```

In line 2, we load the data and select the fourth column. This sets `data` to be an array of `Float64`. In line 4, we compute and print the sample mean using `mean()`. We then compare it to the harmonic mean and geometric mean, computed via `harmmean()` and `geomean()`, respectively. In line 7, we compute the sample variance using `var()` and, then in line 8, the sample standard deviation via `std()`. In lines 9–14, we compute different statistics associated with order statistics including the min, max, median, and quartiles. Finally, in line 16, we use the `summarystats()` function which yields similar output.

Observations in Pairs

When data is configured in the form of pairs, $(x_1, y_1), \dots, (x_n, y_n)$, we often consider the *sample covariance*, which is given by,

$$\widehat{\text{cov}}_{x,y} = \frac{\sum_{i=1}^{n} (x_i - \overline{x})(y_i - \overline{y})}{n-1}, \tag{4.1}$$

where $\overline{x}$ and $\overline{y}$ are the sample means of $(x_1, \dots, x_n)$ and $(y_1, \dots, y_n)$, respectively. A positive covariance indicates a *positive linear relationship* meaning that when x is larger than its mean, we expect y to be larger than its mean, and similarly when x is small, y is small. A negative covariance indicates a *negative linear relationship* meaning that when x is large then y is small, and when x is small then y is large. If the covariance is 0 or near 0, it is an indication that no such relationship holds.

However, like the variance, the covariance is not a normalized quantity and hence depends on the units of measurement. For example, say, we were measuring x and y using kilograms and meters, respectively, and obtain $\widehat{\text{cov}}_{x,y} = 0.003$. Assume that we then decided to modify the data and represent x in grams by multiplying the original x values by 1000. From (4.1), you can observe that the covariance would change to $\widehat{\text{cov}}_{x,y} = 3$. If one was to naively interpret these numbers, in the first case it may appear that there is almost no positive linear relationship, while in the second case, it appears that a positive linear relationship holds. However, nothing changed in the data except for the units of measurement and any relationship existing in the first dataset (kilograms vs. meters) should also exist in the modified one (grams vs. meters).

For this reason, we define another useful statistic, the *sample correlation (coefficient)*:

$$\hat{\rho}_{x,y} := \frac{\widehat{\text{cov}}_{x,y}}{s_x\, s_y},$$

where s_x and s_y are the sample standard deviations of the samples $x_1, \ldots, x_n$ and $y_1, \ldots, y_n$, respectively. Using the *Cauchy-Schwartz* inequality, we can show that $\hat{\rho}_{x,y} \in [-1, 1]$. The sign of $\hat{\rho}_{x,y}$ agrees with the sign of $\widehat{\text{cov}}_{x,y}$, however importantly its magnitude is meaningful. Having $|\hat{\rho}_{x,y}|$ near 0 implies little or no linear relationship, while $|\hat{\rho}_{x,y}|$ closer to 1 implies a stronger linear relationship, which is either positive or negative depending on the sign of $\hat{\rho}_{x,y}$. Also note that that if $(x_1, \ldots, x_n) = (y_1, \ldots, y_n)$ then the sample covariance is simply the sample variance. That is $\widehat{\text{cov}}_{x,x} = s_x^2$.

It is often useful to represent the variances and covariances in the *sample covariance matrix* as follows:

$$\hat{\Sigma} = \begin{bmatrix} \widehat{\text{cov}}_{x,x} & \widehat{\text{cov}}_{x,y} \\ \widehat{\text{cov}}_{x,y} & \widehat{\text{cov}}_{y,y} \end{bmatrix} = \begin{bmatrix} s_x^2 & \hat{\rho}_{x,y}\, s_x\, s_y \\ \hat{\rho}_{x,y}\, s_x\, s_y & s_y^2 \end{bmatrix}. \tag{4.2}$$

In Listing 4.12, we import a weather observation dataset containing pairs of temperature observations (see Section 3.7). We then estimate the elements of the covariance matrix and then store the results in the file `mvParams.jl`. Note that this file is used as input to Listing 3.34 at the end of Chapter 3.

Listing 4.12: Estimating elements of a covariance matrix

```
using DataFrames, CSV, Statistics

data = CSV.read("../data/temperatures.csv", copycols=true)
brisT = data.Brisbane
gcT = data.GoldCoast

sigB = std(brisT)
sigG = std(gcT)
covBG = cov(brisT, gcT)

meanVect = [mean(brisT) , mean(gcT)]
covMat = [sigB^2  covBG
          covBG   sigG^2]

outfile = open("../data/mvParams.jl","w")
write(outfile,"meanVect = $meanVect \ncovMat = $covMat")
close(outfile)
print(read("../data/mvParams.jl", String))
```

```
meanVect = [27.1554, 26.1638]
covMat = [16.1254 13.047; 13.047 12.3673]
```

In lines 3–5, we import the data and store the temperatures for Brisbane and Gold Coast as the arrays `brisT` and `gcT` respectively. In lines 7–8, the standard deviations of our temperature observations are calculated, and in line 9 the `cov()` function is used to estimate the covariance. In line 11, the means of our temperatures are calculated and stored as the array `meanVect`. In lines 12–13, the covariance matrix is calculated and assigned to the variable `covMat`. In lines 15–17, we save `meanVect` and `covMat` to the new Julia file, `mvParams.jl`. Note that this file is used as input for our calculations in Listing 3.34. First, in line 15, the `open()` function is used (with the argument `w`) to create the file `mvParams.jl` in write mode. Note that `open()` creates an input-output stream, `outfile`, which can then be written to. Then in line 16 `write()` is used to write to the input-output stream `outfile`. In line 17, the input-output stream `outfile` is closed. In line 18, the content of the file `mvParams.jl` is printed via the `read()` and `print()` functions.

Observations in Vectors

We now consider data that consists of n vectors. The i'th data vector represents a tuple of values, $(x_{i1}, \ldots, x_{ip})$. In this case, the data can be represented by a $n \times p$ *data matrix*, X, where the rows are observations (data vectors) and each column represents a different *variable*, *feature*, or *attribute*. Such a layout is natural if considering the data as part of a data frame, see Section 4.1. However, in other cases, you may see this data matrix transposed such that each observation is a column vector of features.

In summarizing the data matrix X, a few basic objects arise. These include the *sample mean vector*, *sample standard deviation vector*, *sample covariance matrix*, and the *sample correlation matrix*. We now describe these.

The sample mean vector is simply a vector of length p where the j'th entry, $\overline{x}_j$ is the sample mean of $(x_{1j}, \ldots, x_{nj})$, based on the j'th column of X. Similarly, the sample standard deviation vector has a j'th entry, s_j, which is the sample standard deviation of $(x_{1j}, \ldots, x_{nj})$.

With these we often *standardize* (also called *normalize*) the data by creating a new $n \times p$ matrix Z, with entries,

$$z_{ij} = \frac{x_{ij} - \overline{x}_j}{s_j}, \quad i = 1, \ldots, n, \quad j = 1, \ldots, p, \tag{4.3}$$

sometimes called *z-scores*. It can be created via, $Z = (X - \mathbf{1}\overline{x}^\top)\,\text{diag}(s)^{-1}$, where $\mathbf{1}$ is a column vector of 1's, $\overline{x}$ is the mean vector, s is the standard deviation vector, and diag$(\cdot)$ creates a diagonal matrix from a vector. The standardized data has the attribute that each column, $(z_{1j}, \ldots, z_{nj})^\top$, has a 0 sample mean and a unit standard deviation. Hence, first- and second-order information of the j'th feature is lost when moving from the data matrix X to the standardized matrix Z. Nevertheless, relationships between features are still captured in Z and can be easily calculated. Most notably, the sample correlation between feature j_1 and feature j_2, denoted by $\hat{\rho}_{j_1,j_2}$ is simply calculated via

$$\hat{\rho}_{j_1 j_2} = \frac{\sum_{i=1}^{n} z_{ij_1} z_{ij_2}}{n-1} = \Big[\frac{1}{n-1} Z^\top Z\Big]_{j_1 j_2}. \tag{4.4}$$

Here, the second expression means taking the j_1, j_2 entry from the $p \times p$ sample correlation matrix $Z^\top Z/(n-1)$. In Julia this can be performed via the `cor()` function.

Without resorting to standardization, it is often of interest to calculate the $p \times p$ *sample covariance matrix* $\hat{\Sigma}$ generalizing (4.2). Here, the j_1, j_2 entry is the covariance between the j_1 and j_2 variables. The matrix can be computed in several ways. For example,

$$\hat{\Sigma} = \frac{1}{n-1}(X - \mathbf{1}\overline{x}^\top)^\top (X - \mathbf{1}\overline{x}^\top) = \frac{1}{n-1} X^\top (I - n^{-1}\mathbf{1}\mathbf{1}^\top) X. \tag{4.5}$$

In the second expression, I is the identity matrix and as before $\mathbf{1}$ is a vector of 1's and hence $\mathbf{1}\mathbf{1}^\top$ is a matrix of 1's. Note that $(I - n^{-1}\mathbf{1}\mathbf{1}^\top)X$ is the *de-meaned* data. Also note that the symmetric matrix $(I - n^{-1}\mathbf{1}\mathbf{1}^\top)$ multiplied by itself is itself. In Julia, this calculation can be performed via the `cov()` function. We now illustrate several alternative ways for computing the sample covariance and sample correlation in Listing 4.13 below. In addition to the `cov()`, `cor()`, `mean()`, and `std()` functions, the listing also illustrates the use of the `zscore()` function from `StatsBase`.

Listing 4.13: Sample covariance

```
using Statistics, StatsBase, LinearAlgebra, DataFrames, CSV
df = CSV.read("../data/3featureData.csv",header=false)
n, p = size(df)
println("Number of features: ", p)
println("Number of observations: ", n)
X = convert(Array{Float64,2},df)
println("Dimensions of data matrix: ", size(X))

xbarA = (1/n)*X'*ones(n)
xbarB = [mean(X[:,i]) for i in 1:p]
xbarC = sum(X,dims=1)/n
println("\nAlternative calculations of (sample) mean vector: ")
@show(xbarA), @show(xbarB), @show(xbarC)

Y = (I-ones(n,n)/n)*X
println("Y is the de-meaned data: ", mean(Y,dims=1))

covA = (X .- xbarA')'*(X .- xbarA')/(n-1)
covB = Y'*Y/(n-1)
covC = [cov(X[:,i], X[:,j]) for i in 1:p, j in 1:p]
covD = [cor(X[:,i], X[:,j])*std(X[:,i])*std(X[:,j]) for i in 1:p, j in 1:p]
covE = cov(X)
println("\nAlternative calculations of (sample) covariance matrix: ")
@show(covA), @show(covB), @show(covC), @show(covD), @show(covE)

ZmatA = [(X[i,j] - mean(X[:,j]))/std(X[:,j]) for i in 1:n, j in 1:p ]
ZmatB = hcat([zscore(X[:,j]) for j in 1:p]...)
println("\nAlternate computation of Z-scores yields same matrix: ",
        maximum(norm(ZmatA-ZmatB)))
Z = ZmatA

corA = covA ./ [std(X[:,i])*std(X[:,j]) for i in 1:p, j in 1:p]
corB = covA ./ (std(X,dims = 1)'*std(X,dims = 1))
corC = [cor(X[:,i],X[:,j]) for i in 1:p, j in 1:p]
corD = Z'*Z/(n-1)
corE = cov(Z)
corF = cor(X)
println("\nAlternative calculations of (sample) correlation matrix: ")
@show(corA), @show(corB), @show(corC), @show(corD), @show(corE), @show(corF);
```

```
Number of features: 3
Number of observations: 7
Dimensions of data matrix: (7, 3)

Alternative calculations of (sample) mean vector:
xbarA = [1.05714, 2.08571, 3.5]
xbarB = [1.05714, 2.08571, 3.5]
xbarC = [1.05714, 2.08571, 3.5]
Y is the de-meaned data: [6.74064e-17 3.24889e-16 2.85486e-16]

Alternative calculations of (sample) covariance matrix:
covA = [0.119524 -0.087381 0.44; -0.087381 0.121429 -0.715; 0.44 -0.715 8.03333]
covB = [0.119524 -0.087381 0.44; -0.087381 0.121429 -0.715; 0.44 -0.715 8.03333]
covC = [0.119524 -0.087381 0.44; -0.087381 0.121429 -0.715; 0.44 -0.715 8.03333]
covD = [0.119524 -0.087381 0.44; -0.087381 0.121429 -0.715; 0.44 -0.715 8.03333]
covE = [0.119524 -0.087381 0.44; -0.087381 0.121429 -0.715; 0.44 -0.715 8.03333]

Alternate computation of Z-scores yields same matrix: 2.220446049250313e-16

Alternative calculations of (sample) correlation matrix:
corA = [1.0 -0.725319 0.449032; -0.725319 1.0 -0.723932; 0.449032 -0.723932 1.0]
corB = [1.0 -0.725319 0.449032; -0.725319 1.0 -0.723932; 0.449032 -0.723932 1.0]
corC = [1.0 -0.725319 0.449032; -0.725319 1.0 -0.723932; 0.449032 -0.723932 1.0]
corD = [1.0 -0.725319 0.449032; -0.725319 1.0 -0.723932; 0.449032 -0.723932 1.0]
corE = [1.0 -0.725319 0.449032; -0.725319 1.0 -0.723932; 0.449032 -0.723932 1.0]
corF = [1.0 -0.725319 0.449032; -0.725319 1.0 -0.723932; 0.449032 -0.723932 1.0]
```

In line 2, we read the data with `header = false` since there isn't a line in the csv for the variable (or feature) names. In line 3, we use the `size()` function to set the number of observations, `n`, and number of features `p`. The `convert()` function is used in line 6 to extract a data matrix out of the data frame `df`. Lines 9–13 show alternative ways of computing the sample mean vector. Note the use of `dims=1` in the `sum()` function in line 11, indicating to sum over columns. In line 15, we create the de-meaned data Y and show in line 16 that the mean is 0 (effectively 0 in the output). Lines 18–24 illustrate a variety of ways to calculate the sample covariance matrix using several forms of (4.5). Lines 26–30 deal with standardized data as in (4.3). The printout of the maximum of the norm in line 29 is a way for seeing that the two matrices `ZmatA` and `ZmatB` are identical. Finally, lines 32–39 compute the correlation matrix in a variety of ways. Observe line 35 implementing (4.4). Also observe that the covariance of Z is the correlation of X as shown in lines 36–37.

4.3 Plots for Single Samples and Time Series

In this section, we deal with plots focused on a single collection of observations (numbers), $x_1, \ldots, x_n$, where in certain cases we plot several such single collections jointly for comparison. If the observations are obtained by randomly sampling a population, then the order of the observations is inconsequential. In this case, we say the data is a single sample and in general, plotting the observations one after the other isn't particularly useful. However, if the observations represent measurement over time then we call the dataset a *time-series*, and in this case, plotting the observations one after the other is the standard way for considering temporal patterns in the data.

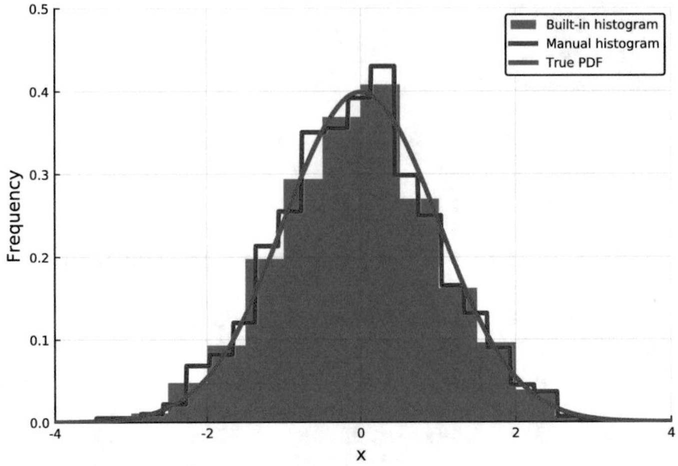

Figure 4.1: A manually created histogram with the same number of bins as `histogram()`. Both are compared to original PDF.

Histograms

In both the single sample and time-series case, considering frequencies of occurrences is generally an insightful way to visualize the data. The most standard mechanism for plotting frequencies is the *histogram*, already used extensively in previous chapters (see for example, Listing 1.11). Mathematically, a histogram can be defined as follows. First denote the support of the observations via $[\ell, m]$, where ℓ is the minimal observation and m is the maximal observation. Then the interval $[\ell, m]$ is partitioned into a finite set of *bins* $\mathcal{B}_1, \ldots, \mathcal{B}_L$, and the frequency in each bin is recorded via

$$f_j = \frac{1}{n} \sum_{i=1}^{n} \mathbf{1}\{x_i \in \mathcal{B}_j\}, \qquad \text{for} \quad j = 1, \ldots, L. \tag{4.6}$$

Here $\mathbf{1}\{\cdot\}$ is 1 if $x_i \in \mathcal{B}_j$ or 0 if $x_i \notin \mathcal{B}_j$. We have that $\sum f_j = 1$ and hence $f_1, \ldots, f_L$ is a discrete probability distribution.

A histogram is then just a visual representation of this discrete probability distribution (the frequencies). One way to plot the frequencies is via a *stem plot* (see for example, Figure 3.10 illustrating a binomial distribution). However, such a plot would not represent the widths of the bins. Hence, an alternative representation is via a *histogram function* $h(x)$ which is a scaled plot of the frequencies $f_1, \ldots, f_L$. The function $h(x)$ is defined for any $x \in [\ell, m]$. It is constructed by staying constant on all values $x \in \mathcal{B}_j$, at a height of $f_j/|\mathcal{B}_j|$, where $|\mathcal{B}_j|$ is the width of bin j. This ensures the total area under the plot is 1. Hence $h(x)$ is actually a probability density function, and can hence be compared to probability densities. Mathematically,

$$h(x) = \sum_{j=1}^{L} \frac{f_j}{|\mathcal{B}_j|} \mathbf{1}\{x \in \mathcal{B}_j\} = \frac{f_{b(x)}}{|\mathcal{B}_{b(x)}|}, \tag{4.7}$$

where $b(x)$ is the bin of x, that is $x \in \mathcal{B}_{b(x)}$.

Note that there are a multitude of ways for choosing the number of bins L and the actual bins $\mathcal{B}_1, \ldots, \mathcal{B}_L$. Different histogram implementations will use different *bin selection heuristics*. We don't discuss these methods here. Throughout this book, we use the `histogram()` function from `Plots` and it contains a default bin selection heuristic. In certain cases, we specify the number of bins using the keyword argument, `bins`, often making a judicious choice based on the appearance of a specific dataset. It is important to keep in mind that histograms are clearly not unique representations of the data because there is not a unique way to choose the bins.

For demonstration purposes, we implement a manual histogram in Listing 4.14 and compare it to `histogram()` from plots. The results are in Figure 4.1. The demonstration illustrates that a histogram is not a unique representation of the data. Both the manually created histogram and the built-in histogram have `L` bins, however, a different choice of actual bins creates different histograms. Note that if in line 21, `L` is replaced with `first.(bins)`, then the two histograms will agree because they use the exact same bins. Also note that our implementation is not an efficient one, but rather aims to illustrate the use of the above equations directly. A related classic plot that we do not survey here is the *stem and leaf plot.*

Listing 4.14: Creating a manual histogram

```
1  using Plots, Distributions, Random; pyplot()
2  Random.seed!(0)
3
4  n = 2000
5  data = rand(Normal(),n)
6  l, m = minimum(data), maximum(data)
7
8  delta = 0.3;
9  bins = [(x,x+delta) for x in l:delta:m-delta]
10 if last(bins)[2] < m
11     push!(bins,(last(bins)[2],m))
12 end
13 L = length(bins)
14
15 inBin(x,j) = first(bins[j]) <= x && x < last(bins[j])
16 sizeBin(j) = last(bins[j]) - first(bins[j])
17 f(j) = sum([inBin(x,j) for x in data])/n
18 h(x) = sum([f(j)/sizeBin(j) * inBin(x,j) for j in 1:L])
19
20 xGrid = -4:0.01:4
21 histogram(data,normed=true, bins=L,
22     label="Built-in histogram",
23     c=:blue, la=0, alpha=0.6)
24 plot!(xGrid,h.(xGrid), lw=3, c=:red, label="Manual histogram",
25     xlabel="x",ylabel="Frequency")
26 plot!(xGrid,pdf.(Normal(),xGrid),label="True PDF",
27     lw=3, c=:green, xlims=(-4,4), ylims=(0,0.5))
```

In lines 4–6, we deal with the data. It is artificially sampled from a standard normal distribution. Lines 8–13 detail our choice of bins. In this case, `L` is implicitly defined based on the bin width `delta`. The statement in line 11 is executed when `l-m` is not a multiple of `delta` and adds an additional final bin (potentially smaller than the rest of the bins). The function `inBin()` implements $\mathbf{1}\{x \in \mathcal{B}_j\}$. The function `sizeBin()` implements $|\mathcal{B}_j|$. The function `f()` implements f_j as in (4.6). We then use these in line 18 to implement $h(x)$ as in the first representation of (4.7). Lines 21–23 plot the histogram using `histogram()` where we specify `L` bins. Lines 24–25 plot our manual implementation of the histogram via `h()`. For comparison, we also plot the PDF of the data in lines 26–27.

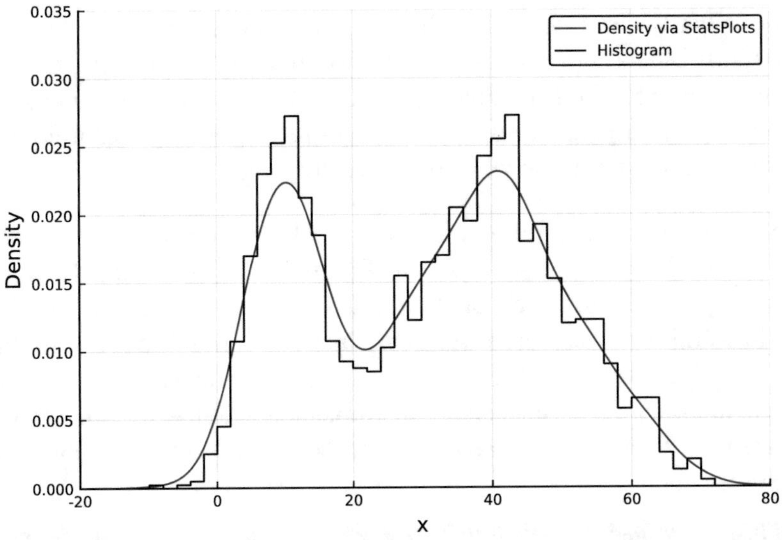

Figure 4.2: Histogram of the underlying data, and the KDE, as generated from the `density()` function in `StatsPlots`.

Density Plots and Kernel Density Estimation

A more modern and visually appealing alternative to histograms is the *smoothed histogram*, also known as a *density plot*, often generated via a *kernel density estimate*. Before we describe and detail kernel density estimation, let's see how to use it to create a smoothed histogram as in Figure 4.2. The figure is generated by Listing 4.15, using the `density()` function from the `StatsPlots` package. The plot is compared to a histogram. Its usage is similar to the `histogram()` or `stephist()` functions from `Plots`, however the result is strikingly different, yielding a smooth curve.

This example and the next are based on synthetic data from a *mixture model*. Such models are useful for situations where we sample from populations made up of heterogeneous sub-populations. Each sub-population has its own probability distribution and these are "mixed" in the process of sampling. At first a *latent* (un-observed) random variable determines which sub-population is used, and then a sample is taken from that sub-population. In terms of random variable generation, creating a mixture simply involves first randomly selecting which probability distribution is used, and then generating an observation from it. Also, the probability density function of the mixture is a convex combination of the probability density functions of each of the sub-populations. That is, if the M sub-populations have densities $g_1(x), \ldots, g_M(x)$ with weights, $p_1, \ldots, p_M$ and $\sum p_i = 1$, then the density of the mixture is

$$f(x) = \sum_{i=1}^{M} p_i \, g_i(x).$$

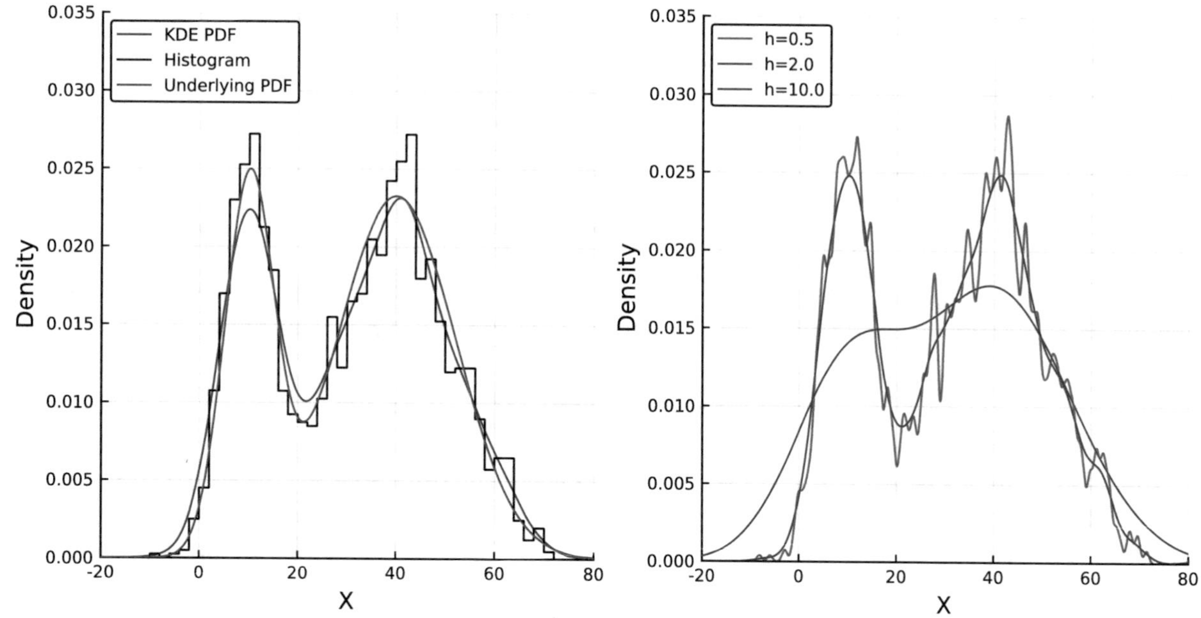

Figure 4.3: Left: KDE compared to the actual underlying PDF as well as a histogram. Right: KDE obtained via several bandwidths settings.

Listing 4.15: Classic vs. smooth histograms

```
1  using Random, Distributions, StatsPlots; pyplot()
2  Random.seed!(0)
3
4  mu1, sigma1 = 10, 5
5  mu2, sigma2 = 40, 12
6  dist1, dist2 = Normal(mu1,sigma1), Normal(mu2,sigma2)
7  p = 0.3
8  mixRv() = (rand() <= p) ? rand(dist1) : rand(dist2)
9
10 n = 2000
11 data = [mixRv() for _ in 1:n]
12
13 density(data, c=:blue, label="Density via StatsPlots",
14           xlims=(-20,80), ylims=(0,0.035))
15 stephist!(data, bins=50, c=:black, norm=true,
16      label="Histogram", xlabel="x", ylabel = "Density")
```

Lines 4–8 deal with the mixture random variable. It is a mixture of two normal distributions, each with parameters as specified in lines 4–5. The mixture places a probability of `p = 0.3` of being from the first distribution and hence a probability of 0.7 of being from the second. Line 8 defines the function that generates the mixture random variable. It evaluates to `rand(dist1)` with probability 0.3 and `rand(dist2)` with probability 0.7. Lines 10–11 generate data samples from this mixture distribution. Lines 13–14 create the density plot. Lines 15–16 plot a histogram for comparison.

How is a density plot created? The typical way is via *kernel density estimation* (*KDE*), which is a way of fitting a probability density function to data. When we used the `density()` function from `StatsPlots` above, KDE was implicitly invoked. However, in certain cases, we may wish to have access to the estimated density. For this we use the `kde()` function from the `KernelDensity` package. Let us first explain how kernel density estimation works.

Given a set of observations, $x_1, \ldots, x_n$, the KDE is the function,

$$\hat{f}(x) = \frac{1}{n} \sum_{i=1}^{n} \frac{1}{h} K\Big(\frac{x - x_i}{h}\Big), \tag{4.8}$$

where $K(\cdot)$ is some specified *kernel function* and $h > 0$ is the *bandwidth* parameter. The kernel function is a function that satisfies the properties of a PDF. A typical example is the Gaussian kernel.

$$K(x) = \frac{1}{\sqrt{2\pi}} e^{-x^2/2}.$$

With such a kernel (or any other) the estimator $\hat{f}(x)$ is a PDF because it is a weighted superposition of scaled kernel functions centered about each of the observations. Like histograms, KDEs are not unique as they depend on the type of kernel function used and more importantly on the bandwidth parameter. A very small bandwidth implies that the density,

$$\frac{1}{h} K\Big(\frac{x - x_i}{h}\Big), \tag{4.9}$$

is very concentrated around x_i. This in turn implies that the KDE (4.8) is comprised of a superposition of very concentrated functions, one for each observation. In contrast, a very large bandwidth implies that the density (4.9) around each observation has a very wide spread. This will make the KDE "smear" over a wide range. Hence ideally, the bandwidth is not too small nor too large. The right hand plot of Figure 4.3 illustrates KDE with different choices of bandwidth. As can be seen when $h = 0.5$ the KDE appears to have multiple spikes. At the other extreme, when $h = 10$ the KDE is very "smeared".

For any value of h, it can be proved under general conditions that if the data is distributed according to some density $f(\cdot)$, then $\hat{f}(\cdot)$ converges to $f(\cdot)$ when the sample size grows. Nevertheless, in practice, choosing the bandwidth is a key issue in the application of KDE. A default classic rule is *Silverman's rule*, which is based on the sample standard deviation of the sample, s. The rule is

$$h = \Big(\frac{4}{3}\Big)^{1/6} s\, n^{-1/5} \approx 1.06\, s\, n^{-1/5}.$$

There is some theory justifying this h in certain cases, and in other cases, more advanced rules such as that in [BGK10] perform better. Listing 4.16 carries out KDE for the same synthetic data as the previous example. It generates Figure 4.3 where the left plot compares the KDE to the underlying PDF of the mixture and the right plot presents the effect of changing the bandwidths.

Listing 4.16: Kernel density estimation

```
1  using Random, Distributions, KernelDensity, Plots; pyplot()
2  Random.seed!(0)
3
4  mu1, sigma1 = 10, 5
5  mu2, sigma2 = 40, 12
6  dist1, dist2 = Normal(mu1,sigma1), Normal(mu2,sigma2)
7  p = 0.3
8  mixRv() = (rand() <= p) ? rand(dist1) : rand(dist2)
9  mixPDF(x) = p*pdf(dist1,x) + (1-p)*pdf(dist2,x)
10
11 n = 2000
12 data = [mixRv() for _ in 1:n]
13
14 kdeDist = kde(data)
15
16 xGrid = -20:0.1:80
17 pdfKDE = pdf(kdeDist,xGrid)
18
19 plot(xGrid, pdfKDE, c=:blue, label="KDE PDF")
20 stephist!(data, bins=50, c=:black, normed=:true, label="Histogram")
21 p1 = plot!(xGrid, mixPDF.(xGrid), c=:red, label="Underlying PDF",
22     xlims=(-20,80), ylims=(0,0.035), legend=:topleft,
23     xlabel="X", ylabel = "Density")
24
25 hVals = [0.5,2,10]
26 kdeS = [kde(data,bandwidth=h) for h in hVals]
27 plot(xGrid, pdf(kdeS[1],xGrid), c = :green, label= "h=$(hVals[1])")
28 plot!(xGrid, pdf(kdeS[2],xGrid), c = :blue, label= "h=$(hVals[2])")
29 p2 = plot!(xGrid, pdf(kdeS[3],xGrid), c = :purple, label= "h=$(hVals[3])",
30     xlims=(-20,80), ylims=(0,0.035), legend=:topleft,
31     xlabel="X", ylabel = "Density")
32 plot(p1,p2,size = (800,400))
```

The first 12 lines are similar to the previous code example with an exception of line 9 that defines the function `mixPDF()` which is the PDF of the mixture distribution. In line 14, we invoke the function `kde()` to generate a KDE-type object `kdeDist`, based on `data`. The `KernelDensity` package supplies methods for the `pdf()` function that can be applied to `UnivariateKDE` objects such as `kdeDist`. This is used in line 17 to create the array `pdfKDE` over `xGrid`. Lines 19–23 plot the KDE, a histogram of the data, and the actual PDF. These plots make up `p1` which is the left hand of the figure. The right hand side `p2` is created in lines 25–32.

Empirical Cumulative Distribution Function

While KDE is a useful way to estimate the PDF of the unknown underlying distribution given some sample data, the *Empirical Cumulative Distribution Function* (ECDF) may be viewed as an estimate of the underlying CDF. In contrast to histograms and KDEs, ECDFs provide a unique representation of the data not dependent on tuning parameters, such as the bins for histograms, or the bandwidth and kernel function for KDE.

The ECDF is a stepped function which, given n data points, increases by $1/n$ at each point.

Mathematically, given the sample, $x_1, \dots, x_n$ the ECDF is given by

$$\hat{F}(t) = \frac{1}{n}\sum_{i=1}^{n} \mathbf{1}\{x_i \leqslant t\} \qquad \text{where } \mathbf{1}\{\cdot\} \text{ is the indicator function.}$$

In the case of i.i.d. data from an underlying distribution with CDF $F(\cdot)$, the *Glivenko-Cantelli theorem* ensures that the ECDF $\hat{F}(\cdot)$ approaches $F(\cdot)$ as the sample size grows.

Constructing an ECDF is possible in Julia through the `ecdf()` function from the `StatsBase` package. In Listing 4.17, we use synthetic data from the same mixture distribution as in the two previous examples. We obtain the ECDF for a sample of size $n = 30$ and then again for $n = 100$. We compare the ECDFs to the underlying actual CDF of the mixture distribution. See Figure 4.4.

Listing 4.17: Empirical cumulative distribution function

```
1   using Random, Distributions, StatsBase, Plots; pyplot()
2   Random.seed!(0)
3
4   mu1, sigma1 = 10, 5
5   mu2, sigma2 = 40, 12
6   dist1, dist2 = Normal(mu1,sigma1), Normal(mu2,sigma2)
7   p = 0.3
8   mixRv() = (rand() <= p) ? rand(dist1) : rand(dist2)
9   mixCDF(x) = p*cdf(dist1,x) + (1-p)*cdf(dist2,x)
10
11  n = [30, 100]
12  data1 = [mixRv() for _ in 1:n[1]]
13  data2 = [mixRv() for _ in 1:n[2]]
14
15  empiricalCDF1 = ecdf(data1)
16  empiricalCDF2 = ecdf(data2)
17
18  xGrid = -10:0.1:80
19  plot(xGrid,empiricalCDF1.(xGrid), c=:blue, label="ECDF with n = $(n[1])")
20  plot!(xGrid,empiricalCDF2.(xGrid), c=:red, label="ECDF with n = $(n[2])")
21  plot!(xGrid, mixCDF.(xGrid), c=:black, label="Underlying CDF",
22      xlims=(-10,80), ylims=(0,1),
23      xlabel="x", ylabel="Probability", legend=:topleft)
```

The first few lines of the code block are similar to the previous examples using a mixture distribution. A difference is that, in line 9, we define the function `mixCDF()` which is the CDF of the mixture distribution. We then generate two samples in lines 12–13, of varying sample sizes. In lines 15–16, we invoke the `ecdf()` function from `StatsBase`. The returned object can then be used as a function, evaluating $\hat{F}(\cdot)$ at any point. This is done in lines 19–20 where we plot the ECDFs evaluated on `xGrid`. Then lines 21–23 plot the actual CDF.

Normal Probability Plot

We now introduce the *normal probability plot*. This plot can be used to indicate if it is likely that a dataset has come from a population that is normally distributed or not. It works by plotting the quantiles of the dataset in question against the theoretical quantiles that one would expect if

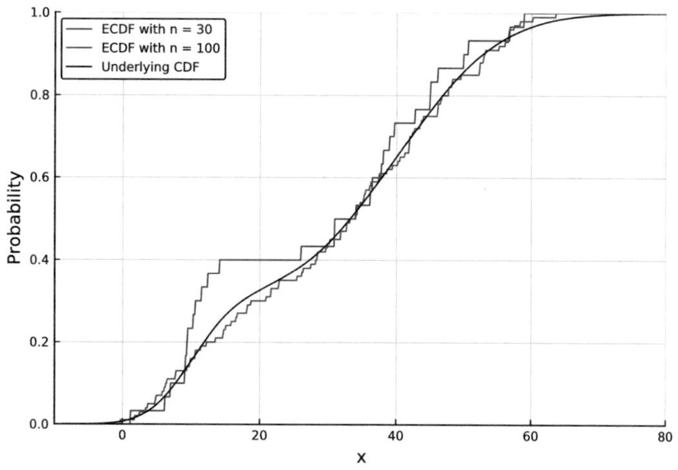

Figure 4.4: The ECDF from a sample compared to the population CDF.

the sample data came from a normal distribution, and checking if the plot is linear. The normal probability plot is actually a special case of the more generalized *Q-Q plot*, or *quantile-quantile plot* that is described in more detail in the next section.

In order to create a normal probability plot, the data points are first sorted in ascending order, $x_1, \ldots, x_n$, then the quantiles of each data point are calculated. Finally, n equally spaced quantiles of the standard normal distribution are calculated, and each ascending quantile pair is then plotted. If the data comes from a normal distribution then we can expect the normal probability plot to follow a straight line, otherwise not. An alternative view of the normal probability plot is to think of the ECDF of the data, plotted with the y-axis stretched according to the inverse of the CDF of the normal distribution.

In Listing 4.18, we create Figure 4.5 which presents two normal probability plots. It is based on two synthetic datasets that have a similar mean and a similar variance. The first comes from a normal distribution and the second from an exponential distribution. As can be seen, the "non-normality" of the data due to the exponential distribution is very apparent.

Listing 4.18: Normal probability plot

```
using Random, Distributions, StatsPlots, Plots; pyplot()
Random.seed!(0)

mu = 20
d1, d2 = Normal(mu,mu), Exponential(mu)

n = 100
data1 = rand(d1,n)
data2 = rand(d2,n)

qqnorm(data1, c=:blue, ms=3, msw=0, label="Normal Data")
qqnorm!(data2, c=:red, ms=3, msw=0, label="Exponential Data",
        xlabel="Normal Theoretical Quantiles",
        ylabel="Data Quantiles", legend=true)
```

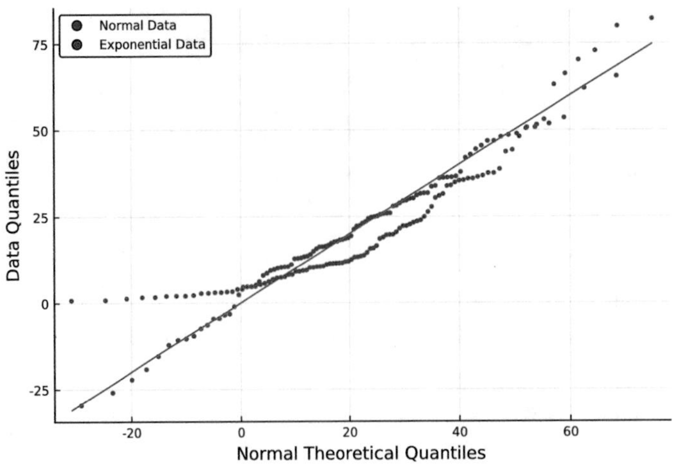

Figure 4.5: Comparing two normal probability plots. One from a normal population and one from an exponential population.

The distributions for the two synthetic datasets are defined in lines 4–5. You can check that they have the same theoretical mean and variance by using `mean()` and `var()` on `d1` and `d2`. The samples are then generated in lines 7–9. Lines 11–14 then plot the normal probability plots via the `qqnorm()` and `qqnorm!()` functions from `StatsPlots`. The second function has a `!` in the name similar to other plotting functions that add onto an existing plot.

Visualizing Time series

Moving from single sample data to time series, we now illustrate a basic example. We create time-series plots of two time series together with an associated histogram. Later, we also show how a radial plot can help visualize cyclic temporal patterns on the same data. In general, when confronted with time-series data, simply plotting a histogram of the data can be misleading because the frequencies in the histogram can be greatly affected by trends or cyclic patterns in the data. For example, the gross domestic product (GDP) of China has risen in the past 20 years from around a trillion US dollars (USD) to roughly 12.5 trillion USD. It does not make sense to plot a histogram of this data, because it would not capture the distribution of the GDP.

Nevertheless, in cases where the time-series data appears to be *stationary*, then a histogram is immediately insightful. Broadly speaking, a stationary sequence is one in which the distributional law of observations does not depend on the exact time. This means that there isn't an apparent trend nor a cyclic component. To illustrate these concepts, we present Figure 4.6. The top left plot presents two time series of temperature data in the adjacent locations of Brisbane and Gold Coast Australia. As apparent from the plot, the sequences are clearly non-stationary. This is because of seasonality. The top right plot shows a zoomed-in view of a specific fortnight. The bottom left plot is a time series of the differences in temperatures between Brisbane and Gold Coast. On initial inspection, this time series appears to be stationary. Hence, for the difference, we plot a histogram in the bottom right. The code for generating Figure 4.6 is in Listing 4.19.

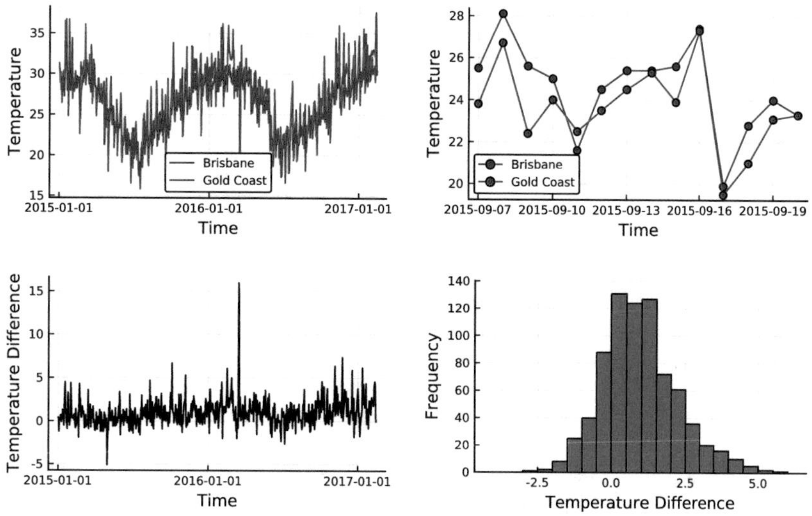

Figure 4.6: Simple plots for time-series data. Top left: Temperatures over time. Top Right: Zooming in on a specific week. Bottom left: Differences in temperature. Bottom right: Histogram of the difference in temperature.

Listing 4.19: Multiple simple plots for a time-series

```
using DataFrames, CSV, Statistics, Dates, Plots, Measures; pyplot()

data = CSV.read("../data/temperatures.csv")
brisbane = data.Brisbane
goldcoast = data.GoldCoast

diff = brisbane - goldcoast
dates = [Date(
            Year(data.Year[i]),
            Month(data.Month[i]),
            Day(data.Day[i])
        ) for i in 1:nrow(data)]

fortnightRange = 250:263
brisFortnight = brisbane[fortnightRange]
goldFortnight = goldcoast[fortnightRange]

default(xlabel="Time", ylabel="Temperature")
default(label=["Brisbane" "Gold Coast"])

p1 = plot(dates, [brisbane goldcoast],
        c=[:blue :red])
p2 = plot(dates[fortnightRange], [brisFortnight goldFortnight],
        c=[:blue :red], m=(:dot, 5, Plots.stroke(1)))
p3 = plot(dates, diff,
        c=:black, ylabel="Temperature Difference",legend=false)
p4 = histogram(diff, bins=-4:0.5:6,
        ylims=(0,140), legend = false,
        xlabel="Temperature Difference", ylabel="Frequency")
plot(p1,p2,p3,p4, size = (800,500), margin = 5mm)
```

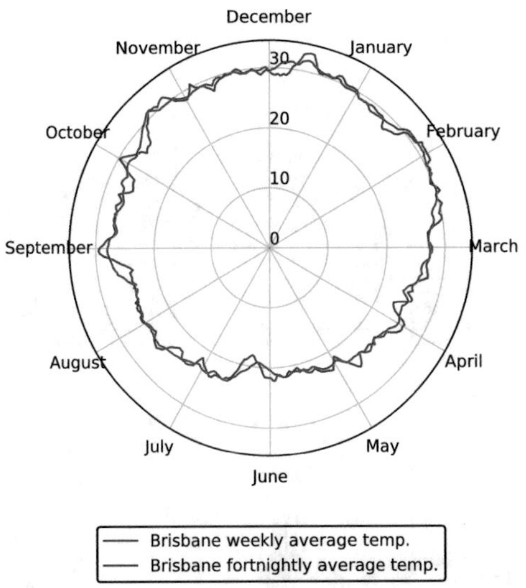

Figure 4.7: A radial plot of (time) averaged weekly and fortnightly temperatures for Brisbane in 2015.

In lines 3-5, we read the data and create the arrays `brisbane` and `goldcoast` describing the temperatures in these respective locations. In line 7, we create the array `diff` made up of temperature differences. In lines 8-12, we create the array `dates` which contains `Date` objects mapped to the days of temperature measurement. It is constructed based on the `Year`, `Month`, and `Day` columns of the data frame by using the respective functions from the `Dates` package. In line 14, we define a range of days spanning a fortnight, `fortnightRange`. This is then used to splice that fortnight from the temperature data into `brisFortnight` and `goldFortnight`. In this plotting example, we use the `default()` function from `Plots` to set some default argument for each subplot. This is in lines 18–19. We then create the plots in lines 21–30, overriding the defaults in certain cases.

Radial Plot

It is often useful to plot time-series data, or cyclic data, on a so-called *radial plot*. Such a plot involves plotting data on a polar coordinate system. See Figure 4.7. This plot can be used to help visualize the nature of a dataset by comparing the distances of each data point radially from the origin. A variation of the radial plot is the *radar plot*, which is often used to visualize the levels of different categorical variables on the one plot.

For our example of a radial plot, we use the Brisbane temperature data in 2015, similar to the data plotted in the previous listing. This time, we present the effect of different forms of *smoothing* on the time-series data. For this, we carry out a *moving average* on the data. Roughly, this transforms the original data sequence $x_1, \ldots, x_n$ into a smoother sequence $\tilde{x}_1, \ldots, \tilde{x}_n$ via

$$\tilde{x}_i = \frac{1}{L} \sum_{j=0}^{L-1} x_{i-j}. \tag{4.10}$$

Hence, each $\tilde{x}_i$ is the average of the L observations $x_{i-L+1}, \ldots, x_i$ neighboring time i in the original sequence. A critical parameter is the *window size* L which determines "how much smoothing" is to be performed. With $L = 1$ no smoothing takes place and as L is increased more smoothing takes place. There are also minor details that we don't specify here associated with shifting the smoothing window and with edge effects.

In Listing 4.20, we use the `TimeSeries` package to carry out such smoothing, comparing two window size values, $L = 7$ (weekly) and $L = 14$ (fortnightly). This package contains a variety of more advanced time-series manipulation functions. Refer to the documentation for more examples. The listing then plots the smoothed data in Figure 4.7. Since Brisbane is in the southern hemisphere, you can observe that September-March temperatures are significantly higher than the temperatures during April-August.

Listing 4.20: Radial plot

```
1  using DataFrames, CSV, Dates, StatsBase, Plots, TimeSeries; pyplot()
2
3  data = CSV.read("../data/temperatures.csv",copycols = true)
4  brisbane = data.Brisbane
5  dates = [Date(
6              Year(data.Year[i]),
7              Month(data.Month[i]),
8              Day(data.Day[i])
9          ) for i in 1:nrow(data)]
10
11 window1, window2 = 7, 14
12 d1 = values(moving(mean,TimeArray(dates,brisbane),window1))
13 d2 = values(moving(mean,TimeArray(dates,brisbane),window2))
14
15 grid = (2pi:-2pi/365:0) .+ pi/2
16 monthsNames = Dates.monthname.(dates[1:31:365])
17
18 plot(grid, d1[1:365],
19     c=:blue, proj=:polar, label="Brisbane weekly average temp.")
20 plot!(grid, d2[1:365],
21     xticks=([mod.((11pi/6:-pi/6:0) .+ pi/2,2pi) ;], monthsNames),
22     c=:red, proj=:polar,
23     label="Brisbane fortnightly average temp.", legend=:outerbottom)
```

Lines 3–9 are similar to the previous listing setting up `brisbane` as an array of temperature readings and `dates` as an array of dates. In line 11, we define `window1` and `window2` which specify the width of the moving average smoothing to be performed. Then lines 12–13 use several functions from the `TimeSeries` package to perform moving average smoothing. We first create `TimeArray` objects, we then use the `moving()` function with first argument `mean`, we then extract the values using the `values()` function. The results are in the arrays `d1` and `d2`. In line 15, we specify the polar plotting grid. Note the use of `.+ pi/2` shifting the range by 90 degrees. In line 16, we use the `monthname()` function from package `Dates` to get an array of month names for labels. The radial plots are generated in lines 18–23 using the argument `proj=:polar`. Notice the specification of `xticks` in line 21 where we broadcast the `mod()` function with second argument `2pi`. This ensures all angles are standardized to lie in the interval $[0, 2\pi]$.

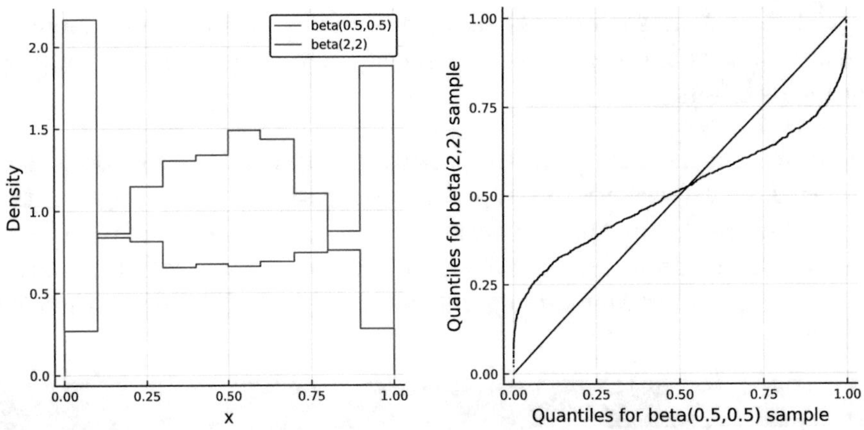

Figure 4.8: Left: Samples from two beta distributions. Right: A Q-Q plot comparing the samples.

4.4 Plots for Comparing Two or More Samples

Having covered plots for single sample data, we now introduce plots that are primarily designed for comparing two or more samples. As described at the start of this chapter, in the case of two samples, we denote the data $x_1, \ldots, x_n$ and $y_1, \ldots, y_m$.

Quantile-Quantile (Q-Q) Plot

The *Quantile-Qunatile* or *Q-Q plot* checks if the distributional shape of two samples is the same or not. For this plot, we require that the sample sizes are the same. Then the ranked quantiles of the first sample are plotted against the ranked quantile of the second sample. While mathematically a Q-Q plot is a parametric curve, in practice since sample sizes are finite, the points plotted are according to the points of the first sample—on the horizontal axis. In the case where the samples have a similar distributional shape, the resulting plot appears like a collection of increasing points along a straight line. However, in cases where the distributional shape varies, other patterns appear. Hence, Q-Q plots serve as a mechanism to compare the distributional shapes of two samples.

A different variant of Q-Q plots is when the quantiles of a single sample are plotted against the quantiles of a theoretical distribution. One such plot is the normal probability plot covered in the previous section. In general, one can create such a plot of a single sample against any theoretical distribution. Refer to the documentation of `qqplot()` from `StatsPlots` for more information. Another variant is the *Probability-Probability* or *P-P plot* where cumulative probabilities are used on the axes instead of quantiles.

In Listing 4.21, we generate Figure 4.8 which considers two synthetic samples from beta distributions. The left plot presents histograms and the right a Q-Q plot. You may modify the parameters in the code and see how this affects the plot.

Listing 4.21: Q-Q Plots

```
1  using Random, Distributions, StatsPlots, Plots, Measures; pyplot()
2  Random.seed!(0)
3
4  b1, b2 = 0.5 , 2
5  dist1, dist2, = Beta(b1,b1), Beta(b2,b2)
6
7  n = 2000
8  data1 = rand(dist1,n)
9  data2 = rand(dist2,n)
10
11 stephist(data1, bins=15, label = "beta($b1,$b1)", c = :red, normed = true)
12 p1 = stephist!(data2, bins=15, label = "beta($b2,$b2)",
13         c = :blue, xlabel="x", ylabel="Density",normed = true)
14
15 p2 = qqplot(data1, data2, c=:black, ms=1, msw =0,
16         xlabel="Quantiles for beta($b1,$b1) sample",
17         ylabel="Quantiles for beta($b2,$b2) sample",
18         legend=false)
19
20 plot(p1, p2, size=(800,400), margin = 5mm)
```

Lines 4–5 define the distributions of the synthetic data and their parameters. Lines 7–9 create the sample data. Lines 11–13 create the histograms. Lines 15–18 call the `qqplot()` function from `StatsPlots` and create the Q-Q plot.

Box Plot

The *box plot*, also known as a *box and whisker plot*, is commonly used to visually draw conclusions of, and to compare two or more single-sample datasets. It displays the first and third quartiles along with the median, i.e. the "box", along with calculated upper and lower bounds of the data, i.e. the "whiskers", hence the name. The location of the whiskers is typically given by

$$\text{minimum} = Q1 - 1.5IQR, \qquad \text{maximum} = Q3 + 1.5IQR,$$

where IQR is the inter-quartile range (see Section 4.2). Observations that lie outside this range are called *outliers*.

In Listing 4.22, we present an example of the box plot, where we compare three datasets. The files `machine1.csv`, `machine2.csv`, and `machine3.csv` represent sample measurements of the diameter of identical pipes produced by three different machines. The diameters of the pipes vary not only due to the imprecision of each machine but also potentially due to the variability between the machines. Statistical analysis of this example via ANOVA is presented in Chapter 7, Section 7.3. The listing produces Figure 4.9, and from this figure, we can visually compare the three sample populations.

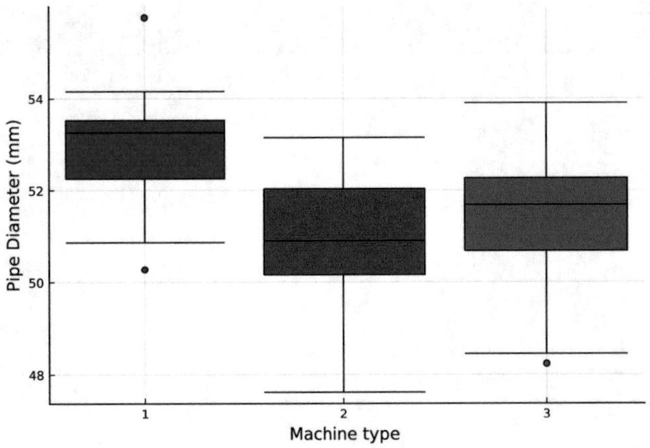

Figure 4.9: Box plots of pipe diameters associated with machines 1, 2, and 3.

Listing 4.22: Box plots of data

```
1  using CSV, StatsPlots; pyplot()
2
3  data1 = CSV.read("../data/machine1.csv", header=false)[:,1]
4  data2 = CSV.read("../data/machine2.csv", header=false)[:,1]
5  data3 = CSV.read("../data/machine3.csv", header=false)[:,1]
6
7  boxplot([data1,data2,data3], c=[:blue :red :green], label="",
8          xticks=([1:1:3;],["1", "2", "3"]), xlabel="Machine type",
9          ylabel="Pipe Diameter (mm)")
```

In lines 3-5, the data files for each of the machines are loaded and the data stored as separate arrays. In lines 7-9, the boxplot is created via the `boxplot()` function from the `StatsPlots` package.

Violin Plot

The *violin plot* is another plot that can be used to compare multiple sample populations. It is similar to the box plot, however, the shape of each sample is represented by a mirrored kernel density estimate of the data. Listing 4.23 creates an example of this plot as shown in Figure 4.10. Note this example uses the `iris` dataset from the `RDatasets` package. This dataset is further explored in the next section.

Listing 4.23: Violin plot

```
using RDatasets, StatsPlots

iris = dataset("datasets", "iris")
@df iris violin(:Species, :SepalLength,
        fill=:blue, xlabel="Species", ylabel="Sepal Length", legend=false)
```

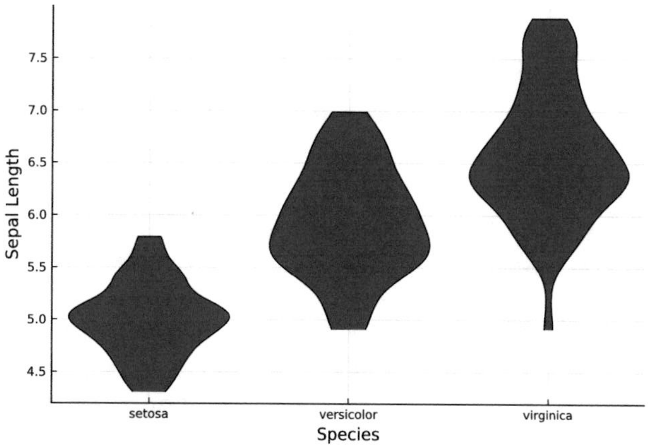

Figure 4.10: An example of a violin plot.

In line 3, the `iris` dataset from the `RDatasets` package is loaded as a `DataFrame` via the `dataset` function. The first argument, `"datasets"`, is the package in `RDatasets` which contains the `"iris"` dataset, which is the second argument. In line 4, the `@df` macro is used to plot the data from the dataframe directly, with the first argument `:Species` the horizontal axis, and the second argument `:SepalLength` the vertical axis.

4.5 Plots for Multivariate and High-Dimensional Data

We now consider vectors of observations, $(x_{11}, \ldots, x_{1p}), \ldots, (x_{n1}, \ldots, x_{np})$, where n is the number of observations and p is the number of variables, or features. In cases where p is large the data is called *high dimensional*. In such cases, analysis of the data can be both challenging and rewarding. Such analysis is the focus of Chapters 8 and 9 where we focus on linear regression and machine learning. Analysis of multivariate data often goes hand in hand with visualization. Here the natural constraint is the fact that images (or plots) are limited to lie on the two-dimensional plane, while in practice p is often much greater than 2 (denoted $p \gg 2$).

We have already explored several basic plots associated with multivariate data. These include the surface plot and heat map first introduced in Figure 1.8, the contour plot introduced in Figure 3.26, and the scatter plot first introduced in Figure 1.12. We augment these by introducing the *scatter plot matrix*, *heat map with marginals* plot, and *Andrews plot*. Also related are plots generated by PCA as that presented in Figure 9.14 in Chapter 9.

Note that with the exception of a basic animation example presented in Listing 1.12 in Chapter 1, we don't cover advanced animation methods, sound generation, interactive graphics, or 3D printing. Nevertheless, the reader should keep in mind that when properly used, all these forms of media allow one to better visualize high-dimensional data. This is still an emerging field and is bound to take on a more prominent role in the coming years. To this end, you may be interested in exploring a growing and diverse set of Julia packages including `PlotLy.jl`, `VegaLite.jl`, and others.

Scatter Plot Matrix

The basic *scatter plot* is very common in data visualization. In its simplest form, when considering pairs of observations $(x_1, y_1), \ldots, (x_n, y_n)$ it is a plot of these coordinates on the *Cartesian plain*. If in addition, the observations are labeled where each pair (x_i, y_i) has a label from a small finite set, then each point can be colored or marked with a symbol, matching the label. See for example Figure 3.25 from Chapter 3 as one of many examples of this type of plot.

When moving from pairs to higher dimensions, each observation is represented as the vector or tuple $(x_{i1}, \ldots, x_{ip})$. If $p = 3$ one may still try to illustrate a *point cloud*, however for higher dimensions this isn't possible. In this case, one of the most popular plots for visualizing relationships is the *scatter plot matrix*. It consists of taking each possible pair of variables and plotting a scatter plot for that pair. This allows one to understand relationships between pairs of variables. With p variables, there are p^2 total plots, where p of the plots are redundant because they plot a variable against itself (on the diagonal), and the other $p^2 - p$ plots each contain a duplicate of the plots (with the axis reversed). Hence, for example, if $p = 4$, there are $(p^2 - p)/2 = 6$ important plots in the scatter plot matrix even though the 4×4 matrix has 16 plots in total.

As an example, Listing 4.24 creates Figure 4.11 where we consider the `iris` dataset that consists of four measurements for each flower: "sepal width", "sepal length", "petal width", and "petal length". Hence each flower can be considered as a tuple $(x_{i1}, x_{i2}, x_{i3}, x_{i4})$. As with any scatter plot, data in scatter plot matrices can also be colored or labeled. In this example, there are three species, "setosa", "versicolor", and "virginica", and each tuple is associated with a species. The listing output also summarizes basic information about the `iris` dataset. Inspection of Figure 4.11 can yield insight and conjectures about the population of flowers.

Listing 4.24: Scatterplot matrix

```
using RDatasets, Plots, Measures; pyplot()

data = dataset("datasets", "iris")
println("Number of rows: ", nrow(data))

insertSpace(name) = begin
    i = findlast(isuppercase,name)
    name[1:i-1]*" "*name[i:end]
end

featureNames = insertSpace.(string.(names(data)))[1:4]
println("Names of features:\n\t", featureNames)

speciesNames = unique(data.Species)
speciesFreqs = [sn => sum(data.Species .== sn) for sn in speciesNames]
println("Frequency per species:\n\t", speciesFreqs)

default(msw = 0, ms = 3)

scatters = [
    scatter(data[:,i], data[:,j], c=[:blue :red :green], group=data.Species,
        xlabel=featureNames[i], ylabel=featureNames[j], legend = i==1 && j==1)
    for i in 1:4, j in 1:4 ]

plot(scatters..., size=(1200,800), margin = 4mm)
```

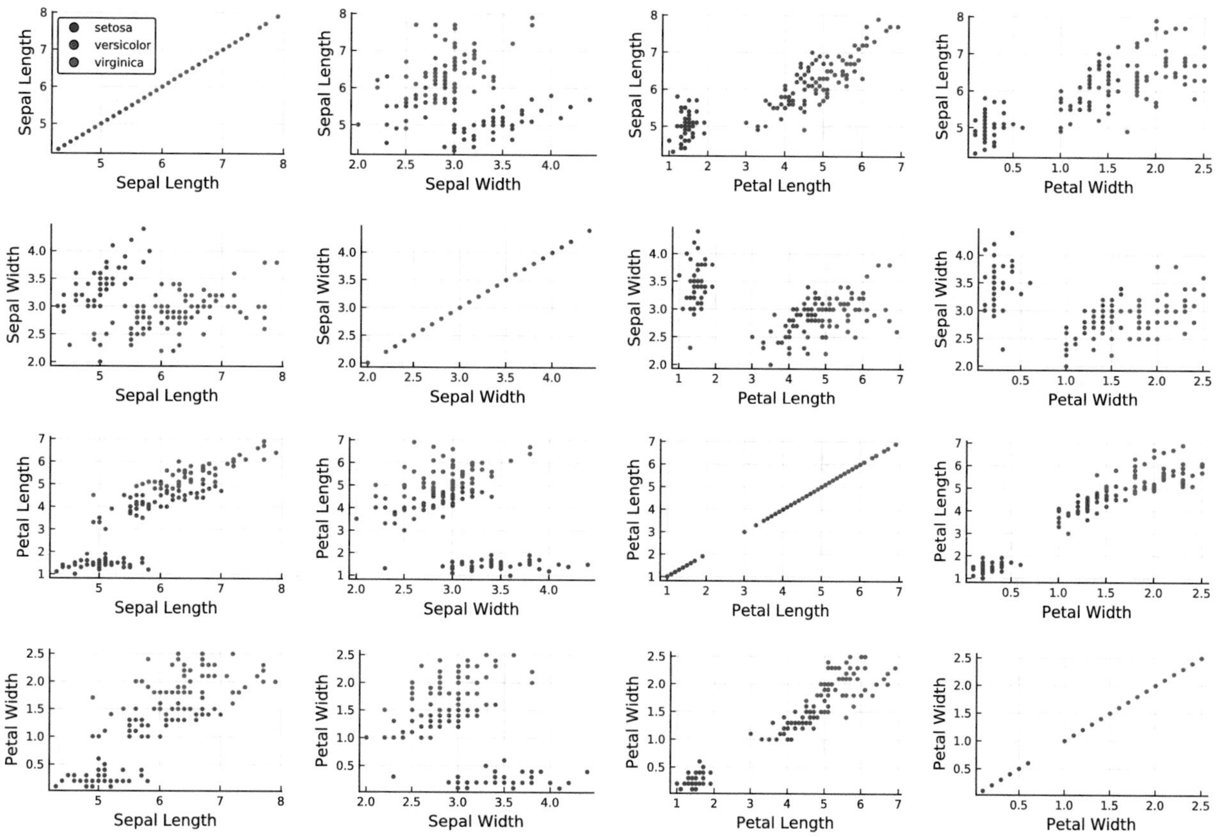

Figure 4.11: A scatter plot matrix of the iris dataset with observations grouped by species. Blue is Setosa, red is versicolor, green is virginica.

```
Number of rows: 150
Names of features:
        ["Sepal Length", "Sepal Width", "Petal Length", "Petal Width", " Species"]
Frequency per species:
        Pair{String,Int64}["setosa"=>50, "versicolor"=>50, "virginica"=>50]
```

In line 3, we create the data frame and, in line 4, we print the number of rows in it. In lines 6–9, we define a function that takes a string, `name`, that is assumed to be of the form `"SepalWidth"` as an example. Such are the names of columns in the `iris` dataset. The function then inserts white space prior to the last capital letter so as to convert the string to `"Sepal Width"`. Notice the use of string concatenation using `*` in line 8. We then use this function in line 11 to create `featureNames`, an array of strings that is later used to label the variables. Note the use of `names()` in line 11, yielding an array of symbols. Lines 14–16 deal with the species and their frequency. The names of species are obtained in line 14 and their frequency is obtained in line 15. This is simply for purposes of summarizing these results in the output generated in line 16. In line 18, we use the `default()` function from `Plots` to set parameters used by all scatter plots. In line 20, we create a matrix of scatter plots. Note the use of `group=` in line 21 based on species. Also note the condition in line 22 for presenting a legend only in the top left plot. The plots are then presented in a figure in line 25.

Heat Map with Marginals

The *heat map*, first seen in Figure 1.8, consists of a grid of shaded cells. Another name for it is a *matrix plot.* The colors of the cells indicate the magnitude, where typically, the "warmer" the color, the higher the value. This is in a sense nothing but a monochrome image.

In cases of pairs of observations $(x_1, y_1), \dots (x_n, y_n)$, the bivariate data can be constructed into a *bivariate histogram* in a manner similar to the (univariate) histogram implemented in Listing 4.14. In the bivariate case, we partition the Cartesian plain (or the subset containing the data) $\mathbb{R}^2$, into a grid of bins $\mathcal{B}_{ij}$ for $i = 1, \dots, L_1$ and $j = 1, \dots, L_2$. Then we count the frequency of observations per bin via

$$f_{ij} = \frac{1}{n} \sum_{k=1}^{n} \mathbf{1}\{x_k \in \mathcal{B}_{ij}\}, \qquad \text{for} \quad i = 1, \dots, L_1, \qquad j = 1, \dots, L_2. \tag{4.11}$$

Compare this with (4.6) dealing with the univariate case. Now the $L_1 \times L_2$ matrix composed of f_{ij} can be plotted as a heat map to yield a bivariate histogram.

The `marginalhist()` function from `StatsPlots` implements this and goes even further to create and present *marginal histograms.* These are two separate univariate histograms, one for $x_1, \dots, x_n$ and the other for $y_1, \dots, y_n$. Then, as shown in Figure 4.12, these histograms are presented on the margins of the heat maps, estimating the marginal distributions. See also Section 3.7.

In Listing 4.25, we create Figure 4.12 that presents two variants of a heat map with marginals. The left plot is for the Brisbane and Gold Coast temperature data, also used in Listings 4.12, Listing 3.34, as well as others. The right plot is for synthetic data based on a bivariate normal distribution fitted to that data, with the actual parameters fit in Listing 4.12. Note that this data is also plotted as a time series in Figure 4.6. Hence in interpreting it via a histogram, one needs to exercise caution due to the cyclic nature of the data.

Listing 4.25: Heatmap and marginal histograms

```
using StatsPlots, Distributions, CSV, DataFrames, Measures; pyplot()

realData = CSV.read("../data/temperatures.csv")

N = 10^5
include("../data/mvParams.jl")
biNorm = MvNormal(meanVect,covMat)
syntheticData = DataFrame(rand(MvNormal(meanVect,covMat),N)')
rename!(syntheticData,[:x1=>:Brisbane, :x2 => :GoldCoast])

default(c=cgrad([:blue, :red]),
    xlabel="Brisbane Temperature",
    ylabel="Gold Coast Temperature")

p1 = marginalhist(realData.Brisbane, realData.GoldCoast, bins=10:45)
p2 = marginalhist(syntheticData.Brisbane, syntheticData.GoldCoast, bins=10:.5:45)

plot(p1,p2, size = (1000,500), margin = 10mm)
```

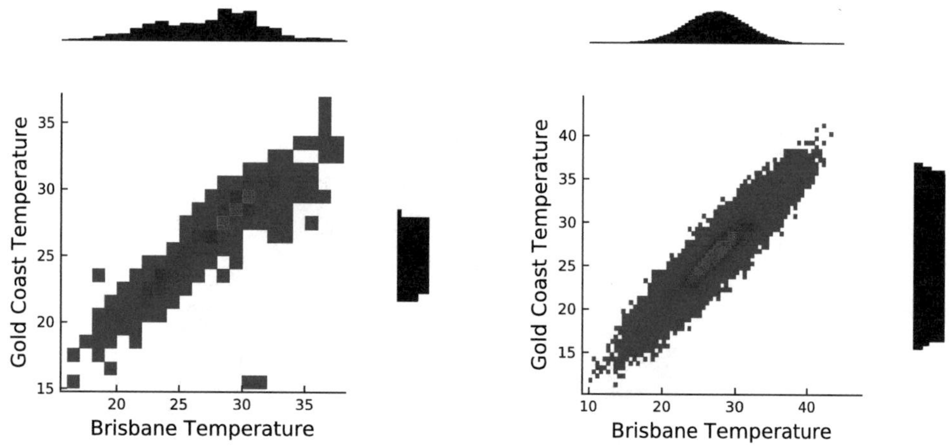

Figure 4.12: Heat map with marginals comparing Brisbane and Gold Coast temperatures. Left: actual data. Right: synthetic multivariate normal data.

In line 3, we create `realData` based on the Brisbane and Gold Coast temperature file. Lines 5–9 create the `syntheticData` data frame with `N` observations based on the bivariate normal distribution `biNorm` using the parameters in `mvParams.jl` similarly to Listing 3.34. The actual creation of the `DataFrame` object in line 8 creates default column names, `x1` and `x2`. We then rename these in line 9. The remainder of the code creates the two heat maps with marginals plots using `marginalhist()` in lines 15–16. Observe that, for the synthetic data, we are able to use a much larger number of bins. Note the use of the `cgrad()` function in line 11, setting the color gradient as part of the default parameters.

Andrews Plot

We now introduce a completely different way to visualize high-dimensional data. The idea is to represent a data vector $(x_{i1}, \ldots, x_{ip})$ via a real-valued function. For any individual vector, such a transformation cannot be generally useful, however, when comparing groups of vectors, it may yield a way to visualize structural differences in the data.

The specific transformation rule that we present here creates a plot known as *Andrews plot*. Here, for the i'th data vector $(x_{i1}, \ldots, x_{ip})$, we create the function $f_i(\cdot)$ defined on $[-\pi, \pi]$ via,

$$f_i(t) = \frac{x_{i1}}{\sqrt{2}} + x_{i2}\sin(t) + x_{i3}\cos(t) + x_{i4}\sin(2t) + x_{i5}\cos(2t) + x_{i6}\sin(3t) + x_{i7}\sin(3t) + \cdots,$$

with the last term involving a sin() if p is even and a cos() if p is odd. The for $i = 1, \ldots, n$, the functions $f_1(\cdot), \ldots, f_n(\cdot)$ are plotted. In cases where each i has an associated label from a small finite set, different colors or line patterns can be used. An example of this plot is shown in Figure 4.13, where the results hint at similarities within species and differences between species.

In Listing 4.26 below, we present a standard example of Andrews plot based on the `iris` dataset. The resulting Figure 4.13 indicates that differences exist between each of the species.

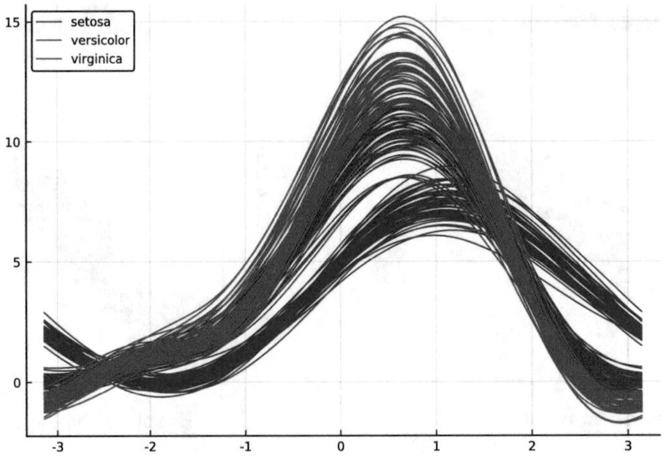

Figure 4.13: Andrews plot, showing an underlying structure in the Iris dataset.

Listing 4.26: Andrews plot

```
1  using RDatasets, StatsPlots; pyplot()
2
3  iris = dataset("datasets", "iris")
4  @df iris andrewsplot(:Species, cols(1:4),
5          line=(fill=[:blue :red :green]), legend=:topleft)
```

In line 4, the `andrewsplot()` function from `StatsPlots` is used to plot the data. Note the `@df` macro is used in a similar format to that of Listing 4.23. The first argument, `:Species`, determines how the data should be grouped, while the second argument determines what variables should be included in the calculation, in this case columns 1–4.

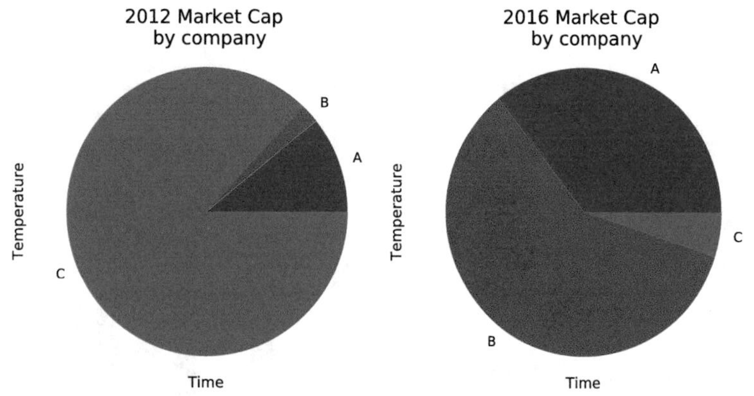

Figure 4.14: Two pie charts.

4.6 Plots for the Board Room

In this section, we introduce more simple plots, such as those that one may typically see in business summaries, or news reports. We show how to create *pie charts*, *bar charts*, and *stack plots*. Although the plots covered here are not as technical as those covered previously, they are still useful as they can quickly convey information in a very clear manner. The examples in this section rely on data for three fictitious companies, stored in `companyData.csv`.

Pie Chart

We first look at the *pie chart*, which is a simple plot that conveys relative proportions. In Listing 4.27, we construct two pie charts which show the relative market capitalization of each company A, B, and C for the years 2012 and 2016. The results are shown in Figure 4.14.

Listing 4.27: A pie chart

```
1  using CSV, CategoricalArrays, Plots; pyplot()
2
3  df = CSV.read("../data/companyData.csv")
4  companies = levels(df.Type)
5
6  year2012 = df[df.Year .== 2012, :MarketCap]
7  year2016 = df[df.Year .== 2016, :MarketCap]
8
9  p1 = pie(companies, year2012, title="2012 Market Cap \n by company")
10 p2 = pie(companies, year2016, title="2016 Market Cap \n by company")
11 plot(p1, p2, size=(800, 400))
```

In line 4, `levels()` from the `CategoricalArrays` package is used to extract the name of each company as a level, and store them in the array `companies`. In lines 6–7, the market capitalization for each company is stored as arrays `year2012` and `year2016` for the years 2012 and 2016, respectively. In lines 9–10, the `pie()` function is used to create the pie charts.

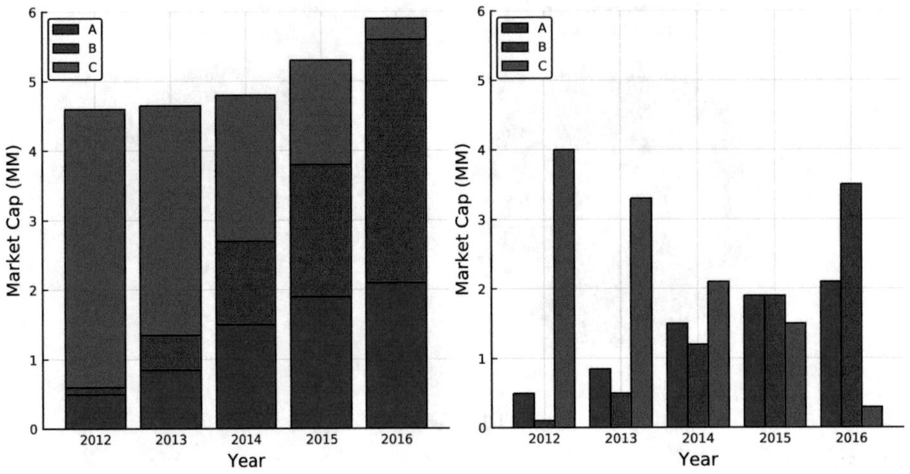

Figure 4.15: A stacked bar plot (left), and non-stacked bar plot (right).

Bar Plot

The *bar plot*, or *bar chart*, is another useful plot which conveys proportions through the use of vertical bars. This type of plot was first seen in Figure 3.13, and we present another example of this plot here. Listing 4.28 summarizes the data from `companyData.csv`, and presents the total market capitalization of each company for each year through a *stacked bar plot* and a *grouped bar plot*. The results are shown in Figure 4.15.

Listing 4.28: Two different bar plots

```
1  using CSV, CategoricalArrays, StatsPlots; pyplot()
2
3  df = CSV.read("../data/companyData.csv")
4  years = levels(df.Year)
5  data  = reshape(df.MarketCap, 5, 3)
6
7  p1 = groupedbar(years, data, bar_position=:stack)
8  p2 = groupedbar(years, data, bar_position=:dodge)
9  plot(p1, p2, bar_width=0.7, fill=[:blue :red :green], label=["A" "B" "C"],
10         ylims=(0,6), xlabel="Year", ylabel="Market Cap (MM)",
11         legend=:topleft, size=(800,400))
```

In line 5, `reshape()` is used to reshape the market capitalization data from a single column to a 5×3 array, with the rows representing years and columns companies. In lines 7–11, the `groupedbar()` function from `StatsPlots` is used to create the bar plots. By setting `bar_position=:stack`, a stackplot is created, while `bar_position=:dodge` creates a grouped bar plot instead.

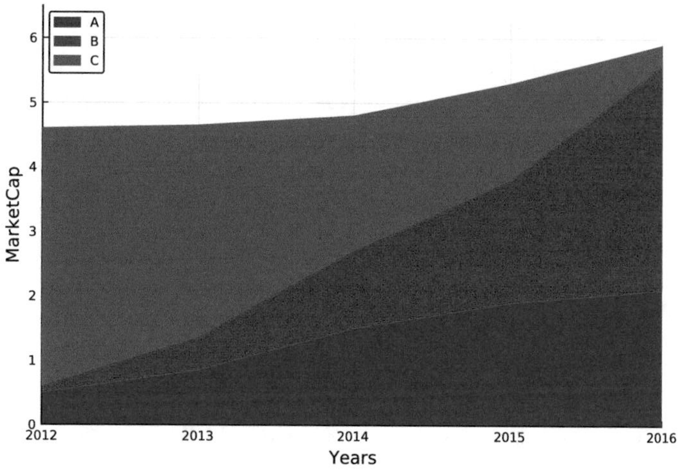

Figure 4.16: A stack plot showing the change in market capitalization of several companies over time.

Stack Plot

The *stack plot* is a commonly used plot which shows how constituent amounts of a metric change over time. In Listing 4.29, we present an example, where we consider the changing total market capitalization of the three companies A, B, and C over several years (Figure 4.16).

Listing 4.29: A stack plot

```
1  using CSV, CategoricalArrays, Plots; pyplot()
2
3  df = CSV.read("../data/companyData.csv")
4  mktCap = reshape(df.MarketCap, 5, 3)
5  years  = levels(df.Year)
6
7  areaplot(years, mktCap,
8          c=[:blue :red :green], labels=["A" "B" "C"],
9          xlims=(minimum(years),maximum(years)), ylims=(0,6.5),
10         legend=:topleft, xlabel="Years", ylabel="MarketCap")
```

In line 4, the data in the `MarketCap` column is reshaped into a 5×3 array via the `reshape()` function. In line 5, `levels()` is used to store the unique years of the dataset in the array `years` in ascending order. In lines 7–10, `areaplot()` is used to create the plot, with the horizontal values given as the first argument, and the data to be plotted as the second argument, with rows treated as individual years.

4.7 Working with Files and Remote Servers

The ability to work with files is an important skill that the modern data scientist is expected to have. Often one will be required to perform various operations with files programmatically, such as create new files, open files, interact with their content, and save information to existing files. Julia provides various methods to interact with and work with files via *input/output streams* (I/O). In this section, we provide two simple examples which involve working with files programmatically. In addition, at the end of this section, we provide a brief pseudo-code example of how one might connect to a remote server, query a database, and save the results to a locally stored file.

Searching a File for Keywords

For a first example, we show how one might programmatically search a file for a specific keyword or content, and then save that content to a separate file. In Listing 4.30, we create a function which searches a text document for a given keyword and then saves every line of text containing this keyword to a new text file, along with the associated line number.

Listing 4.30: Filtering an input file

```
1   function lineSearch(inputFilename, outputFilename, keyword)
2       infile  = open(inputFilename, "r")
3       outfile = open(outputFilename,"w")
4
5       for (index, line) in enumerate(split(read(infile, String), "\n"))
6           if occursin(keyword, line)
7               println(outfile, "$index: $line")
8           end
9       end
10      close(infile)
11      close(outfile)
12  end
13
14  lineSearch("../data/earth.txt", "../data/waterLines.txt", "water")
```

```
17: 71% of Earth's surface is covered with water, mostly by oceans. The
19: have many lakes, rivers and other sources of water that contribute to the
```

In lines 1–12, the function `lineSearch()` is defined, which searches an input file, `inputFilename`, for a `keyword`, and saves the lines and line numbers where it appears to an output file, `outputFilename`. Line 2 uses `open()` with "`r`" to open the file to be searched in read mode. It creates an `IOStream` object, which can be used as arguments to other functions. We define this as the variable `infile`. Line 3 uses `open()` with "`w`" to create and open a file in write mode, with the given file name `outputFilename`. This file is created on disk ready to have information written to it. Lines 5–9 contain a for loop, which is used to search through the input file for the given `keyword`. Line 5 reads the file as a `String` via `read()`, and `split()` is used along with "`\n`" to convert the single string into an array of strings, where the content of each line is stored in a separate index of the array. Line 6 uses `occursin()` to check if the given `line` contains our given `keyword`. If it does, then we proceed to line 7, where `println()` is used to write both the `index` and the `line` content to the `outfile`. Lines 10 and 11 close both our input file and output file. In line 14 `lineSearch` is used to search the file "`earth.txt`", for the keyword "`water`", with the line numbers and text saved to the file `waterLines.txt`".

Searching for Files in a Directory

In Listing 4.31, we present our next example, where we create a function which searches a directory for all filenames which contain a particular string. It then saves a list of these files to a file, `fileList`. Note that this function does not behave recursively and only searches the directory given.

Listing 4.31: Searching files in a directory

```
1  function directorySearch(directory, searchString)
2      outfile  = open("../data/fileList.txt","w")
3      fileList = filter(x->occursin(searchString, x), readdir(directory))
4
5      for file in fileList
6          println(outfile, file)
7      end
8      close(outfile)
9  end
10
11 directorySearch(pwd(),".jl")
```

In lines 1–9, we define the function `directorySearch`. As arguments, it takes a `directory` to search through, and a `searchString`. Line 2 uses `open` with "w" to create our output file `fileList.txt`, which we will write to. In line 3, we create a string array of all filenames in our specified `directory` that contain our `searchString`. This string array is defined as the variable `fileList`. The `readdir()` function is used to list all files in the given `directory`, and `filter()` is used, along with `occursin()` to check each element contains the `searchString`. Lines 5–7 loop through each element in `fileList` and print them to theoutput file `outfile`. Line 8 closes the `IOStream outfile`. Line 11 provides an example of the use of our `directorySearch` function, where we use it to obtain a shortlist of all files whose extensions contain "`.jl`" within our current working directory, i.e. `pwd()`.

Connecting to a Remote Server

One may not always work with data stored locally on their machine or network. For example, sometimes a dataset is too large to be stored on a workstation, and therefore must be stored remotely in a datacentre, or on a *remote server*. In this scenario, one must first connect to the server before working with the data. A typical workflow involves connecting to the remote database, submitting a query, and then saving the result locally. There are different types of databases, including Oracle, MySQL, PostgreSQL, MongoDB, and many others. There are several Julia packages for connecting to remote servers including `LibPQ.jl`, which is a wrapper for the PostgreSQL libpq C library, `SQLite.jl`, as well as `ODBC.jl` and several others. Once a connection is established, one will typically submit a so-called *SQL* query to the server. SQL stands for structured query language and is a common syntax used to query remote databases in order to extract a subsect of data from the database.

In this section, we do not expand on the details of databases, nor the syntax of SQL queries. Instead, in Listing 4.32, we present a simple pseudocode example of how a user may connect to a remote PostgreSQL database, submit a SQL query, and then save the results.

Listing 4.32: Pseudocode for a remote database query

```
1  using LibPQ, DataFrames, CSV
2
3  host     = "remoteHost"
4  dbname   = "db1"
5  user     = "username"
6  password = "userPwd"
7  port     = "1111"
8
9  conStr= "host=" *host *
10         " port=" *port *
11         " dbname=" *dbname *
12         " user=" *user *
13         " password=" *password
14 conn = LibPQ.Connection(conStr)
15
16 df = DataFrame(execute(conn, "SELECT * FROM S1.T1"))
17 close(conn)
18
19 CSV.write("example.csv", df);
```

In line 1, the `LibPQ` package is included. It is a wrapper for the libpq postgreSQL library, and contains methods to remotely connect to postgreSQL servers and submit queries. In lines 3–7, the details of the connection are specified and stored as strings, they include the host name, database name, username, password, and specific port to connect on. Lines 9–13 concatenate these details together into the string `conStr`. In line 14, a connection to the remote server is established via the `Connection()` function from the `LibPQ` package. The details in the string `conStr` are used to establish the connection. Note that if the password is not given in the connection string, then the server will prompt for a password. In line 16, a SQL query is submitted to the server via the `execute()` function. It takes two arguments, the first is the connection to the server, and the second is the SQL query. The query submitted here is simple: `SELECT *` is used to select all columns `FROM` the `T1` table, from the `S1` schema, from database `db1`. The results are stored as the `DataFrame` `df`. The connection to the server is closed in line 17 via `close()`. In line 19, the data in `df` is written to the CSV file `example.csv`, in the current working directory.

Chapter 5

Statistical Inference Concepts

This chapter introduces statistical inference concepts with the goal of establishing a theoretical footing of key concepts that follow in later chapters. The approach is that of classical statistics as opposed to machine learning, covered in Chapter 9. The action of *statistical inference* involves using mathematical techniques to make conclusions about unknown *population* parameters based on collected data. The field of statistical inference employs a variety of stochastic models to analyze and put forward efficient methods for carrying out such analyses.

In broad generality, analysis and methods of statistical inference can be categorized as either *frequentist* (also known as classical) or *Bayesian*. The former is based on the assumption that population parameters of some underlying distribution, or probability law, exist and are fixed, but are yet unknown. The process of statistical inference then deals with making conclusions about these parameters based on sampled data. In the latter Bayesian case, it is only assumed that there is a *prior distribution* of the parameters. In this case, the key process deals with analyzing a *posterior distribution* (of the parameters)—an outcome of the inference process. In this book, we focus almost solely on the classical frequentist approach with the exception of Section 5.7 where we explore Bayesian statistics briefly.

In general, a statistical inference process involves *data*, a *model*, and *analysis*. The data is assumed to be comprised of random samples from the model. The goal of the analysis is then to make informed statements about population parameters of the model based on the data. Such statements typically take one of the following forms:

Point estimation - Determination of a single value (or vector of values) representing a best estimate of the parameter/parameters. In this case, the notion of "best" can be defined in different ways.

Confidence intervals - Determination of a range of values where the parameter lies. Under the model and the statistical process used, it is guaranteed that the parameter lies within this range with a pre-specified probability.

Hypothesis tests - The process of determining if the parameter lies in a given region, in the complement of that region, or fails to take on a specific value. Such tests often represent a scientific hypothesis in a very natural way.

Y. Nazarathy and H. Klok, *Statistics with Julia*, Springer Series in the Data Sciences,
https://doi.org/10.1007/978-3-030-70901-3_5

Most of the point estimation, confidence intervals, and hypothesis tests that we introduce and carry out in this book are elementary. Chapter 6 is devoted to covering elementary confidence intervals in detail, and Chapter 7 is devoted to covering elementary hypothesis tests in detail. We now begin to explore key ideas and concepts of statistical inference.

This chapter is structured as follows: In Section 5.1, we present the concept of a random sample together with the distribution of statistics, such as the distribution of the sample mean and the sample variance. In Section 5.2, we focus on random samples of normal random variables. In this common case, certain statistics have well-known distributions that play a central role. In Section 5.3, we explore the central limit theorem, providing justification for the ubiquity of the normal distribution. In Section 5.4, we explore basics of point estimation. In Section 5.5, we explore the concept of a confidence interval. In Section 5.6, we explore concepts of hypothesis testing. Finally, in Section 5.7, we explore the basics of Bayesian statistics.

5.1 A Random Sample

When carrying out (frequentist) statistical inference, we assume there is some underlying distribution $F(x\,;\,\theta)$ from which we are sampling, where θ is the scalar or vector-valued unknown parameter we wish to know. Importantly, we assume that each observation is statistically independent and identically distributed as the rest. That is, from a probabilistic perspective, the observations are taken as *independent and identically distributed (i.i.d.)* random variables. In mathematical statistics language, this is called a *random sample*. We denote the random variables of the observations by $X_1, \ldots, X_n$ and their respective values by $x_1, \ldots, x_n$.

Typically, we compute *statistics* from the random sample. For example, two common standard statistics include the *sample mean* and *sample variance*, introduced in Section 4.2 in the context of data summary. However, we can model these *statistics* as random variables

$$\overline{X} = \frac{1}{n}\sum_{i=1}^{n} X_i, \qquad \text{and} \qquad S^2 = \frac{1}{n-1}\sum_{i=1}^{n}(X_i - \overline{X})^2. \tag{5.1}$$

Note that for S^2, the denominator is $n-1$ (as opposed to n as one might expect). This makes S^2 an *unbiased estimator*. We discuss this property further in Section 5.4.

In general, the phrase *statistic* implies a quantity calculated based on the sample. When working with data, the sample mean and sample variance are nothing but numbers computed from our sample observations. However, in the statistical inference paradigm, we associate random variables to these values, since they themselves are functions of the random sample. We look at properties of such statistics and see how they play a role in estimating the unknown underlying distribution parameter θ.

To illustrate the fact that $\overline{X}$ and S^2 are random variables, assume we have sampled data from an exponential distribution with $\lambda = 4.5^{-1}$ (a mean of 4.5 and a variance of 20.25). If we collect $n = 10$ observations, then the sample mean and sample variance are random variables. In Listing 5.1, we investigate their distribution through Monte Carlo simulation and create Figure 5.1. The point to see is that $\overline{X}$ and S^2 are themselves random variables with underlying distributions.

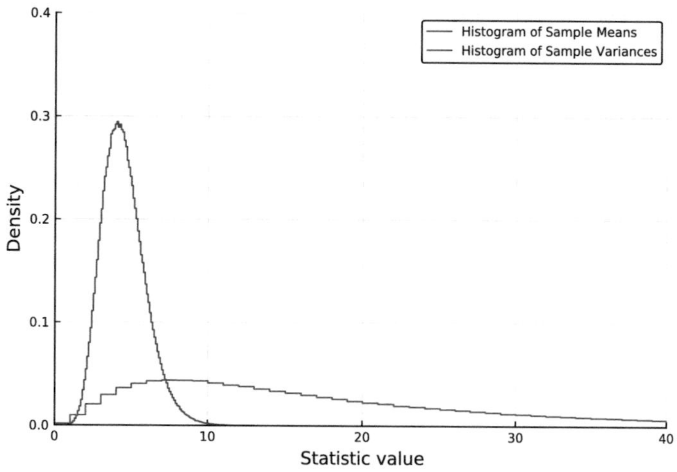

Figure 5.1: Histograms of the sample mean and sample variance of an exponential distribution.

Listing 5.1: Distributions of the sample mean and sample variance

```
using Random, Distributions, Plots; pyplot()
Random.seed!(0)

lambda  = 1/4.5
expDist = Exponential(1/lambda)
n, N    = 10, 10^6

means     = Array{Float64}(undef, N)
variances = Array{Float64}(undef, N)

for i in 1:N
    data = rand(expDist,n)
    means[i] = mean(data)
    variances[i] = var(data)
end

println("Actual mean: ",mean(expDist),
                "\nMean of sample means: ",mean(means))
println("Actual variance: ",var(expDist),
        "\nMean of sample variances: ",mean(variances))

stephist(means, bins=200, c=:blue, normed=true,
        label="Histogram of Sample Means")
stephist!(variances, bins=600, c=:red, normed=true,
        label="Histogram of Sample Variances", xlims=(0,40), ylims=(0,0.4),
        xlabel = "Statistic value", ylabel = "Density")
```

```
Actual mean: 4.5
Mean of sample means: 4.500154606762812
Actual variance: 20.25
Mean of sample variances: 20.237117004185237
```

In lines 8–9, we initialize the empty arrays `means` and `variances`, respectively. In lines 11–15, we create `N` random samples, each of length `n`. For each sample, we calculate the sample mean and sample variance. In lines 17–20, we calculate the `mean()` of both arrays `means` and `variances`. It can be seen that the estimated expected value of our simulated data is good approximations to the mean and variance parameters of the underlying exponential distribution. That is, for an exponential distribution with rate λ, the mean is λ^{-1} and the variance is λ^{-2}. In lines 22–26, we generate histograms of the sample means and sample variances, using `200` and `600` bins, respectively.

5.2 Sampling from a Normal Population

It is often assumed that the distribution $F(x\,;\,\theta)$ is a normal distribution, and hence $\theta = (\mu, \sigma^2)$. This assumption is called the *normality assumption*, and is sometimes justified due to the central limit theorem, which we cover in Section 5.3. Under the normality assumption, the distribution of the random variables $\overline{X}$ and S^2 as well as transformations of them are well known. The following three distributional relationships play a key role:

$$\begin{aligned} \overline{X} &\sim \text{Normal}(\mu, \sigma^2/n), \\ (n-1)S^2/\sigma^2 &\sim \chi^2_{n-1}, \\ T := \frac{\overline{X}-\mu}{S/\sqrt{n}} &\sim t_{n-1}. \end{aligned} \tag{5.2}$$

Here, "$\sim$" denotes "*distributed as*", and implies that the statistics on the left hand side of the "$\sim$" symbols are distributed according to the distributions on the right hand side. The notation χ^2_{n-1} and t_{n-1} denotes a *chi-squared* and *student T-distribution*, respectively, each with $n-1$ degrees of freedom. The chi-squared distribution is a gamma distribution (see Section 3.6) with parameters $\lambda = 1/2$ and $\alpha = n/2$. The student T-distribution is discussed later in this section.

Importantly, these distributional properties of the statistics from a normal sample theoretically support the statistical procedures that are presented in Chapters 6 and 7.

We now look at an example in Listing 5.2, where we sample data from a normal distribution and compute the statistics, $\overline{X}$, T, and S^2. As seen in Figure 5.2, the distribution of sample means, sample variances and T-statistics (T) indeed follow the distributions given by (5.2).

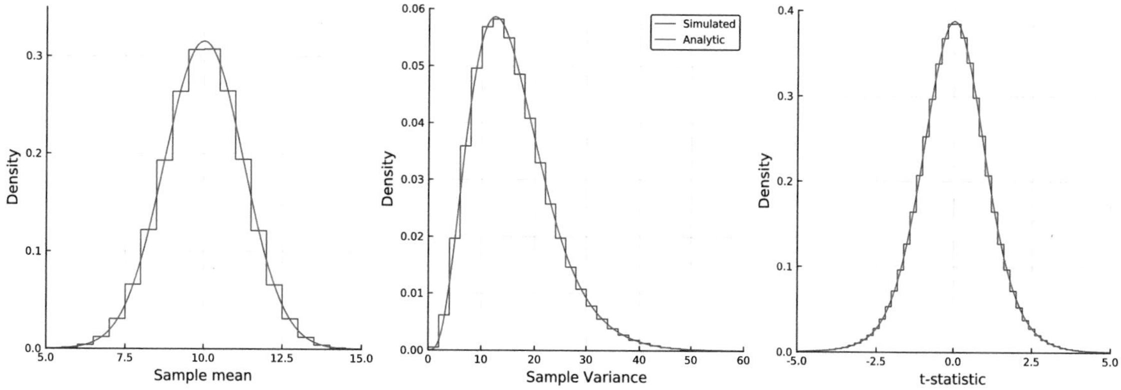

Figure 5.2: Histograms of the simulated sample means, sample variances, and T-statistics, against their analytic counterparts.

Listing 5.2: Friends of the normal distribution

```
using Distributions, Plots; pyplot()

mu, sigma = 10, 4
n, N = 10, 10^6

sMeans = Array{Float64}(undef, N)
sVars  = Array{Float64}(undef, N)
tStats = Array{Float64}(undef, N)

for i in 1:N
    data       = rand(Normal(mu,sigma),n)
    sampleMean = mean(data)
    sampleVars = var(data)
    sMeans[i]  = sampleMean
    sVars[i]   = sampleVars
    tStats[i]  = (sampleMean - mu)/(sqrt(sampleVars/n))
end

xRangeMean = 5:0.1:15
xRangeVar = 0:0.1:60
xRangeTStat = -5:0.1:5

p1 = stephist(sMeans, bins=50, c=:blue, normed=true, legend=false)
p1 = plot!(xRangeMean, pdf.(Normal(mu,sigma/sqrt(n)), xRangeMean),
    c=:red, xlims=(5,15), ylims=(0,0.35), xlabel="Sample mean",ylabel="Density")

p2 = stephist(sVars, bins=50, c=:blue, normed=true, label="Simulated")
p2 = plot!(xRangeVar, (n-1)/sigma^2*pdf.(Chisq(n-1), xRangeVar*(n-1)/sigma^2),
    c=:red, label="Analytic", xlims=(0,60), ylims=(0,0.06),
    xlabel="Sample Variance",ylabel="Density")

p3 = stephist(tStats, bins=100, c=:blue, normed=true, legend=false)
p3 = plot!(xRangeTStat, pdf.(TDist(n-1), xRangeTStat),
    c=:red, xlims=(-5,5), ylims=(0,0.4), xlabel="t-statistic",ylabel="Density")

plot(p1, p2, p3, layout = (1,3), size=(1200, 400))
```

In line 3, we specify the parameters of the underlying normal distribution from which we sample our data. In line 4, we specify the number of samples in each group `n` and the total number of Monte Carlo repetitions `N`. In lines 6–8, we initialize three arrays which will be used to store our sample means, variances, and T-statistics. In lines 10–17, we conduct our numerical simulation by taking `n` sample observations from the underlying normal distribution and calculating the sample mean, sample variance, and T-statistic. This process is repeated `N` times, and the values are stored in the arrays `sMeans`, `sVars`, and `tStats` respectively. The remainder of the code creates the histograms of the sample means, sample variances, and T-statistics alongside the analytic PDF's given by (5.2). Observe the PDF of the sample mean in line 24. Observe the PDF of a scaled chi-squared distribution through the use of the `pdf()` and `Chisq()` functions in line 28. Note that the values on the x-axis and the density are both normalized by `(n-1)/sigma^2` to reflect the fact we are interested in the PDF of a scaled chi-squared distribution. Finally, observe the PDF of the T-statistic (T), which is described by a T-distribution, is plotted via the use of the `TDist()` function in line 33.

Independence of the Sample Mean and Sample Variance

We now look at a key property of the sample mean and sample variance. Consider a random sample, $X_1, \ldots, X_n$. In general, one would not expect the sample mean, $\overline{X}$ and the sample variance S^2 to be independent random variables—since both of these statistics rely on the same underlying values. For example, consider a random sample where $n = 2$, and let each X_i be Bernoulli distributed, with parameter p. The joint distribution of $\overline{X}$ and S^2 can then be computed as follows.

If both X_i's are 0, which happens with probability $(1-p)^2$, then,

$$\overline{X} = 0 \qquad \text{and} \qquad S^2 = 0.$$

If both X_i's are 1, which happens with probability p^2, then,

$$\overline{X} = 1 \qquad \text{and} \qquad S^2 = 0.$$

If one of the X_i's is 0, and the other is 1, which happens with probability $2p(1-p)$, then,

$$\overline{X} = \frac{1}{2} \qquad \text{and} \quad S^2 = 1 - 2\left(\frac{1}{2}\right)^2 = \frac{1}{2}.$$

.

Hence, as shown in Figure 5.3 the joint PMF of $\overline{X}$ and S^2 is

$$\mathbb{P}(\overline{X} = \overline{x}, S^2 = s^2) = \begin{cases} (1-p)^2, & \text{for } \overline{x} = 0 \text{ and } s^2 = 0, \\ 2p(1-p), & \text{for } \overline{x} = 1/2 \text{ and } s^2 = 1/2, \\ p^2, & \text{for } \overline{x} = 1 \text{ and } s^2 = 0. \end{cases} \tag{5.3}$$

Furthermore, the (marginal) PMF of $\overline{X}$ is

$$\mathbb{P}_{\overline{X}}(0) = (1-p)^2, \qquad \mathbb{P}_{\overline{X}}\left(\frac{1}{2}\right) = 2p(1-p), \qquad \mathbb{P}_{\overline{X}}(1) = p^2.$$

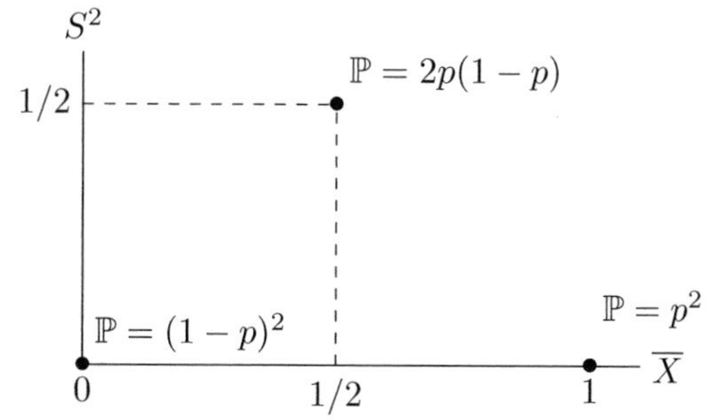

Figure 5.3: PMF of the sample mean and sample variance.

And the (marginal) PMF of S^2 is

$$\mathbb{P}_{S^2}(0) = (1-p)^2 + p^2, \qquad \mathbb{P}_{S^2}\left(\frac{1}{2}\right) = 2p(1-p), \qquad \mathbb{P}_{S^2}(1) = 0.$$

We now see that $\overline{X}$ and S^2 are not independent because the joint distribution

$$\tilde{\mathbb{P}}(i,j) = \mathbb{P}_{\overline{X}}(i)\,\mathbb{P}_{S^2}(j) \qquad \text{for} \qquad i,j \in \{0, \frac{1}{2}, 1\},$$

constructed by the product of the marginal distributions does not equal the joint distribution in (5.3).

The example above demonstrates dependence between $\overline{X}$ and S^2. This is in many ways unsurprising. However importantly, in the special case where the samples, $X_1, \ldots, X_n$ are from a normal distribution, independence between $\overline{X}$ and S^2 does hold. In fact, this property characterizes the normal distribution—that is, this property only holds for the normal distribution, see [L42].

We now explore this concept further in Listing 5.3. In it, we compare a standard normal distribution to what we call a standard uniform distribution—a uniform distribution on $[-\sqrt{3}, \sqrt{3}]$ which exhibits zero mean and unit variance. For both distributions, we consider a random sample of size $n = 3$, and from this, we obtain the pair $(\overline{X}, S^2)$. We then plot points of these pairs against points of pairs where $\overline{X}$ and S^2 are each obtained from two separate sample groups.

From Figure 5.4, it can be seen that for the normal distribution, regardless of whether the pair $(\overline{X}, S^2)$ is calculated from the same sample group, or from two different sample groups, the points appear to behave similarly. This is because they have the same joint distribution. However, for the standard uniform distribution, it can be observed that the points behave in a completely different manner. If the sample mean and variance are calculated from the same sample group, then all pairs of $\overline{X}$ and S^2 fall within a specific bounded region. The envelope of this blue region can be clearly observed and represents the region of all possible combinations of $\overline{X}$ and S^2 when calculated based on the same sample data. On the other hand, if $\overline{X}$ and S^2 are calculated from two separate samples, then we observe scattering of data, shown by the points in red. This difference in behavior shows that in this case $\overline{X}$ and S^2 are not independent, but rather the outcome of one imposes some restriction on the outcome of the other. By comparison, in the case of the standard normal distribution, regardless of how the pair $(\overline{X}, S^2)$ are calculated, (from the same sample group or from two different groups) the same scattering of points is observed, supporting the fact that $\overline{X}$ and S^2 are independent.

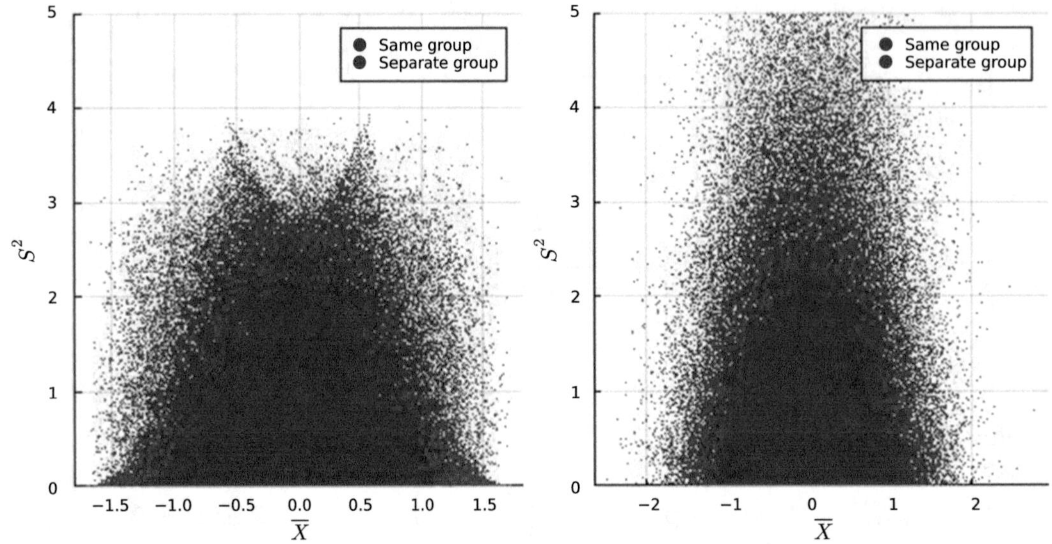

Figure 5.4: Pairs of $\overline{X}$ and S^2 for standard uniform (left) and standard normal (right). Blue points are for statistics calculated from the same sample, and red for statistics calculated from separate samples.

Listing 5.3: Are the sample mean and variance independent?

```
using Distributions, Plots, LaTeXStrings; pyplot()

function statPair(dist,n)
    sample = rand(dist,n)
    [mean(sample),var(sample)]
end

stdUni = Uniform(-sqrt(3),sqrt(3))
n, N = 3, 10^5

dataUni     = [statPair(stdUni,n) for _ in 1:N]
dataUniInd  = [[mean(rand(stdUni,n)),var(rand(stdUni,n))] for _ in 1:N]
dataNorm    = [statPair(Normal(),n) for _ in 1:N]
dataNormInd = [[mean(rand(Normal(),n)),var(rand(Normal(),n))] for _ in 1:N]

p1 = scatter(first.(dataUni), last.(dataUni),
        c=:blue, ms=1, msw=0, label="Same group")
p1 = scatter!(first.(dataUniInd), last.(dataUniInd),
        c=:red, ms=0.8, msw=0, label="Separate group",
        xlabel=L"\overline{X}", ylabel=L"S^2")

p2 = scatter(first.(dataNorm), last.(dataNorm),
        c=:blue, ms=1, msw=0, label="Same group")
p2 = scatter!(first.(dataNormInd), last.(dataNormInd),
        c=:red, ms=0.8, msw=0, label="Separate group",
        xlabel=L"\overline{X}", ylabel=L"$S^2$")

plot(p1, p2, ylims=(0,5), size=(800, 400))
```

In lines 3–6, the function `statPair()` is defined. It takes a distribution and integer `n` as input, generates a random sample of size `n`, and then returns the sample mean and sample variance of this random sample as an array. In line 8, we define the standard uniform distribution, which has a mean of zero and a standard deviation of 1. In line 9, we set the number of observations for each sample `n`, along with the total number of sample groups `N`. In line 11, the function `statPair()` is used along with a comprehension to calculate `N` pairs of sample means and variances from `N` sample groups. Note that the observations are all sampled from the standard uniform distribution `stdUni` and that the output is an array of arrays. In line 12, a similar approach to line 11 is used. However, in this case, rather than calculating the sample mean and variance from the same sample group each time, they are calculated from two separate sample groups `N` times. As before, the data is sampled from the standard uniform distribution `stdUni`. Lines 13 and 14 are identical to lines 11–12, however, in this case, observations are sampled from a standard normal distribution `Normal()`.

More on the T-Distribution

Having explored the fact that $\overline{X}$ and S^2 are independent in the case of a normal sample, we now elaborate on the *Student T-distribution* and focus on the distribution of the *T-statistic*, that appeared earlier in (5.2). This random variable is given by

$$T = \frac{\overline{X} - \mu}{S/\sqrt{n}}.$$

Denoting the mean and variance of the normally distributed observations by μ and σ^2, respectively, we can represent the T-statistic as

$$T = \frac{\sqrt{n}(\overline{X} - \mu)/\sigma}{\sqrt{(n-1)S^2/\sigma^2(n-1)}} = \frac{Z}{\sqrt{\frac{\chi^2_{n-1}}{n-1}}}. \tag{5.4}$$

Here, the numerator Z is a standard normal random variable and in the denominator the random variable, $\chi^2_{n-1} = (n-1)S^2/\sigma^2$ is chi-squared distributed with $n-1$ degrees of freedom, as claimed in (5.2). Furthermore, the numerator and denominator random variables are independent because they are based on the sample mean and sample variance.

One can show that a ratio of a standard normal random variable and the square root of a scaled independent chi-squared random variable (scaled by its degrees of freedom parameter) is distributed according to a T-distribution with the same number of degrees of freedom as the chi-squared random variable. Hence, $T \sim t(n-1)$. This means a "T-distribution with $n-1$ degrees of freedom". The T-distribution is a symmetric distribution with a "bell-curved" shape similar to that of the normal distribution, with "heavier tails" for non-large n. A t-distribution with k degrees of freedom can be shown to have a density function,

$$f(x) = \frac{\Gamma\left(\frac{k+1}{2}\right)}{\sqrt{k\pi}\,\Gamma\left(\frac{k}{2}\right)} \left(1 + \frac{x^2}{k}\right)^{-\frac{k+1}{2}}.$$

Note the presence of the gamma function, $\Gamma(\cdot)$, which is defined in Section 3.6.

To gain further insight from the representation (5.4), note that $\mathbb{E}[\chi^2] = (n-1)$ and $\text{Var}(\chi^2) = 2(n-1)$. Thus, the variance of $\chi^2/(n-1)$ is $2/(n-1)$, and hence one may expect that as $n \to \infty$, the random variable $\chi^2/(n-1)$ gets more and more concentrated around 1, with the same holding for $\sqrt{\chi^2/(n-1)}$. Hence, for large n, one may expect the distribution of T to be similar to the distribution of Z, which is indeed the case. This plays a role in the confidence intervals and hypothesis tests in the chapters that follow.

In practice, when carrying out elementary statistical inference using the T-distribution (as presented in the following chapters), the most commonly used attribute is the quantile, covered in Section 3.3. It is typically denoted by $t_{k,\alpha}$ where the *degrees of freedom* (DOF), k, define the specific T-distribution. Such quantiles are often tabulated in standard statistical tables.

In Listing 5.4 below, we first illustrate the validity of the representation (5.4) by generating T-distributed random variables by using a standard normal and a chi-squared random variable. We then plot the PDFs of several T-distributions, illustrating that as the degrees of freedom increase, the PDF converges to the standard normal PDF. See Figure 5.5.

Listing 5.4: Student's T-distribution

```
1  using Distributions, Random, Plots; pyplot()
2  Random.seed!(0)
3
4  n, N, alpha = 3, 10^7, 0.1
5
6  myT(nObs) = rand(Normal())/sqrt(rand(Chisq(nObs-1))/(nObs-1))
7  mcQuantile = quantile([myT(n) for _ in 1:N],alpha)
8  analyticQuantile = quantile(TDist(n-1),alpha)
9
10 println("Quantile from Monte Carlo: ", mcQuantile)
11 println("Analytic qunatile: ", analyticQuantile)
12
13 xGrid = -5:0.1:5
14 plot(xGrid, pdf.(Normal(), xGrid), c=:black, label="Normal Distribution")
15 scatter!(xGrid, pdf.(TDist(1) ,xGrid),
16         c=:blue, msw=0, label="DOF = 1")
17 scatter!(xGrid, pdf.(TDist(3), xGrid),
18         c=:red, msw=0, label="DOF = 3")
19 scatter!(xGrid, pdf.(TDist(100),xGrid),
20         c=:green, msw=0, label="DOF = 100",
21         xlims=(-4,4), ylims=(0,0.5), xlabel="X", ylabel="Density")
```

```
Quantile from Monte Carlo: -1.8848554309670498
Analytic qunatile: -1.8856180831641265
```

In line 6, we specify the function `myT()` which generates a T-distributed random variable by using a standard normal and a chi-squared random variable, just as in (5.4). In line 7, we use `N` replications of `myT()` to estimate the `alpha` quantile. Then in line 8, we compute the quantile analytically for a corresponding T-distribution represented by `TDist(n-1)`. The estimated quantile and computed quantile are then printed in lines 10–11. The remainder of the code plots three T-distributions, generating Figure 5.5.

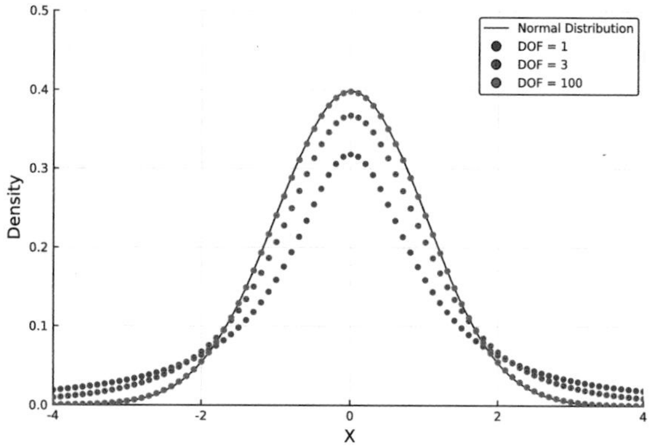

Figure 5.5: As the number of degrees of freedom (DOF) increases, the T-distribution approaches that of the normal distribution.

Two Samples and the F-Distribution

Many statistical procedures involve the ratio of sample variances, or similar quantities, for two or more samples. For example, if $X_1, \ldots, X_{n_1}$ is one sample and $Y_1, \ldots, Y_{n_2}$ is another sample, and both samples are distributed normally with the same parameters, one can look at the ratio of the two sample variances

$$F_{\text{statistic}} = \frac{S_X^2}{S_Y^2}.$$

It turns out that such a statistic is distributed according to what is called the *F-distribution*, with density given by

$$f(x) = K(a,b)\,\frac{x^{a/2-1}}{(b+ax)^{(a+b)/2}} \qquad \text{with} \qquad K(a,b) = \frac{\Gamma\Big(\dfrac{a+b}{2}\Big)a^{a/2}b^{b/2}}{\Gamma\Big(\dfrac{a}{2}\Big)\Gamma\Big(\dfrac{b}{2}\Big)}.$$

Here, the parameters a and b are the *numerator degrees of freedom* and *denominator degrees of freedom*, respectively. In the case of $F_{\text{statistic}}$, we set $a = n_1 - 1$ and $b = n_2 - 1$.

In agreement with (5.2), an alternative view is that the random variable F is obtained by the ratio of two independent chi-squared random variables, normalized by their degrees of freedom. The F-distribution plays a key role in the popular Analysis of Variance (ANOVA) procedures, further explored in Section 7.3.

We now briefly explore the F-distribution in Listing 5.5 by simulating two sample sets of data with n_1 and n_2 observations, respectively, from a normal distribution. The ratio of the sample variances from the two distributions is then compared to the PDF of an F-distribution with parameters $n_1 - 1$ and $n_2 - 1$. The listing generates Figure 5.6.

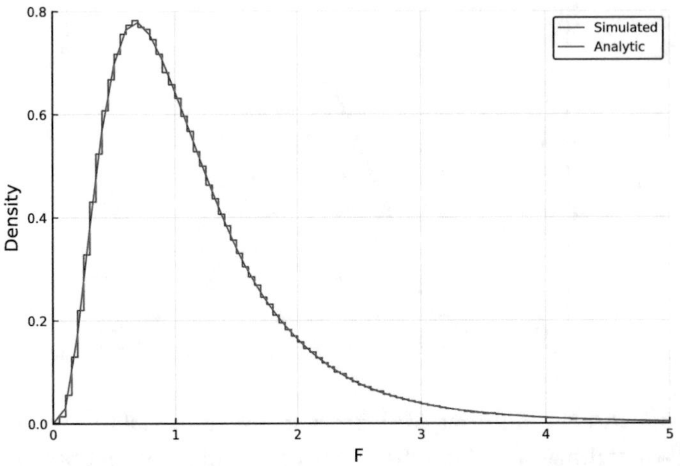

Figure 5.6: Histogram of the ratio of two sample variances against the PDF of an F-distribution.

Listing 5.5: Ratio of variances and the F-distribution

```
1  using Distributions, Plots; pyplot()
2
3  n1, n2 = 10, 15
4  N = 10^6
5  mu, sigma = 10, 4
6  normDist = Normal(mu,sigma)
7
8  fValues = Array{Float64}(undef, N)
9
10 for i in 1:N
11     data1 = rand(normDist,n1)
12     data2 = rand(normDist,n2)
13     fValues[i] = var(data1)/var(data2)
14 end
15
16 fRange = 0:0.1:5
17 stephist(fValues, bins=400, c=:blue, label="Simulated", normed=true)
18 plot!(fRange, pdf.(FDist(n1-1, n2-1), fRange),
19         c=:red, label="Analytic", xlims=(0,5), ylims=(0,0.8),
20         xlabel = "F", ylabel = "Density")
```

In lines 3–4, we define the total number of observations for our two sample groups, `n1` and `n2`, as well as the total number of F-statistics we will generate, `N`. In lines 10–14, we simulate two separate sample groups, `data1` and `data2`, by randomly sampling from the same underlying normal distribution. A single F-statistic is then calculated from the ratio of the sample variances of these two groups. The remainder of the code creates the figure where in line 18 the constructor `FDist()` is used to create an F-distribution with the parameters `n1-1` and `n2-2`.

5.3 The Central Limit Theorem

In the previous section, we assumed sampling from a normal population, and this assumption gave rise to a variety of properties of statistics associated with the sampling. However, why would such an assumption hold? A key lies in one of the most fundamental results of probability and statistics, *the Central Limit Theorem* (CLT).

While the CLT has several versions and many generalizations, they all have one thing in common: summations of a large number of random quantities, each with finite variance, yield a sum that is approximately normally distributed. This is the main reason that the normal distribution is ubiquitous in nature and present throughout the universe.

We now develop this more formally. Consider an i.i.d. sequence $X_1, X_2, \ldots$ where all X_i are distributed according to some distribution $F(x_i\ ;\ \theta)$ with mean μ and finite variance σ^2. Consider now the random variable

$$Y_n := \sum_{i=1}^{n} X_i.$$

It is clear that $\mathbb{E}[Y_n] = n\mu$ and $\mathrm{Var}(Y_n) = n\sigma^2$. Hence, we may consider a random variable

$$\tilde{Y}_n := \frac{Y_n - n\mu}{\sqrt{n}\sigma}.$$

Observe that $\tilde{Y}_n$ is zero mean and unit variance. The CLT states that as $n \to \infty$, the distribution of $\tilde{Y}_n$ converges to a standard normal distribution. That is, for every $x \in \mathbb{R}$,

$$\lim_{n\to\infty} \mathbb{P}(\tilde{Y}_n \le x) = \int_{-\infty}^{x} \frac{1}{\sqrt{2\pi}} e^{-\frac{u^2}{2}}\, du.$$

Alternatively, this may be viewed as indicating that for non-small n

$$Y_n \underset{\text{approx}}{\sim} N\big(n\mu, n\sigma^2\big),$$

where N is a the normal distribution with mean $n\mu$ and variance $n\sigma^2$.

In addition, by dividing the numerator and denominator of $\tilde{Y}_n$ by n, we see an immediate consequence of the CLT. That is, for non-small n, the sample mean of n observations denoted by $\overline{X}_n$ satisfies

$$\overline{X}_n \underset{\text{approx}}{\sim} N\left(\mu, \left(\frac{\sigma}{\sqrt{n}}\right)^2\right).$$

Hence, the CLT states that sample means from i.i.d. samples with finite variances are asymptotically distributed according to a normal distribution as the sample size grows. This ubiquity of the normal distribution justifies the normality assumption employed when using many of the statistical procedures that we cover in Chapters 6 and 7.

To illustrate the CLT, consider three different distributions below, noting that each has a mean and variance both equal 1:

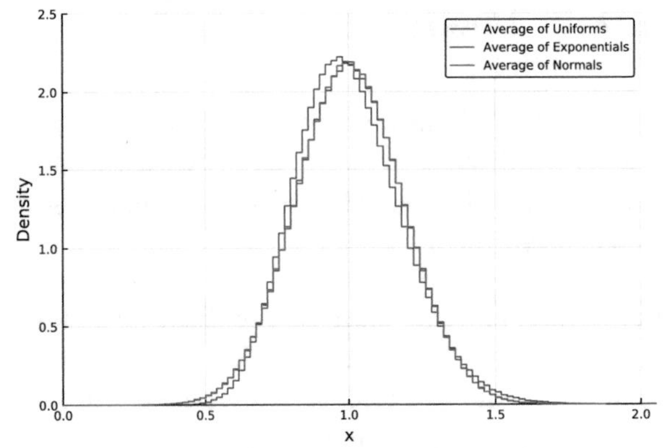

Figure 5.7: Histograms of sample means for different underlying distributions.

1. A uniform distribution, on $[1-\sqrt{3},\ 1+\sqrt{3}]$.

2. An exponential distribution with $\lambda = 1$.

3. A normal distribution with both a mean and variance of 1.

In Listing 5.6, we illustrate the central limit theorem, by generating a histogram of `N` sample means for each of the three different distributions mentioned above. Although each of the underlying distributions is very different, i.e. uniform, exponential and normal, the sampling distribution of the sample means all approach that of the normal distribution centered about 1 with standard deviation $1/\sqrt{n}$. Notice that in the case of the exponential distribution, $n = 30$ isn't "enough" to get a "perfect fit" to a normal distribution (Figure 5.7).

Listing 5.6: The central limit theorem

```
using Distributions, Plots; pyplot()

n, N = 30, 10^6

dist1 = Uniform(1-sqrt(3),1+sqrt(3))
dist2 = Exponential(1)
dist3 = Normal(1,1)

data1 = [mean(rand(dist1,n)) for _ in 1:N]
data2 = [mean(rand(dist2,n)) for _ in 1:N]
data3 = [mean(rand(dist3,n)) for _ in 1:N]

stephist([data1 data2 data3], bins=100,
    c=[:blue :red :green], xlabel = "x", ylabel = "Density",
    label=["Average of Uniforms" "Average of Exponentials" "Average of Normals"],
    normed=true, xlims=(0,2), ylims=(0,2.5))
```

In lines 5–7, we define three different distribution type objects: a continuous uniform distribution over the domain $[1-\sqrt{3},\ 1+\sqrt{3}]$, an exponential distribution with a mean of 1, and a normal distribution with mean and standard deviation both 1. In lines 9–11, we generate `N` sample means, each consisting of `n` observations, for each distribution defined above. In lines 13–16, we plot three separate histograms based on the sample mean vectors previously generated. It can be observed that for large `N`, these histograms approach that of a normal distribution, and in addition, the mean of the data approaches the mean of the underlying distribution from which the samples were taken.

5.4 Point Estimation

Given a random sample, $X_1, \ldots, X_n$, a common task of statistical inference is to estimate a parameter θ, or a function of it, say $h(\theta)$. The process of designing an estimator, analyzing its performance, and carrying out the estimation is called *point estimation*.

Although we can never know the underlying parameter θ, or $h(\theta)$ exactly, we can arrive at an estimate for it via an *estimator* $\hat{\theta} = f(X_1, \ldots, X_n)$. Here, the design of the estimator is embodied by $f(\cdot)$, a function that specifies how to construct the estimate from the sample.

An important question to ask is how close is $\hat{\theta}$ to the actual unknown quantity θ or $h(\theta)$. In this section, we first describe several ways of quantifying and categorizing this "closeness", and then present two common methods for designing estimators; the *method of moments* and *maximum likelihood estimation* (MLE).

The design of (point) estimators is a central part of statistics. However, in elementary statistics courses for science students, engineers, or social studies researchers, point estimation is often not explicitly mentioned. The reason for this is that one can estimate the mean and variance via, $\overline{X}$ and S^2, respectively, see (5.1). That is, in the case of $h(\cdot)$ being either the mean or the variance of the distribution, the estimator given by the sample mean or sample variance, respectively, is a natural candidate and performs exceptionally well. However, in other cases, choosing an estimation procedure is less straightforward.

Consider, for example, the case of a uniform distribution on the range $[0, \theta]$, and say we are interested in estimating θ based on a random sample, $X_1, \ldots, X_n$. In this case, one could construct an estimator in many different ways. For example, here are a few alternative estimators:

$$\begin{aligned}
\hat{\theta}_1 &= f_1(X_1, \ldots, X_n) := \max\{X_i\}, \\
\hat{\theta}_2 &= f_2(X_1, \ldots, X_n) := 2\,\overline{X}, \\
\hat{\theta}_3 &= f_3(X_1, \ldots, X_n) := 2\,\mathrm{median}(X_1, \ldots, X_n), \\
\hat{\theta}_4 &= f_4(X_1, \ldots, X_n) := \sqrt{12 S^2}.
\end{aligned} \tag{5.5}$$

Each of these makes some sense in their own right; $\hat{\theta}_1$ is based on the fact that θ is an upper bound of the observations, $\hat{\theta}_2$ and $\hat{\theta}_3$ utilize the fact that the sample mean and sample median are both expected to fall on $\theta/2$, and finally $\hat{\theta}_4$ utilizes the fact that the variance of the distribution is given by $S^2 = \theta^2/12$. Given that there are various possible estimators, we require a methodology for comparing them and perhaps developing others, with the aim of choosing a suitable one. In the remainder of this section, we describe some methods for analyzing the performance of such estimators and others.

Describing the Performance and Behavior of Estimators

When analyzing the performance of an estimator $\hat{\theta}$, it is important to understand that it is a random variable. One common measure of its performance is the *Mean Squared Error* (MSE),

$$MSE_\theta(\hat{\theta}) := \mathbb{E}\big[(\hat{\theta} - \theta)^2\big] = \text{Var}(\hat{\theta}) + (\mathbb{E}[\hat{\theta}] - \theta)^2 := \text{variance} + \text{bias}^2. \tag{5.6}$$

The second equality arises naturally from adding and subtracting $\mathbb{E}[\hat{\theta}]$, expanding and collecting terms. In this representation, we see that the MSE can be decomposed into the variance of the estimator, and it's bias squared. Low variance is clearly a desirable performance measure. The same applies to the *bias*, which is a measure of the expected difference between the estimator and the true parameter value. Note that in machine learning the "Bias variance tradeoff" is often considered as a tradeoff between model complexity and model generalizability. The idea is similar to the decomposition in (5.6), however it is conceptually different because the setting is different. More details are in Chapter 9.

One question that arises with regards to estimation is: are there cases where estimators are *unbiased*—that is, they have a bias of 0, or alternatively $\mathbb{E}[\hat{\theta}] = \theta$? The answer is yes. We show this now using the sample mean as a simple example.

Consider $X_1, \ldots, X_n$ distributed according to any distribution with a finite mean μ. In this case, say we are interested in estimating μ (note that $\mu = h(\theta)$ for some function h). It is easy to see that the sample mean $\overline{X}$ is itself a random variable with mean μ and is hence unbiased. Furthermore, the variance of this estimator is σ^2/n, where σ^2 is the original variance of X_i. Since the estimator is unbiased, the MSE equals the variance, i.e. σ^2/n. In fact, it can be shown that the sample mean is the estimator of θ with minimal mean square error over all other estimators.

Now consider a case where the population mean μ is known, but the population variance σ^2 is unknown, and that we wish to estimate it. As a sensible estimator consider

$$\hat{\sigma^2} := \frac{1}{n}\sum_{i=1}^{n}(X_i - \mu)^2. \tag{5.7}$$

Computing the mean of $\hat{\sigma^2}$ yields:

$$\mathbb{E}[\hat{\sigma^2}] = \frac{1}{n}\mathbb{E}\left[\sum_{i=1}^{n}(X_i - \mu)^2\right] = \frac{1}{n}\sum_{i=1}^{n}\mathbb{E}\big[(X_i - \mu)^2\big] = \frac{1}{n}n\sigma^2 = \sigma^2.$$

Hence, $\hat{\sigma^2}$ is an unbiased estimator for σ^2. However, say we are now also interested in estimating the (population) standard deviation, σ. In this case, it is natural to use the estimator

$$\hat{\sigma} := \sqrt{\hat{\sigma^2}} = \sqrt{\frac{1}{n}\sum_{i=1}^{n}(X_i - \mu)^2}.$$

Interestingly, while this is a perfectly sensible estimator, it is not unbiased. We illustrate this via simulation in Listing 5.7. In it, we consider a uniform distribution over $[0,1]$, where the population mean, variance, and standard deviation are 0.5, 1/12, and $\sqrt{1/12}$, respectively. We then estimate the bias of $\hat{\sigma^2}$ and $\hat{\sigma}$ via Monte Carlo simulation. The output shows that $\hat{\sigma}$ is not unbiased. However, as the numerical results illustrate, it is *asymptotically unbiased*. That is, the bias tends to 0 as the sample size n grows.

Listing 5.7: A biased estimator

```julia
1  using Random, Statistics
2  Random.seed!(0)
3
4  trueVar, trueStd = 1/12, sqrt(1/12)
5
6  function estVar(n)
7      sample = rand(n)
8      sum((sample .- 0.5).^2)/n
9  end
10
11 N = 10^7
12 for n in 5:5:30
13     biasVar = mean([estVar(n) for _ in 1:N]) - trueVar
14     biasStd = mean([sqrt(estVar(n)) for _ in 1:N]) - trueStd
15     println("n = ",n, " Var bias: ", round(biasVar, digits=5),
16                 "\t Std bias: ", round(biasStd, digits=5))
17 end
```

```
n = 5 Var bias: 1.0e-5          Std bias: -0.00642
n = 10 Var bias: 1.0e-5         Std bias: -0.00303
n = 15 Var bias: 0.0            Std bias: -0.00199
n = 20 Var bias: -1.0e-5        Std bias: -0.00148
n = 25 Var bias: -1.0e-5        Std bias: -0.00117
n = 30 Var bias: 0.0            Std bias: -0.00098
```

In lines 6–8, the function `estVar()` is defined, which implements (5.7). In lines 12–17, we loop over sample sizes $n = 5, 10, 15, \dots, 30$, and for each we repeat `N` sampling experiments, for which we estimate the biases for $\hat{\sigma^2}$ and $\hat{\sigma}$, respectively. The biases are then estimated and the values stored in `biasVar` and `biasStd`.

Having explored an estimator for σ^2 with μ known, as well as briefly touching an estimator for σ for the same case, we now ask the question: What would be a sensible estimator for σ^2 for the more realistic case where μ is not known? A natural first suggestion would be to replace μ in (5.7) with $\overline{X}$ to obtain

$$\tilde{S}^2 := \frac{1}{n}\sum_{i=1}^{n}(X_i - \overline{X})^2.$$

With a few lines of computations involving expectations, one can verify that

$$\mathbb{E}[\tilde{S}^2] = \frac{n-1}{n}\sigma^2.$$

Hence, it is biased, albeit asymptotically unbiased. This is the reason that the preferred estimator, S^2 is actually,

$$S^2 = \frac{n}{n-1}\tilde{S}^2.$$

as in (5.1). This yields an unbiased estimator.

There are other important qualitative properties of estimators that one may explore. One such property is *consistency*. Roughly, we say that an estimator is consistent if it converges to the true value as the number of observations grows to infinity. More can be found in mathematical statistics references such as [DS11] and [CB01]. The remainder of this section presents two common methodologies for estimating parameters; method of moments and maximum likelihood estimation. A comparison of these two methodologies is presented.

Method of Moments

The *method of moments* is a methodological way to obtain parameter estimates for a distribution. The key idea is based on moment estimators for the k'th moment, $\mathbb{E}[X_i^k]$,

$$\hat{m}_k = \frac{1}{n}\sum_{i=1}^{n} X_i^k. \tag{5.8}$$

As a simple example, consider a uniform distribution on $[0, \theta]$. An estimator for the first moment ($k = 1$) is then, $\hat{m}_1 = \overline{X}$. Now we denote by X a typical random variable from this sample. For such a distribution, $\mathbb{E}[X^1] = \theta/2$. We can then equate the moment estimator with the first-moment expression to arrive at the equation,

$$\frac{\theta}{2} = \hat{m}_1.$$

Notice that this equation involves the unknown parameter θ and the moment estimator obtained from the data. Then trivially solving for θ yields the estimator,

$$\hat{\theta} = 2\hat{m}_1,$$

which is exactly $\hat{\theta}_2$ from (5.5).

In cases where there are multiple unknown parameters, say K, we use the first K moment estimates to formulate a system of K equations and K unknowns. This system of equations can be written as

$$\mathbb{E}[X^k \; ; \; \theta_1, \ldots, \theta_K] = \hat{m}_k \qquad \text{for} \qquad k = 1, \ldots, K. \tag{5.9}$$

For many textbook examples (such as the uniform distribution case described above), we are able to solve this system of equations analytically, yielding a solution,

$$\hat{\theta}_k = g_k(\hat{m}_1, \ldots, \hat{m}_K) \qquad \text{for} \qquad k = 1, \ldots, K. \tag{5.10}$$

Here the functions $g_k(\cdot)$ describe the solution of the system of equations. However, it is often not possible to obtain explicit expressions for $g_k(\cdot)$. In these cases, numerical techniques are typically used to solve the corresponding system of equations.

As an example, consider the triangular distribution with density,

$$f(x) = \begin{cases} 2\dfrac{x-a}{(b-a)(c-a)}, & x \in [a, c), \\ 2\dfrac{b-x}{(b-a)(b-c)}, & x \in [c, b]. \end{cases}$$

This distribution has support $[a, b]$, and a maximum at c with $a \le c \le b$ and $a < b$. Note that the `Julia` triangular distribution function uses this same parameterization: `TriangularDist (a,b,c)`.

Now straightforward (yet tedious) computation yields the first three moments, $\mathbb{E}[X^1]$, $\mathbb{E}[X^2]$, $\mathbb{E}[X^3]$, as well as the system of equations for the method of moments:

$$\begin{aligned}\hat{m}_1 &= \frac{1}{3}(a+b+c), \\ \hat{m}_2 &= \frac{1}{6}(a^2+b^2+c^2+ab+ac+bc), \\ \hat{m}_3 &= \frac{1}{10}(a^3+b^3+c^3+a^2b+a^2c+b^2a+b^2c+c^2a+c^2b+abc).\end{aligned} \tag{5.11}$$

Generally, this system of equations is not analytically solvable. Hence, the method of moments estimator is given by a numerical solution to (5.11). In Listing 5.8, given a series of observations, we numerically solve this system of equations through the use of the `NLsolve` package and arrive at estimates for the values of a, b and c. Observe that the Eq. (5.11) are symmetric in terms of a, b, and c in the sense that permuting these values does not change the equations. Hence, when using a numerical solver, there is a possibility that it will return an arbitrary permutation of the solutions. We remedy this by sorting the solutions and picking estimators for a, b, and c according to the sorted order.

Listing 5.8: Point estimation via the method of moments using a numerical solver

```
using Random, Distributions, NLsolve
Random.seed!(0)

a, b, c = 3, 5, 4
dist = TriangularDist(a,b,c)
n = 2000
samples = rand(dist,n)

m_k(k,data) = 1/n*sum(data.^k)
mHats = [m_k(i,samples) for i in 1:3]

function equations(F, x)
    F[1] = 1/3*( x[1] + x[2] + x[3] ) - mHats[1]
    F[2] = 1/6*( x[1]^2 + x[2]^2 + x[3]^2 + x[1]*x[2] + x[1]*x[3] +
                 x[2]*x[3] ) - mHats[2]
    F[3] = 1/10*( x[1]^3 + x[2]^3 + x[3]^3 + x[1]^2*x[2] + x[1]^2*x[3] +
                 x[2]^2*x[1] + x[2]^2*x[3] + x[3]^2*x[1] + x[3]^2*x[2] +
                 x[1]*x[2]*x[3] ) - mHats[3]
end

nlOutput = nlsolve(equations, [ 0.1; 0.1; 0.1])
sol = sort(nlOutput.zero)
aHat, bHat, cHat = sol[1], sol[3], sol[2]
println("Found estimates for (a,b,c) = ", (aHat, bHat, cHat) , "\n" )
println(nlOutput)
```

```
Found estimates for (a,b,c) = (3.002706152232, 5.003033254712, 3.999191608726)

Results of Nonlinear Solver Algorithm
 * Algorithm: Trust-region with dogleg and autoscaling
 * Starting Point: [0.1, 0.1, 0.1]
 * Zero: [5.00303, 3.99919, 3.00271]
 * Inf-norm of residuals: 0.000000
 * Iterations: 14
 * Convergence: true
```

```
    * |x - x'| < 0.0e+00: false
    * |f(x)| < 1.0e-08: true
  * Function Calls (f): 15
  * Jacobian Calls (df/dx): 13
```

In line 1, we specify using the `NLsolve` package. This package contains numerical methods for solving non-linear systems of equations. In lines 4–7, we specify the parameters of the triangular distribution and the distribution itself `dist`. We also specify the total number of samples `n` and generate our sample set of observations `samples`. In line 9, the function `m_k()` is defined, which implements (5.8), and in line 10, this function is used to estimate the first three moments, given our observations `samples`. In line 12–19, we set up the system of simultaneous equations within the function `equations()`. This specific format is used as it is a requirement of the `nlsovle()` function which is used later. The `equations()` function takes two arrays as input, `F` and `x`. The elements of `F` represent the left hand side of the series of equations (which are later solved for zero), and the elements of `x` represent the corresponding constants of the equations. Note that in setting up the equations from (5.11), the moment estimators are moved to the right hand side, so that the zeros can be found. In line 21, the `nlsolve()` function from the `NLsolve` package is used to solve the zeros of the function `equations()`, given starting coefficient estimates of `[0.1; 0.1; 0.1]`. In this example, since the Jacobian was not specified, it is computed by finite differences. In lines 22–23, we sort the solution and set the estimates of the parameters based on the sorted order. In line 24, the zeros of our function are printed as output through the use of `.zero`, which is used to return just the zero field of the `nlsolve()` output. In line 25, the complete output from the function `nlsolve()` is printed as output.

Maximum Likelihood Estimation (MLE)

Maximum likelihood estimation is another commonly used technique for creating point estimators. In fact, in the study of mathematical statistics, it is probably the most popular method used. The key principle is to consider the *likelihood* of the parameter θ having a specific value given observations $x_1, \ldots, x_n$. That is, what is the most likely parameter value based on the observations. This is done via the likelihood function, which is presented below for the i.i.d. case of continuous probability distributions,

$$L(\theta \; ; \; x_1, \ldots, x_n) = f_{X_1,\ldots,X_n}(x_1, \ldots, x_n \; ; \; \theta) = \prod_{i=1}^{n} f(x_i \; ; \; \theta). \tag{5.12}$$

In the second equality, the joint probability density of $X_1, \ldots, X_n$ is represented as the product of the individual probability densities since the observations are assumed i.i.d.

A key observation is that the likelihood, $L(\cdot)$, in (5.12) is a function of the parameter θ, influenced by the sample, $x_1, \ldots, x_n$. Now given the likelihood, the *maximum likelihood estimator* is a value θ that maximizes $L(\theta \; ; \; x_1, \ldots, x_n)$. The rational behind using this as an estimator is that it chooses the parameter value θ that is most plausible, given the observed sample.

As an example, consider the continuous uniform distribution on $[0, \theta]$. In this case, it is useful to consider the PDF for an individual observation as

$$f(x \; ; \; \theta) = \frac{1}{\theta} \mathbf{1}\{x \in [0, \theta]\} \qquad \text{for} \qquad x \in \mathbb{R}.$$

Here the *indicator function* $\mathbf{1}\{\cdot\}$ explicitly constrains the support of the random variable to $[0, \theta]$. Now using (5.12), it follows that

$$L(\theta \; ; \; x_1, \ldots, x_n) = \frac{1}{\theta^n} \prod_{i=1}^{n} \mathbf{1}\{x_i \in [0, \theta]\} = \frac{1}{\theta^n} \mathbf{1}\{0 \leq \min_i x_i\} \mathbf{1}\{\max_i x_i \leq \theta\}.$$

From this, we see that for any sample $x_1, \ldots, x_n$ with non-negative values, this function (of θ) is maximized at $\hat{\theta} = \max_i x_i$. Hence, as you can see the MLE for this case is exactly $\hat{\theta}_1$ from (5.5).

Many textbooks present constructed examples of MLEs, where the likelihood is a differentiable function of θ. In such cases, these MLEs can be solved explicitly, by carrying out the optimization of the likelihood function analytically (for example, see [CB01] and [DS11]). However, this is not always possible, and often numerical optimization of the likelihood function is carried out instead.

As an example, consider the case where we have n random samples from what we know to be a gamma distribution, with PDF,

$$f(x) = \frac{\lambda^\alpha}{\Gamma(\alpha)} x^{\alpha-1} e^{-\lambda x}.$$

and parameters, $\lambda > 0$ and $\alpha > 0$. In such a case where λ and α are both unknown, there is not an explicit solution to the MLE optimization problem, and hence we resort to numerical methods instead. In Listing 5.9, we use MLE to construct a plot of the likelihood function. That is, given synthetic data, we calculate the likelihood function for various combinations of α and λ. Note that directly after this example, we present an elegant approach for this numerical problem.

Listing 5.9: The likelihood function for a gamma distribution's parameters

```
using Random, Distributions, Plots, LaTeXStrings; pyplot()
Random.seed!(0)

actualAlpha, actualLambda = 2,3
gammaDist = Gamma(actualAlpha,1/actualLambda)
n = 10^2
sample = rand(gammaDist, n)

alphaGrid = 1:0.02:3
lambdaGrid = 2:0.02:5

likelihood = [prod([pdf.(Gamma(a,1/l),v) for v in sample])
                        for l in lambdaGrid, a in alphaGrid]

surface(alphaGrid, lambdaGrid, likelihood, lw=0.1,
        c=cgrad([:blue, :red]), legend=:none, camera = (135,20),
        xlabel=L"\alpha", ylabel=L"\lambda", zlabel="Likelihood")
```

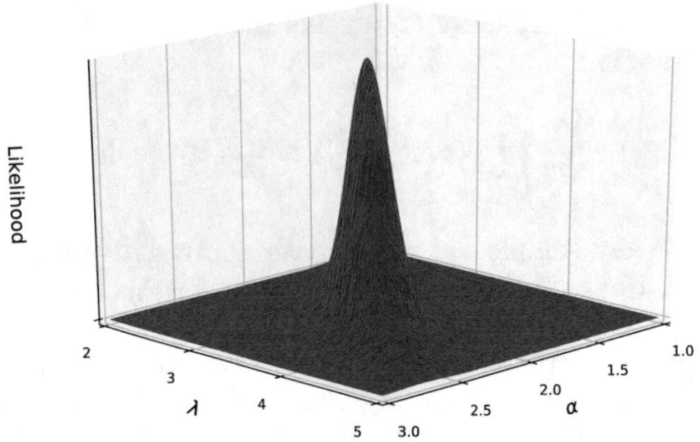

Figure 5.8: Likelihood function on different combinations of α and λ for a gamma distribution.

In lines 4–5, we specify the parameters α and λ, as well as the underlying distribution, `gammaDist`. Note that the gamma distribution in Julia, `Gamma()`, uses a different parameterization to what is outlined in Chapter 3 (i.e. `Gamma()` uses α, and $1/\lambda$). In lines 6–7, we generate `n` sample observations, `sample`. In lines 9–10, we specify the grid of values over which we will calculate the likelihood function, based on various combinations of α and λ. In lines 12–13, we first evaluate the likelihood function, (5.12), through the use of the `prod()` function on an array of all PDF values, evaluated for each sample observation, `v`. Through the use of a two-way comprehension, this process is repeated for all possible combinations of `a` and `l` in `alphaGrid` and `lambdaGrid`, respectively. This results in a two-dimensional array of evaluated likelihood functions for various combinations of α and λ, denoted `likelihood`. Lines 15–17 create the plot.

The likelihood function plotted in Figure 5.8 embodies the data. An MLE is then the maximizer of the likelihood. We now investigate this optimization problem further, and in the process present further insight. First observe that any maximizer, $\hat{\theta}$, of $L(\theta \; ; \; x_1, \ldots, x_n)$ will also maximize its logarithm. Practically, both from an analytic and numerical perspective, considering this *log-likelihood function* is often more attractive:

$$\ell(\theta \; ; \; x_1, \ldots, x_n) := \log L(\theta \; ; \; x_1, \ldots, x_n) = \sum_{i=1}^{n} \log \big(f(x_i \; ; \; \theta)\big).$$

Hence, given a sample from a gamma distribution as before, the log-likelihood function is

$$\ell(\theta \; ; \; x_1, \ldots, x_n) = n\alpha \log(\lambda) - n \log(\Gamma(\alpha)) + (\alpha - 1) \sum_{i=1}^{n} \log(x_i) - \lambda \sum_{i=1}^{n} x_i.$$

We may then divide by n (without compromising the optimizer) to obtain the following function that needs to be maximized:

$$\tilde{\ell}(\theta \; ; \overline{x}, \; \overline{x}_\ell) = \alpha \log(\lambda) - \log(\Gamma(\alpha)) + (\alpha - 1)\overline{x}_\ell - \lambda \overline{x},$$

where $\overline{x}$ is the sample mean and,

$$\overline{x}_\ell := \frac{1}{n} \sum_{i=1}^{n} \log(x_i).$$

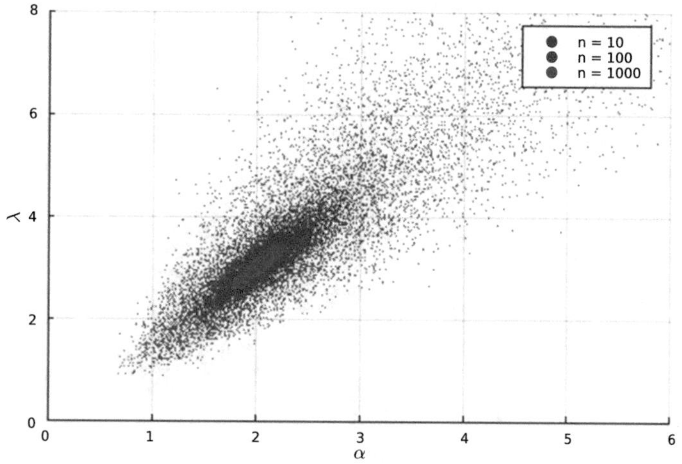

Figure 5.9: Repetitions of MLE for a gamma(2, 3) distribution with $n = 10, 100, 1000$. For $n = 100$ and $n = 1000$, asymptotic normality is visible.

Further simplification is possible by removing the stand-alone $-\overline{x}_\ell$ term, as it does not affect the optimal value. Hence, our optimization problem is then,

$$\max_{\lambda>0,\ \alpha>0} \quad \alpha\big(\log(\lambda) + \overline{x}_\ell\big) - \log(\Gamma(\alpha)) - \lambda\overline{x}. \tag{5.13}$$

As is typical in such cases, the function actually depends on the sample only through the two *sufficient statistics* $\overline{x}$ and $\overline{x}_\ell$. Now in optimizing (5.13), we aren't able to obtain an explicit expression for the maximizer. However, taking α as fixed, we may consider the derivative with respect to λ, and equate this to 0:

$$\frac{\alpha}{\lambda} - \overline{x} = 0.$$

Hence, for any optimal α^*, we have that $\lambda^* = \alpha^*/\overline{x}$. This allows us to substitute λ^* for λ in (5.13) to obtain

$$\max_{\alpha>0} \quad \alpha(\log(\alpha) - \log(\overline{x}) + \overline{x}_\ell) - \log(\Gamma(\alpha)) - \alpha. \tag{5.14}$$

Now by taking the derivative of (5.14) with respect to α, and equating this to 0, we obtain

$$\log(\alpha) + 1 - \log(\overline{x}) + \overline{x}_\ell - \psi(\alpha) - 1 = 0,$$

where $\psi(z) := \frac{d}{dz}\log(\Gamma(z))$ is the well-known *digamma function*. Hence, we find that α^* must satisfy

$$\log(\alpha) - \psi(\alpha) - \log(\overline{x}) + \overline{x}_\ell = 0. \tag{5.15}$$

In addition, since $\lambda^* = \alpha^*/\overline{x}$, our optimal MLE solution is given by (α^*, λ^*). In order to find this value, (5.15) must be solved numerically.

In Listing 5.10, we do just this. In fact, we repeat the act of numerically solving (5.15) many times, and in the process illustrate the distribution of the MLE in terms of λ and α. Note that there are many more properties of the MLE that we do not discuss here, including the asymptotic distribution of the MLE, which happens to be a multivariate normal. However, through this example, we provide an intuitive illustration of the distribution of the MLE, which is bivariate in this case and can be observed in Figure 5.9.

Listing 5.10: MLE for the gamma distribution

```
1  using SpecialFunctions, Distributions, Roots, Plots, LaTeXStrings; pyplot()
2  
3  eq(alpha, xb, xbl) = log(alpha) - digamma(alpha) - log(xb) + xbl
4  
5  actualAlpha, actualLambda = 2, 3
6  gammaDist = Gamma(actualAlpha,1/actualLambda)
7  
8  function mle(sample)
9      alpha  = find_zero( (a)->eq(a,mean(sample),mean(log.(sample))), 1)
10     lambda = alpha/mean(sample)
11     return [alpha,lambda]
12 end
13 
14 N = 10^4
15 
16 mles10   = [mle(rand(gammaDist,10)) for _ in 1:N]
17 mles100  = [mle(rand(gammaDist,100)) for _ in 1:N]
18 mles1000 = [mle(rand(gammaDist,1000)) for _ in 1:N]
19 
20 scatter(first.(mles10), last.(mles10),
21         c=:blue, ms=1, msw=0, label="n = 10")
22 scatter!(first.(mles100), last.(mles100),
23         c=:red, ms=1, msw=0, label="n = 100")
24 scatter!(first.(mles1000), last.(mles1000),
25         c=:green, ms=1, msw=0, label="n = 1000",
26         xlims=(0,6), ylims=(0,8), xlabel=L"\alpha", ylabel=L"\lambda")
```

In line 1, we specify usage of the `SpecialFunctions` and `Roots` packages, as they contain the `digamma()` and `find_zero()` functions, respectively. In line 3, the `eq()` function implements equation (5.15). Note it takes three arguments, an alpha value `alpha`, a sample mean `xb`, and the mean of the log of each observation `xbl`, which is calculated element-wise via `log.()`. This allows us to apply `eq()` on vectors. In lines 5–6, we specify the actual parameters of the underlying gamma distribution, as well as the distribution itself. In lines 8–12, the function `mle()` is defined, which in line 9 takes an array of sample observations, and solves the value of `alpha` which satisfies the zero of `eq()`. This is done through the use of the `find_zero()` function, and the anonymous function `(a)->eq(a,mean(sample),mean(log.(sample)))`. Note the trailing `1` in line 9, which is used as the initial value of the iterative solver. In line 10, the corresponding `lambda` value is calculated, and both alpha and lambda are returned as an array of values. In line 16, ten random samples are made from our gamma distribution, and then the function `mle()` is used to solve for the corresponding values of alpha and lambda and an array of arrays. This experiment is repeated through a comprehension `N` times total and the resulting array of arrays stored as `mles10`. Lines 17–18 repeat the same procedure as that in line 16, however in these two cases, the experiments are conducted for 100 and 1000 random samples, respectively. In lines 20–22, a scatterplot of the resulting pairs of $\hat{\alpha}$ and $\hat{\lambda}$ are plotted, for the cases of the sample size being equal to 10, 100, and 1000. Note the use of the `first()` and `last()` functions, which are used to return the values of alpha and lambda respectively. Note that the bivariate distribution of alpha and lambda can be observed. In addition, for a larger number of observations, it can be seen that the data is centered on the true underlying parameters alpha and lambda values of 2 and 3. This agrees with the fact that the MLE is asymptotically unbiased.

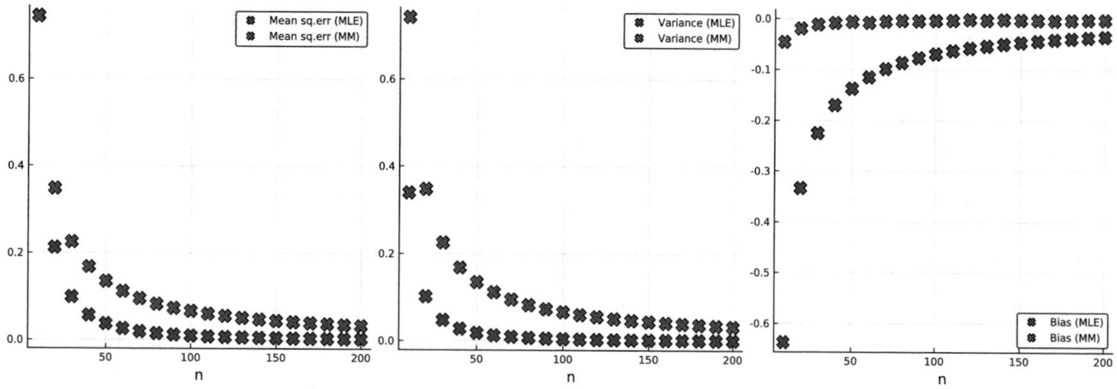

Figure 5.10: Comparing the method of moments and MLE in terms of MSE, variance, and bias.

Comparing the Method of Moments and MLE

We now carry out an illustrative comparison between a method of moments estimator and an MLE estimator on a specific example. Consider a random sample $x_1, \ldots, x_n$ from a uniform distribution on the interval (a, b). The MLE for the parameter $\theta = (a, b)$, can be shown to be

$$\hat{a} = \min\{x_1, \ldots, x_n\}, \qquad \hat{b} = \max\{x_1, \ldots, x_n\}. \tag{5.16}$$

For the method of moments estimator, since $X \sim \text{uniform}(a, b)$, it follows that

$$\mathbb{E}[X] = \frac{a+b}{2}, \qquad \text{Var}(X) = \frac{(b-a)^2}{12}.$$

Hence, by solving for a and b, and replacing $\mathbb{E}[X]$ and $\text{Var}(X)$ with $\overline{x}$ and s^2 respectively, we obtain,

$$\hat{a} = \overline{x} - \sqrt{3}\, s, \qquad \hat{b} = \overline{x} + \sqrt{3}\, s. \tag{5.17}$$

Observe that here we are actually using the second central moment (variance) as opposed to the second moment to construct the estimator. This is a slight variation on the method of moments method described above and yields a nicer expression.

Now we can compare how the estimators (5.16) and (5.17) perform based on MSE, specifically the variance and bias. In Listing 5.11, we use Monte Carlo simulation to compare the estimates of $\hat{b}$ using both the method of moments and MLE, for different cases of n. The code creates Figure 5.10 analyzing MSE, bias, and variance. As can be seen, the MSE of maximum likelihood is lower than the MSE of the method of moments and this is due to the variance of maximum likelihood being lower. However, maximum likelihood exhibits more significant bias than the method of moments. Nevertheless, observe that after squaring, the bias contribution to the MSE is not as significant as the variance. The reader should keep in mind that these conclusions about MSE/variance/bias are specific to this example. However, there is more supporting theory for the usefulness of maximum likelihood estimation as $n \to \infty$. See for example [CB01].

Listing 5.11: MSE, bias and variance of estimators

```
1   using Distributions, Plots; pyplot()
2
3   N = 10^5
4   nMin, nStep, nMax = 10, 10, 200
5   nn = Int(nMax/nStep)
6   sampleSizes = nMin:nStep:nMax
7   trueB = 5
8   trueDist = Uniform(-2, trueB)
9
10  MLEest(data) = maximum(data)
11  MMest(data)  = mean(data) + sqrt(3)*std(data)
12
13  res = Dict{Symbol,Array{Float64}}(
14      ((sym) -> sym => Array{Float64}(undef,nn)).(
15          [:MSeMLE,:MSeMM, :VarMLE,:VarMM,:BiasMLE,:BiasMM]))
16
17  for (i, n) in enumerate(sampleSizes)
18      mleEst, mmEst = Array{Float64}(undef, N), Array{Float64}(undef, N)
19      for j in 1:N
20          sample    = rand(trueDist,n)
21          mleEst[j] = MLEest(sample)
22          mmEst[j]  = MMest(sample)
23      end
24      meanMLE, meanMM = mean(mleEst), mean(mmEst)
25      varMLE, varMM = var(mleEst), var(mmEst)
26
27      res[:MSeMLE][i] = varMLE + (meanMLE - trueB)^2
28      res[:MSeMM][i] = varMM + (meanMM - trueB)^2
29      res[:VarMLE][i] = varMLE
30      res[:VarMM][i] = varMM
31      res[:BiasMLE][i] = meanMLE - trueB
32      res[:BiasMM][i] = meanMM - trueB
33  end
34
35  p1 = scatter(sampleSizes, [res[:MSeMLE] res[:MSeMM]], c=[:blue :red],
36      label=["Mean sq.err (MLE)" "Mean sq.err (MM)"])
37  p2 = scatter(sampleSizes, [res[:VarMLE] res[:VarMM]], c=[:blue :red],
38      label=["Variance (MLE)" "Variance (MM)"])
39  p3 = scatter(sampleSizes, [res[:BiasMLE] res[:BiasMM]], c=[:blue :red],
40      label=["Bias (MLE)" "Bias (MM)"])
41
42  plot(p1, p2, p3, ms=10, shape=:xcross, xlabel="n",
43      layout=(1,3), size=(1200, 400))
```

In line 4, the minimum, maximum and step size for sample size observations are specified. These are used to define the number of sample size groups `nn`. In lines 7 and 8, the true parameter, `trueB` is specified. Lines 10 and 11 specify the two estimators in the functions `MLEest()` and `MMest()`. Line 13–15 create a dictionary mapping symbols (type `Symbol`) to arrays (type `Array{Float64}`). The dictionary is initialized with symbol keys `:MSeMLE`,...,`:BiasMM`, and with values that are empty arrays. The main simulation loop is in lines 17–33 where we use `enumerate` to loop over tuples `(i,n)`, with `i` the index of the iteration and `n` a value from the range `sampleSizes`. In each iteration, we initialize empty arrays for parameter estimates in line 18. We then repeat the experiment `N` times in the loop of lines 19–23. Lines 24–32 record performance measures in the dictionary `res`.

5.5 Confidence Interval as a Concept

Now that we have dealt with the concept of a point estimator, we consider how confident we are about our estimate. The previous section included an analysis of such confidence in terms of the mean squared error and its variance and bias components. However, given a single sample, $X_1, \ldots, X_n$, how does one obtain an indication about the accuracy of the estimate? Here the concept of a *confidence interval* comes as an aid.

Consider the case where we are trying to estimate the parameter θ. A confidence interval is then an interval $[L, U]$ obtained from our sample data, such that

$$\mathbb{P}(L \leq \theta \leq U) = 1 - \alpha, \tag{5.18}$$

where $1 - \alpha$ is called the *confidence level*. Knowing this range $[L, U]$ in addition to θ is useful, as it indicates some level of certainty in regards to the unknown value. Much of elementary classical statistics involves explicit formulas for L and U, based on the sample $X_1, \ldots, X_n$. Most of Chapter 6 is dedicated to this, however in this section, we simply introduce the concept through an elementary non-standard example.

Consider a case of a single observation X ($n = 1$) taken from a symmetric triangular distribution, with a spread of 2 and an unknown center (mean) μ. In this case, we would set

$$L = X + q_{\alpha/2}, \qquad U = X + q_{1-\alpha/2},$$

where q_u is the u'th quantile of a triangular distribution centered at 0, and having a spread of 2. Setting L and U in this manner ensures that (5.18) holds. Note that this is not the only possible construction of a confidence interval, however it makes sense due to the symmetry of the problem. For such a triangular distribution, calculating quantiles using integration or areas of triangles, it holds that $q_{\alpha/2} = -1 + \sqrt{\alpha}$ and $q_{1-\alpha/2} = 1 - \sqrt{\alpha}$.

Now, given observations, (a single observation in this case), we can compute L and U. A demonstration of this is performed in Listing 5.12 below.

Listing 5.12: A confidence interval for a symmetric triangular distribution

```julia
using Random, Distributions
Random.seed!(0)

alpha = 0.05
L(obs) = obs - (1-sqrt(alpha))
U(obs) = obs + (1-sqrt(alpha))

mu = 5.57
observation = rand(TriangularDist(mu-1,mu+1,mu))
println("Lower bound L: ", L(observation))
println("Upper bound U: ", U(observation))
```

```
Lower bound L: 5.1997170907797585
Upper bound U: 6.7525034952798
```

In lines 5–6, the functions `L()` and `U()` implement the formulas above. In this simple example, the actual (unknown) parameter value μ is set in line 8. Then the sample, a single observation in this case, is obtained in line 9. The virtue of the example is in presenting the 95% confidence interval, as output by lines 10 and 11. Based on the output (after rounding), we know that with probability 0.95, the unknown parameter lies in the range $[5.2, 6.75]$.

Let us now further explore the meaning of a confidence interval by considering (5.18). The key point is that there is a $1-\alpha$ chance that the actual parameter θ lies in the interval $[L, U]$. This means that if the sampling experiment is repeated say N times, then on average, $N \times (1-\alpha)\%$ of the time the actual parameter θ is covered by the interval.

In Listing 5.13, we present an example where we repeat the previous sampling process $N = 100$ times. Each time we take a single sample (a single observation in this case) and construct the corresponding confidence interval. We observe that about $\alpha \times 100$ times the confidence interval, $[L, U]$, does not include the parameter in question, μ. The results are presented in Figure 5.11.

Listing 5.13: Repetitions of a confidence interval

```
1   using Random, Distributions, StatsPlots; pyplot()
2   Random.seed!(2)
3
4   alpha   = 0.05
5   L(obs)  = obs - (1-sqrt(alpha))
6   U(obs)  = obs + (1-sqrt(alpha))
7
8   mu = 5.57
9   triDist = TriangularDist(mu-1,mu+1,mu)
10
11  N = 100
12  hitBounds, missBounds = zeros(N, 2), zeros(N,2)
13  for i in 1:N
14      observation = rand(triDist)
15      LL, UU = L(observation), U(observation)
16      if LL <= mu && mu <= UU
17          hitBounds[i,:] = [LL  UU-LL]
18      else
19          missBounds[i,:] = [LL  UU-LL]
20      end
21  end
22
23  groupedbar(hitBounds, bar_position=:stack,
24      c=:blue, la=0, fa=[0 1], label="", ylims=(3,8))
25  groupedbar!(missBounds, bar_position=:stack,
26      c=:red, la=0, fa=[0 1], label="", ylims=(3,8))
27  plot!([0,N+1],[mu,mu],
28      c=:black, xlims=(0,N+1),
29      ylims=(3,8), label="Parameter value", ylabel="Value Estimate")
```

At the heart of this example, we repeat the experiment `N = 100` times and create the matrices `hitBounds` and `missBounds`. These are plotted via the `groupedbar()` function from package `StatsPlots`. The main loop in lines 13–21 records a confidence interval as a "hit" in line 17 or alternatively as a "miss" in line 19.

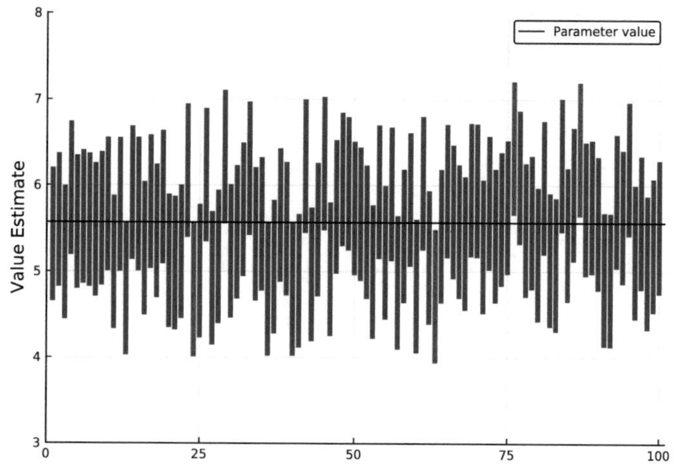

Figure 5.11: 100 confidence intervals. The blue confidence interval bars contain the unknown parameter, while the red ones do not.

5.6 Hypothesis Tests Concepts

Having explored point estimation and confidence intervals, we now consider ideas associated with *hypothesis testing*. The approach involves partitioning the parameter space Θ into Θ_0 and Θ_1, and then, based on the sample, concluding whether one of two hypotheses, H_0 or H_1, holds. Here,

$$H_0 : \theta \in \Theta_0, \qquad H_1 : \theta \in \Theta_1. \tag{5.19}$$

The hypothesis H_0 is called the *null hypothesis* and H_1 the *alternative hypothesis*. The former is the default hypothesis, and in carrying out hypothesis testing our general aim (or hope) is to reject this hypothesis. This is because in typical situations we wish to demonstrate that the alternative hypothesis holds, as opposed to some well-established status quo captured by the null hypothesis.

		Decision	
		Do not reject H_0	Reject H_0
Reality	H_0 is true	Correct (1-α) “true negative”	Type I error (α) “false positive”
	H_0 is false	Type II error (β) “false negative’	Correct ($1-\beta$) “true positive”

Table 5.1: Type I and Type II errors with their probabilities α and β respectively.

Since our decision is based on a random sample, there is always a chance of making a mistakenly false conclusion. As summarized in Table 5.1, the two types of errors that can be made are a *type I error*: Rejecting H_0 falsely, sometimes called a “false positive”, or a *type II error*: Failing to correctly reject H_0, sometimes called a “false negative”. The probability α quantifies the likelihood of making a type I error, while the probability of making a type II error is denoted by β. Note that $1-\beta$ is known as the *power* of the hypothesis test, and this concept of power is covered in more detail in

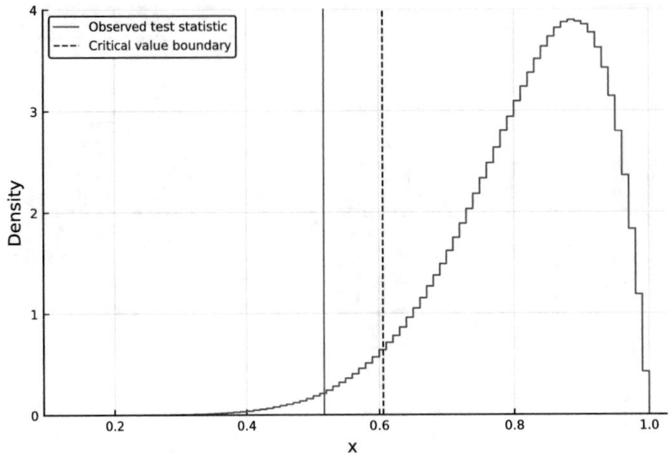

Figure 5.12: The distribution of the test statistic, X^*, under H_0. With $\alpha = 0.05$ the rejection region is to the left of the black dashed line. In a specific sample, the test statistic is on the red line and we reject H_0.

Section 7.5. Note that in carrying out a hypothesis test, α is typically specified, while power is not directly controlled, but rather is influenced by the sample size and other factors.

An important point in terminology is that we don't use the phrase "accept" for the null hypothesis, rather we "fail to reject it" (if we stick with H_0) or "reject it" (if we choose H_1). This is because when we fail to reject H_0, we typically don't know the actual value of β, hence we aren't able to put a level of certainty on H_0 being the case. However if we do reject H_0, then by the design of hypothesis tests we can say that our error probability is bounded by α.

We now present some elementary examples which illustrate the basic concepts involved. Standard hypothesis tests are discussed in depth in Chapter 7.

The Test Statistic, Rejection Region and p-Values

In general, the key objects in hypothesis testing are the *test statistic*, the *rejection region* and *p-values.* Once the scientific question is formulated as a hypothesis by partitioning the parameter space according to (5.19), the next step is to calculate the test statistic. For this, we define the test statistic, denoted X^*, as a function of the data. An example can be the sample mean, the sample variance, or other statistics. Importantly, with probabilistic assumptions on the sample data, the test statistic is a random variable itself.

Since the test statistic follows some distribution under H_0, the next step is to consider how likely it is to observe the specific value calculated from our sample data. To this end, in setting up the hypothesis test, we typically choose a significance level α, at 0.05, 0.01, or a similar value. It quantifies our level of tolerance for enduring a type I error. For example, setting $\alpha = 0.01$ implies we wish to design a test where the probability of type I error is at most 0.01 if H_0 holds. Clearly, a low α is desirable, however, there are tradeoffs involved since seeking a very low α will imply a high β (low power).

With the test statistic and α at hand, we are able to determine the rejection region which we denote by $\mathcal{R}$. It is a subset of the real line where $\mathbb{P}(X^* \in \mathcal{R}) \leq \alpha$, under H_0. The idea is then to calculate the test statistic X^* and reject H_0 if $X^* \in \mathcal{R}$, and otherwise not to reject H_0. Typically, $\mathcal{R}$ is selected at one or both extremes of the support, depending on the distribution of the test statistic and the hypothesis (5.19).

To illustrate these concepts, we now present a simple yet non-standard example. Consider that we have a series of sample observations distributed as continuous uniform between 0 and some unknown upper bound, m. Say that we set

$$H_0: \quad m = 1, \qquad H_1: \quad m < 1.$$

With observations $X_1, \ldots, X_n$, one possible test statistic is the sample range:

$$X^* = \max(X_1, \ldots, X_n) - \min(X_1, \ldots, X_n).$$

As is always the case, the test statistic is a random variable. Under H_0, we expect the distribution of X^* to have support $[0, 1]$ with the most likely value being close to 1. This is because low values of X^* are less plausible under H_0 since we can expect the minimum to be near 0 and the maximum to be near 1. The explicit form of the distribution of X^* can be analytically obtained however for simplicity we use a Monte Carlo simulation to estimate it and present the density in Figure 5.12 for $n = 10$ observations.

For this case, it is sensible to reject H_0 if X^* is small. Hence, denoting quantiles of this distribution by $q_0(u)$, we set the rejection region as $\mathcal{R} = [0, q_0(\alpha)]$. Using Monte Carlo, we also compute the rejection region and present it in the figure where the *critical value* is the upper boundary, $q_0(\alpha)$, of the rejection region. Note that computing the rejection region does not require any sample data as it is based on model assumptions and not the sample. Still, it is computed via Monte Carlo in this specific example. The *decision rule* for this hypothesis test is simple: Compare the observed value of the test statistic, x^*, to the critical value $q_0(\alpha)$ and reject H_0 if $x^* \leq q_0(\alpha)$, otherwise do not reject.

An alternative view of hypothesis tests is to consider the p-value. Here we collect the data and compute the observed value of the test statistic x^*. The p-value is then the maximal α under which the test would be rejected with the observed test statistic. In other words we find p which solves $x^* = q_0(p)$. This is computed via $F_0(x^*)$, where $F_0(\cdot)$ is the CDF of X^*.

Using the p-value approach, reporting a low p-value (e.g. $p = 0.0024$) implies that we are very confident in rejecting H_0, while a high p-value (e.g. $= 0.24$) implies we are not. The p-value approach can be used to decide whether H_0 should be rejected or not with a specified α. For this, simply compare p and α, and reject H_0 if $p \leq \alpha$.

Listing 5.14 creates Figure 5.12 and illustrates the operation of the hypothesis test. In this case, we illustrate a scenario where the unknown parameter m is `muActual = 0.75`. With the specific seed selected, it turns out that $x^* = 0.517$. This corresponds to a p-value of 0.0141, which is rejected for $\alpha = 0.05$, however would not be rejected if $\alpha = 0.01$. Keep in mind that the `N` repetitions in this example are simple to obtain the distribution of X^* under H_0 and the critical value, `0.6058`. In the many standard hypothesis tests presented in Chapter 7, the distribution of the test statistic is analytically available, so such Monte Carlo-based computation is not needed.

Notice that with specific sample (depends on the seed), lines 18–19 of the code are not executed. However, if you were to change the seed in line 2, this would simulate a scenario with different data points, and it is possible to not reject H_0 even though H_1 holds (`muActual` < 1).

Listing 5.14: The distribution of a test statistic under H_0

```
1  using Distributions, Random, Statistics, Plots; pyplot()
2  Random.seed!(2)
3
4  n, N, alpha = 10, 10^7, 0.05
5  mActual = 0.75
6  dist0, dist1 = Uniform(0,1), Uniform(0,mActual)
7
8  ts(sample) = maximum(sample) - minimum(sample)
9
10 empiricalDistUnderH0 = [ts(rand(dist0,n)) for _ in 1:N]
11 rejectionValue = quantile(empiricalDistUnderH0,alpha)
12
13 sample = rand(dist1,n)
14 testStat = ts(sample)
15 pValue = sum(empiricalDistUnderH0 .<= testStat)/N
16
17 if testStat > rejectionValue
18     print("Didn't reject: ", round(testStat,digits=4))
19     print(" > ", round(rejectionValue,digits=4))
20 else
21     print("Reject: ", round(testStat,digits=4))
22     print(" <= ", round(rejectionValue,digits=4))
23 end
24 println("\np-value = $(round(pValue,digits=4))")
25
26 stephist(empiricalDistUnderH0, bins=100, c=:blue, normed=true, label="")
27 plot!([testStat, testStat], [0,4], c=:red, label="Observed test statistic")
28 plot!([rejectionValue, rejectionValue], [0,4], c=:black, ls=:dash,
29         label="Critical value boundary", legend=:topleft, ylims=(0,4),
30         xlabel = "x", ylabel = "Density")
```

```
Reject: 0.517 <= 0.6058
p-value = 0.0141
```

In line 8, we define the function `ts()` which calculates the test statistic from a sample. We use it in lines 10–11 to obtain `N` (many) samples under H_0 and calculate the `rejectionValue`, $q_0(\alpha)$. The actual testing procedure begins in line 13 when we collect our sample, simulating a point in H_1 (since `mActual = 0.75`). The test statistic is calculated in line 14 and the p-value in line 15. The decision rule is then executed in lines 17–23 and the p-value is also presented. The remainder of the code creates Figure 5.12.

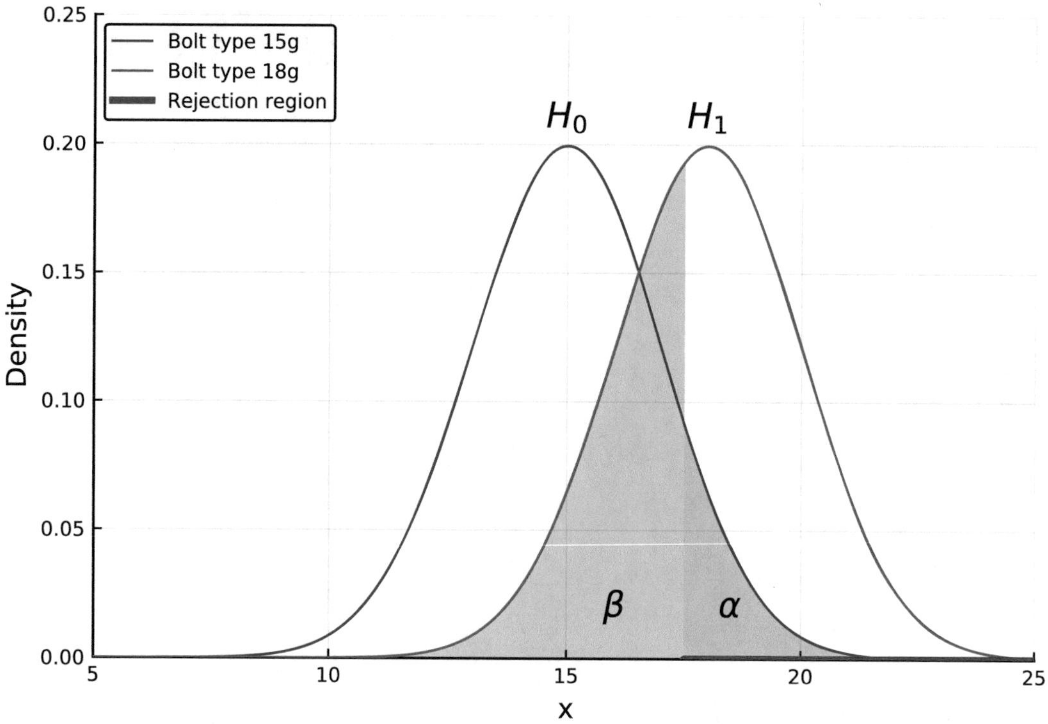

Figure 5.13: Type I (blue) and Type II (green) errors. The rejection region based from $\tau = 17.5$ to the right is colored with red on the horizontal axis.

Simple Hypothesis Tests

When the alternative parameter spaces Θ_0 and Θ_1 are only comprised of a single point each, the hypothesis test is called a *simple hypothesis test.* Such a test is often not of great practical use, but we introduce it here for pedagogical purposes. Specifically, by analyzing such tests we can understand how type I and type II errors interplay.

As an introductory example, consider a container that contains two identical types of pipes, except that one type weighs 15 grams on average and the other 18 grams on average. The standard deviation of the weights of both pipe types is 2 grams. Imagine now that we sample a single pipe, and wish to determine its type. Denote the weight of this pipe by the random variable X. For this example, we devise the following statistical hypothesis test: $\Theta_0 = \{15\}$ and $\Theta_1 = \{18\}$. Now, given a threshold τ, we reject H_0 if $X > \tau$, otherwise we retain H_0.

In this circumstance, we can explicitly analyze the probabilities of both the type I and type II errors, α and β respectively. Listing 5.15 generates Figure 5.13, which illustrates this graphically for $\tau = 17.5$. You may try to modify the value of `tau` in the code to see how the probabilities for type I and type II errors vary.

Listing 5.15: A simple hypothesis test

```
1  using Distributions, StatsBase, Plots, LaTeXStrings; pyplot()
2
3  mu0, mu1, sd, tau  = 15, 18, 2, 17.5
4  dist0, dist1 = Normal(mu0,sd), Normal(mu1,sd)
5  grid = 5:0.1:25
6  h0grid, h1grid = tau:0.1:25, 5:0.1:tau
7
8  println("Probability of Type I error: ", ccdf(dist0,tau))
9  println("Probability of Type II error: ", cdf(dist1,tau))
10
11 plot(grid, pdf.(dist0,grid),
12         c=:blue, label="Bolt type 15g")
13 plot!(h0grid, pdf.(dist0, h0grid),
14         c=:blue, fa=0.2, fillrange=[0 1], label="")
15 plot!(grid, pdf.(dist1,grid),
16         c=:green, label="Bolt type 18g")
17 plot!(h1grid, pdf.(dist1, h1grid),
18         c=:green, fa=0.2, fillrange=[0 1], label="")
19 plot!([tau, 25],[0,0],
20         c=:red, lw=3, label="Rejection region",
21         xlims=(5, 25), ylims=(0,0.25) , legend=:topleft,
22     xlabel="x", ylabel="Density")
23 annotate!([(16, 0.02, text(L"\beta")),(18.5, 0.02, text(L"\alpha")),
24             (15, 0.21, text(L"H_0")),(18, 0.21, text(L"H_1"))])
```

```
Probability of Type I error: 0.10564977366685525
Probability of Type II error: 0.4012936743170763
```

In line 3, we set the parameters of the example. In line 4, we define the distributions under H_0 and H_1. Line 6 sets girds of values that are used for plotting type I and type II error ranges. In lines 8-9, we compute α and β using `ccdf()` and `cdf()` on `dist0` and `dist1`, respectively. The remainder of the code creates the figure using the `pdf()` function. Notice the calls to `plot!()` in lines 13–14 and 17–18 using the `fillrange` argument.

The Receiver Operating Curve

In the previous example, $\tau = 17.5$ was arbitrarily chosen. Clearly, if τ was increased the probability of making a Type I error, α, would decrease, while the probability of making a type II error, β, would increase. Conversely, if we decreased τ the reverse would occur. We now introduce the *Receiver Operating Curve* (ROC), also sometime called the *receiver operating characteristic curve*. It is a tool that helps to visualize the tradeoff between type I and type II errors. It allows one to visualize the error tradeoffs for all possible τ values simultaneously, for a particular alternative hypothesis H_1.

We look at three different scenarios for μ_1: $16, 18,$ and 20. Clearly, the bigger the difference between μ_0 and μ_1, the easier it should be to make a decision without errors. In Listing 5.16, we consider each scenario and shift τ, and in the process plot the analytic coordinates of $\big(\alpha(\tau), 1-\beta(\tau)\big)$. This is the ROC. It is a *parametric plot* of the probability of a type I error and power. The results are in Figure 5.14. ROCs are also a way of comparing different sets of hypotheses simultaneously. By plotting several different ROCs on the same figure, we can compare the likelihood of making errors for various scenarios of different μ_1's.

To better understand how the ROCs are generated, consider also Figure 5.13 and imagine the effect of sliding τ. In this figure, the shaded blue area represents α, while 1 minus the green area represents power. If one considers $\tau = 25$, then both α and power are almost zero, and this corresponds (approximately) to $(0, 0)$ in Figure 5.14. Now, as the τ threshold is slowly decreased, it can be seen that the power increases at a much faster rate than α, and this behavior is observed in the ROC. In addition, as the difference in means between the null and alternative hypotheses are greater, the ROC curves are shown to be pushed "further out" from the diagonal dashed line, reflecting the fact that such alternative sets of hypotheses are easier to detect.

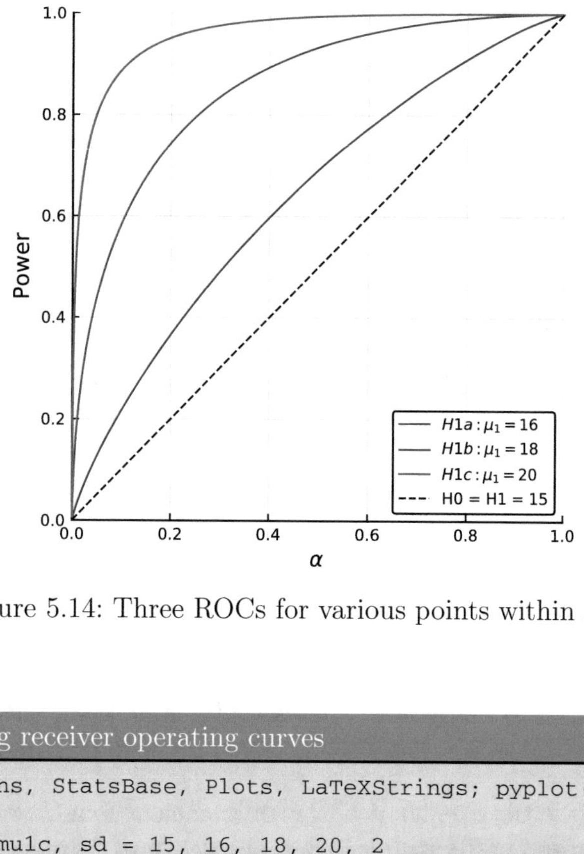

Figure 5.14: Three ROCs for various points within H_1.

Listing 5.16: Comparing receiver operating curves

```
using Distributions, StatsBase, Plots, LaTeXStrings; pyplot()

mu0, mu1a, mu1b, mu1c, sd = 15, 16, 18, 20, 2
tauGrid = 5:0.1:25

dist0 = Normal(mu0,sd)
dist1a, dist1b, dist1c  = Normal(mu1a,sd), Normal(mu1b,sd), Normal(mu1c,sd)

falsePositive = ccdf.(dist0,tauGrid)
truePositiveA, truePositiveB, truePositiveC =
    ccdf.(dist1a,tauGrid), ccdf.(dist1b,tauGrid), ccdf.(dist1c,tauGrid)

plot(falsePositive, [truePositiveA truePositiveB truePositiveC],
    c=[:blue :red :green],
    label=[L"H1a: \mu_1 = 16" L"H1b: \mu_1 = 18" L"H1c: \mu_1 = 20"])
plot!([0,1], [0,1], c=:black, ls=:dash, label="H0 = H1 = 15",
    xlims=(0,1), ylims=(0,1), xlabel=L"\alpha", ylabel="Power",
    ratio=:equal, legend=:bottomright)
```

The range `tauGrid` presents the range of possible values for τ that are used. The distribution `dist0` is for H_0 and the distributions `dist1a`, `dist1b`, and `dist1c` are for three variants of H_1. The plots are then plots of `falsePositives` vs. `truePositiveA`, `truePositiveB` or `truePositiveC`. The essence of the plotting code is to use `falsePositives` for arguments of the horizontal coordinate and `truePositiveA`, `truePositiveB`, or `truePositiveC` as arguments of the vertical coordinate. This creates a parametric plot. Lines 16–18 plot a diagonal dashed line. This line represents the extreme case of the distributions of H_0 and H_1 directly overlapping. In this case, the probability of a Type I error is the same as the power.

A Randomized Hypothesis Test

We now investigate the concept of a *randomization test*, which is a type of *non-parametric test*, i.e. a statistical test which does not require that we know what type of distribution the data comes from. A virtue of non-parametric tests is that they do not impose a specific model. Consider the following example, where a farmer wants to test whether a new fertilizer is effective at increasing the yield of her tomato plants. As an experiment, she took 20 plants, kept 10 as controls, and treated the remaining 10 with fertilizer. After 2 months, she harvested the plants and recorded the yield of each plant (in kg) as shown in Table 5.2.

Control	4.17	5.58	5.18	6.11	4.5	4.61	5.17	4.53	5.33	5.14
Fertilizer	6.31	5.12	5.54	5.5	5.37	5.29	4.92	6.15	5.8	5.26

Table 5.2: Yield in kg for 10 plants with, and 10 plants without fertilizer (control).

It can be observed that the group of plants treated with fertilizer have an average yield 0.494 kg greater than that of the control group. One could argue that this difference is due to the effects of the fertilizer. We now investigate if this is a reasonable assumption. Let us assume for a moment that the fertilizer had no effect on plant yield (H_0) and that the result was simply due to random chance. In such a scenario, we actually have 20 observations from the same group, and regardless of how we arrange our observations, we would expect to observe similar results.

Hence, we can investigate the likelihood of this outcome occurring by random chance, by considering all possible *combinations* of 10 samples from our group of 20 observations, and counting how many of these combinations result in a difference in sample means greater than or equal to 0.494 kg. The proportion of times this occurs is analogous to the likelihood that the difference we observe in our sample means was purely due to random chance. It is in a sense the p-value.

Before proceeding, we calculate the number of ways one can sample $r = 10$ unique items from $n = 20$ total, which is given by

$$\binom{20}{10} = 184,756.$$

Hence, the number of possible combinations in our example is computationally manageable. Note that in a different situation where n and r would be bigger, e.g. $n = 40$ and $r = 20$, the number of combinations would be too big for an exhaustive search (about 137 billion). In such a case, a viable alternative is to randomly sample combinations for estimating the p-value.

In Listing 5.17, we use Julia's `Combinatorics` package to enumerate the difference in sample means for every possible combination. From the output, we observe that only 2.39% of all possible combinations result in a sample mean greater than or equal to our treated group, i.e. a difference greater than or equal to 0.494 kg. Therefore, there is significant statistical evidence that the fertilizer increases the yield of the tomato plants since under H_0, there is only a 2.39% chance of obtaining this value or greater by random chance.

Listing 5.17: A randomized hypothesis test

```
1  using Combinatorics, Statistics, DataFrames, CSV
2
3  data = CSV.read("../data/fertilizer.csv")
4  control = data.Control
5  fertilizer = data.FertilizerX
6
7  subGroups = collect(combinations([control;fertilizer],10))
8
9  meanFert = mean(fertilizer)
10 pVal = sum([mean(i) >= meanFert for i in subGroups])/length(subGroups)
11 println("p-value = ", pVal)
```

```
p-value = 0.023972157873086666
```

We use the `Combinatorics` package for the `combinations()` function. In line 3–5, we import our data and store the data for the control and fertilized groups in the arrays `control` and `fertilizer`. In line 7, all observations are concatenated into one array via the use of `[ ; ]`. Following this, the `combinations()` function is used to generate an iterator object for all combinations of 10 elements from our 20 observations. The `collect()` function then converts this iterator into an array of all possible combinations of 10 objects, sampled from 20 total. This array of all combinations is stored as `subGroups`.In line 9, the mean of the fertilizer group is calculated and assigned to the variable `meanFert`. In line 10, the mean of each combination in the array `x` is calculated and compared against `meanFert`. The proportion of means which are greater than or equal to `meanFert` is then calculated through the use of a comprehension, and the functions `sum()` and `length()`.

5.7 A Taste of Bayesian Statistics

In this section, we briefly explore the Bayesian approach to statistical inference as an alternative to the frequentist view of statistics which was introduced in Sects. 5.4, 5.5 and 5.6, and used throughout the remainder of this book. In the Bayesian paradigm, the (scalar or vector) parameter θ is not assumed to exist as some fixed unknown quantity but instead is assumed to follow a distribution. That is, the parameter itself is a random variable, and the act of *Bayesian inference* is the process of obtaining more information about the distribution of θ. Such a setup is useful in many practical situations since it allows one to incorporate prior beliefs about the parameter before experience from new observations is taken into consideration. It also allows one to carry out repeated inference in a very natural manner by allowing inference in future periods to rely on past experience or past data.

The key objects at play are the *prior distribution* of the parameter and the *posterior distribution* of the parameter. The former is postulated beforehand, or exists as a consequence of previous inference, while the latter captures the distribution of the parameter after observations are taken into account. The relationship between the prior and the posterior is

$$\text{posterior} = \frac{\text{likelihood} \times \text{prior}}{\text{evidence}} \qquad \text{or} \qquad f(\theta \mid x) = \frac{f(x \mid \theta) \times f(\theta)}{\int f(x \mid \theta) f(\theta)\, d\theta}. \tag{5.20}$$

This is nothing but Bayes' rule applied to densities. Here the prior distribution (density) is $f(\theta)$, and the posterior distribution (density) is $f(\theta \mid x)$. Observe that the denominator, known as *evidence* or *marginal likelihood*, is constant with respect to the parameter θ. This allows the equation to be written as

$$f(\theta \mid x) \propto f(x \mid \theta) \times f(\theta), \tag{5.21}$$

where the symbol "$\propto$" denotes "proportional to". Hence, the posterior distribution can be easily obtained up to the normalizing constant (the evidence) by multiplying the prior with the likelihood, $f(x \mid \theta)$.

In general, carrying out *Bayesian inference* involves the following steps:

1. Assume some distributional model for the data based on the parameter θ which is a random variable.

2. Use previous inference experience, elicit an expert, or make an educated guess to determine a prior distribution, $f(\theta)$. The prior distribution might be parameterized by its own parameters, called *hyper-parameters.*

3. Collect data x and create an expression or a computational mechanism for the likelihood $f(x \mid \theta)$ based on the distributional model chosen.

4. Use the relationship (5.20) to obtain the posterior distribution of the parameters, $f(\theta \mid x)$. In most cases, the evidence (denominator of (5.20)) is not easily computable. Hence the posterior distribution is only available up to a normalizing constant. In some special cases, the form of the posterior distribution is the same as the prior distribution. In such cases, *conjugacy* holds, the prior is called a *conjugate prior*, and the hyper-parameters are updated from prior to posterior.

5. The posterior distribution can then be used to make conclusions about the model. For example, if a single specific parameter value is needed to make the model concrete, a *Bayes estimate* based on the posterior distribution, for example, the *posterior mean*, may be computed:

   $$\hat{\theta} = \int \theta f(\theta \mid x)\, d\theta. \tag{5.22}$$

   Further analyses such as obtaining *credible intervals*, similar to confidence intervals, may also be carried out. See a brief discussion in Section 6.7.

6. The model with $\hat{\theta}$ can then be used for making conclusions. Alternatively, a whole class of models based on the posterior distribution $f(\hat{\theta} \mid x)$ can be used. This often goes hand in hand with simulation as one is able to generate Monte Carlo samples from the posterior distribution.

Bayesian inference has gained significant popularity over the past few decades and has evolved together with the whole field of *computational statistics*. Unless conjugacy holds, there is typically not an explicit expression for the evidence (the integral in (5.20)), and hence a computational challenge is to make use of the posterior available only up to a normalizing constant. We now elaborate on the details through variants of a very simple example in order to understand the main concepts. For a general treatment of Bayesian inference, we recommend [R07].

A Simple Poisson Example

Consider an example where an insurance company models the number of weekly fires in a city using a Poisson distribution with parameter λ. Here, λ is also the expected number of fires per week. Assume that the following data is collected over a period of 16 weeks,

$$x = (x_1, \dots, x_{16}) = (2, 1, 0, 0, 1, 0, 2, 2, 5, 2, 4, 0, 3, 2, 5, 0).$$

Each data point indicates the number of fires per week. In this case, the MLE is $\hat{\lambda} = 1.8125$ simply obtained by the sample mean. Hence, in a frequentist approach, after 16 weeks the distribution of the number of fires per week is modeled by a Poisson distribution with $\lambda = 1.8125$. One can then obtain estimates for say, the probability of having more than 5 fires in a given week as follows:

$$\mathbb{P}(\text{fires per week} > 5) = 1 - \sum_{k=0}^{5} e^{-\lambda} \frac{\lambda^k}{k!} \approx 0.0107. \tag{5.23}$$

However, the drawback of such an approach in estimating λ is that it didn't make use of previous information. By comparison, in a Bayesian approach, the estimate would allow one to incorporate information from previous years, or alternatively from adjacent geographical areas. Say that for example, further knowledge comes to light that the number of fires per week ranges between 0 and 10 and that the typical number is 3 fires per week. In this case, one can assign a prior distribution to λ that captures this belief. Here is where some critics claim that such use of Bayesian statistics turns into somewhat of an "inexact science" since we have an infinite number of options to choose for the prior. Still, having some prior is often more useful than no incorporation of prior information at all. A simple justification is that you also have an infinite number of choices for the model no matter what approach you use. Hence, almost any paradigm relies on some sort of subjective prior information.

In our example, assume that we decide to use a triangular distribution as shown in blue in Figure 5.15. Such a triangular distribution captures prior beliefs about the parameter λ well because it has a defined range and a defined mode.

With the prior assigned and the data collected, we can now use the machinery of Bayesian inference of (5.20). In this specific case, the prior distribution of the parameter λ is the triangular distribution with the PDF,

$$f(\lambda) = \begin{cases} \frac{1}{15}\lambda, & \lambda \in [0, 3], \\ \frac{1}{35}(10 - \lambda), & \lambda \in (3, 10]. \end{cases}$$

With the 16 observations, $x_1, \dots, x_{16}$, the likelihood is

$$f(\lambda \mid x) = \prod_{k=1}^{16} e^{-\lambda} \frac{\lambda^{x_k}}{x_k!}.$$

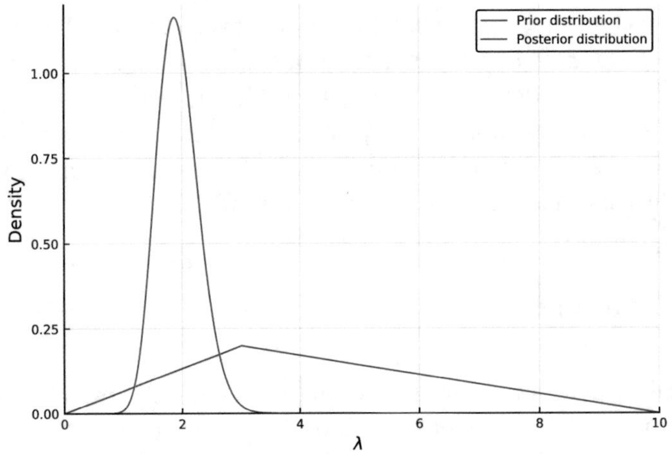

Figure 5.15: The prior distribution in blue and the posterior in red for Bayesian estimation of a Poisson distribution.

Hence the posterior is proportional to $f(x \mid \lambda)f(\lambda)$. However, normalization of this function in λ requires dividing it by the evidence, given by,

$$\int_0^{10} f(x \mid \lambda)f(\lambda)\, d\lambda.$$

Typically, this integral isn't easy to evaluate analytically, hence numerical methods are often used. For illustration purposes, we carry out this numerical integration as part of Listing 5.18 where we also plot the resulting posterior distribution (red curve in Figure 5.15). To appreciate potential problems with such a numerical solution, imagine cases where the parameter θ is not just the scalar λ but rather consists of multiple dimensions. The integral of the evidence cannot be efficiently computed in such cases.

In Listing 5.18, once the prior distribution is obtained, we compute its mean to obtain a Bayes estimate for λ. The value obtained differs from the MLE obtained above and hence probability estimates using the model, such as (5.23) would also vary. Importantly, by employing the Bayesian perspective, we were able to incorporate prior knowledge into the inference procedure.

Listing 5.18: Bayesian inference with a triangular prior

```
1  using Distributions, Plots, LaTeXStrings; pyplot()
2
3  prior(lam) = pdf(TriangularDist(0, 10, 3), lam)
4  data = [2,1,0,0,1,0,2,2,5,2,4,0,3,2,5,0]
5
6  like(lam) = *([pdf(Poisson(lam),x) for x in data]...)
7  posteriorUpToK(lam) = like(lam)*prior(lam)
8
9  delta = 10^-4.
10 lamRange = 0:delta:10
11 K = sum([posteriorUpToK(lam)*delta for lam in lamRange])
12 posterior(lam) = posteriorUpToK(lam)/K
13
14 bayesEstimate = sum([lam*posterior(lam)*delta for lam in lamRange])
15 println("Bayes estimate: ",bayesEstimate)
16
17 plot(lamRange, prior.(lamRange),
18         c=:blue, label="Prior distribution")
19 plot!(lamRange, posterior.(lamRange),
20         c=:red, label="Posterior distribution",
21         xlims=(0, 10), ylims=(0, 1.2),
22         xlabel=L"\lambda",ylabel="Density")
```

```
Bayes estimate: 1.9371887551439297
```

In line 3, we define the prior. In line 4, we set the data values. In line 6, the likelihood function is defined. Notice that the * operator is used as a function and that the splat operator `...` is applied inside the brackets. Equation (5.21) is implemented in Line 7, while lines 9–11 are used to numerically compute the evidence. The actual posterior is defined in line 12. In line 14, a Bayes estimate from the prior is calculated, according to (5.22) and printed in line 15. The remainder of the code creates Figure 5.15.

Conjugate Priors

Following on from the previous example, a natural question arises: why use the specific form of the prior distribution? After all, the results would vary if we were to choose a different prior. While in generality Bayesian statistics doesn't supply a complete answer, there are cases where certain families of prior distributions work very well with certain (other) families of statistical models (likelihoods).

For example, in our case of a Poisson probability distribution model, it turns out that assuming a gamma prior distribution works nicely. This is because the resulting posterior distribution is also guaranteed to be gamma. In such a case, the gamma distribution is said to be a *conjugate prior* to the Poisson distribution. The parameters of the prior/posterior distribution are called *hyper-parameters*, and by exhibiting a conjugate prior distribution relationship, the hyper-parameters typically have a simple update law from prior to posterior. This relieves a huge computational burden.

To see this in the case of a gamma-Poisson conjugate pair, assume the hyper-parameters of the prior to have α (shape parameter) and β (rate parameter). Now by using the Poisson likelihood and the gamma PDF, we obtain

$$\begin{aligned}
\text{posterior} &\propto \left(\prod_{k=1}^{n} e^{-\lambda}\frac{\lambda^{x_k}}{x_k!}\right)\frac{\beta^\alpha}{\Gamma(\alpha)}\lambda^{\alpha-1}e^{-\beta\lambda} \\
&\propto e^{-n\lambda}\lambda^{\sum_{k=1}^{n} x_k}\lambda^{\alpha-1}e^{-\beta\lambda} \\
&= \lambda^{\alpha+(\sum_{k=1}^{n} x_k)-1}e^{-\lambda(\beta+n)} \\
&\propto \text{gamma density with shape parameter } \alpha+\sum x_i \text{ and rate parameter } \beta+n.
\end{aligned} \tag{5.24}$$

This shows us the gamma-Poisson conjugacy and implies a slick update rule for the hyper-parameters: The hyper-parameter α is updated to $\alpha + \sum x_i$ and the hyper-parameter β is updated to $\beta + n$.

In Listing 5.19, we use a gamma prior with prior parameters of $\alpha = 8$ and $\beta = 2$. For illustration, we compute the posterior using both the brute force method of the previous listing and using the simple hyper-parameter update rule due to conjugacy. The posterior and prior are plotted in Figure 5.16.

Listing 5.19: Bayesian inference with a gamma prior

```
using Distributions, Plots; pyplot()

alpha, beta = 8, 2
prior(lam) = pdf(Gamma(alpha, 1/beta), lam)
data = [2,1,0,0,1,0,2,2,5,2,4,0,3,2,5,0]

like(lam) = *([pdf(Poisson(lam),x) for x in data]...)
posteriorUpToK(lam) = like(lam)*prior(lam)

delta = 10^-4.
lamRange = 0:delta:10
K = sum([posteriorUpToK(lam)*delta for lam in lamRange])
posterior(lam) = posteriorUpToK(lam)/K

bayesEstimate = sum([lam*posterior(lam)*delta for lam in lamRange])

newAlpha, newBeta = alpha + sum(data), beta + length(data)
closedFormBayesEstimate = mean(Gamma(newAlpha, 1/newBeta))

println("Computational Bayes Estimate: ", bayesEstimate)
println("Closed form Bayes Estimate: ", closedFormBayesEstimate)

plot(lamRange, prior.(lamRange),
        c=:blue, label="Prior distribution")
plot!(lamRange, posterior.(lamRange),
        c=:red, label="Posterior distribution",
        xlims=(0, 10), ylims=(0, 1.2),
        xlabel=L"\lambda",ylabel="Density")
```

```
Computational Bayes Estimate: 2.0555555555555556
Closed form Bayes Estimate: 2.0555555555555554
```

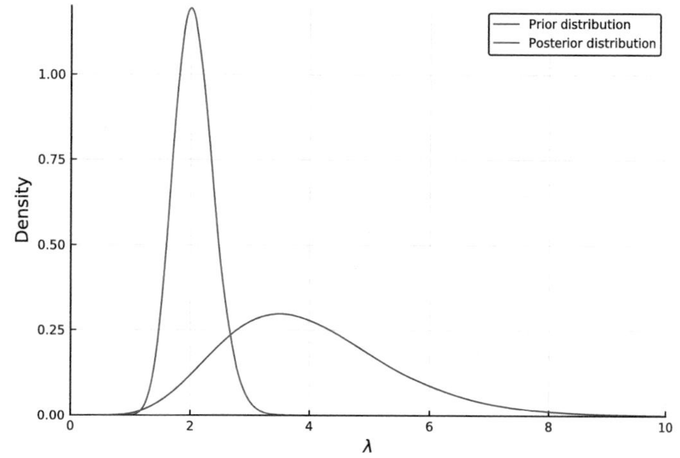

Figure 5.16: The prior and posterior for Bayesian estimation of a Poisson distribution using gamma conjugacy.

In lines 3, the prior hyper-parameters are defined, and in line 4, the prior distribution is defined. In lines 7–13, the posterior is calculated in the brute force same manner as listing 5.18. Similarly in line 15, we compute the Bayes estimate in the same manner. Line 17 is where the simplicity of conjugacy comes about, the hyper-parameters are updated according to the conjugacy rule. Then in line 18 `closedFormBayesEstimate` is computed just using the formula for the mean of a gamma distribution (using `mean()` from `Distributions.jl`). The Bayes estimates are printed in lines 20–21 and the remaining code lines create Figure 5.16.

Markov Chain Monte Carlo

In many applicative cases of Bayesian statistics, convenient situations of conjugate priors are not available, yet computation of posterior distributions and Bayes estimates are needed. In cases where the dimension of the parameter space is high, carrying out straightforward integration as done in Listing 5.18 is not possible. However, there are other ways of carrying out Bayesian inference. One such popular way is by using algorithms that fall under the category known as *Markov Chain Monte Carlo*, MCMC, also known as *Monte Carlo Markov Chain* (with a different word order).

The *Metropolis–Hastings* algorithm is one such popular MCMC algorithm. It produces a series of samples $\theta(1), \theta(2), \theta(3), \ldots$, where it is guaranteed that for large t, $\theta(t)$ is distributed according to the posterior distribution. Technically, the random sequence $\{\theta(t)\}_{t=1}^{\infty}$ is a Markov chain (see Chapter 9 for more details about Markov chains) and it is guaranteed that the stationary distribution of this Markov chain is the specified posterior distribution. That is, the posterior distribution is an input parameter to the algorithm.

The major benefit of Metropolis-Hastings and similar MCMC algorithms is that they only use ratios of the posterior on different parameter values. For example, for parameter values θ_1 and θ_2, the algorithm only uses the posterior distribution via the ratio

$$L(\theta_1, \theta_2) = \frac{f(\theta_1 \mid x)}{f(\theta_2 \mid x)}.$$

This means that the normalizing constant (evidence) is not needed as it is implicitly canceled out. Thus, using the posterior in the proportional form (5.21) suffices.

Further to the posterior distribution, an additional input parameter to Metropolis-Hastings is the so-called *proposal density*, denoted by $q(\cdot \mid \cdot)$. This is a family of probability distributions where given a certain value of θ_1 taken as a parameter, the new value, say θ_2, is distributed with PDF,

$$q(\theta_2 \mid \theta_1).$$

The idea of Metropolis-Hastings is to walk around the parameter space by randomly generating new values using $q(\cdot \mid \cdot)$. At each step, some new values are 'accepted' while others are not, all in a manner which ensures the desired limiting behavior. The algorithm specification is to accept with probability,

$$H = \min\left\{1, \ \ L\big(\theta^*, \theta(t)\big) \ \frac{q\big(\theta(t) \mid \theta^*\big)}{q\big(\theta^* \mid \theta(t)\big)}\right\},$$

where θ^* is the new proposed value, generated via $q\big(\cdot \mid \theta(t)\big)$, and $\theta(t)$ is the current value. With each such iteration, the new value is accepted with probability H and rejected otherwise. With certain technical requirements on the posterior and proposal densities, the theory of Markov chains then guarantees that the stationary distribution of the sequence $\{\theta(t)\}$ is the posterior distribution.

Different variants of the Metropolis-Hastings algorithm employ different types of proposal densities. There are also generalizations and extensions that we don't discuss here, such as *Gibbs Sampling* and *Hamiltonian Monte Carlo*, for example.

To help illustrate some of these concepts, we now implement a simple version of Metropolis-Hastings where we use the *folded normal distribution* as a proposal density. This distribution is achieved by taking a normal random variable X with mean μ and variance σ^2 and considering $Y = |X|$. In this case, the PDF of Y is

$$f(y) = \frac{1}{\sigma\sqrt{2\pi}}\Big(e^{-\frac{(y-\mu)^2}{2\sigma^2}} + e^{-\frac{(y+\mu)^2}{2\sigma^2}}\Big). \tag{5.25}$$

Our choice of this specific density is purely for simplicity of implementation, and in addition, it suits the case that we demonstrate, where the support of the parameter in question is non-negative.

In Listing 5.20, we implement Metropolis-Hastings for the same data and prior as the previous example, Listing 5.19. In such an example, one would not use MCMC since conjugacy is much more efficient, however we do so here for purposes of comparison. Our results show that we obtain the same numerical results as we did using gamma conjugacy. The histogram of the samples is plotted in Figure 5.17.

Listing 5.20: Bayesian inference using MCMC

```
using Distributions, Plots; pyplot()

alpha, beta = 8, 2
prior(lam) = pdf(Gamma(alpha, 1/beta), lam)
data = [2,1,0,0,1,0,2,2,5,2,4,0,3,2,5,0]

like(lam) = *([pdf(Poisson(lam),x) for x in data]...)
posteriorUpToK(lam) = like(lam)*prior(lam)

sig = 0.5
foldedNormalPDF(x,mu) = (1/sqrt(2*pi*sig^2))*(exp(-(x-mu)^2/2sig^2)
                                + exp(-(x+mu)^2/2sig^2))
foldedNormalRV(mu) = abs(rand(Normal(mu,sig)))

function sampler(piProb,qProp,rvProp)
    lam = 1
    warmN, N = 10^5, 10^6
    samples = zeros(N-warmN)

    for t in 1:N
        while true
            lamTry = rvProp(lam)
            L = piProb(lamTry)/piProb(lam)
            H = min(1,L*qProp(lam,lamTry)/qProp(lamTry,lam))
            if rand() < H
                lam = lamTry
                if t > warmN
                    samples[t-warmN] = lam
                end
                break
            end
        end
    end
    return samples
end

mcmcSamples = sampler(posteriorUpToK,foldedNormalPDF,foldedNormalRV)
println("MCMC Bayes Estimate: ",mean(mcmcSamples))

stephist(mcmcSamples, bins=100,
        c=:black, normed=true, label="Histogram of MCMC samples")

lamRange = 0:0.01:10
plot!(lamRange, prior.(lamRange),
        c=:blue, label="Prior distribution")

closedFormPosterior(lam)=pdf(Gamma(alpha + sum(data),1/(beta+length(data))),lam)
plot!(lamRange, closedFormPosterior.(lamRange),
        c=:red, label="Posterior distribution",
        xlims=(0, 10), ylims=(0, 1.2),
    xlabel=L"\lambda",ylabel="Density")
```

```
MCMC Bayes Estimate: 2.065756632471559
```

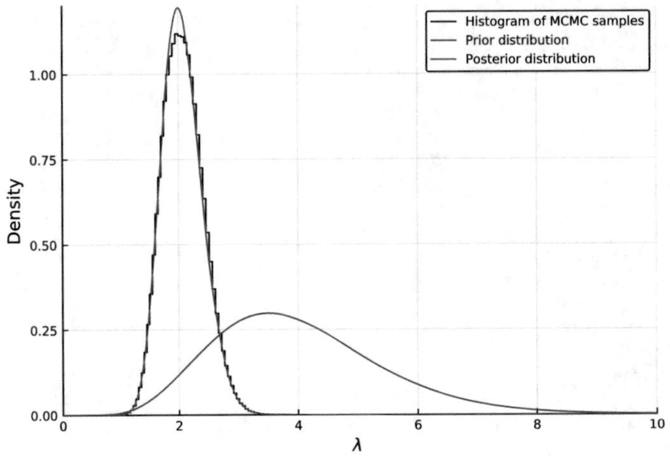

Figure 5.17: The prior and the posterior for Monte Carlo Markov Chain samples generated using Metropolis-Hastings.

Lines 3–8 are similar to the previous listings 5.18 and 5.19. In lines 10–13, the proposal density `foldedNormalPDF()` is defined in accordance with (5.25), along with a function for generating a proposal random variable, `foldedNormalRV()`. Lines 15–35 define the function `sampler()`. It operates on a desired (non-normalized) density, `piProb`, and runs the Metropolis-Hastings algorithm for sampling from that density. The argument `qProp` is the proposal density, and the argument `rvProp` is for generating from the proposal. All three arguments are assumed to be functions which `sampler()` invokes. Our implementation uses a *warm up sequence* with a length specified by `warmN` in line 17. The idea here is to let the algorithm run for a while to remove any bias introduced by initial values. Lines 20–33 constitute the main loop over N samples generated by the algorithm. In our implementation, we set up an internal loop (lines 21–32) that iterates until a proposal is accepted (and breaks in line 30). Line 38 prints the Bayes estimate. As can be seen, it agrees with the estimate of Listing 5.19.

Chapter 6

Confidence Intervals

In this chapter, we cover a variety of confidence intervals used in standard statistical procedures. As introduced in Section 5.5, a confidence interval with a confidence level $1-\alpha$ is an interval $[L, U]$ resulting from the observations. When considering confidence intervals in the setting of symmetric sampling distributions (as is the case for most of this chapter), a typical formula for $[L, U]$ is of the form,

$$\hat{\theta} \pm K_\alpha \ s_{\text{err}}. \tag{6.1}$$

Here, $\hat{\theta}$ is typically the point estimate for the parameter in question, s_{err} is some measure or estimate of the variability (e.g. standard error), and K_α is a constant which depends on the model at hand and on α. Typically by decreasing $\alpha \to 0$, we have that K_α increases, implying a wider confidence interval. For the examples in this chapter, common values for K_α are in the range of $[1.5, 3.5]$ for values of α in the range of $[0.01, 0.1]$. Most of the confidence intervals presented in this chapter follow the form of (6.1), with the specific form of K_α often depending on conditions such as

Sample size: Is the sample size small or large.

Variance: Is the variance σ^2 known or unknown.

Distribution: Is the population assumed normally distributed or not.

In exploring confidence intervals with Julia, we compute the confidence intervals using standard statistical formulas and then illustrate how they can be obtained using the `HypothesisTests` package. This package includes various functions that generate objects resulting from specific statistical procedures. We can either look at the output of these objects or query them using other functions, specifically the `confint()` function. This package is also used extensively in Chapter 7.

The individual sections of this chapter focus on specific confidence intervals and general concepts. In Section 6.1, we cover confidence intervals for the mean of a single population. In Section 6.2, we present comparisons of means of two populations. In Section 6.3, we cover confidence intervals for proportions. In Section 6.4, we gain a better understanding of model assumptions via the example of a confidence interval for the variance. In Section 6.5, we present the bootstrap method, a general methodology for creating confidence intervals. In Section 6.6, we present *prediction intervals*, a concept dealing with prediction of future observations based on previous ones. We close with Section 6.7 which deals with credible intervals from Bayesian statistics.

Y. Nazarathy and H. Klok, *Statistics with Julia*, Springer Series in the Data Sciences,
https://doi.org/10.1007/978-3-030-70901-3_6

6.1 Single Sample Confidence Intervals for the Mean

Let us first consider the case where we wish to estimate the population mean μ using a random sample, $X_1, \ldots, X_n$. As covered previously, a point estimate for the mean is the sample mean $\overline{X}$. A typical formula for the confidence interval of the mean is then

$$\overline{X} \pm K_\alpha \frac{S}{\sqrt{n}}. \tag{6.2}$$

Here, the bounds around the point estimator $\overline{X}$ are defined by the addition and subtraction of a multiple, K_α, of the standard error, $S/\sqrt{n}$, first introduced in Section 4.2. The multiple K_α takes on different forms depending on the specific case at hand.

Population Variance Known

If we assume that the population variance, σ^2, is known and the data is normally distributed, then the sample mean $\overline{X}$ is normally distributed with mean μ and variance σ^2/n. This yields

$$\mathbb{P}\Big(\mu - z_{1-\frac{\alpha}{2}} \frac{\sigma}{\sqrt{n}} \le \overline{X} \le \mu + z_{1-\frac{\alpha}{2}} \frac{\sigma}{\sqrt{n}}\Big) = 1 - \alpha, \tag{6.3}$$

where $z_{1-\frac{\alpha}{2}}$ is the $1 - \frac{\alpha}{2}$ quantile of the standard normal distribution. In Julia, this is computed via `quantile(Normal(),1-alpha/2)`. If we denote the actual sample mean estimate obtained from data by $\overline{x}$, then by rearranging the inequalities inside the probability statement above, we obtain the following confidence interval formula:

$$\overline{x} \pm z_{1-\frac{\alpha}{2}} \frac{\sigma}{\sqrt{n}}. \tag{6.4}$$

In practice, σ^2 is rarely known; hence, it is tempting to replace σ by s (sample standard deviation), in the formula above. Such a replacement is generally fine for large samples. However, in the case of small samples, one should confidence intervals assuming population variance unknown, covered at the end of this section.

The validity of the normality assumption should also be considered. In cases where the data is not normally distributed, the probability statement (6.3) only approximately holds. However, as $n \to \infty$, it quickly becomes precise due to the central limit theorem. Hence the confidence interval (6.4) may be used for non-small samples.

In Julia, computation of confidence intervals is done using functions from the `HypothesisTests` package (even when we don't carry out a hypothesis test). The code in Listing 6.1 illustrates computation of the confidence interval (6.4) using both the package and by evaluating the formula directly. It can be observed that both the direct computation and the use of the `confint()` function yield the same result.

Listing 6.1: CI for single sample population, variance assumed known

```
1  using CSV, Distributions, HypothesisTests
2
3  data = CSV.read("../data/machine1.csv", header=false)[:,1]
4  xBar, n = mean(data), length(data)
5  sig = 1.2
6  alpha = 0.1
7  z = quantile(Normal(),1-alpha/2)
8
9  println("Calculating formula: ", (xBar - z*sig/sqrt(n), xBar + z*sig/sqrt(n)))
10 println("Using confint() function: ", confint(OneSampleZTest(xBar,sig,n),alpha))
```

```
Calculating formula:       (52.51484557853184, 53.397566664027984)
Using confint() function: (52.51484557853184, 53.397566664027984)
```

Line 3 loads the data. Note the use of the `header=false` argument, and also the trailing `[:,1]` which is used to select all rows of the data. In line 4, we calculate the sample mean and the number of observations. In line 5, we stipulate the standard deviation as 1.2, as this scenario is one in which the population standard deviation, or population variance, is assumed known. In line 7, we calculate the value of z for $1 - \alpha/2$. This quantity does not depend on the sample but is a fixed number. As is well known from statistical tables it equals approximately 1.65 when $\alpha = 10\%$. In line 9, the formula for the confidence interval (6.4) is evaluated directly. In line 10, the function `OneSampleZTest()` is first used to conduct a one-sample z-test given the parameters `xBar`, `sig`, and `n`. The `confint()` function is then applied to this output, for the specified value of `alpha`. It can be observed that the two methods are in agreement. Note that hypothesis tests are covered further in Chapter 7.

Population Variance Unknown

A celebrated procedure in elementary statistics is the confidence interval based on the T-distribution. Here, we relax the assumptions of the previous confidence interval by allowing σ^2 to be an unknown quantity. In this case, if we replace σ by the sample standard deviation s, then the probability statement (6.3) no longer holds. However, by using the T-distribution (see Section 5.2), we are able to correct the confidence interval to

$$\overline{x} \pm t_{1-\frac{\alpha}{2},n-1} \, \frac{s}{\sqrt{n}}. \tag{6.5}$$

Here, $t_{1-\frac{\alpha}{2},n-1}$ is the $1 - \frac{\alpha}{2}$ quantile of a T-distribution with $n-1$ degrees of freedom. This can be calculated in Julia via `quantile(TDist(n-1),1-alpha/2)`.

For small samples, the replacement of $z_{1-\frac{\alpha}{2}}$ from (6.4) by $t_{1-\frac{\alpha}{2},n-1}$ in (6.5) significantly affects the width of the confidence interval, as for the same value of α, the T case is wider. However, as $n \to \infty$, we have, $t_{1-\frac{\alpha}{2},n-1} \to z_{1-\frac{\alpha}{2}}$, as illustrated in Figure 5.5. Hence for non-small samples, the confidence interval (6.5) is very close to the confidence interval (6.4) with s replacing σ. Note that the T-confidence interval hinges on the normality assumption of the data. In fact for small samples, cases that deviate from normality imply imprecision of the confidence intervals. However, for larger samples, these confidence intervals serve as a good approximation. Still in these larger sample cases, one might as well use $z_{1-\frac{\alpha}{2}}$ instead of $t_{1-\frac{\alpha}{2},n-1}$.

The code in Listing 6.2 calculates the confidence interval (6.5), where it is assumed that the population variance is unknown.

Listing 6.2: CI for single sample population with variance assumed unknown

```
1   using CSV, Distributions, HypothesisTests
2
3   data = CSV.read("../data/machine1.csv", header=false)[:,1]
4   xBar, n = mean(data), length(data)
5   s = std(data)
6   alpha = 0.1
7   t = quantile(TDist(n-1),1-alpha/2)
8
9   println("Calculating formula: ", (xBar - t*s/sqrt(n), xBar + t*s/sqrt(n)))
10  println("Using confint() function: ", confint(OneSampleTTest(xBar,s,n),alpha))
```

```
Calculating formula:       (52.49989385779555, 53.412518384764276)
Using confint() function: (52.49989385779555, 53.412518384764276)
```

This example is very similar to Listing 6.1, however, there are several differences. In line 5, since the population variance is assumed unknown, the population standard deviation `sig` of Listing 6.1 is replaced with the sample standard deviation `s`. In line 7, the quantile `t` is calculated on a T-distribution, `TDist(n-1)`, with $n-1$ degrees of freedom. Previously, the quantile `z` was calculated on a standard normal distribution `Normal()`. Lines 9 and 10 are very similar to those in the previous listing, but `z` and `sig` are replaced with `t` and `s`, respectively. It can be seen that the outputs of lines 9 and 10 are in agreement and that the confidence interval is wider than that calculated in the previous Listing 6.1.

6.2 Two Sample Confidence Intervals for the Difference in Means

We now consider cases in which there are two populations involved. As an example, consider two separate machines, 1 and 2, which are designed to make pipes of the same diameter. In this case, due to small differences and tolerances in the manufacturing process, the distribution of pipe diameters from each machine will differ. In such cases where two populations are involved, it is often of interest to estimate the difference between the population means, $\mu_1 - \mu_2$.

In order to do this we first collect two random samples, $x_{11}, \ldots, x_{n_1 1}$ and $x_{12}, \ldots, x_{n_2 2}$. For each sample $i = 1, 2$ we have the sample mean $\overline{x}_i$, and sample standard deviation s_i. In addition, the difference in sample means, $\overline{x}_1 - \overline{x}_2$ serves as a point estimate for the difference in population means, $\mu_1 - \mu_2$.

A confidence interval for $\mu_1 - \mu_2$ around the point estimate $\overline{x}_1 - \overline{x}_2$ is then constructed via the same process seen previously,

$$\overline{x}_1 - \overline{x}_2 \pm K_\alpha s_{\text{err}}. \tag{6.6}$$

We now elaborate on the values of K_α and s_{err} based on model assumptions.

Population Variances Known

In the (unrealistic) case that the population variances are known, we may explicitly compute

$$\text{Var}(\overline{X}_1 - \overline{X}_2) = \frac{\sigma_1^2}{n_1} + \frac{\sigma_2^2}{n_2}.$$

Hence, the standard error is given by

$$s_{\text{err}} = \sqrt{\frac{\sigma_1^2}{n_1} + \frac{\sigma_2^2}{n_2}}. \tag{6.7}$$

When combined with the assumption that the data is normally distributed, we can derive the following confidence interval

$$\overline{x}_1 - \overline{x}_2 \pm z_{1-\frac{\alpha}{2}} \sqrt{\frac{\sigma_1^2}{n_1} + \frac{\sigma_2^2}{n_2}}. \tag{6.8}$$

While this case is not often applicable in practice, it is useful to cover for pedagogical reasons. Due to the fact that the population variances are almost always unknown, the `HypothesisTests` package in Julia does not have a function for this case. However, for completeness, we evaluate equation (6.8) manually in Listing 6.3.

Listing 6.3: CI for difference in population means with variances known

```
1  using CSV, Distributions, HypothesisTests
2
3  data1 = CSV.read("../data/machine1.csv", header=false)[:,1]
4  data2 = CSV.read("../data/machine2.csv", header=false)[:,1]
5  xBar1, xBar2 = mean(data1), mean(data2)
6  n1, n2 = length(data1), length(data2)
7  sig1, sig2 = 1.2, 1.6
8  alpha = 0.05
9  z = quantile(Normal(),1-alpha/2)
10
11 println("Calculating formula: ", (xBar1 - xBar2 - z*sqrt(sig1^2/n1+sig2^2/n2),
12                                   xBar1 - xBar2 + z*sqrt(sig1^2/n1+sig2^2/n2)))
```

```
Calculating formula:     (1.1016568035908845, 2.9159620096069574)
```

This listing is similar to those previously covered in this chapter. The sample means and number of observations are calculated in lines 5–6. In line 7, we stipulate the standard deviations of both populations 1 and 2, as 1.2 and 1.6, respectively (since in this scenario the population variances are assumed known). In lines 11–12 the confidence interval (6.8) is evaluated manually and printed as output.

Population Variances Unknown and Assumed Equal

Typically, when considering cases consisting of two populations, the population variances are unknown. In such cases, a common and practical assumption is that the variances are equal, denoted by σ^2. Based on this assumption, it is sensible to use both sample variances to estimate σ^2. This estimated variance using both samples is known as the *pooled sample variance*, and is given by

$$S_p^2 = \frac{(n_1 - 1)S_1^2 + (n_2 - 1)S_2^2}{n_1 + n_2 - 2}.$$

Upon closer inspection, it can be observed that the above is in fact a weighted average of the sample variances of the individual samples.

In this case, it can be shown that

$$T = \frac{\overline{X}_1 - \overline{X}_2 - (\mu_1 - \mu_2)}{S_{\text{err}}} \tag{6.9}$$

is distributed according to a T-distribution with $n_1 + n_2 - 2$ degrees of freedom, where the standard error is,

$$S_{\text{err}} = S_p \sqrt{\frac{1}{n_1} + \frac{1}{n_2}}.$$

Hence, we arrive at the following confidence interval:

$$\overline{x}_1 - \overline{x}_2 \pm t_{1-\frac{\alpha}{2},n-2}\ s_p \sqrt{\frac{1}{n_1} + \frac{1}{n_2}}, \tag{6.10}$$

where s_p is the square root of the observed pooled sample variance and $t_{1-\frac{\alpha}{2},n-2}$ is a quantile of a T-distribution with $n - 2$ degrees of freedom.

The code in Listing 6.4 calculates the confidence interval (6.10), where it is assumed that the population variance is unknown. This is compared with the result from the `HypothesisTests` package using `EqualVarianceTTest()`. It can be observed that the results are in agreement.

Listing 6.4: CI for difference in means, variance unknown, assumed equal

```
1   using CSV, Distributions, HypothesisTests
2
3   data1 = CSV.read("../data/machine1.csv", header=false)[:,1]
4   data2 = CSV.read("../data/machine2.csv", header=false)[:,1]
5   xBar1, xBar2 = mean(data1), mean(data2)
6   n1, n2 = length(data1), length(data2)
7   alpha = 0.05
8   t = quantile(TDist(n1+n2-2),1-alpha/2)
9
10  s1, s2 = std(data1), std(data2)
11  sP = sqrt(((n1-1)*s1^2 + (n2-1)*s2^2) / (n1+n2-2))
12
13  println("Calculating formula: ", (xBar1 - xBar2 - t*sP* sqrt(1/n1 + 1/n2),
14                                    xBar1 - xBar2 + t*sP* sqrt(1/n1 + 1/n2)))
15  println("Using confint(): ", confint(EqualVarianceTTest(data1,data2),alpha))
```

```
Calculating formula:        (1.1127539574575822, 2.90486485574026)
Using confint() function: (1.1127539574575822, 2.90486485574026)
```

In line 8, a T-distribution with `n1+n2-2` degrees of freedom is used. In line 10, the sample standard deviations are calculated. In line 11, the pooled sample variance `sP` is calculated. In lines 13–14, (6.10) is evaluated manually, while in line 15 the `confint()` function is used.

Population Variances Unknown and not Assumed Equal

In certain cases, it may be appropriate to relax the assumption of equal population variances. In this case, the estimate for S_{err} is given by

$$S_{\text{err}} = \sqrt{\frac{S_1^2}{n_1} + \frac{S_2^2}{n_2}}.$$

This is due to the fact that the variance of the difference of two independent sample means is the sum of the variances of each of the means. Hence, in this case the statistic (6.9) is adapted to

$$T = \frac{\overline{X}_1 - \overline{X}_2 - (\mu_1 - \mu_2)}{\sqrt{\frac{S_1^2}{n_1} + \frac{S_2^2}{n_2}}}. \tag{6.11}$$

It turns out that (6.11) is only T-distributed if the variances are equal, otherwise, it isn't. Nevertheless, an approximate confidence interval is commonly used by approximating the distribution of (6.11) with a T-distribution. This is called the *Satterthwaite approximation.*

The approximation suggests a T-distribution with a parameter (degrees of freedom) given via

$$v = \frac{\left(\frac{s_1^2}{n_1} + \frac{s_2^2}{n_2}\right)^2}{\frac{\left(s_1^2/n_1\right)^2}{n_1 - 1} + \frac{\left(s_2^2/n_2\right)^2}{n_2 - 1}}. \tag{6.12}$$

Now it holds that

$$T \underset{\text{approx}}{\sim} t(v). \tag{6.13}$$

That is, the random variable T from (6.11) is approximately distributed according to a T-distribution with v degrees of freedom. Note that v does not need to be an integer. We investigate this approximation further in Listing 6.6 later.

Using the Satterthwaite approximation, following steps similar to previous confidence intervals, and given (6.13), we arrive at the following confidence interval formula:

$$\overline{x}_1 - \overline{x}_2 \pm t_{1-\frac{\alpha}{2},v} \sqrt{\frac{s_1^2}{n_1} + \frac{s_2^2}{n_2}}. \tag{6.14}$$

In Listing 6.5, we calculate the confidence interval (6.14), where it is assumed that the population variances are unknown and not assumed equal. We then compare the result to those resulting from the use of `UnequalVarianceTTest()` from the `HypothesisTests` package. It can be observed that the results are in agreement.

Listing 6.5: CI for difference in means, variance unknown and unequal

```
1   using CSV, Distributions, HypothesisTests
2
3   data1 = CSV.read("../data/machine1.csv", header=false)[:,1]
4   data2 = CSV.read("../data/machine2.csv", header=false)[:,1]
5   xBar1, xBar2 = mean(data1), mean(data2)
6   s1, s2 = std(data1), std(data2)
7   n1, n2 = length(data1), length(data2)
8   alpha = 0.05
9
10  v = (s1^2/n1 + s2^2/n2)^2 / ( (s1^2/n1)^2 / (n1-1) + (s2^2/n2)^2 / (n2-1) )
11
12  t = quantile(TDist(v),1-alpha/2)
13
14  println("Calculating formula: ", (xBar1 - xBar2 - t*sqrt(s1^2/n1 + s2^2/n2),
15                                    xBar1 - xBar2 + t*sqrt(s1^2/n1 + s2^2/n2)))
16  println("Using confint(): ",    confint(UnequalVarianceTTest(data1,data2),alpha))
```

```
Calculating formula: (1.0960161148824918, 2.9216026983153505)
Using confint():     (1.0960161148824918, 2.9216026983153505)
```

The main difference in this code block from the previous code block is the calculation of the degrees of freedom, `v`, which is performed in line 10. In line 12, `v` is then used to derive the T-statistic `t`. In lines 14–15, equation (6.14) is evaluated manually, while in line 16 the `confint()` function is used.

Exploring the Satterthwaite Approximation

We now investigate the approximate distribution stated in (6.13). Observe that both sides of the "distributed as" ($\sim$) symbol are random variables which depend on the same random experiment. Hence, the statement can be presented generally, as a case of the following format:

$$X(\omega) \sim F_{h(\omega)}, \tag{6.15}$$

where ω is a point in the sample space (see Chapter 2). Here, $X(\omega)$ is a random variable, and F is a distribution that depends on a parameter h, which itself depends on ω. In our case of the Satterthwaite approximation, h is given by (6.12). That is, h can be thought of as v, which itself depends on the specific observations made for our two sample groups (a function of s_1 and s_2).

Now by recalling the inverse probability transform from Section 3.4, we have that (6.15) is equivalent to

$$F^{-1}_{h(\omega)}\big(X(\omega)\big) \ \sim \ \text{uniform}(0,1). \tag{6.16}$$

Hence, in the case of the Satterthwaite approximation, we expect that (6.16) hold approximately. This distributional relationship would not hold with the naive alternative of treating h as simply dependent on the number of observations made (n_1 and n_2). Hence, in this naive case, the distribution is not expected to be uniform.

We investigate this in Listing 6.6, where we construct Figure 6.1, a Q-Q plot comparing T-values calculated from the Satterthwaite approximation (6.13) and T-values calculated via the naive equal variance case. See Section 4.4 for a description of Q-Q plots.

The results in Figure 6.1, indicate that the Satterthwaite approximation is a better approximation than simply using the degrees of freedom. It can be observed that the data from the fixed v case deviates further from the 1:1 slope in comparison to the case where v was calculated based on each experiment's sample observations, i.e. calculated from equation (6.12). Hence, the distribution of T-statistics from Satterthwaite calculated v's yields better results than the constant v case.

Listing 6.6: Analyzing the Satterthwaite approximation

```
using Distributions, Statistics, Plots, Random; pyplot()
Random.seed!(0)

mu1, sig1, n1 = 0, 2, 8
mu2, sig2, n2 = 0, 30, 15
dist1 = Normal(mu1, sig1)
dist2 = Normal(mu2, sig2)

N = 10^6
tdArray = Array{Tuple{Float64,Float64}}(undef,N)

df(s1,s2,n1,n2) =
    (s1^2/n1 + s2^2/n2)^2 / ( (s1^2/n1)^2/(n1-1) + (s2^2/n2)^2/(n2-1) )

for i in 1:N
    x1Data = rand(dist1, n1)
    x2Data = rand(dist2, n2)
    x1Bar,x2Bar = mean(x1Data),mean(x2Data)
    s1,s2 = std(x1Data),std(x2Data)
    tStat = (x1Bar - x2Bar) / sqrt(s1^2/n1 + s2^2/n2)
    tdArray[i] = (tStat , df(s1,s2,n1,n2))
end
sort!(tdArray, by = first)

invVal(v,i) = quantile(TDist(v),i/(N+1))

xCoords  = Array{Float64}(undef,N)
yCoords1 = Array{Float64}(undef,N)
yCoords2 = Array{Float64}(undef,N)

for i in 1:N
    xCoords[i] = first(tdArray[i])
    yCoords1[i] = invVal(last(tdArray[i]), i)
    yCoords2[i] = invVal(n1+n2-2, i)
end

scatter(xCoords, yCoords1, c=:blue, label="Calculated v", msw=0)
scatter!(xCoords, yCoords2, c=:red, label="Fixed v", msw=0)
plot!([-10,10], [-10,10],
        c=:black, lw=0.3, xlims=(-8,8), ylims=(-8,8), ratio=:equal, label="",
        xlabel="Theoretical t-distribution quantiles",
        ylabel="Simulated t-distribution quantiles", legend=:topleft)
```

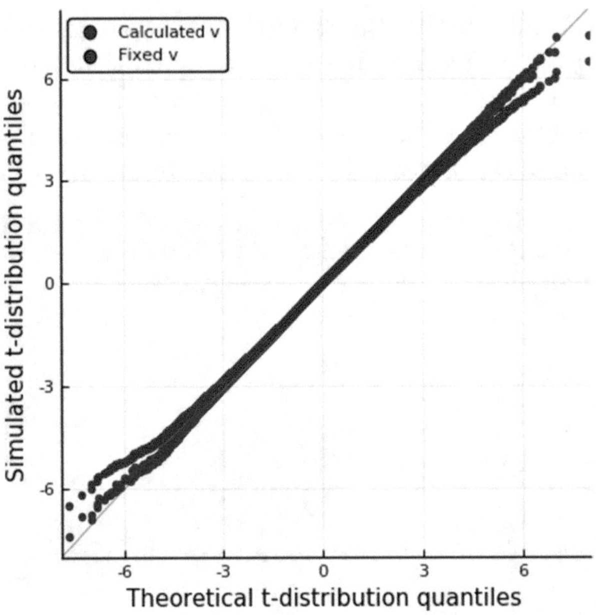

Figure 6.1: Q-Q plots of T-statistics from identical experiments, given Satterthwaite calculated v's, along with T-statistics given constant v.

In lines 4–5, we set the means and standard deviations of the two underlying distributions, and the number of observations that will be made for each group. In line 8, we set the number of times we repeat the experiment, `N`. In line 9, we pre-allocate the array `tdArray`, in which each element is a tuple. The first element of each tuple will be the T-statistic calculated via (6.11), while the second will be the corresponding degrees of freedom calculated via (6.12). In lines 12–13, we define the function `df()`, which implements (6.12). In lines 15–22, we conduct `N` experiments, where for each, we calculate the T-statistic and the degrees of freedom. In line 23, `sort!()` is used to re-order `tdArrray` in ascending order according to the T-statistics via `by = first`. This is done so that we can construct the Q-Q plot. In line 25, the function `invVal()` is defined, which uses the `quantile()` function to perform the inverse probability transform on the degrees of freedom associated with each T-statistic for each experiment. Note that the number of quantiles is one more than the number of experiments, i.e. `N+1`. In lines 31-35, the quantiles of our data are calculated. Here, `xCoords` represents the T-statistic quantiles, and `yCoords1` represents the quantiles of a T-distribution with v degrees of freedom, where v is calculated via (6.12). The array `yCoords2`, on the other hand, represents the quantiles of a T-distribution with v degrees of freedom, where $v = n_1 + n_2 - 2$. Lines 37–42 plot the Q-Q plots creating Listing 6.6.

6.3 Confidence Intervals for Proportions

In certain inference settings the parameter of interest is a *population proportion*. Examples include the proportion of females within an animal population, the proportion of customers that own two or more cars, or the proportion of baby turtles that survive the first day after hatching. In all such cases, one may view the proportion as either a characteristic of the population or alternatively as the probability of some event happening when randomly sampling from the population.

When carrying out inference for a proportion we assume that there exists some unknown population proportion $p \in (0,1)$. We then sample an i.i.d. sample of observations $I_1, \ldots, I_n$, where for the i'th observation, $I_i = 0$ if the event in question does not happen, and $I_i = 1$ if the event occurs. For example, dealing with the proportion of females, we set $I_i = 0$ if the i'th sample is not a female and $I_i = 1$ if the i'th sample is a female.

A natural estimator for the proportion is then the sample mean of $I_1, \ldots, I_n$ which we denote

$$\hat{p} = \frac{\sum_{i=1}^{n} I_i}{n}. \tag{6.17}$$

In this case, since the summands in the numerator are indicator variables, the sum is simply a count of the number of observations for which the event occurred. Hence, we also call $\hat{p}$, the *proportion estimator.*

Now observe that each I_i is a Bernoulli random variable with success probability p. Under the i.i.d. assumption this means that the numerator of (6.17) is binomially distributed with parameters n and p (see Section 3.5 to review the binomial distribution). Hence,

$$\mathbb{E}\left[\sum_{i=1}^{n} I_i\right] = np, \qquad \text{and} \qquad \text{Var}\left(\sum_{i=1}^{n} I_i\right) = np(1-p).$$

By combining (6.17) with the above, we have that

$$\mathbb{E}[\hat{p}] = p, \qquad \text{and} \qquad \text{Var}(\hat{p}) = \frac{p(1-p)}{n}. \tag{6.18}$$

Hence, $\hat{p}$ is an unbiased and consistent estimator of p. That is, on average $\hat{p}$ estimates p perfectly, and if more observations are collected the variance of the estimator vanishes and $\hat{p} \to p$. Furthermore, we can use the central limit theorem to create a normal approximation for the distribution of $\hat{p}$ and yield an approximate confidence interval. To do so denote

$$\tilde{Z}_n = \frac{\hat{p} - p}{\sqrt{p(1-p)/n}}.$$

This is a random variable that approximately follows a standard normal distribution. The approximation becomes exact as n grows. The same also holds for a slightly different random variable

$$\hat{Z}_n = \frac{\hat{p} - p}{\sqrt{\hat{p}(1-\hat{p})/n}}. \tag{6.19}$$

This is because $\hat{p}$ is an unbiased and consistent estimator of p and thus replacing the p's in the denominator of $\tilde{Z}_n$ with $\hat{p}$ to yield $\hat{Z}_n$ does not significantly affect the distribution for large n. We now use the approximate normality of $\hat{Z}_n$ to create a confidence interval for p. First observe that as a consequence of the approximate standard normal distribution,

$$\mathbb{P}(z_{\alpha/2} \le \hat{Z}_n \le z_{1-\alpha/2}) \approx 1 - \alpha. \tag{6.20}$$

We now use (6.19) in (6.20), along with the fact that $z_{\alpha/2} = -z_{1-\alpha/2}$ as follows,

$$\begin{aligned}
1-\alpha &\approx \mathbb{P}(-z_{1-\alpha/2} \le \hat{Z}_n \le z_{1-\alpha/2}) \\
&= \mathbb{P}\Big(-z_{1-\alpha/2} \le \frac{\hat{p}-p}{\sqrt{\hat{p}(1-\hat{p})/n}} \le z_{1-\alpha/2}\Big) \\
&= \mathbb{P}(-z_{1-\alpha/2}\sqrt{\hat{p}(1-\hat{p})/n} \;\le\; \hat{p}-p \;\le\; z_{1-\alpha/2}\sqrt{\hat{p}(1-\hat{p})/n}) \\
&= \mathbb{P}(-\hat{p}-z_{1-\alpha/2}\sqrt{\hat{p}(1-\hat{p})/n} \;\le\; -p \;\le\; -\hat{p}+z_{1-\alpha/2}\sqrt{\hat{p}(1-\hat{p})/n}) \\
&= \mathbb{P}(\hat{p}-z_{1-\alpha/2}\sqrt{\hat{p}(1-\hat{p})/n} \;\le\; p \;\le\; \hat{p}+z_{1-\alpha/2}\sqrt{\hat{p}(1-\hat{p})/n}).
\end{aligned}$$

We thus arrive at the following (approximate) *confidence interval for proportions* formula:

$$\hat{p} \pm z_{1-\alpha/2}\sqrt{\frac{\hat{p}(1-\hat{p})}{n}}. \tag{6.21}$$

Observe that this confidence interval formula agrees with the general form of (6.1), where the standard error depends only on the statistic $\hat{p}$ and is represented as

$$s_{\text{err}} = \sqrt{\frac{\hat{p}(1-\hat{p})}{n}}. \tag{6.22}$$

Similar more complex confidence interval formulas also exist for the case of two populations. Say one is interested in comparing the proportion of females in two different sub-species populations of crocodiles. By sampling n_1 crocodiles from one sub-species and n_2 crocodiles for the other subspecies, one can form the point estimators $\hat{p}_1$ and $\hat{p}_2$, each in the same manner as (6.17). The point estimator for the difference in proportions is then simply $\hat{p}_1 - \hat{p}_2$. An approximate $1-\alpha$ confidence interval for this parameter is

$$\hat{p}_1 - \hat{p}_2 \pm z_{1-\alpha/2}\sqrt{\frac{\hat{p}_1(1-\hat{p}_1)}{n_1} + \frac{\hat{p}_2(1-\hat{p}_2)}{n_2}}. \tag{6.23}$$

Compare this formula with the general form (6.6) and other formulas for two populations presented in the previous section. You can see that (6.23) follows a similar structure. We now return to examples and discussions of (6.21) and don't discuss (6.23) further.

In Listing 6.7, we demonstrate basic usage of the confidence interval for proportions formula (6.21). We consider the `Grade` column of `purchaseData.csv`. As the code demonstrates, here the possible grades are "`A`"–"`E`". We obtain a point estimate and a confidence interval for the proportion of observations with level "`E`". You may modify line 11 of the code to carry out inference for other levels. Note that this code also deals with missing observations by culling the missing observations and only uses observations for which `Grade` is not `missing`.

Listing 6.7: Confidence interval for a proportion

```
1  using CSV, DataFrames, CategoricalArrays
2
3  data = CSV.read("../data/purchaseData.csv", copycols = true)
4  println("Levels of Grade: ", levels(data.Grade))
5  println("Data points: ", nrow(data))
6
7  n = sum(.!(ismissing.(data.Grade)))
8  println("Non-missing data points: ", n)
9  data2 = dropmissing(data[:,[:Grade]],:Grade)
10
11 gradeInQuestion = "E"
12 indicatorVector = data2.Grade .== gradeInQuestion
13 numSuccess = sum(indicatorVector)
14 phat = numSuccess/n
15 serr = sqrt(phat*(1-phat)/n)
16
17 alpha = 0.05
18 confidencePercent = 100*(1-alpha)
19 zVal = quantile(Normal(),1-alpha/2)
20 confInt = (phat - zVal*serr, phat + zVal*serr)
21
22 println("\nOut of $n non-missing observations, "*
23         "$numSuccess are at level $gradeInQuestion.")
24 println("Hence a point estimate for the proportion "*
25         "of grades at level $gradeInQuestion is $phat.")
26 println("A $confidencePercent% confidence interval for "*
27         "the proprotion of level $gradeInQuestion is:\n$confInt.")
```

```
Levels of Grade: ["A", "B", "C", "D", "E"]
Data points: 200
Non-missing data points: 187

Out of 187 non-missing observations, 61 are at level E.
Hence a point estimate for the proportion of grades at level E is 0.3262.
A 95.0% confidence interval for the proprotion of level E is:
(0.2590083767381328, 0.3933980403741667).
```

Lines 3–5 load and describe the data, focusing on the `Grade` column. Note the use of `levels()` from `CategoricalArrays`. Lines 7–9 handle missing values. Note the use of "`.!()`" to broadcast negation on the output of a broadcasted `ismissing()`. Summing this yields the number of (non-missing) observations `n`. The new data frame, `data2` is comprised of a single variable `Grade` after `dropmissing()` is applied with `:Grade` as a second argument. In line 11, we choose to carry out proportion estimation for `"E"`. Line 12 creates $I_1, \ldots, I_n$. Line 14 calculates $\hat{p}$ and line 15 calculates s_{err} as in (6.22). Lines 17–20 determine the confidence interval (6.21), with `confInt` represented as a `Tuple`. The remainder of the code prints the output describing the results.

Sample Size Planning

Denote the confidence interval (6.21) as $\hat{p} \pm E$ where E is the *margin of error* or half the *width of the confidence interval*, denoted by

$$E = z_{1-\alpha/2}\sqrt{\frac{\hat{p}(1-\hat{p})}{n}}.$$

You may often want to plan an experiment or a sampling scheme such that E is not too wide. For example, "not more than 0.1". For this, you need to choose a sample size n prior to sampling. We now illustrate a crude yet effective way for such *sample size planning.*

First observe that for typical values of α we have that $z_{1-\alpha/2} \approx 2$. In fact, when $\alpha = 0.0455$ we have that $z_{1-\alpha/2} = 2$ almost exactly. Values ranging between 1.5 and 2.5 are common for most chosen confidence levels in practice. Hence, in general, crudely taking $z_{1-\alpha/2}$ as "2" helps simplify expressions.

Say we want $E \leq \varepsilon$, e.g. with the maximal margin of error $\varepsilon = 0.1$, or any other similar value. Then taking $z_{1-\alpha/2} = 2$ for simplicity, we get

$$2\sqrt{\frac{\hat{p}(1-\hat{p})}{n}} \leq \varepsilon, \qquad \text{or} \qquad 4\frac{\hat{p}(1-\hat{p})}{\varepsilon^2} \leq n. \tag{6.24}$$

Now also observe that $x(1-x)$ is maximized at $x = 1/2$ with a maximal value of $1/4$. Hence,

$$4\frac{\hat{p}(1-\hat{p})}{\varepsilon^2} \leq \frac{1}{\varepsilon^2}.$$

This means that by taking $n \geq \varepsilon^{-2}$ we ensure (6.24). For seeking a whole number of observations we use the $\lceil \cdot \rceil$ "ceiling" (rounding up) operator, to get the *proportions sample size formula*

$$n^* = \left\lceil \frac{1}{\varepsilon^2} \right\rceil. \tag{6.25}$$

In Listing 6.8, we create a simple table implementing (6.25) to get a sense for the magnitude of samples needed. As you can see for $\varepsilon = 0.1$ we need 100 observations. However, if we seek a more accurate confidence interval with $\varepsilon = 0.01$ then 10,000 observations are needed! Again, keep in mind that these calculations are assuming $\alpha = 0.0455$.

Listing 6.8: Sample size planning for proportions

```
1 for eps in [0.1, 0.05, 0.02, 0.01]
2     n = ceil(1/eps^2)
3     println("For epsilon = $eps set n = $n")
4 end
```

```
For epsilon = 0.1 set n = 100.0
For epsilon = 0.05 set n = 400.0
For epsilon = 0.02 set n = 2500.0
For epsilon = 0.01 set n = 10000.0
```

The listing is a straightforward implementation of (6.25). Observe the use of `ceil()` in line 3.

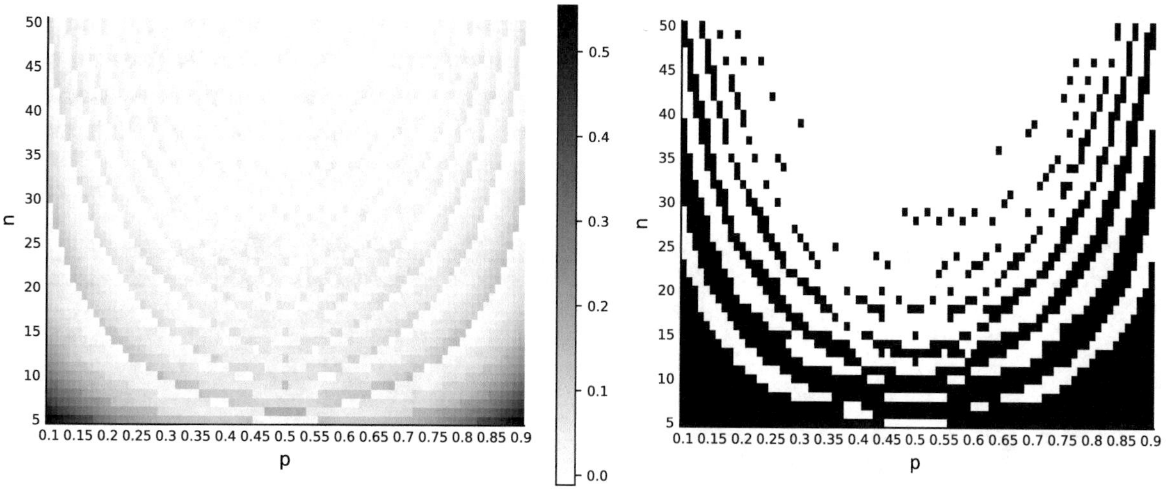

Figure 6.2: Approximation error of a confidence interval for proportion. The heatmaps are for various n and p combinations for $\alpha = 0.05$.

Validity of the Approximation

The key to the derivation of (6.21) is the distributional approximation in (6.20). In many cases, this approximation works well, however for small sample sizes n or values of p near 0 or 1, this is often too crude of an approximation. A consequence is that one may obtain a confidence interval that isn't actually a $1-\alpha$ confidence interval, but rather has a different coverage probability.

One common rule of thumb used to decide if (6.21) is valid is to require that both the product np and the product $n(1-p)$ be at least 10. For $p = 0.5$, this rule specifies a minimal sample size of $n = 20$, and for other values of p higher values of n are required.

How does such a rule come about? To explore this we now present a computational experiment, aiming to asses when (6.21) is valid. For this, we explore a grid of n ranging from 5 to 50 and p in the interval $[0.1, 0.9]$. For each combination, we repeat $N = 5{,}000$ Monte Carlo experiments and calculate the following:

$$(1-\alpha) \quad - \quad \frac{1}{N}\sum_{k=1}^{N} \mathbf{1}\left\{ p \in \left[\hat{p} - z_{1-\alpha/2}\sqrt{\frac{\hat{p}(1-\hat{p})}{n}},\ \ \hat{p} + z_{1-\alpha/2}\sqrt{\frac{\hat{p}(1-\hat{p})}{n}}\right]\right\}. \tag{6.26}$$

This estimated difference of the actual coverage probability of the confidence interval (6.21) and the desired confidence level $1-\alpha$ is a measure of the accuracy of the confidence level. We expect this difference to be almost 0 if the approximation is "good". Otherwise, a higher absolute difference is observed.

Listing 6.9 creates Figure 6.2 which presents the results of this simulation experiment for $\alpha = 0.05$. The left plot illustrates the estimated difference between the actual coverage probability and $1-\alpha$. Observe that, in general for p values around 0.5, there is less error than p values closer to 0 or 1. Also, as expected when n is increased the error probabilities drop. There is also a periodic effect due to the fact that for small n, the proportion estimator $\hat{p}$ only falls on a small finite set of values.

The right hand plot of Figure 6.2 compares the absolute value of (6.26) to 0.04 (an ad hoc tolerance that we selected). For (n, p) where the absolute error is less than 0.04, we can say the confidence error is "small" and conclude that using the confidence interval formula is satisfactory. As seen from the plot, this occurs when p is closer to 0.5 and n is at around 20 or more. This may give some insight into the heuristic rule described above and generally agrees with it.

Listing 6.9: Coverage accuracy of a confidence interval for proportions

```
1  using Random, Plots, Distributions, Measures; pyplot()
2
3  N = 5*10^3
4  alpha = 0.05
5  confLevel = 1 - alpha
6  z = quantile(Normal(),1-alpha/2)
7
8  function randCI(n,p)
9      sample = rand(n) .< p
10     pHat = sum(sample)/n
11     serr = sqrt(pHat*(1-pHat)/n)
12     (pHat - z*serr, pHat + z*serr)
13 end
14 cover(p,ci) = ci[1] <= p && p <= ci[2]
15
16 pGrid = 0.1:0.01:0.9
17 nGrid = 5:1:50
18 errs = zeros(length(nGrid),length(pGrid))
19
20 for i in 1:length(nGrid)
21     for j in 1:length(pGrid)
22         Random.seed!(0)
23         n, p = nGrid[i], pGrid[j]
24         coverageRatio = sum([cover(p,randCI(n,p)) for _ in 1:N])/N
25         errs[i,j] = confLevel - coverageRatio
26     end
27 end
28
29 default(xlabel = "p", ylabel = "n",
30     xticks =([1:5:length(pGrid);], minimum(pGrid):0.05:maximum(pGrid)),
31     yticks =([1:5:length(nGrid);], minimum(nGrid):5:maximum(nGrid)))
32
33 p1 = heatmap(errs, c=cgrad([:white, :black]))
34 p2 = heatmap(abs.(errs) .<= 0.04, legend = false, c=cgrad([:black, :white]))
35 plot(p1,p2, size = (1000,400), margin = 5mm)
```

In line 3, we set the number of Monte Carlo repetitions, `N`. Lines 4–6 define constants for the confidence interval based on `alpha` from line 4. In lines 8–13, we define the function `randCI()` for generating a random sample and an associated confidence interval. The function `cover()` that we define in line 14 checks if `p` is covered by the given confidence interval, `ci`. Lines 16–17 define the grid of p values and n values on which we estimate (6.26). In line 18, we initialize the matrix `errs` using `zeros()`. The simulation repetitions are in lines 20–27 where, after resetting the seed in line 22, for each (n, p) combination we compute the sum in (6.26) into `coverageRatio` in line 24 by composing `cover()` on `randCI()`. Then in line 25 we record the estimated difference in the matrix `errs`. Lines 29–35 create Figure 6.2.

6.4 Confidence Interval for the Variance of a Normal Population

We now consider confidence intervals when the parameter in question is the variance. We also use this as an example to show how model assumptions may strongly affect the accuracy of the confidence interval. Consider sampling from a population that follows a normal distribution. A point estimator for the population variance is the sample variance,

$$S^2 = \frac{1}{(n-1)} \sum_{i=1}^{n} (X_i - \overline{X})^2.$$

As illustrated in Section 5.2, when multiplied by the constant $(n-1)/\sigma^2$, the sample variance follows a chi-squared distribution with $n-1$ degrees of freedom

$$\frac{(n-1)S^2}{\sigma^2} \sim \chi^2_{n-1}.$$

Therefore, denoting the γ-quantile of this distribution via $\chi^2_{\gamma,n-1}$, we have

$$\mathbb{P}\left(\chi^2_{\frac{\alpha}{2},n-1} < \frac{(n-1)S^2}{\sigma^2} < \chi^2_{1-\frac{\alpha}{2},n-1} \right) = 1-\alpha. \tag{6.27}$$

Hence, we can re-arrange to obtain a two-sided $100(1-\alpha)\%$ confidence interval for the variance of a normal population where we denote the observed estimator by s^2:

$$\frac{(n-1)s^2}{\chi^2_{1-\frac{\alpha}{2},n-1}} < \sigma^2 < \frac{(n-1)s^2}{\chi^2_{\frac{\alpha}{2},n-1}}. \tag{6.28}$$

Note that (6.27) only holds when sampling from data that is normally distributed. If the data is not normally distributed, then our confidence intervals will be inaccurate. Such sensitivity to assumptions is explored later in this section. However, first we demonstrate a simple example for using the confidence interval (6.28) in Listing 6.10.

Listing 6.10: Confidence interval for the variance

```
1 using CSV, Distributions, HypothesisTests
2
3 data = CSV.read("../data/machine1.csv", header=false)[:,1]
4 n, s, alpha = length(data), std(data), 0.1
5 ci = (   (n-1)*s^2/quantile(Chisq(n-1),1-alpha/2),
6          (n-1)*s^2/quantile(Chisq(n-1),alpha/2)   )
7
8 println("Point estimate for the variance: ", s^2)
9 println("Confidence interval for the variance: ", ci)
```

```
Point estimate for the variance: 1.3928282706110504
Confidence interval for the variance: (0.8779243703322502, 2.6157658366723124)
```

The code is similar to Listing 6.2 and uses the same dataset. Lines 5–6 implement (6.28) based on the sample variance standard deviation `s` and a chi-squared distribution, `Chisq()`.

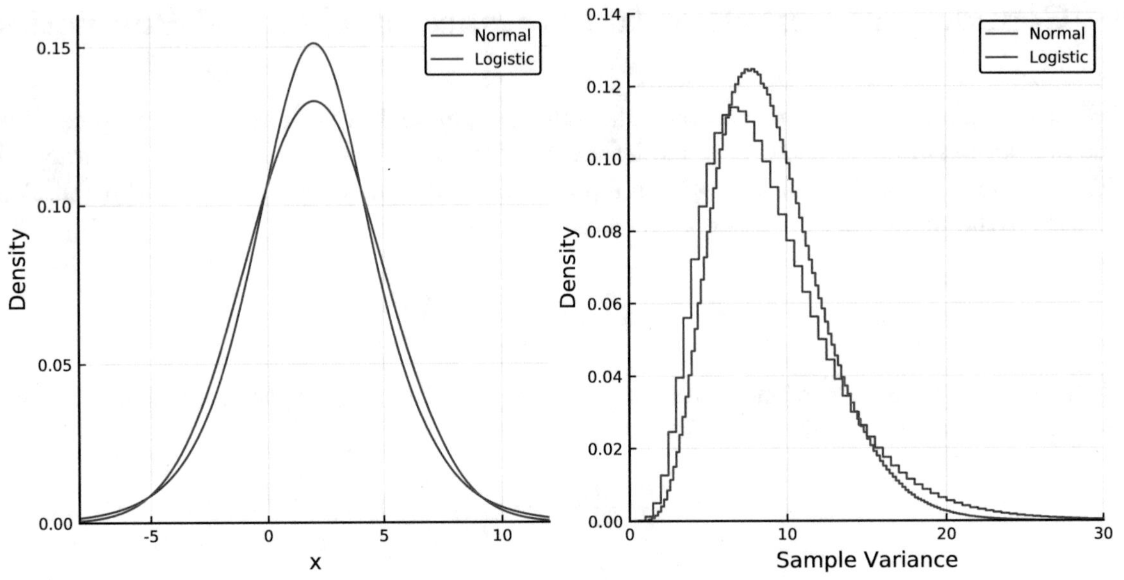

Figure 6.3: PDF's of the normal and logistic distributions, along with histograms of the sample variances from the corresponding distributions.

Sensitivity of the Normality Assumption

We now look at the sensitivity of the normality assumption on the confidence interval for the variance. As part of this example we first introduce the *logistic distribution* which has a "bell curved" shape somewhat similar to the normal distribution. It is defined by the location and scale parameters, μ and η. The PDF of the logistic distribution is

$$f(x) = \frac{e^{-\frac{x-\mu}{\eta}}}{\eta\left(1+e^{-\frac{x-\mu}{\eta}}\right)^2}, \tag{6.29}$$

with the variance given by $\eta^2\pi^2/3$.

In Listing 6.11, we create Figure 6.3 where in the left plot, the PDF of a normal distribution with mean $\mu = 2$ and standard deviation $\sigma = 3$ is plotted against that of a logistic distribution with the same mean and variance. To achieve that we require $\eta^2\pi^2/3 = \sigma^2$ and hence,

$$\eta = \frac{\sqrt{3}}{\pi}\sigma. \tag{6.30}$$

While both of these symmetric (about the mean) distributions share the same mean and variance and hence are somewhat similar, in the right plot, we show via Monte Carlo that the distributions of their sample variances with $n = 15$ are actually significantly different. This gives a first hint at the fact that the confidence interval formula (6.28) may be very sensitive to the normality assumption. Later, in the example that follows, we investigate the effect of this on the confidence interval.

Listing 6.11: Comparison of sample variance distributions

```
1  using Distributions, Plots; pyplot()
2
3  mu, sig = 2, 3
4  eta = sqrt(3)*sig/pi
5  n, N = 15, 10^7
6  dNormal   = Normal(mu, sig)
7  dLogistic = Logistic(mu, eta)
8  xGrid = -8:0.1:12
9
10 sNormal   = [var(rand(dNormal,n)) for _ in 1:N]
11 sLogistic = [var(rand(dLogistic,n)) for _ in 1:N]
12
13 p1 = plot(xGrid, pdf.(dNormal,xGrid), c=:blue, label="Normal")
14 p1 = plot!(xGrid, pdf.(dLogistic,xGrid), c=:red, label="Logistic",
15         xlabel="x",ylabel="Density", xlims= (-8,12), ylims=(0,0.16))
16
17 p2 = stephist(sNormal, bins=200, c=:blue, normed=true, label="Normal")
18 p2 = stephist!(sLogistic, bins=200, c=:red, normed=true, label="Logistic",
19         xlabel="Sample Variance", ylabel="Density", xlims=(0,30), ylims=(0,0.14))
20
21 plot(p1, p2, size=(800, 400))
```

In line 3, we define the mean and standard deviation that will be used for both distributions. Then in line 4, we calculate `eta` according to (6.30). In line 5, the number of sample observations, `n`, and total number of experiments, `N`, are specified. In lines 6–7, we define the two distributions with matched moments and variance. Note that the Julia `Logistic()` function uses the same parametrization as that of equation (6.29). In lines 10–11, comprehensions are used to generate `N` sample variances from the normal and logistic distributions `dNormal` and `dLogistic`, with the values assigned to the arrays `sNormal` and `sLogistic`, respectively. The remainder of the code creates the plots with lines 13–15 creating the left plot by using `pdf()` broadcasted over `xGrid` and lines 17–19 creating histograms of the sample variances.

Having seen that the distribution of the sample variance heavily depends on the shape of the actual distribution, we now investigate the effect that this has on the accuracy of the confidence interval. Specifically, we show that while usage of the confidence interval formula (6.28) yields $1-\alpha$ coverage for normally distributed data, it strongly deviates from the logistic distribution case.

In Listing 6.12, we cycle through different values of α from 0.001 to 0.1, and for each value, we perform N of the following identical experiments: calculate the sample variance of n observations and evaluate the confidence interval (6.28). We then calculate the proportion of times that the actual (unknown) variance of the distribution is contained within the confidence interval in a similar manner to what we did in the context of proportions in (6.26). The effective α values are then plotted against the actual values in Figure 6.4.

It can be observed that in the case of the normal distribution, the simulated α values align with those of the actual α used. However, in the case of logistic distribution, there is a strong discrepancy. This illustrates that model assumptions are critical for the correctness of the confidence interval (6.28). Note that in general, confidence intervals for the mean such as (6.5), would be less sensitive to model assumptions than the confidence interval for the variance.

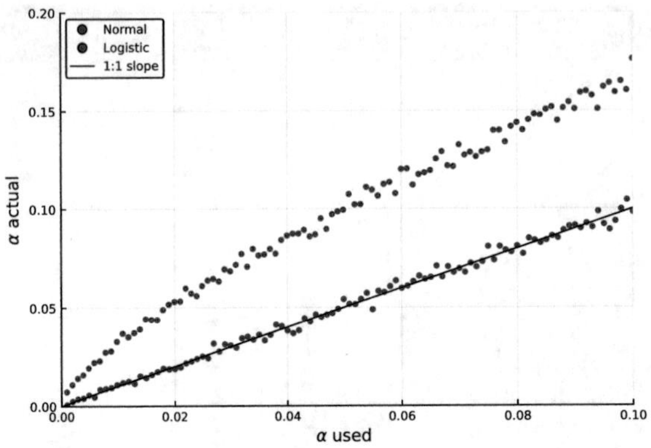

Figure 6.4: Actual α values vs. α values used in confidence intervals.

Listing 6.12: Actual α vs. α used in variance confidence intervals

```
1   using Distributions, Plots, LaTeXStrings; pyplot()
2
3   mu, sig = 2, 3
4   eta = sqrt(3)*sig/pi
5   n, N = 15, 10^4
6   dNormal   = Normal(mu, sig)
7   dLogistic = Logistic(mu, eta)
8   alphaUsed = 0.001:0.001:0.1
9
10  function alphaSimulator(dist, n, alpha)
11      popVar        = var(dist)
12      coverageCount = 0
13      for _ in 1:N
14          sVar = var(rand(dist, n))
15          L = (n - 1) * sVar / quantile(Chisq(n-1),1-alpha/2)
16          U = (n - 1) * sVar / quantile(Chisq(n-1),alpha/2)
17          coverageCount +=  L < popVar && popVar < U
18      end
19      1 - coverageCount/N
20  end
21
22  scatter(alphaUsed, alphaSimulator.(dNormal,n,alphaUsed),
23          c=:blue, msw=0, label="Normal")
24  scatter!(alphaUsed, alphaSimulator.(dLogistic, n, alphaUsed),
25          c=:red, msw=0, label="Logistic")
26  plot!([0,0.1],[0,0.1],c=:black, label="1:1 slope",
27          xlabel=L"\alpha"*" used", ylabel=L"\alpha"*" actual",
28          legend=:topleft, xlim=(0,0.1), ylims=(0,0.2))
```

Lines 3–7 are identical to the previous listing. In line 8 we define a grid of α values over which we carry out the experiment. Lines 10–20 define the function `alphaSimulator()`. This function takes a distribution, the total number of sample observations, and a value of alpha as input. It then generates `N` separate confidence intervals for the variance via equation (6.28) and returns the corresponding proportion of times the confidence intervals do not contain the actual variance of the distribution. We apply `alphaSimulator()` directly in lines 22 and 24 as part of the scatter plots. Lines 26-28 plot the 1:1 line on which α used equals α actual.

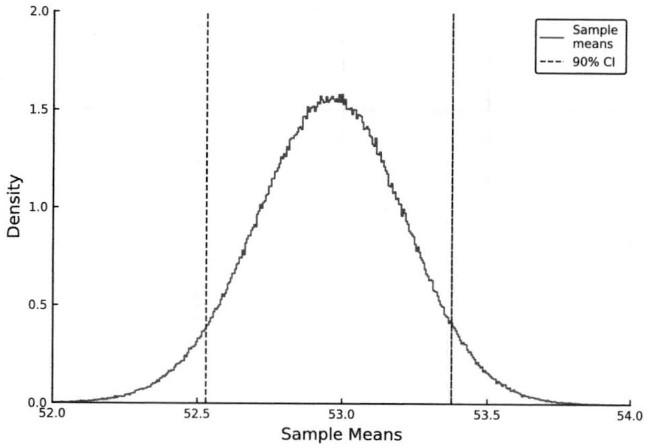

Figure 6.5: A single confidence interval for the mean generated by bootstrapped data.

6.5 Bootstrap Confidence Intervals

When developing confidence intervals, the main goal is to make some sort of inference about the population based on sample data. However, in some cases ,a statistical model may not be readily available, or as we saw in Listing 6.9 and Listing 6.12, model error may cause inaccuracies. Hence, it is useful to have an alternative method for finding confidence intervals. One such general method is the method of *bootstrap confidence intervals.*

Bootstrap, also called *empirical bootstrap*, is a useful technique which relies on resampling from the observed data $x_1, \dots, x_n$ in order to empirically construct the distribution of the point estimator. One way in which this resampling can be conducted is to apply the inverse probability transform on the empirical cumulative distribution function. However, from an implementation perspective, a simpler alternative is to consider the data points $x_1, \dots, x_n$,and then randomly sample n discrete uniform indexes, $j_1, \dots, j_n$ each in the range $\{1, \dots, n\}$. The *resampled data* denoted via $x^* = (x_1^*, \dots, x_n^*)$ is then

$$x^* = (x_{j_1}, \dots, x_{j_n}).$$

That is, each point in the resampled data is a random observation from the original data, where we allow to "sample with replacement". In Julia, if the sample is represented by an array called `sampleData`, say of length `n`, we create an instance of x^* by executing `rand(sampleData,n)`. This method of the `rand()` function will uniformly sample `n` random copies of elements from `sampleData` with replacement.

The idea of empirical bootstrap is now to repeat the resampling a large number of times, say N, and for each resampled data vector, $x^*(1), \dots, x^*(N)$ to compute the parameter estimate. If the parameter estimate is denoted by the function $h : \mathbb{R}^n \mapsto \mathbb{R}$, then we end up with values,

$$\begin{aligned} h^*(1) &= h\big(x_1^*(1), \dots, x_n^*(1)\big), \\ h^*(2) &= h\big(x_1^*(2), \dots, x_n^*(2)\big), \\ &\vdots \\ h^*(N) &= h\big(x_1^*(2), \dots, x_n^*(N)\big). \end{aligned}$$

A bootstrap confidence interval is then determined by computing the respective lower and upper $(\frac{\alpha}{2}, 1-\frac{\alpha}{2})$ quantiles of the sequence $h^*(1), \ldots, h^*(N)$. The beauty of this method is that if n is not too small, the resulting quantiles are quite close to the actual quantiles of the distribution of the point estimate. Hence, in general, bootstrap confidence intervals provide a very generic and general method to obtain confidence intervals for parameters in question.

In Listing 6.13, we generate a bootstrap confidence interval for the mean, using the same data that was used in Section 6.1. The listing also creates Figure 6.5 which illustrates the empirical distribution of $h^*(1), \ldots, h^*(N)$, where $h(\cdot)$ is the sample mean. The 90% confidence interval is $(52.53, 53.38)$. Compare this with the output of Listing 6.2 where the 90% confidence interval using the T-distribution and formula (6.5) for the population mean was $(52.5, 53.41)$. Clearly, the results are similar. While bootstrap requires more computational effort and doesn't come with a neat simple formula as (6.5), it is useful because we can use it to generate confidence intervals for other point estimators. For example, as observed in the output, we also generate a 90% confidence interval for the median.

Listing 6.13: Bootstrap confidence interval

```
1  using Random, CSV, Distributions, Plots; pyplot()
2  Random.seed!(0)
3
4  sampleData = CSV.read("../data/machine1.csv", header=false)[:,1]
5  n, N = length(sampleData), 10^6
6  alpha = 0.1
7
8  bootstrapSampleMeans = [mean(rand(sampleData, n)) for i in 1:N]
9  Lmean = quantile(bootstrapSampleMeans, alpha/2)
10 Umean = quantile(bootstrapSampleMeans, 1-alpha/2)
11
12 bootstrapSampleMedians = [median(rand(sampleData, n)) for i in 1:N]
13 Lmed = quantile(bootstrapSampleMedians, alpha/2)
14 Umed = quantile(bootstrapSampleMedians, 1-alpha/2)
15
16 println("Bootstrap confidence interval for the mean: ", (Lmean, Umean) )
17 println("Bootstrap confidence interval for the median: ", (Lmed, Umed) )
18
19 stephist(bootstrapSampleMeans, bins=1000, c=:blue,
20     normed=true, label="Sample \nmeans")
21 plot!([Lmean, Lmean],[0,2], c=:black, ls=:dash, label="90% CI")
22 plot!([Umean, Umean],[0,2],c=:black, ls=:dash, label="",
23     xlims=(52,54), ylims=(0,2), xlabel="Sample Means", ylabel="Density")
```

```
Bootstrap confidence interval for the mean: (52.530497748, 53.376643266)
Bootstrap confidence interval for the median: (52.373195891, 53.49007500)
```

In line 4, we load our sample observations and store them in the array `sampleData`. In line 5, the total number of sample observations is assigned to `n` and the number of repetitions of the bootstrap, `N`, is specified. In line 6 we specify the level of our confidence interval `alpha`. In line 8, we generate `N` bootstrapped sample means which are assigned to the array `bootstrapSampleMeans`. In lines 9–10, the lower and upper quantiles of our bootstrapped sample data are calculated and stored as the variables `Lmean` and `Umean`, respectively. Then lines 12–14 repeat the process for the sample median using `median()`. The resulting confidence intervals are printed in lines 16–17. Lines 19–23 plot a histogram of bootstrapped sample means with an illustration of the resulting confidence interval.

One may ask, how accurate are bootstrap confidence intervals? We now carry out a computational experiment and see that if the number of sample observations is not very large, then the *coverage probability* of bootstrapped confidence interval is only approximately $1-\alpha$, but not exactly. However, as the sample size n grows, the coverage probability converges to the desired $1-\alpha$.

In Listing 6.14, we create a series of confidence intervals based on different numbers of sample observations from an exponential distribution with $\lambda = 0.1$. Our confidence intervals are for the median which theoretically equals $\log(2)/\lambda = 6.931$. By increasing the sample size n and repeating many sampling scenarios, $M = 10^3$, each with a bootstrap computation using $N = 10^4$, we estimate the coverage probability. We see that as n increases the coverage probability approaches $1-\alpha$.

Listing 6.14: Coverage probability for bootstrap confidence intervals

```
1  using Random, Distributions
2  Random.seed!(0)
3
4  lambda = 0.1
5  dist = Exponential(1/lambda)
6  actualMedian = median(dist)
7
8  M = 10^3
9  N = 10^4
10 nRange = 2:2:10
11 alpha = 0.05
12
13 for n in nRange
14     coverageCount = 0
15     for _ in 1:M
16         sampleData = rand(dist, n)
17         bootstrapSampleMeans = [median(rand(sampleData, n)) for _ in 1:N]
18         L = quantile(bootstrapSampleMeans, alpha/2)
19         U = quantile(bootstrapSampleMeans, 1-alpha/2)
20         coverageCount += L < actualMedian && actualMedian < U
21     end
22     println("n = ",n,"\t coverage = ", coverageCount/M)
23 end
```

```
n = 2     coverage = 0.483
n = 4     coverage = 0.881
n = 6     coverage = 0.936
n = 8     coverage = 0.939
n = 10    coverage = 0.949
```

In line 4, we specify λ. In line 5 we create `dist` remembering that in Julia the exponential distribution is parameterized by the inverse of λ. The actual median is then computed in line 6 via `median()` method implemented in the `Distributions` package. In line 8, we specify the number of repetitions we make to evaluate the coverage probability, `M`. Then in line 9 the number of bootstrap repetitions, `N`, is specified. In line 10, we specify the range of number of observations to consider, `nRange`. We then loop over this range in lines 13-23, where, in each iteration, we count `coverageCount`, counting the number of times the bootstrap confidence interval contained `actualMedian`. Each actual bootstrap confidence interval procedure is in lines 17–19. Note the boolean value to the right of `+=` in line 20 which evaluates to `true` if the median is covered by `(L,U)`.

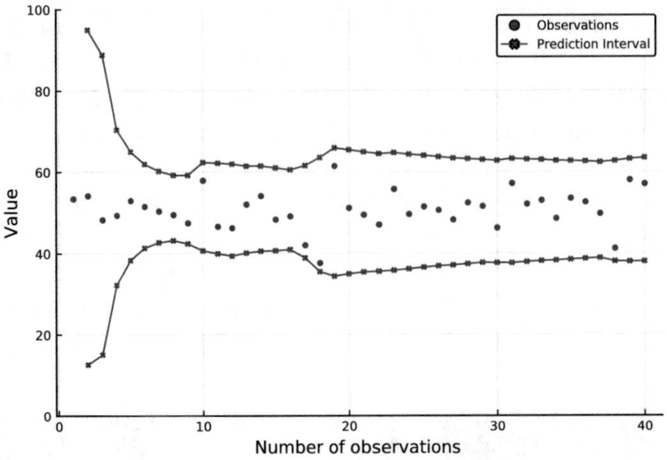

Figure 6.6: As the number of observations increases, the width of the prediction interval decreases to a constant.

6.6 Prediction Intervals

We now look at the concept of a *prediction interval* which is somewhat related to confidence intervals, however has a different meaning. A prediction interval tells us a predicted range that a single next observation of data is expected to fall within. This differs from a confidence interval which indicates how confident we are of a particular parameter that we are trying to estimate. For a given distributional model, the bounds of a prediction interval are always wider than those of a confidence interval, as the prediction interval must account for the uncertainty in knowing the population mean, as well as the spread of the data due to variance.

The example that we use is for a sequence of data points $x_1, x_2, x_3, \ldots$, which come from a normal distribution and are assumed i.i.d. Further assume that we observed $x_1, \ldots, x_n$ but have not yet observed X_{n+1}. Note that we use "little" x for values observed and "upper case" X for (yet) unobserved random variables.

In this case, a $100(1-\alpha)\%$ prediction interval for the single future observation, X_{n+1}, is given by

$$\overline{x} - t_{1-\frac{\alpha}{2},n-1} s\sqrt{1+\frac{1}{n}} \le X_{n+1} \le \overline{x} + t_{1-\frac{\alpha}{2},n-1} s\sqrt{1+\frac{1}{n}}, \tag{6.31}$$

where, $\overline{x}$ and s are respectively, the sample mean and sample standard deviation computed from $x_1, \ldots, x_n$. Note that as the number of observations, n, increases, the bounds of the prediction interval decreases towards,

$$\overline{x} - z_{1-\frac{\alpha}{2}} s \le X_{n+1} \le \overline{x} + z_{1-\frac{\alpha}{2}} s. \tag{6.32}$$

In Listing 6.15, we illustrate the use of prediction intervals based on a series of observations made from a normal distribution. We start with $n = 2$ observations and calculate the corresponding prediction interval for the next observation. The sample size n is then progressively increased, and the prediction interval for each next observation is calculated for each subsequent case. The listing creates Figure 6.6 which illustrates that as the number of observations increases, the prediction interval width decreases. Ultimately, it follows (6.32). and has an expected width close to $2\, z_{1-\frac{\alpha}{2}}\, \sigma$.

Listing 6.15: Prediction interval with unknown population mean and variance

```
1  using Random, Statistics, Distributions, Plots; pyplot()
2  Random.seed!(0)
3
4  mu, sig = 50, 5
5  dist = Normal(mu, sig)
6  alpha = 0.01
7  nMax = 40
8
9  observations = rand(dist,1)
10 piLarray, piUarray = [], []
11
12 for _ in 2:nMax
13     xNew = rand(dist)
14     push!(observations,xNew)
15
16     xbar, sd = mean(observations), std(observations)
17     n = length(observations)
18     tVal = quantile(TDist(n-1),1-alpha/2)
19     delta = tVal * sd * sqrt(1+1/n)
20     piL, piU = xbar - delta, xbar + delta
21
22     push!(piLarray,piL); push!(piUarray,piU)
23 end
24
25 scatter(1:nMax, observations,
26         c=:blue, msw=0, label="Observations")
27 plot!(2:nMax, piUarray,
28         c=:red, shape=:xcross, msw=0, label="Prediction Interval")
29 plot!(2:nMax, piLarray,
30         c=:red, shape=:xcross, msw=0, label="",
31         ylims=(0,100), xlabel="Number of observations", ylabel="Value")
```

In lines 4–7, we set up the distributional parameters, choose α, and also set the limiting number of observations we will make, `nMax`. In line 9, we sample the first sample observation and store it in the array `observations`. In line 10, we create the arrays `piLarray` and `piUarray`, which will be used to store the lower and upper bounds of the prediction interval. Lines 12–23 contain the main logic of this example where, in lines 13–14, a new data point is obtained and stored, in lines 16–18 updated prediction interval is calculated, and in line 22, the prediction interval is stored for plotting afterwards. Lines 25–31 create Figure 6.6. Observe the use of `shpae=:xcross` for setting tick marks,

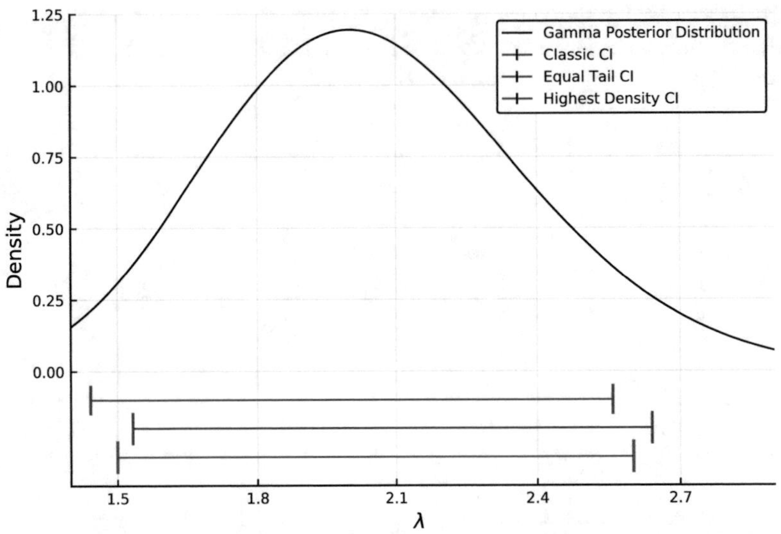

Figure 6.7: The posterior distribution and three forms of 90% credible intervals.

6.7 Credible Intervals

This section presents two related concepts, *credible intervals* and *intervals on asymmetric distributions*. The concept of credible intervals comes from the field of Bayesian statistics which we overviewed in Section 5.7 of Chapter 5. It is an analog of a confidence interval in the Bayesian setting. Before we explain this concept further, we first deal with various ways of finding intervals on asymmetric distributions.

In general, we often need to find an interval $[\ell, u]$ such that given some probability density $f(x)$, the interval satisfies,

$$\int_{\ell}^{u} f(x)\, dx = 1 - \alpha. \tag{6.33}$$

This is needed for confidence intervals which were the focus for most of this chapter, for prediction intervals which were discussed in the previous section, or for credible intervals which are discussed below. However, as long as $\alpha < 1$, there is never a single unique interval $[\ell, u]$ that satisfies (6.33).

In certain cases, there is a "natural" interval. For example, for the normal distribution, using equal tail quantiles is natural. We do this by choosing $\ell = z_{\alpha/2}$ and $u = z_{1-\alpha/2}$ as was used throughout this chapter. Similarly, for a T-distribution with $n-1$ degrees of freedom, we used $\ell = t_{\alpha/2,n-1}$ and $u = t_{1-\alpha/2,n-1}$. Such choices for $[\ell, u]$ are natural because in both the case of the normal and the T-distribution, the density $f(x)$ is symmetric about the mean. In such cases, the mean is also the median and further since the density is *unimodal* (has a single maximum) the mean and median are also the mode. With such symmetry and unimodality, while there exist other choices of ℓ and u, they don't appeal to applications.

However, consider *asymmetric distributions* such as the one presented in Figure 6.7. In this case, there isn't an immediate "natural" choice for ℓ and u. Without going into the actual meaning of the density in the figure just yet, you can already observe three different plotted intervals. They all satisfy (6.33) A blue *classic CI*, a red *equal tail CI*, and a green *highest density CI*. We now describe each of these intervals.

Classic interval—This type of interval has the mode of the density (assuming the density is unimodal), at its center between ℓ and u. An alternative is to use mean or median at the center. That is, assuming the centrality measure (mode, median or mean) is m, we have, $[\ell, u] = [m - E, m + E]$. One way to define E is via

$$E = \max\{\varepsilon \geq 0 \ : \ \int_{m-\varepsilon}^{m+\varepsilon} f(x)\, dx \leq 1 - \alpha\}. \tag{6.34}$$

That is, we can search for the highest ε such that the integral over $[m-\varepsilon, m+\varepsilon]$ doesn't exceed $1 - \alpha$. This is crudely implemented in Listing 6.16 in the function `classicalCI()`.

Equal tail interval—This type of interval simply sets ℓ and u as the $\alpha/2$ and $1 - \alpha/2$ quantiles respectively. Namely,

$$\frac{\alpha}{2} = \int_{-\infty}^{\ell} f(x)\, dx, \qquad \text{and} \qquad \frac{\alpha}{2} = \int_{u}^{\infty} f(x)\, dx. \tag{6.35}$$

Such an interval was implicitly used with the asymmetric chi-squared distribution in Section 6.4. See for example formula (6.27).

Highest density interval—This type of interval seeks to cover the part of the support that is most probable. Define the smallest probability densities falling over an interval $[\ell, u]$ via $M(\ell, u) = \min\{f(x) \ : \ x \in [\ell, u]\}$. Then the highest density interval seeks for an interval $[\ell, u]$ that satisfies (6.33) while maximizing $M(\ell, u)$. A consequence is that if the density is unimodal then this highest density interval is also the narrowest possible confidence interval.

There are multiple computational ways to find such a confidence interval. In Listing 6.16, the function `highestDensityCI()` crudely does so by starting with a high-density value and decreasing it gradually while seeking for the associated interval $[\ell, u]$. An alternative would be to gradually increment ℓ each time finding a corresponding u that satisfies (6.33) and within this search to choose the interval that minimizes the width $u - \ell$.

For a symmetric and unimodal distribution such as the Normal distribution or T-distribution, all three of these confidence intervals agree. However, in general, they differ. In a Bayesian context, as we describe below, one often prefers the highest density interval. However, in other settings, equal tail intervals are common. For example, when considering confidence intervals for the variance, we used equal tail intervals in (6.27) because it yielded the simple confidence interval formula (6.28) which uses tabulated quantiles (of chi-squared distributions).

We now explain the *credible intervals*. These come instead of confidence intervals in a Bayesian setting. Recall that in the Bayesian setting, we treat the unknown parameter, θ, as a random variable. As described in Section 5.7, the process of inference is based on observing data, $x = (x_1, \ldots, x_n)$ and fusing it with the prior distribution $f(\theta)$ to obtain the posterior distribution $f(\theta \mid x)$. Here too, as in the frequentist case, we may wish to describe an interval where it is likely that our parameter lies. Then for a fixed confidence level, $1 - \alpha$, seek $[\ell, u]$ such that,

$$\int_{\ell}^{u} f(\theta \mid x)\, d\theta = 1 - \alpha.$$

Compare this with (5.18) of Chapter 5 where $[L, U]$ denotes the confidence interval. In the Bayesian case, ℓ and u are deterministic values determined from the prior distribution of the random θ, whereas in the frequentist case, θ is deterministic with L and U random. This is a conceptual difference between confidence intervals and credible intervals. However practically there isn't really a difference in how they are used.

Nevertheless, in a Bayesian context, unless dealing with special cases of conjugacy (see Listing 5.19), the posterior distribution of the parameter is often only available numerically via MCMC or similar methods (see Listing 5.20). Hence there is no general motivation to use equal tail intervals or classical intervals. Instead, the highest density intervals are often a prime choice.

In Listing 6.16, we generate Figure 6.7 and compute alternative credible intervals using classical, equal tail, and highest density methods. We deal with the same small data set that was used in Section 5.7 of Chapter 5. Here the unknown parameter is λ, the mean of a Poisson distribution. In this simple example, for simplicity, we use gamma-Poisson conjugacy and thus update hyper-parameters of with simple rules as in (5.24) of Chapter 5.

Observe that by design all three intervals have the same $1 - \alpha = 0.9$ coverage probability. Slight differences only appear due to numerical inaccuracy. Also note that the width of the highest density interval is lowest, at 1.102. Indeed, there isn't a 90% confidence interval over this prior distribution narrower than 1.102. We finally note that our implementation aims to be simple and readable but not the most time-efficient nor the most numerically precise.

Listing 6.16: Credible intervals on a posterior distribution

```
using Distributions, Plots, LaTeXStrings; pyplot()

alpha, beta = 8, 2
data = [2,1,0,0,1,0,2,2,5,2,4,0,3,2,5,0]

newAlpha, newBeta = alpha + sum(data), beta + length(data)
post = Gamma(newAlpha, 1/newBeta)

xGrid = quantile(post,0.01):0.001:quantile(post,0.99)
significance = 0.9; halfAlpha = (1-significance)/2

coverage(l,u) = cdf(post,u) - cdf(post,l)

function classicalCI(dist)
    l, u = mode(dist),mode(dist)
    bestl, bestu = l, u
    while  coverage(l,u) < significance
        l -= 0.00001; u += 0.00001
    end
    (l,u)
end
equalTailCI(dist) = (quantile(dist,halfAlpha), quantile(dist,1-halfAlpha))
function highestDensityCI(dist)
    height = 0.999 * maximum(pdf.(dist,xGrid))
    l,u = mode(dist),mode(dist)
    while coverage(l,u) <= significance
        range = filter(theta -> pdf(dist,theta) > height, xGrid)
        l,u = minimum(range), maximum(range)
        height -= 0.00001
    end
    (l,u)
end

l1, u1 = classicalCI(post)
l2, u2 = equalTailCI(post)
l3, u3 = highestDensityCI(post)
println("Classical: ", (l1,u1), "\tWidth: ",u1-l1,
        "\tCoverage: ", coverage(l1,u1))
println("Equal tails: ", (l2,u2), "\tWidth: ",u2-l2,
        "\tCoverage: ", coverage(l2,u2))
println("Highest density: ", (l3,u3), "\tWidth: ",u3-l3,
        "\tCoverage: ", coverage(l3,u3))

plot(xGrid,pdf.(post,xGrid),  yticks=(0:0.25:1.25),
        c=:black, label="Gamma Posterior Distribution",
        xlims=(1.4, 2.9), ylims=(-0.4,1.25))
plot!([l1,u1],[-0.1,-0.1], label="Classic CI",
        c=:blue, shape=:vline, ms=16)
plot!([l2,u2],[-0.2,-0.2], label="Equal Tail CI",
        c=:red, shape=:vline, ms=16)
plot!([l3,u3],[-0.3,-0.3], label="Highest Density CI",
        c=:green, shape=:vline, ms=16, xlabel=L"\lambda", ylabel="Density")
```

```
Classical: (1.44, 2.56)         Width: 1.1146    Coverage: 0.90000
Equal tails: (1.53, 2.64)       Width: 1.1081    Coverage: 0.89999
Highest density: (1.51, 2.60)   Width: 1.1020    Coverage: 0.90018
```

In line 3, we define the prior hyper-parameters, `alpha` and `beta`. Line 4 defines the observations. In line 6, the posterior hyper-parameters are calculated using the Poisson-gamma conjugacy update rule. In line 7, we create an object, `post`, for the posterior distribution. In line 9, we define a fine grid for carrying out computation. In line 10, we define the confidence level $1 - \alpha$ via `significance` and also use `halfAlpha` to denote $\alpha/2$. In line 12, the function `coverage()` is defined to compute the coverage probability in the posterior distribution on the interval ranging between `l` and `u`. Lines 14–32 define the three main functions of this code block, `classicalCI()`, `equalTailCI()`, and `highestDensityCI()`, each returning the respective confidence/credible interval, classical, equal tail, and highest density. The implementation of `classicalCI()` in lines 14–21 starts with `(l,u)` at the mode of the distribution and then expands outwards in small steps until the desired coverage is reached. The implementation of `equalTailCI()` in line 22 simply returns the $\alpha/2$ and $1 - \alpha/2$ quantiles. The implementation of `highestDensityCI()` in lines 23–32 works by starting with `height` at the maximal possible value and decreasing it until coverage is satisfied. In this implementation, in each iteration we use `filter()` in line 27 with the anonymous function `theta -> pdf(dist,theta) > height`. This sets `range` to be the array of all values of `theta` for which the density exceeds `height`. In lines 34–36, we use our confidence/credible interval functions to obtain credible intervals for the posterior distribution. The intervals, their widths and coverage probabilities are printed in lines 37–42. Then lines 44–52 generate Figure 6.7. Note the use of `:vline` with a specification of `ms` to create the confidence intervals in the figure.

Chapter 7

Hypothesis Testing

In this chapter, we explore hypothesis testing through a few specific practical hypothesis tests. Recall the general hypothesis test formulation first introduced in Section 5.6, where we partition the parameter space Θ as follows:

$$H_0 : \theta \in \Theta_0, \qquad H_1 : \theta \in \Theta_1.$$

One of the most common cases for a single population is to consider θ as μ, the population mean, in which case $\Theta = \mathbb{R}$. Often, we wish to test if the population mean is equal to some value, μ_0. This allows us to construct a *two-sided hypothesis test* as follows:

$$H_0 : \mu = \mu_0, \qquad H_1 : \mu \neq \mu_0. \tag{7.1}$$

However, one could instead chose to construct a *one-sided hypothesis test*, as

$$H_0 : \mu \leq \mu_0, \qquad H_1 : \mu > \mu_0, \tag{7.2}$$

or alternatively, in the opposite direction,

$$H_0 : \mu \geq \mu_0, \qquad H_1 : \mu < \mu_0, \tag{7.3}$$

where the choice of setting up (7.1), (7.2), or (7.3) depends on the context of the problem.

As covered in Section 5.6, once the hypothesis is established, the general approach involves calculating the *test statistic*, along with the corresponding *p-value*, and then finally making some statement about the null hypothesis based on some chosen level of significance. In this chapter, we present some specific common examples often used in practice.

In Section 7.1, we introduce hypothesis testing via several examples involving a single population. In Section 7.2, we present extensions of these concepts and related ideas by looking at inference for the difference in means of two populations. In Section 7.3, we focus on methods of *Analysis of Variance (ANOVA)*. Then in Section 7.4 we explore *chi-squared tests* and *Kolmogorov-Smirnov tests*. These latter two procedures are often used to assess goodness of fit, independence, or both. We then close with Section 7.5, where we illustrate how power curves can aid in experimental design.

Y. Nazarathy and H. Klok, *Statistics with Julia*, Springer Series in the Data Sciences,
https://doi.org/10.1007/978-3-030-70901-3_7

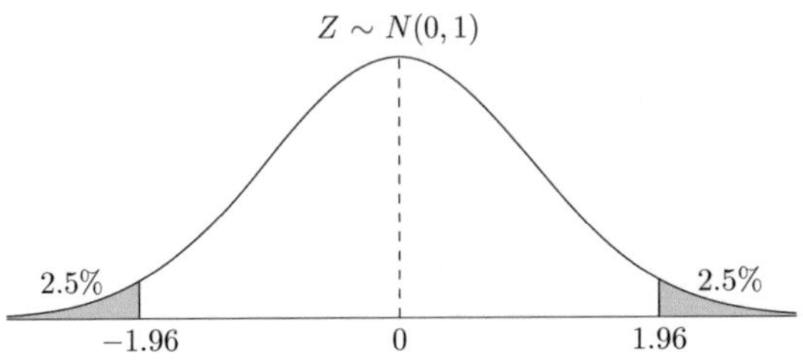

Figure 7.1: The standard normal distribution and rejection regions for the two-sided hypothesis test (7.4) at significance level $\alpha = 5\%$.

As in the previous chapters, we try to strike a balance between use of `HypothesisTests.jl` and reproducing results from fundamental calculations, with the purpose of highlighting key phenomena. Several of the examples make use of the datasets `machine1.csv` and `machine2.csv`, which represent the diameter (in mm) of pipes produced via two separate machines.

7.1 Single Sample Hypothesis Tests for the Mean

As an introduction, consider the case where we wish to make inference on the mean diameter of pipes produced by a machine. Specifically, assume that we are interested in checking if the machine is producing pipes of the specified diameter $\mu_0 = 52.2$ mm. In this case, using a hypothesis testing methodology, we may wish to set up the hypothesis as

$$H_0 : \mu = 52.2, \qquad H_1 : \mu \neq 52.2. \tag{7.4}$$

Here, H_0 represents the situation where the machine is functioning properly, and deviation from H_0 in either the positive or negative direction is captured by H_1, which represents that the machine is malfunctioning. Alternatively, one could have treated $\mu_0 = 52.2$ as a specified upper limit on the pipe diameter, in which case the hypothesis would be formulated as

$$H_0 : \mu \leq 52.2, \qquad H_1 : \mu > 52.2. \tag{7.5}$$

Similarly, one could envision a case where (7.3) was used instead. In most of this chapter, we do not dive deeply into the aspects of formulating the hypotheses themselves but rather the hypotheses are introduced and treated as given. If you are interested in the *experimental design* aspect of hypothesis testing, you may consult [M17] or similar texts.

Once the hypothesis is formulated, the next step is the collection of data, which in this section is taken from `machine1.csv`. We now separate the inference of the mean of a single population into the two cases of variance known and unknown, similarly to what was done in Section 6.1. Note also that, similarly to Chapter 6, it is assumed that the observations $X_1, \ldots, X_n$ are normally distributed. Finally, at the end of this section, we consider a simple *non-parametric* test, where we make no assumptions about the distribution of the observations.

Population Variance Known

Consider the case where we wish to test whether a single machine in a factory is producing pipes of a specified diameter. For this example, we set up the hypothesis as two-sided according to (7.4), and assume that σ is a known value. Recall from Section 5.2 that, under H_0, $\overline{X}$ follows a normal

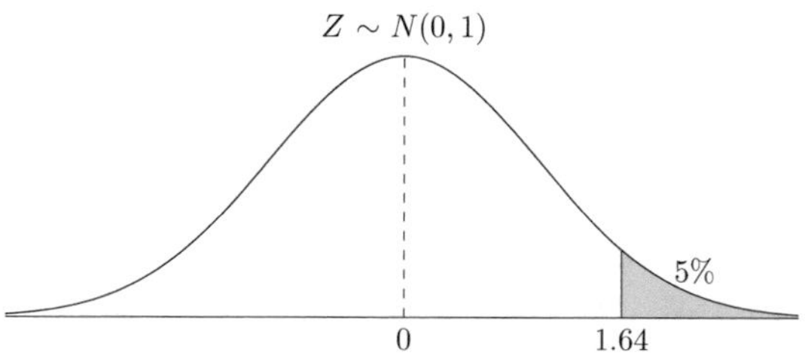

Figure 7.2: The standard normal distribution and rejection region for the one-sided hypothesis test (7.5) at significance level $\alpha = 5\%$.

distribution with mean μ_0 and variance σ^2/n. Hence, it holds that under H_0 the *Z-statistic*,

$$Z = \frac{\overline{X} - \mu_0}{\sigma/\sqrt{n}}, \tag{7.6}$$

follows a standard normal distribution.

As we will see through the various examples in this chapter, the test statistic often follows a general form similar to that of (7.6). In this case, under the null hypothesis, the random variable Z is normally distributed, and hence to carry out a hypothesis test we observe its position relative to a standard normal distribution. Specifically, we check if it lies within the *rejection region* or not, and if it does, we reject the null hypothesis, otherwise we don't. In Figure 7.1, we present the rejection region corresponding to $\alpha = 5\%$. It is obtained by considering the $\alpha/2$ and $1 - \alpha/2$ quantiles of the standard normal distribution.

With the hypothesis test and rejection region specified, we are ready to collect data, calculate the test statistic, and make a conclusion. For this example, the data is taken from `machine1.csv` where we assume $\sigma = 1.2$ (i.e. is known). After collecting the data, the observed Z-statistic is calculated via

$$z = \frac{\overline{x} - \mu_0}{\sigma/\sqrt{n}}. \tag{7.7}$$

Since this is a two-sided test, we aim for a symmetric rejection region. We then reject H_0 if $|z| > z_{1-\alpha/2}$ where $z_{1-\alpha/2}$ is the quantile of the standard normal distribution for a specific confidence level α. We may also compute the p-value of the test via

$$p = 2\,\mathbb{P}(Z > |z|). \tag{7.8}$$

That is, we consider the observed test statistic, z, and determine the maximal significance level for which we would reject H_0. Hence, for a fixed significance level α, if $p < \alpha$, we reject H_0 and otherwise not. The calculation of critical values such as $z_{1-\alpha/2}$ and p-values is typically done via software, or more traditionally via *statistical tables*, which list the area under the normal curve along with different quantiles of a standard normal. For example, $z_{0.025} = -1.96$, or $z_{0.975} = 1.96$. For $\alpha = 0.05$, we reject the null hypothesis if $z > 1.96$ or $z < -1.96$, otherwise we don't reject.

If the null hypothesis is rejected then we conclude the test by stating, "*there is sufficient evidence to reject the null hypothesis at the* 5% *significance level*". Otherwise, we conclude by stating, "*there is insufficient evidence to reject the null hypothesis at the* 5% *significance level*". Note that if a different hypothesis test setup was used, such as (7.5), then the rejection region would not be symmetric as shown in Figure 7.1, but rather would cover only one tail of the distribution. This is illustrated in Figure reffig:hypothesis2, where $z_{0.95} = 1.645$ is used to determine the boundary of the rejection region. In such a case, the p-value is calculated via $p = \mathbb{P}(Z > z)$.

In Listing 7.1, we present an example containing two hypothesis tests (using the same data) where the first (`mu0A`) is rejected and the second (`mu0B`) is not rejected. For the `mu0A` case, the test statistic is first calculated via (7.7) along with the corresponding p-value via (7.8). Then the `HypothesisTests` package is used to perform the same hypothesis test for both `mu0A` and `mu0B`. The default test assumes $\alpha = 5\%$. Observe that the p-value in the `mu0A` case is less than 0.05 and hence H_0 is rejected. In comparison, for the `mu0B` case, the p-value is greater than 0.05 and hence H_0 is not rejected.

Listing 7.1: Inference with single sample, population variance is known

```
using CSV, Distributions, HypothesisTests

data = CSV.read("../data/machine1.csv", header=false)[:,1]
xBar, n = mean(data), length(data)
sigma = 1.2
mu0A, mu0B = 52.2, 53

testStatistic = ( xBar - mu0A ) / ( sigma/sqrt(n) )
pVal = 2*ccdf(Normal(), abs(testStatistic))

testA = OneSampleZTest(xBar, sigma, n, mu0A)
testB = OneSampleZTest(xBar, sigma, n, mu0B)

println("Results for mu0 = ", mu0A,":")
println("Manually calculated test statistic: ", testStatistic)
println("Manually calculated p-value: ", pVal,"\n")
println(testA)

println("\n In case of  mu0 = ", mu0B, " then p-value = ", pvalue(testB))
```

```
Results for mu0 = 52.2:

Manually calculated test statistic: 2.8182138203055467
Manually calculated p-value: 0.004829163880878602

One sample z-test
-----------------
Population details:
    parameter of interest:   Mean
    value under h_0:         52.2
    point estimate:          52.95620612127991
    95% confidence interval: (52.4303, 53.4821)

Test summary:
    outcome with 95% confidence: reject h_0
    two-sided p-value:           0.0048

Details:
    number of observations:   20
    z-statistic:              2.8182138203055467
    population standard error: 0.2683281572999747


 In case of  mu0 = 53 then p-value = 0.870352975060586
```

In lines 3-6, we load the data, calculate the sample mean, and specify the values of `mu0A` and `mu0B` under H_0 (there are two separate tests in this example). Note that importantly, the standard deviation, `sigma`, is specified as `1.2`, as it is "known". In line 8, we calculate the test statistic for case `mu0A` according to (7.7). In line 9, we calculate the p-value according to (7.8). Note that the `ccdf()` function is used to find the area to the right of the absolute value of the test statistic. In lines 11 and 12, `OneSampleZTest()` from `HypothesisTests` is used to perform the same calculations for both the `mu0A` and `mu0B` cases. The results are stored in `testA` and `testB`. These objects can then be printed or queried. Note that `OneSampleZTest()` was called with four arguments. If the last argument (`mu0A` or `mu0B`) was excluded, then the function would have performed the one sample z-test assuming $\mu_0 = 0$. There is also an additional method for `OneSampleZTest()`, which simply takes a single argument of an array of values. In this case, it will use the sample standard deviation as the population standard deviation, and will assume $\mu_0 = 0$. Further information is available in the documentation of the `HypothesisTests` package. Lines 14–17 print the results for the `mu0A` case. The p-value of 0.0048 merits rejection of H_0 for $\alpha = 5\%$. The output from line 17 also lists the value of the parameter under H_0, the point estimate of the parameter (`xBar`), as well as the corresponding 95% confidence interval. Line 19 prints the p-value for the second hypothesis test which uses `mu0B`, and since the p-value is greater than 0.05, we do not reject H_0. Note the use of the `pvalue()` function applied to `testB`. This way of using the `HypothesisTests` package is based on creating an object (`testB` in this case) and then querying it using a function like `pvalue()`.

Population Variance Unknown

Having covered the case of variance known, we now consider the more realistic scenario where the population variance is unknown. Informally called the *T-test*, this is perhaps the most famous and widely used hypothesis test in elementary statistics. Here the test statistic is the *T-statistic*,

$$T = \frac{\overline{X} - \mu_0}{S/\sqrt{n}}. \tag{7.9}$$

Notice that it is similar to (7.6); however, the sample standard deviation, S, is used instead of the population standard deviation σ, since σ is unknown. As presented in Section 5.2, in the case where the data is normally distributed with mean μ_0, the random variable T follows a T-distribution with $n-1$ degrees of freedom and this is the basis for the *T-test*. The procedure is the same as the Z-test presented above, except that a T-distribution is used instead of a normal distribution. Note that for non-small n, the T-distribution is almost identical to a standard normal distribution.

The observed test statistic from the data is then

$$t = \frac{\overline{x} - \mu_0}{s/\sqrt{n}}, \tag{7.10}$$

and the corresponding p-value for a two-sided test is

$$p = 2\,\mathbb{P}(T_{n-1} > |t|), \tag{7.11}$$

where T_{n-1} is a random variable distributed according to a T-distribution with $n-1$ degrees of freedom. Note that standardized tables present *critical values* of the T-distribution, namely, t_γ where γ is typically $0.9, 0.95, 0.975, 0.99$, and 0.995. These are typically presented in detail for degrees of freedom ranging from $n = 2$ to $n = 30$, after which the T-distribution is very similar to a normal distribution. These values are then compared to the T-statistic (7.10) where $\gamma = 1 - \alpha$ in the one-sided case or $\gamma = 1 - \alpha/2$ in the two-sided case. However, for precise calculation of p-values, software must be used.

In Listing 7.2, we first calculate the test statistic via (7.10), and then use this to manually calculate the corresponding p-value via (7.11). Then `OneSampleTTest()` from the `HypothesisTests` package is used to perform the same hypothesis. The output from our manual calculation matches that of `OneSampleTTest()`.

Listing 7.2: Inference with single sample, population variance unknown

```
1  using CSV, Distributions, HypothesisTests
2
3  data = CSV.read("../data/machine1.csv", header=false)[:,1]
4  xBar, n = mean(data), length(data)
5  s = std(data)
6  mu0 = 52.2
7
8  testStatistic = ( xBar - mu0 ) / ( s/sqrt(n) )
9  pVal = 2*ccdf(TDist(n-1), abs(testStatistic))
10
11 println("Manually calculated test statistic: ", testStatistic)
12 println("Manually calculated p-value: ", pVal,"\n")
13 OneSampleTTest(data, mu0)
```

```
Manually calculated test statistic: 2.86553950269453
Manually calculated p-value: 0.009899631865162935

One sample t-test
-----------------
Population details:
    parameter of interest:   Mean
    value under h_0:         52.2
    point estimate:          52.95620612127991
    95% confidence interval: (52.4039, 53.5085)

Test summary:
    outcome with 95% confidence: reject h_0
    two-sided p-value:           0.0099

Details:
    number of observations:   20
    t-statistic:              2.86553950269453
    degrees of freedom:       19
    empirical standard error: 0.2638965962845154
```

Lines 1–9 are similar to Listing 7.1. In line 5, the sample standard deviation is calculated and stored as `s`. In lines 8 and 9, the test statistic and p-value are calculated according to (7.10) and (7.11), respectively. Here the `ccdf()` function is used on a T-distribution with $n - 1$ degrees of freedom, `TDist(n-1)`. The manual calculations are output in lines 11 and 12. In line 13, `OneSampleTTest()` is used to perform the same hypothesis test on the data. Note that in this case we only specify two arguments, the array of our data and the value of μ_0, `mu0`.

A Non-parametric Sign Test

The validity of the T-test relies heavily on the assumption that the sample $X_1, \ldots, X_n$ is comprised of independent normal random variables. This is because only under this assumption does the T-statistic follows a T-distribution. This assumption may often be safely made; however, in certain cases, we cannot assume a normal population and we need an alternative test.

Here we present a particular type of *non-parametric test* known as the *sign test.* The phrase "non-parametric" implies that the distribution of the test statistic does not depend on any particular distributional assumption for the population.

For the sign test, we begin by denoting the random variables,

$$X^+ = \sum_{i=1}^{n} \mathbf{1}\{X_i > \mu_0\} \qquad \text{and} \qquad X^- = \sum_{i=1}^{n} \mathbf{1}\{X_i < \mu_0\} = n - X^+, \tag{7.12}$$

where $\mathbf{1}\{\cdot\}$ is the indicator function. The variable X^+ is a count of the number of observations that exceed μ_0, and similarly X^- is a count of the number of observations that are below μ_0.

Observe that under $H_0 : \mu = \mu_0$, it holds that $\mathbb{P}(X_i > \mu_0) = \mathbb{P}(X_i < \mu_0) = 1/2$. Note that here we are actually taking μ_0 as the median of the distribution and assuming that $\mathbb{P}(X_i = \mu_0) = 0$ as is the case for a continuous distribution. Hence, under H_0 the random variables X^+ and X^- both follow a binomial $(n, 1/2)$ distribution (see Section 3.5). Given the symmetry of this binomial distribution we define the test statistic to be

$$U = \max\{X^+, X^-\}. \tag{7.13}$$

Hence, with observed data, and an observed test statistic u, the p-value can be calculated via

$$p = 2\,\mathbb{P}(B > u), \tag{7.14}$$

where B is a binomial$(n, 1/2)$ random variable. Here, under H_0, p is the probability of getting an extreme number of signs greater than u (either too many via X^+ or a very small number via X^-). The test procedure is then to reject H_0 if $p < \alpha$.

In Listing 7.3, we present an example where we calculate the value of the test statistic and its corresponding p-value manually. We then use these to make conclusions about the null hypothesis at the 5% significance level. As was done in Listing 7.1, we compare two hypothetical cases. In the first case $\mu_0 = 52.2$, and the second $\mu_0 = 53.0$. As can be observed from the output, the former case is significant (H_0 is rejected) while the latter is not, as the test statistic of 11 is not non-plausible under H_0.

Listing 7.3: Non-parametric sign test

```
1  using CSV, Distributions, HypothesisTests
2
3  data = CSV.read("../data/machine1.csv", header=false)[:,1]
4  n = length(data)
5  mu0A, mu0B = 52.2, 53
6
7  xPositiveA = sum(data .> mu0A)
8  testStatisticA = max(xPositiveA, n-xPositiveA)
9
10 xPositiveB = sum(data .> mu0B)
11 testStatisticB = max(xPositiveB, n-xPositiveB)
12
13 binom = Binomial(n,0.5)
14 pValA = 2*ccdf(binom, testStatisticA)
15 pValB = 2*ccdf(binom, testStatisticB)
16
17 println("Binomial mean: ", mean(binom))
18
19 println("Case A: mu0: ", mu0A)
20 println("\tTest statistc: ", testStatisticA)
21 println("\tP-value: ", pValA)
22
23 println("Case B: mu0: ", mu0B)
24 println("\tTest statistc: ", testStatisticB)
25 println("\tP-value: ", pValB)
```

```
Binomial mean: 10.0

Case A: mu0: 52.2
        Test statistc: 15
        p-value: 0.011817932128906257
Case B: mu0: 53
        Test statistc: 11
        p-value: 0.5034446716308596
```

In line 5, the value of the population mean under the null hypothesis for both cases, `mu0A` and `mu0B`, is specified. In lines 7–11, the observed test statistics for both cases are calculated via (7.13). Note the use of `.>` for comparing the array `data` element-wise with the scalars `mu0A` and `mu0B`. In lines 13–15, `Binomial()` and `ccdf()` are used used to compute the p-values for both cases via (7.14). The results are printed in lines 19–25. As there are $n = 20$ observations, the binomial mean is 10.

Sign Test versus T-Test

With the sign test presented as a robust alternative to the T-test, one may ask why not simply always use the sign test. After all, the validity of the T-test rests on the assumption that $X_1, \ldots, X_n$ are normally distributed. Otherwise, T of (7.9) does not follow a T-distribution, and conclusions drawn from the test may be potentially imprecise.

One answer is due to the statistical power of the test. As we show in the example below, the T-test is often a more powerful test than the sign test when the normality assumption holds. That is, for a fixed α, the probability of detecting H_1 is higher for the T-test than for the sign test. This makes it a more effective test to use, if the data can be assumed normally distributed. The concept of power was first introduced in Section 5.6, and is further investigated in Section 7.5.

In Listing 7.4, we perform a two-sided hypothesis test for $H_0 : \mu = 53$ vs. $H_1 : \mu \neq 53$ via both the T-test and sign test. We consider a range of scenarios where we let the actual μ vary over $[51.0, 55.0]$. When $\mu = 53$, H_0 is the case; however, all other cases fall in H_1. On a grid of such cases, we use Monte Carlo to estimate the power of the tests (for $\sigma = 1.2$). The resulting curves in Figure 7.3 show that the T-test is more powerful than the sign test.

Listing 7.4: Comparison of sign test and T-test

```
using Random, Distributions, HypothesisTests, Plots; pyplot()

muRange = 51:0.02:55
n = 50
N = 10^4
mu0 = 53.0
powerT, powerU = [], []

for muActual in muRange

    dist = Normal(muActual, 1.2)
    rejectT, rejectU = 0, 0
    Random.seed!(1)

    for _ in 1:N
        data = rand(dist,n)
        xBar, stdDev = mean(data), std(data)

        tStatT = (xBar - mu0)/(stdDev/sqrt(n))
        pValT  = 2*ccdf(TDist(n-1), abs(tStatT))

        xPositive = sum(data .> mu0)
        uStat = max(xPositive, n-xPositive)
        pValSign = 2*ccdf(Binomial(n,0.5), uStat)

        rejectT += pValT < 0.05
        rejectU += pValSign < 0.05
    end

    push!(powerT, rejectT/N)
    push!(powerU, rejectU/N)

end

plot(muRange, powerT, c=:blue, label="t test")
plot!(muRange, powerU, c=:red, label="Sign test",
    xlims=(51,55), ylims=(0,1),
    xlabel="Different values of muActual",
    ylabel="Proportion of times H0 rejected", legend=:bottomleft)
```

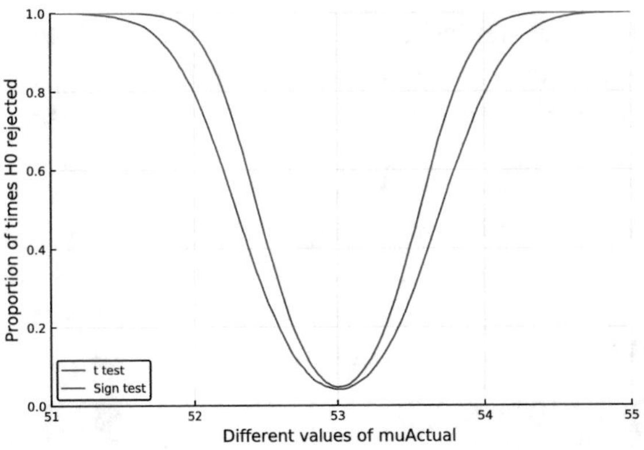

Figure 7.3: Power of the T-test vs. power of the sign test.

In lines 3–6, we set up the basic parameters. The sample size `n` is 50. The range `muRange` represents the range $[51.0, 55.0]$ in discrete steps of 0.02. The number `N` is the number of simulation repetitions to carry out for each value of μ in that range. The value `mu0` is the value under H_0. In line 7, we initialize empty arrays that are to be populated with power estimates. Lines 9–33 contain the main loop where each discrete step in `muRange` is tested. In line 11, for each value, a distribution `dist` is created with the same standard deviation, `1.2`, but with a different mean, `muActual`. The inner loop of lines 15–28 is a repetition of the sampling experiment `N` times where for the same `data` we compute `tStat`, `uStat`, and the corresponding p-values. Rejection counts are accumulated in lines 26 and 27, where `pValT` $<$ `0.05` and `pValSign` $<$ `0.05` constitutes a rejection. Note `Random.seed!()` is used in Line 13 so as to obtain a smoother curve via common random numbers. See Section 10.6. Lines 30–31 record the power for the respective `muActual` by appending to the adrrays using `push!()`. Lines 35–39 plot the power curves showing the superiority of the T-test. Observe that at $\mu = 53$ the power is identical to $\alpha = 0.05$.

7.2 Two Sample Hypothesis Tests for Comparing Means

Having dealt with several examples involving one population, we now present some common hypothesis tests for the inference on the difference in means of two populations. As with all hypothesis tests we start by first establishing the testing methodology. Commonly, we wish to investigate if the population difference, Δ_0, takes on a specific value. Hence, we may wish to set up a two-sided hypothesis test as

$$H_0 : \mu_1 - \mu_2 = \Delta_0, \qquad H_1 : \mu_1 - \mu_2 \neq \Delta_0. \tag{7.15}$$

Alternatively, one could formulate a one-sided hypothesis test, such as

$$H_0 : \mu_1 - \mu_2 \leq \Delta_0, \qquad H_1 : \mu_1 - \mu_2 > \Delta_0, \tag{7.16}$$

or the reverse if desired. It is common to consider $\Delta_0 = 0$, in which case (7.15) would be stated as $H_0 : \mu_1 = \mu_2$, and similarly (7.16) as $H_0 : \mu_1 \leq \mu_2$. Once the testing methodology has been established, the approach then follows the same outline as that covered previously, the test statistic is calculated along with its corresponding p-value, which is then used to make some conclusion about the hypothesis for some significance level α.

For the tests introduced in this section, we assume that the observations $X_1^{(1)}, \ldots, X_{n_1}^{(1)}$ from population 1 and $X_1^{(2)}, \ldots, X_{n_2}^{(2)}$ from population 2 are all normally distributed, where $X_i^{(j)}$ has mean μ_j and variance σ_j^2. The testing methodology then differs based on the following three cases:

(I) The population variances σ_1 and σ_2 are known.

(II) The population variances σ_1 and σ_2 are unknown and assumed equal.

(III) The population variances σ_1 and σ_2 are unknown, and not assumed equal.

In each of these cases, the test statistic is given by

$$\frac{\overline{X}_1 - \overline{X}_2 - \Delta_0}{S_{\text{err}}}, \tag{7.17}$$

where $\overline{X}_j$ is the sample mean of $X_1^{(j)}, \ldots, X_{n_j}^{(j)}$, and the standard error S_{err} varies according to the case (**I–III**). In each example, the two datasets `machine1.csv` and `machine2.csv` are used, and it is considered that H_0 implies that both machines are identical (i.e. we use (7.15) with $\Delta_0 = 0$).

Population Variances Known

In case (**I**), where the population variances σ_1^2 and σ_2^2 are known, we set

$$S_{\text{err}} = \sqrt{\frac{\sigma_1^2}{n_1} + \frac{\sigma_2^2}{n_2}}.$$

In this case, S_{err} is not a random quantity, and hence the test statistic (7.17) follows a standard normal distribution under H_0. This is due to the distribution of $\overline{X}_j$ as described in Section 5.2. For this case, the observed test statistic is

$$z = \frac{(\overline{x}_1 - \overline{x}_2) - \Delta_0}{\sqrt{\frac{\sigma_1^2}{n_1} + \frac{\sigma_2^2}{n_2}}}. \tag{7.18}$$

At this point, z is used for hypothesis tests in a manner analogous to the single sample test for the population mean when the variance is known, as described at the start of Section 7.1.

Note that in reality it is highly unlikely that both the population variances would be known, and hence the `HypothesisTests` package does not contain functionality for this test. Nevertheless, in Listing 7.5, we perform this hypothesis test manually for pedagogical completeness. The output shows there is a very significant difference between the machines with the p-value almost zero.

Listing 7.5: Inference on difference of two means with variances known

```julia
1  using CSV, Distributions, HypothesisTests
2
3  data1 = CSV.read("../data/machine1.csv", header=false)[:,1]
4  data2 = CSV.read("../data/machine2.csv", header=false)[:,1]
5  xBar1, n1 = mean(data1), length(data1)
6  xBar2, n2 = mean(data2), length(data2)
7  sig1, sig2 = 1.2, 1.6
8  delta0 = 0
9
10 testStatistic = ( xBar1-xBar2 - delta0 ) / ( sqrt( sig1^2 / n1 + sig2^2 / n2 ) )
11 pVal = 2*ccdf(Normal(), abs(testStatistic))
12
13 println("Sample mean machine 1: ",xBar1)
14 println("Sample mean machine 2: ",xBar2)
15 println("Manually calculated test statistc: ", testStatistic)
16 println("Manually calculated p-value: ", pVal)
```

```
Sample mean machine 1: 52.95620612127991
Sample mean machine 2: 50.94739671468099
Manually calculated test statistc: 4.340167327618076
Manually calculated p-value: 1.423742605667141e-5
```

In lines 3–7, we load our data, calculate the sample means, and specify the values of the population variances. In line 8, we specify the value of our test parameter under the null hypothesis, `delta0`, as 0. In line 10, we calculate the test statistic via (7.18), and in line 11 we calculate the p-value.

Population Variances Unknown and Assumed Equal

We now consider case **(II)** where the population variances are unknown, and assumed equal ($\sigma^2 := \sigma_1^2 = \sigma_2^2$). In this case, the pooled sample variance, S_p^2, is used to estimate σ^2 based on both samples. As covered in Section 6.2, it is given by

$$S_p^2 = \frac{(n_1 - 1)S_1^2 + (n_2 - 1)S_2^2}{n_1 + n_2 - 2}, \tag{7.19}$$

where S_j^2 is the sample variance of sample j. It can be shown that under H_0, if we set

$$S_{\text{err}} = S_p \sqrt{\frac{1}{n_1} + \frac{1}{n_2}},$$

the test statistic is distributed according to a T-distribution with $n_1 + n_2 - 2$ degrees of freedom. For this case, the observed test statistic is

$$t = \frac{(\overline{x}_1 - \overline{x}_2) - \Delta_0}{s_p \sqrt{\dfrac{1}{n_1} + \dfrac{1}{n_2}}}, \tag{7.20}$$

where s_p is the observed pooled sample variance. At this point, the procedure follows similar lines to the single sample T-test described in the previous section. Note that this two sample T-test with equal variance is one of the most commonly used tests in statistics.

In Listing 7.6, we present an example where we perform a two-sided hypothesis test on the difference in means of pipes produced from machines 1 and 2. First, the test is performed manually, and then the `EqualVarianceTTest()` function from the `HypothesisTests` package is used. The resulting output shows that the manually calculated values match those given by `EqualVarianceTTest()`.

Listing 7.6: Inference on difference of means, variances unknown, assumed equal

```
using CSV, Distributions, HypothesisTests

data1 = CSV.read("../data/machine1.csv", header=false)[:,1]
data2 = CSV.read("../data/machine2.csv", header=false)[:,1]
xBar1, s1, n1 = mean(data1), std(data1), length(data1)
xBar2, s2, n2 = mean(data2), std(data2), length(data2)
delta0 = 0

sP = sqrt( ( (n1-1)*s1^2 + (n2-1)*s2^2 ) / (n1 + n2 - 2) )
testStatistic = ( xBar1-xBar2 - delta0 ) / ( sP * sqrt( 1/n1 + 1/n2) )
pVal = 2*ccdf(TDist(n1+n2 -2), abs(testStatistic))

println("Manually calculated test statistic: ", testStatistic)
println("Manually calculated p-value: ", pVal, "\n")
println(EqualVarianceTTest(data1, data2, delta0))
```

```
Manually calculated test statistic: 4.5466542394674425
Manually calculated p-value: 5.9493058655043084e-5

Two sample t-test (equal variance)
----------------------------------
Population details:
    parameter of interest:   Mean difference
    value under h_0:         0
    point estimate:          2.008809406598921
    95% confidence interval: (1.1128, 2.9049)

Test summary:
    outcome with 95% confidence: reject h_0
    two-sided p-value:           <1e-4

Details:
    number of observations:   [20,18]
    t-statistic:              4.5466542394674425
    degrees of freedom:       36
    empirical standard error: 0.44182145832893077
```

Lines 3-7 are similar to Listing 7.5, however note the calculations of the sample standard deviations `s1` and `s2`. In line 7, we specify the value of our test parameter under the null hypothesis as 0. Line 9 calculates the square root of the pooled sample variance, `sP`, via (7.19). Note the use of `sqrt()`, as the test statistic (7.20) makes use of s_p not s_p^2. Lines 10 and 11 calculate the test statistic via (7.20), and the corresponding p-value, respectively. These are printed as output in lines 13 and 14. Line 15 uses the `EqualVarianceTTest()` function to perform the hypothesis test. Note three arguments are given, the two arrays `data1` and `data2`, and the value of the test parameter under H_0. Note the function defaults to $\Delta_0 = 0$ if only two arguments are given, but here we demonstrate the general use of the function.

Population Variances Unknown, and not Assumed Equal

In case **(III)** where the population variances are unknown and not assumed equal ($\sigma_1^2 \neq \sigma_2^2$), we set

$$S_{\text{err}} = \sqrt{\frac{S_1^2}{n_1} + \frac{S_2^2}{n_2}}.$$

Then the observed test statistic is given by

$$t = \frac{(\overline{x}_1 - \overline{x}_2) - \Delta_0}{\sqrt{\frac{s_1^2}{n_1} + \frac{s_2^2}{n_2}}}. \tag{7.21}$$

As covered in Section 6.2, the distribution of the test statistic does not follow an exact T-distribution. Instead, we use the *Satterthwaite approximation*, and determine the degrees of freedom via

$$v = \frac{\left(s_1^2\, n_1 + s_2^2\, n_2\right)^2}{\frac{\left(s_1^2/n_1\right)^2}{n_1 - 1} + \frac{\left(s_2^2/n_2\right)^2}{n_2 - 1}}. \tag{7.22}$$

In Listing 7.7, we perform a two-sided hypothesis test that the difference between the population means is zero ($\Delta_0 = 0$). We first manually calculate the test statistic and p-value, and then make use of the `UnequalVarianceTTest()` function from the `HypothesisTests` package. The output shows both methods are in agreement.

Listing 7.7: Inference on difference of means, variances unknown, not assumed equal

```
1  using CSV, Distributions, HypothesisTests
2
3  data1 = CSV.read("../data/machine1.csv", header=false)[:,1]
4  data2 = CSV.read("../data/machine2.csv", header=false)[:,1]
5  xBar1, s1, n1 = mean(data1), std(data1), length(data1)
6  xBar2, s2, n2 = mean(data2), std(data2), length(data2)
7  delta0 = 0
8
9  v = ( s1^2/n1 + s2^2/n2 )^2 / ( (s1^2/n1)^2/(n1-1) + (s2^2/n2)^2/(n2-1)  )
10 testStatistic = ( xBar1-xBar2 - delta0 )  / sqrt( s1^2/n1 + s2^2/n2 )
11 pVal = 2*ccdf(TDist(v), abs(testStatistic))
12
13 println("Manually calculated degrees of freedom, v: ", v)
14 println("Manually calculated test statistic: ", testStatistic)
15 println("Manually calculated p-value: ", pVal, "\n")
16 println(UnequalVarianceTTest(data1, data2, delta0))
```

```
Manually calculated degrees of freedom, v: 31.82453144280283
Manually calculated test statistic: 4.483705005611673
Manually calculated p-value: 8.936189820683007e-5

Two sample t-test (unequal variance)
------------------------------------
Population details:
    parameter of interest:   Mean difference
    value under h_0:         0
    point estimate:          2.008809406598921
    95% confidence interval: (1.096, 2.9216)

Test summary:
    outcome with 95% confidence: reject h_0
    two-sided p-value:           <1e-4

Details:
    number of observations:   [20,18]
    t-statistic:              4.483705005611673
    degrees of freedom:       31.82453144280282
    empirical standard error: 0.4480244360600785
```

Lines 3–8 are similar to Listing 7.6. In line 9, the degrees of freedom, v, is calculated via (7.22). In line 10, the test statistic is calculated via (7.21). In line 11, these are both used to calculate the p-value. Lines 13-16 output the degrees of freedom, test statistic, and p-value calculated. Line 16 uses the `UnequalVarianceTTest()` function to perform the hypothesis test.

7.3 Analysis of Variance (ANOVA)

The methods presented in Section 7.2 handle the problem of comparing means of two populations. However, what if there are more than two populations that need to be compared? This is often the case in biological, agricultural, and medical trials, among other fields, where it is of interest to see if various "treatments", also known as "groups", have an effect on some mean value or not. In these cases, each type of treatment is considered as a different population.

More formally, assume that there is some *overall mean* μ and there are $L \geq 2$ treatments, where each treatment may potentially alter the mean by τ_i. In this case, the mean of the population under treatment i can be represented by $\mu_i = \mu + \tau_i$, with μ an overall mean and

$$\sum_{i=1}^{L} \tau_i = 0.$$

This condition on the parameters $\tau_1, \dots, \tau_L$ ensures that given $\mu_1, \dots, \mu_L$, the overall mean μ and $\tau_1, \dots, \tau_L$ are well defined.

The question is then: *Do the treatments have any effect or not*? Such a question is presented via the hypothesis formulation:

$$H_0 : \tau_1 = \tau_2 = \ldots = \tau_L = 0, \qquad \text{vs.} \qquad H_1 : \exists\, i \;\mid\; \tau_i \neq 0. \tag{7.23}$$

Note that H_0 is equivalent to the statement that $\mu_1 = \mu_2 = \ldots = \mu_L$, indicating that the treatments do not have an effect. Furthermore, H_1 stating that there exists an i with $\tau_i \neq 0$ is equivalent to the case where there exist at least two treatments, i and j such that $\mu_i \neq \mu_j$. In other words, this means that the choice of treatment has an effect, at least between some treatments.

In conducting hypotheses tests such as (7.23), we collect observations (data) as follows:

$$\begin{aligned}
&\text{Treatment 1:} \quad x_{11},\ x_{12}, \dots,\ x_{1n_1},\\
&\text{Treatment 2:} \quad x_{21},\ x_{22}, \dots,\ x_{2n_2},\\
&\qquad\qquad \vdots\\
&\text{Treatment L:} \quad x_{L1},\ x_{L2}, \dots,\ x_{Ln_L},
\end{aligned}$$

where $n_1, n_2, \dots, n_L$ are the sample sizes for each of the treatments (or groups). If all samples are the same size (say $n_i = n$ for all i) then this is called a *balanced design* problem. However, often different treatments have different sample sizes. It is also convenient to denote the total number of observations via

$$m = \sum_{i=1}^{L} n_i. \tag{7.24}$$

Note that in a balanced design we have $m = L\,n$.

We focus on an example for three treatments where data from `machine1.csv`, `machine2.csv`, and `machine3.csv` represent sample measurements of the diameter of pipes produced by three different machines. Here each machine has a different treatment. The diameters of the pipes vary due to imprecision of machines but also potentially (if H_0 does not hold) due to variability between the machines. A box plot of this dataset was already presented in Figure 4.9 of Chapter 4. Looking at that plot, while there are differences between the groups, it isn't immediately clear if the machine affects the pipe diameter or if the differences are just due to noise in manufacturing within each machine.

Once the data is collected, in addition to displaying a box plot as shown in Figure 4.9, we also consider the values of the sample means for each individual treatment,

$$\overline{x}_i = \frac{1}{n_i}\sum_{j=1}^{n_i} x_{ij}.$$

These values can then be compared with the overall sample mean,

$$\overline{x} = \frac{1}{m}\sum_{i=1}^{L}\sum_{j=1}^{n_i} x_{ij} = \sum_{i=1}^{L}\frac{n_i}{m}\,\overline{x}_i. \tag{7.25}$$

In Listing 7.8, the sample means are computed for three datasets of pipe diameters. Note the use of the broadcast operator in lines 4 and 7.

Listing 7.8: Sample means for ANOVA

```
1  using CSV, Statistics
2
3  rfile(name) = CSV.read(name, header=false)[:,1]
4  data = rfile.(["../data/machine1.csv",
5                 "../data/machine2.csv",
6                 "../data/machine3.csv"])
7  println("Sample means for each treatment: ",round.(mean.(data),digits=2))
8  println("Overall sample mean: ",round(mean(vcat(data...)),digits=2))
```

```
Sample means for each treatment: [52.96, 50.95, 51.43]
Overall sample mean: 51.82
```

Even though the sample mean values are not exactly the same (they are at 52.96, 50.95, and 51.43), without looking at variability we can't conclude if the machine type (treatment) affects the diameter size or not. Here is where ANOVA comes into play.

The typical way to establish whether or not an effect between the treatments exists is to examine the variability of the individual treatment means, and compare these to the overall variability of the observations. If the variability of means significantly exceeds the variability of the individual observations, then H_0 is rejected, otherwise it is not. Such an approach is called *ANOVA*, which stands for *analysis of variance*, and it is based on the decomposition of the sum of squares. In fact, ANOVA is a broad collection of statistical methods, and here we only provide an introduction to ANOVA by covering the *one-way ANOVA* test. In this test, the statistical model assumes that the observations of each treatment group come from an underlying model of the following form:

$$X_i = \mu_i = \mu + \tau_i + \varepsilon \qquad \text{where} \qquad \varepsilon \sim N(0, \sigma^2), \tag{7.26}$$

where X_i is the model for the ith treatment group and ε is some noise term with common unknown variance across all treatment groups, independent across measurements. In this sense, the ANOVA model (7.26) generalizes the assumptions of the T-test applied to case **II** (comparison of two population means with variance unknown and assumed equal), as presented in the previous section.

The process of conducting a *one-way ANOVA test* follows the same general approach as any other hypothesis test. First the test statistic is calculated, then the corresponding p-value is obtained, and finally the p-value is used to make some conclusion about whether or not to reject H_0 at some chosen confidence level α. The test statistic for ANOVA, known as *F-statistic*, is the ratio of the average variance between the groups, divided by the average variance within the groups. Under the null hypothesis, the F-value is distributed according to the *F-distribution*, covered at the end of Section 5.2. Hence, the ANOVA test is sometimes referred to as the *F-test.*

Before we present the details of carrying out an F-test, we present the mathematical motivation used to calculate the variability within the groups (or treatments) and the variability between the groups.

Decomposing Sum of Squares

A key idea of ANOVA is the decomposition of the total variability into two components: the variability between the treatments, and the variability within the treatments. There are explicit expressions for both, and here we show how to derive them by performing what is known as the *decomposition of the sum of squares.*

The total sum of squares, also known as the *sum of squares total* (SS_{Total}), is a measure of the total variability of all observations, and is calculated as follows:

$$SS_{\text{Total}} = \sum_{i=1}^{L}\sum_{j=1}^{n_i} (x_{ij} - \overline{x})^2, \tag{7.27}$$

where $\overline{x}$ is given by (7.25). Now through algebraic manipulation (adding and subtracting treatment means) we can show that SS_{Total} can be decomposed as follows:

$$\begin{aligned}\sum_{i=1}^{L}\sum_{j=1}^{n_i} (x_{ij} - \overline{x})^2 &= \sum_{i=1}^{L}\sum_{j=1}^{n_i} (x_{ij} - \overline{x}_i + \overline{x}_i - \overline{x})^2 \\ &= \sum_{i=1}^{L}\sum_{j=1}^{n_i} \Big((x_{ij} - \overline{x}_i)^2 - 2(x_{ij} - \overline{x}_i)(\overline{x}_i - \overline{x}) + (\overline{x}_i - \overline{x})^2 \Big) \\ &= \underbrace{\sum_{i=1}^{L}\sum_{j=1}^{n_i} (x_{ij} - \overline{x}_i)^2}_{SS_{\text{Error}}} + \underbrace{\sum_{i=1}^{L} n_i(\overline{x}_i - \overline{x})^2}_{SS_{\text{Treatment}}}. \end{aligned} \tag{7.28}$$

Note that on the second line, the middle term reduces to zero, since $\sum_{j=1}^{n_i}(x_{ij} - \overline{x}_i) = 0$. Hence, we have shown that the total sum of squares, SS_{Total}, can be decomposed to

$$SS_{\text{Total}} = SS_{\text{Error}} + SS_{\text{Treatment}}. \tag{7.29}$$

Note that the *sum of squares error*, SS_{Error}, is also known as the sum of the variability within the groups, and that the *sum of squares Treatment*, $SS_{\text{Treatment}}$, is also known as the variability between the groups. The decomposition (7.29) holds under both H_0 and H_1, and hence allows us to construct a test statistic. Intuitively, under H_0, both SS_{Error} and $SS_{\text{Treatment}}$ should contribute to SS_{Total} in the same manner (once properly normalized). Alternatively, under H_1 it is expected that $SS_{\text{Treatment}}$ would contribute more heavily to the total variability.

Before proceeding with the construction of a test statistic, we present Listing 7.9, where the decomposition of (7.29) is demonstrated for the purpose of showing how to compute its individual components in Julia. Note that this verification of the decomposition is not something one would normally carry out in practice as it is already proven in (7.28).

Listing 7.9: Decomposing the sum of squares

```
1  using Random, Statistics
2  Random.seed!(1)
3  allData = [rand(24), rand(15), rand(73)]
4
5  xBarArray = mean.(allData)
6  nArray = length.(allData)
7  xBarTotal = mean(vcat(allData...))
8  L = length(nArray)
9
10 ssBetween=sum([nArray[i]*(xBarArray[i] - xBarTotal)^2 for i in 1:L])
11 ssWithin=sum([sum([(ob - xBarArray[i])^2 for ob in allData[i]]) for i in 1:L])
12 ssTotal=sum([sum([(ob - xBarTotal)^2 for ob in allData[i]]) for i in 1:L])
13
14 println("Sum of squares between groups: ", ssBetween)
15 println("Sum of squares within groups: ", ssWithin)
16 println("Sum of squares total: ", ssTotal)
```

```
Sum of squares between groups: 0.2941847110381936
Sum of squares within groups: 8.50335257006105
Sum of squares total: 8.797537281099242
```

The data is generated in line 3 in an array of arrays, `allData`. In line 5, the mean of each treatment is calculated via the `mean()` function with the broadcast operator "`.`", which performs the operation over each of the three elements of `allData`. In line 6, we retrieve the length of each array via the `length()` function and the broadcast operator. In line 7, the point estimate for the total population mean is calculated and stored as `xBarTotal`. Note that in contrast to line 5, here we first vertically concatenate all the groups into a single array. This is done via the `vcat()` function, and the splat operator `...`. In line 12, the number of treatments is stored as `L`. In line 10, $SS_{\text{Treatment}}$ is calculated. A comprehension is used, and the point estimate of the population mean `xbarTotal` is subtracted from the ith element of each array, and the results squared. These are each multiplied by the length of their respective arrays, and the results for each of the arrays summed together, and stored as `ssBetween`. Note that "between" is sometimes used as an alternative name to "treatments". In line 11, SS_{Error} is calculated. The inner comprehension is used to square the difference between each observation, `ob`, and the group mean `xBarArray[i]`. The outer comprehension is used to repeat this process from the `1:L`-th group. The results for all groups are summed. In line 12, SS_{Total} is calculated via (7.27). The difference between each observation, `ob`, and the point estimate for the population mean, `xBarTotal`, is calculated and each result squared. This is first performed for the i-th array, in the inner comprehension, and then repeated for all arrays via the outer comprehension. Finally, all the squares are summed, via the outer `sum()` function.

Carrying out ANOVA

Having understood the sum of squares decomposition, we now present the F-statistic of ANOVA

$$F = \frac{SS_{\text{Treatment}}/(L-1)}{SS_{\text{Error}}/(m-L)}. \tag{7.30}$$

It is a ratio of the two sum of squares components of (7.29) normalized by their respective *degrees of freedom*, $L-1$ and $m-L$. These normalized quantities are, respectively, denoted by $MS_{\text{Treatment}}$ and MS_{Error} standing for "Mean Squared". Hence $F = MS_{\text{Treatment}}/MS_{\text{Error}}$.

Under H_0 and with the model assumptions presented in (7.26), the ratio F follows an F-distribution (first introduced in Section 5.2) with $L-1$ degrees of freedom for the numerator and $m-L$ degrees of freedom for the denominator. Intuitively, under H_0, we expect the numerator and denominator to have similar values, and hence expect F to be around 1 (indeed most of the mass of F distributions is concentrated around 1). However, if $MS_{\text{Treatment}}$ is significantly larger, then it indicates that H_0 may not hold. Hence, the approach of the F-test is to reject H_0 if the F-statistic is greater than the $1-\alpha$ quantile of the respective F-distribution. Similarly, the p-value for an observed F-statistic f_o is given by

$$p = \mathbb{P}(F_{L-1,m-L} > f_o),$$

where $F_{L-1,m-L}$ is an F-distributed random variable with $L-1$ numerator degrees of freedom and $m-L$ denominator degrees of freedom.

It is often customary to summarize both the intermediate and final results of an ANOVA F-test in an *ANOVA table* as shown in Table 7.1, where "T" and "E" are shorthand for "Treatments" and "Error", respectively. Such tables also generalize to more complex ANOVA procedures not covered here.

Table 7.1: A one-way ANOVA table.

Source of variance:	DOF:	Sum of sq's:	Mean sum of sq's:	F-value:
Treatments (between treatments)	$L-1$	SS_T	$MS_T = \dfrac{SS_T}{L-1}$	$\dfrac{MS_T}{MS_E}$
Error (within treatments)	$m-L$	SS_E	$MS_E = \dfrac{SS_E}{m-L}$	
Total	$m-1$	SS_{Total}		

We now return to the three machine example and carry out a one-way ANOVA F-test. This is carried out in Listing 7.10 where we implement two alternative functions for ANOVA. The first function, `manualANOVA()`, extends the sum of squares code presented in Listing 7.9. The second function, `glmANOVA()`, utilizes the `GLM` package that is described in detail in Chapter 8. Note that `GLM` requires the `DataFrames` package. Both implementations yield identical results, returning a tuple of the F-statistic and the associated p-value. In this example, the p-value is very small and hence under any reasonable α we would reject H_0 and conclude that there is sufficient evidence that the diameter of the pipe depends on the type of machine used. Related is Listing 1.18 in Chapter 1 where we carry out ANOVA for the same data using the R-language.

Listing 7.10: Executing one-way ANOVA

```
using GLM, Distributions, DataFrames

data1 = parse.(Float64, readlines("../data/machine1.csv"))
data2 = parse.(Float64, readlines("../data/machine2.csv"))
data3 = parse.(Float64, readlines("../data/machine3.csv"))

function manualANOVA(allData)
    nArray = length.(allData)
    d = length(nArray)

    xBarTotal = mean(vcat(allData...))
    xBarArray = mean.(allData)

    ssBetween = sum( [nArray[i]*(xBarArray[i] - xBarTotal)^2 for i in 1:d] )
    ssWithin = sum([sum([(ob - xBarArray[i])^2 for ob in allData[i]])
                                for i in 1:d])
    dfBetween = d-1
    dfError = sum(nArray)-d

    msBetween = ssBetween/dfBetween
    msError = ssWithin/dfError
    fStat = msBetween/msError
    pval = ccdf(FDist(dfBetween,dfError),fStat)
    return (fStat,pval)
end

function glmANOVA(allData)
    nArray = length.(allData)
    d = length(nArray)

    treatment = vcat([fill(k,nArray[k]) for k in 1:d]...)
    response = vcat(allData...)
    dataFrame = DataFrame(Response=response, Treatment=categorical(treatment))
    modelH0  = lm(@formula(Response ~ 1), dataFrame)
    modelH1a = lm(@formula(Response ~ 1 + Treatment), dataFrame)
    res = ftest(modelH1a.model, modelH0.model)
    (res.fstat[1],res.pval[1])
end

println("Manual ANOVA: ", manualANOVA([data1, data2, data3]))
println("GLM ANOVA: ", glmANOVA([data1, data2, data3]))
```

```
Manual ANOVA: (10.516968568709117, 0.00014236168817139249)
GLM ANOVA: (10.516968568708988, 0.0001)
```

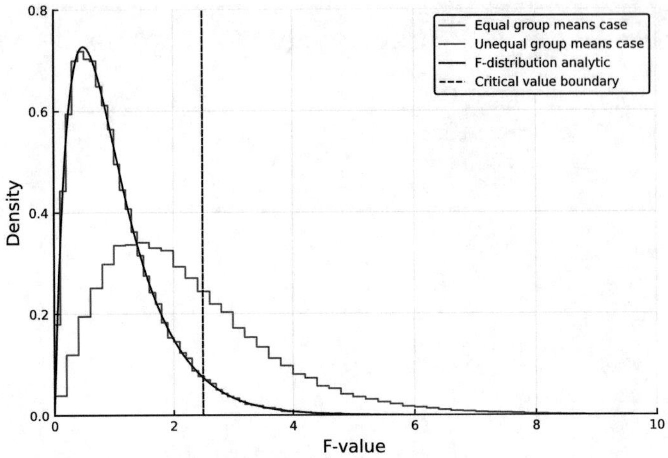

Figure 7.4: Histograms of the F-statistic for the case of equal group means (with analytic F-distribution), and not all equal group means.

In lines 7–25, the function `manualANOVA()` is implemented, which calculates the sum of squares in the same manner as in Listing 7.9. The sums of squares are normalized by their corresponding degrees of freedom `dfBetween` and `dfError`, and then in line 22 the F-statistic `fStat` is calculated. The p-value is then calculated in line 23 via the `ccdf()` function and the F-distribution `FDist()` with the degrees of freedom calculated above. The function returns a tuple of values, comprising the F-statistic and the corresponding p-value. In lines 27–38, the function `glmANOVA()` is defined. This function calculates the F-statistic and p-value, via functionality of the `GLM` package which is heavily discussed in Chapter 8. In lines 30–32, a `DataFrame` (see Chapter 4) is set up in the manner required by the `GLM` package. Then in lines 34–35 two "model objects" are created via the `lm()` function from the `GLM` package. Note that `modelH0` is constructed on the assumption that the machine type has no effect on the response, while `modelH1` is constructed on the assumption that treatment has an effect. Finally, the `ftest()` function from the `GLM` package is used to compare if `modelH1a` fits the data "better" than `modelH0`. Also note that the `model` fields of the model objects are used. Finally, the F-statistic and p-value are returned in line 37. The results of both functions are printed in lines 40–41 and it can be observed that the F-statistics and p-values calculated are identical to within the numerical error expected due to the different implementations.

More on the Distribution of the F-Statistic

Having explored the basics of ANOVA, we now use Monte Carlo simulation to illustrate that under H_0 the F-statistic is indeed distributed according to the F-distribution. In Listing 7.11, we present an example where Monte Carlo simulation is used to empirically generate the distribution of the F-statistic for two different cases where the number of groups is $L = 5$. In the first case, the means of each group are all the same and are at 13.4, but in the second case, the means are not all the same. The first case represents H_0, while the latter is one possibility within H_1. For both cases, the standard deviation of each group is identical (2).

In this example, for each of the two cases, N sample runs are generated, where each run consists of a separate random collection of sample observations for each group. Hence, by using a large number of sample runs N, histograms can be used to empirically represent the theoretical distributions of the F-statistics for both cases. The results presented in Figure 7.4 show that the distribution of the F-statistics for the equal group means case is in agreement with the analytically expected F-distribution, while the F-statistic for the case of unequal group means is not. The figure also illustrates the critical value for rejection with $\alpha = 0.05$. The area under the red curve to the left of that boundary is the power of the test under the specific point in H_1 that is simulated.

Listing 7.11: Monte Carlo based distributions of the ANOVA F-statistic

```
using Distributions, Plots; pyplot()

function anovaFStat(allData)
    xBarArray = mean.(allData)
    nArray = length.(allData)
    xBarTotal = mean(vcat(allData...))
    L = length(nArray)

    ssBetween = sum( [nArray[i]*(xBarArray[i] - xBarTotal)^2 for i in 1:L] )
    ssWithin = sum([sum([(ob - xBarArray[i])^2 for ob in allData[i]])
                            for i in 1:L])
    return (ssBetween/(L-1))/(ssWithin/(sum(nArray)-L))
end

case1 = [13.4, 13.4, 13.4, 13.4, 13.4]
case2 = [12.7, 11.8, 13.4, 12.7, 12.9]
stdDevs = [2, 2, 2, 2, 2]
numObs = [24, 15, 13, 23, 9]
L = length(case1)

N = 10^5

mcFstatsH0 = Array{Float64}(undef, N)
for i in 1:N
    mcFstatsH0[i] = anovaFStat([ rand(Normal(case1[j],stdDevs[j]),numObs[j])
                        for j in 1:L ])
end

mcFstatsH1 = Array{Float64}(undef, N)
for i in 1:N
    mcFstatsH1[i] = anovaFStat([ rand(Normal(case2[j],stdDevs[j]),numObs[j])
                        for j in 1:L ])
end

stephist(mcFstatsH0, bins=100,
        c=:blue, normed=true, label="Equal group means case")
stephist!(mcFstatsH1, bins=100,
        c=:red, normed=true, label="Unequal group means case")

dfBetween = L - 1
dfError = sum(numObs) - 1
xGrid = 0:0.01:10
plot!(xGrid, pdf.(FDist(dfBetween, dfError),xGrid),
           c=:black, label="F-statistic analytic")
critVal = quantile(FDist(dfBetween, dfError),0.95)
plot!([critVal, critVal],[0,0.8],
        c=:black, ls=:dash, label="Critical value boundary",
        xlims=(0,10), ylims=(0,0.8), xlabel="F-value", ylabel="Density")
```

In lines 3–13, we create the function anovaFStat(), which takes an array of arrays as input, calculates the sums of squares and mean sums of squares as per Table 7.1, and returns the F-statistic of the data. It is similar to Listing 7.9. In lines 15 and 16, we create two arrays where case1 represents an array of means for the case of all means being equal, and case2 represents an array of means for the case of all means not equal. In line 17, we create the array of group standard deviations, stdDevs. Note that in both cases of equal group means and unequal group means, the standard deviations of all the groups are equal as per the model assumption. In line 18, we create the array numObs, where each element represents the number of observations of the ith group, or level. In line 21, we specify the total number of Monte Carlo runs to be performed, N. In line 23, we pre-allocate the array mcFStatsH0, which will store N Monte Carlo generated F-statistics, for the case of all group means equal. In lines 24–27, we use a loop to generate N F-statistics via the anovaFStat() function defined earlier. We use the rand() and Normal() functions within a comprehension to generate data for each of the sample groups, using the group means, standard deviations, and number of observations. The comprehension generates an array of arrays, where each of the five elements of the outermost array is another array containing the observations for that group, 1 to5. This array of arrays is then used as the argument for anovaFStat(), which carries out a one-way ANOVA test on the data and outputs the corresponding F-value. Lines 29–33 are similar, using the case2 means. In lines 35–38, histograms of the F-statistics are generated. In lines 40–41, the degrees of freedom of the treatments dfBetween and the degrees of freedom of the error dfError are calculated. In lines 43–44, the analytic PDF of the is plotted. In line 45, the quantile() function is used to calculate the 95th quantile of the F-distribution, FDist() and this is then used in lines 46–48 to plot the critical value.

Extensions

We have only touched on the very basics of ANOVA via the one-way ANOVA case. This stands at the basis of *experimental design*. However, there are many more aspects to ANOVA, and related ideas that one can explore. These include, but are not limited to:

- Extensions to *two-way ANOVA* where there are two treatment categories, for example, "machine type" and "type of lubricant used in the machine", each having multiple treatments.
- Higher dimensional extensions, which are often considered in *block factorial design*.
- Comparison of individual factors to determine which specific treatments have an effect and in which way.
- Using ANOVA for *longitudinal data analysis* using *repeated measures*.
- Aspects of optimal experimental design.

These and many more aspects can be found in design and analysis of experiment texts such as [M17]. At the time of writing, many such procedures are not implemented directly in Julia. However, one alternative is the R software package, which contains many different implementations of these ANOVA extensions, among others. One can call these R packages directly from Julia as in Listing 1.18 of Chapter 1.

7.4 Independence and Goodness of Fit

We now consider a different group of hypothesis tests and associated procedures that deal with *checking for independence* and more generally checking *goodness of fit*. One question often posed is: *Does the population follow a specific distributional form?* We may hypothesize that the distribution is normal, exponential, Poisson, or that it follows any other form (see Chapter 3 for an extensive survey of probability distributions). Checking such a hypothesis is loosely called goodness of fit. Furthermore, in the case of observations over multiple dimensions, we may hypothesize that the different dimensions are independent. Checking for such independence is similar to the goodness-of-fit check.

In order to test for goodness of fit against some hypothesized distribution F_0, we set up the hypothesis test as

$$H_0: \ X \sim F_0, \qquad \text{vs.} \qquad H_1: \text{otherwise}. \tag{7.31}$$

Here X denotes an arbitrary random variable from the population. In this case, we consider the parameter space associated with the test as the space of all probability distributions. The hypothesis formulation then partitions this space into $\{F_0\}$ (for H_0) and all other distributions in H_1.

For the independence case, assume for simplicity that X is a vector of two random variables, say $X = (X_1, X_2)$. Then for this case the hypothesis test setup would be

$$H_0: \ X_1 \text{ independent of } X_2, \qquad \text{vs.} \qquad H_1: \ X_1 \text{ not independent of } X_2. \tag{7.32}$$

This sets the space of H_0 as the space of all distributions of independent random variable pairs, and H_1 as the complement.

To handle hypotheses such as (7.31) and (7.32), we introduce two different test procedures, the *chi-squared test* and the *Kolmogorov-Smirnov test*. The chi-squared test is used for goodness of fit of discrete distributions and for checking independence, while the Kolmogorov-Smirnov test is used for goodness of fit for arbitrary distributions based on the empirical cumulative distribution function. Before we dive into the individual test examples, we explain how to construct the corresponding test statistics.

In the chi-squared case, the approach involves looking at counts of observations that match disjoint categories $i = 1, \ldots, M$. For each category i, we denote O_i as the number of observations that match that category. In addition, for each category, there is also an expected number of observations under H_0, which we denote as E_i. With these, one can express the test statistic as

$$\chi^2 = \sum_{i=1}^{M} \frac{(O_i - E_i)^2}{E_i}. \tag{7.33}$$

Notice that under H_0 of (7.31), we expect that for each category i, both O_i and E_i will be relatively close, and hence it is expected that the sum of relative squared differences, χ^2, will not be too big. Conversely, a large value of χ^2 may indicate that H_0 is not plausible. Later, in this section, we show how to use χ^2 to construct the test to check for both goodness of fit (7.31), and to check for independence (7.32).

In the case of Kolmogorov-Smirnov, a key aspect is the Empirical Cumulative Distribution Function (ECDF), which was introduced in Section 4.3. Recall that for a sample of observations,

$x_1, \ldots, x_n$, the ECDF is

$$\hat{F}(x) = \frac{1}{n} \sum_{i=1}^{n} \mathbf{1}\{x_i \leqslant x\}, \qquad \text{where } \mathbf{1}\{\cdot\} \text{ is the indicator function.} \tag{7.34}$$

The approach of Kolmogorov-Smirnov test is to check the closeness of the ECDF to the CDF hypothesized under H_0 in (7.31). This is done via the *Kolmogorov-Smirnov statistic*,

$$\tilde{S} = \sup_x |\hat{F}(x) - F_0(x)|, \tag{7.35}$$

where $F_0(\cdot)$ is the CDF under H_0 and sup is the supremum over all possible x-values. Similar to the case of chi-squared, under H_0 it is expected that $\hat{F}(\cdot)$ does not deviate greatly from $F_0(\cdot)$, and hence it is expected that $\tilde{S}$ is not very large.

The key to both the chi-squared and Kolmogorov-Smirnov tests is that under H_0 there are tractable known approximations to the distribution of the test statistics of both (7.33) and (7.35). These approximations allow us to obtain an approximate p-value in the standard way via

$$p = \mathbb{P}(W > u), \tag{7.36}$$

where W denotes a random variable distributed according to the approximate distribution and u is the observed test statistic of either (7.33) or (7.35). We now elaborate on the details.

Chi-squared Test for Goodness of Fit

Consider the hypothesis (7.31) and assume that the distribution F_0 can be partitioned into categories $i = 1, \ldots, M$. Such a partition naturally occurs when the distribution is discrete with a finite number of outcomes. It can also be artificially introduced in other cases. With such a partition, having n sample observations, we denote by E_i the expected number of observations satisfying category i. These values are theoretically computed. Then, based on observations $x_1, \ldots, x_n$, we denote by O_i as the number of observations that satisfy category i. Note that

$$\sum_{i=1}^{M} E_i = n, \qquad \text{and} \qquad \sum_{i=1}^{M} O_i = n.$$

Now, based on $\{E_i\}$ and $\{O_i\}$, we can compute the χ^2 test statistic (7.33).

It turns out that under H_0, the χ^2 test statistic of (7.33) approximately follows a chi-squared distribution with $M - 1$ degrees of freedom. Hence, this allows us to approximate the p-value via (7.36), where W is taken as such a chi-squared random variable and u as the test statistic. This is also sometimes called *Pearson's chi-squared test.*

We now present an example where we assume under H_0 that a die is biased, with the probabilities for each side (1 to 6) given by the following vector $\mathbf{p}$:

$$\mathbf{p} = (0.08, \quad 0.12, \quad 0.2, \quad 0.2, \quad 0.15, \quad 0.25). \tag{7.37}$$

Note that if there are then n observations, we have that $E_i = n\, p_i$. For this example $n = 60$, and hence the vector of expected values for each side is

$$\mathbf{E} = (4.8, \quad 7.2, \quad 12, \quad 12, \quad 9, \quad 15).$$

Now imagine that the die is rolled $n = 60$ times, and the following count of outcomes (1 to 6) is observed:

$$\mathbf{O} = (3, \quad 2, \quad 9, \quad 11, \quad 8, \quad 27).$$

In Listing 7.12, we use this data to compute the test statistic and p-value. This is done first manually, and then the `ChisqTest()` function from the `HypothesisTests` package is used. From the output, we see that the p-value is around 0.0105. Hence, at the $\alpha = 0.05$ level, we would reject H_0 and conclude the distribution does not follow (7.37). That is we would conclude there is sufficient evidence to believe the die is weighted differently to the weights $\mathbf{p}$. However, at $\alpha = 0.01$, we will fail to reject H_0.

Listing 7.12: Chi-squared test for goodness of fit

```
using Distributions, HypothesisTests

p = [0.08, 0.12, 0.2, 0.2, 0.15, 0.25]
O = [3, 2, 9, 11, 8, 27]
M = length(O)
n = sum(O)
E = n*p

testStatistic = sum((O-E).^2 ./E)
pVal = ccdf(Chisq(M-1), testStatistic)

println("Manually calculated test statistic: ", testStatistic)
println("Manually calculated p-value: ", pVal,"\n")

println(ChisqTest(O,p))
```

```
Manually calculated test statistic: 14.974999999999998
Manually calculated p-value: 0.010469694843220351

Pearson's Chi-square Test
-------------------------
Population details:
    parameter of interest:   Multinomial Probabilities
    value under h_0:         [0.08, 0.12, 0.2, 0.2, 0.15, 0.25]
    point estimate:          [0.05, 0.0333333, 0.15, 0.183333, 0.133333, 0.45]
    95% confidence interval: Tuple{Float64,Float64}[(0.0, 0.1828), (0.0, 0.1662),
        (0.0333, 0.2828), (0.0667, 0.3162), (0.0167, 0.2662), (0.3333, 0.5828)]

Test summary:
    outcome with 95% confidence: reject h_0
    one-sided p-value:           0.0105

Details:
    Sample size:        60
    statistic:          14.975000000000001
    degrees of freedom: 5
    residuals:          [-0.821584, -1.93793, -0.866025, -0.288675, -0.333333, 3.09839]
    std. residuals:     [-0.85656, -2.06584, -0.968246, -0.322749, -0.361551, 3.57771]
```

In line 3, the array `p` is created, which represents the probabilities of each side occurring under H_0. In line 4, the array `O` is created, which contains the frequencies, or counts, of each side outcome observed. In line 5, the total number of categories (or side outcomes) is stored as `M`. In line 6, the total number of observations is stored as `n`. In line 7, the array of expected number of observed outcomes for each side is calculated by multiplying the vector of expected probabilities under H_0 by the total number of observations `n`. The resulting array is stored as `E`. In line 9, (7.33) is used to calculate the chi-squared test statistic. In line 10, the test statistic is used to calculate the p-value. Since under the null hypothesis the test statistic is asymptotically distributed according to a chi-squared distribution, the `ccdf()` function is used on a `Chisq()` distribution with `M-1` degrees of freedom. In lines 12 and 13, the manually calculated test statistic and p-value are printed. In line 15, the `ChisqTest()` function from the `HypothesisTests` package is used to perform the chi-squared test on the frequency data in array `p`.

Chi-squared Test Used to Check Independence

We now show how a chi-squared statistic can be used to check for independence, as in (7.32). Consider an example where 373 individuals are categorized as Male/Female, and Smoker/Non-smoker, as in the following *contingency table*:

	Smoker	Non-smoker
Male	18	132
Female	45	178

In this example, 18 individuals were recorded as "male" and "smoker", and so forth. Now under H_0, we assume that the smoking or non-smoking behavior of the individual is independent of the gender (male or female). To check for this using a chi-squared statistic, we first set up $\{E_i\}$ and $\{O_i\}$ as in the following table.

Table 7.2: The elements $\{O_{ij}\}$ and $\{E_{ij}\}$ as in the contingency table.

	Smoker	Non-smoker	Total/proportion
Male	$O_{11} = 18$ $E_{11} = 25.34$	$O_{12} = 132$ $E_{12} = 124.67$	150 / 0.402
Female	$O_{21} = 45$ $E_{21} = 37.66$	$O_{22} = 178$ $E_{22} = 185.33$	223 / 0.598
Total/Proportion	63/0.169	310/0.831	373 / 1

Here the observed *marginal distribution* over male versus female is based on the proportions $\mathbf{p} = (0.402, 0.598)$ and the distribution over smoking versus non-smoking is based on the proportions $\mathbf{q} = (0.169, 0.831)$. Then, since independence is assumed under H_0, we multiply the marginal probabilities to obtain the expected observation counts

$$E_{ij} = n\ p_i\ q_j. \tag{7.38}$$

For example, $E_{21} = 373 \times 0.169 \times 0.598 = 37.66$. Now with these values at hand, the chi-squared test statistic can be set up as follows:

$$\chi^2 = \sum_{i=1}^{m} \sum_{j=1}^{\ell} \frac{(O_{ij} - E_{ij})^2}{E_{ij}}, \tag{7.39}$$

where m and ℓ are the respective dimensions of the contingency table ($m = \ell = 2$ in this example).

It turns out that under H_0 the test statistic (7.39) is approximately chi-squared distributed with $(m-1) \times (\ell - 1)$ degrees of freedom. This implies 1 degree of freedom in our example. Hence, (7.36) can be used to determine a (approximate) p-value for this test, just like in the previous example.

Listing 7.13 carries out a chi-squared test in order to check if there is a relationship between gender and smoking. In this example, since the p-value is 0.0387, we conclude by saying there is some evidence that there is a relationship. That is, if $\alpha = 0.05$ we conclude that there is sufficient evidence to reject H_0. However, if $\alpha = 0.01$ we conclude that there is insufficient evidence to reject H_0.

Listing 7.13: Chi-squared for checking independence

```
1  using Distributions
2
3  xObs      = [18 132; 45 178]
4  rowSums   = [sum(xObs[i,:]) for i in 1:2]
5  colSums   = [sum(xObs[:,i]) for i in 1:2]
6  n         = sum(xObs)
7
8  rowProps = rowSums/n
9  colProps = colSums/n
10
11 xExpect   = [colProps[c]*rowProps[r]*n for r in 1:2, c in 1:2]
12 testStat = sum([(xObs[r,c]-xExpect[r,c])^2 / xExpect[r,c] for r in 1:2,c in 1:2])
13 pVal = ccdf(Chisq(1),testStat)
14
15 println("Chi-squared value: ", testStat)
16 println("P-value: ", pVal)
```

```
Chi-squared value: 4.274080056208799
P-value: 0.03869790606536347
```

In line 3, the observation counts from the contingency table are stored as the two-dimensional array, `xObs`. In line 4, the observations in each row are summed via `xObs[i,:]`, and the use of a comprehension. In line 5, the observations in each column are calculated via a similar approach to that in line 4 above. In line 6, the total number of observations is stored as `n`. In lines 8 and 9, the row and column proportions are calculated. In line 11, the expected number of observations, $\{E_{ij}\}$ (shown in Table 7.2), are calculated. Note the use of the comprehension, which calculates (7.38) for each combination of sex and smoker/non-smoker. In line 12, the test statistic is calculated via (7.39) through the use of a comprehension. In line 13, the test statistic is used to calculate the p-value. Since under the null hypothesis the test statistic is asymptotically distributed according to a chi-squared distribution, the `ccdf()` function is used on a `Chisq()` distribution with $(m-1) \times (\ell - 1)$ degrees of freedom, i.e. 1 in this example.

Kolmogorov-Smirnov Test

We now depart from the situations of a finite number of categories as in the chi-squared test, and consider the Kolmogorov-Smirnov test, which is based on the test statistic $\tilde{S}$ from (7.35). The approach is based on the fact that, under H_0 of (7.31), the Empirical Cumulative Distribution (ECDF) $\hat{F}(\cdot)$ is close to the actual CDF $F_0(\cdot)$. To get a feel for this notice that for every value $x \in \mathbb{R}$, the ECDF at that value, $\hat{F}(x)$, is the proportion of the number of observations less than or equal to x. Under H_0, multiplying the ECDF by n yields a binomial random variable with success probability, $F_0(x)$:

$$n\hat{F}(x) \quad \sim \quad \text{Bin}\big(n, F_0(x)\big).$$

Hence,

$$\mathbb{E}[\hat{F}(x)] = F_0(x), \qquad \text{Var}\big(\hat{F}(x)\big) = \frac{F_0(x)\big(1 - F_0(x)\big)}{n}.$$

See the binomial distribution in Section 3.5. Hence, for non-small n, the ECDF and CDF should be close since the variance for every value x is of the order of $1/n$ and diminishes as n grows. The formal statement of this is, taking $n \to \infty$ and considering all values of x simultaneously is known as the *Glivenko Cantelli Theorem.*

For finite n, the ECDF will not exactly align with the CDF. However, the Kolmogorov-Smirnov test statistic (7.35) is useful when it comes to measuring this deviation. This is due to the fact that under H_0 the *stochastic process* in the variable, x,

$$\sqrt{n}\big(\hat{F}(x) - F_0(x)\big) \tag{7.40}$$

is approximately identical in probability law to a standard *Brownian Bridge*, $B(\cdot)$, composed with $F_0(x)$. That is, by denoting $\hat{F}_n(\cdot)$ as the ECDF with n observations, we have that

$$\sqrt{n}\big(\hat{F}_n(x) - F_0(x)\big) \quad \overset{d}{\approx} \quad B\big(F_0(x)\big), \tag{7.41}$$

which asymptotically converges to equality in distribution as $n \to \infty$. Note that a Brownian Bridge, $B(t)$, is a form of a variant of *Brownian motion*, constrained to equal 0 both at $t = 0$ and $t = 1$. It is a type of *diffusion process.* See [K12] for a good introduction to diffusion processes and *stochastic calculus.*

Now consider the supremum as in the Kolmogorov-Smirnov test statistic $\tilde{S}$, as defined in (7.35). It can be shown that, in cases where $F_0(\cdot)$ is a continuous function (distribution), as $n \to \infty$,

$$\sqrt{n}\tilde{S} \quad \overset{d}{=} \quad \sup_{t\in[0,1]} |B(t)|. \tag{7.42}$$

Importantly, notice that the right-hand side does not depend on $F_0(\cdot)$, but rather is the maximal value attained by the absolute value of the Brownian bridge process over the interval $[0, 1]$. It then turns out that (see, for example, [M07] for a derivation) such a random variable, denoted by K, has CDF,

$$F_K(x) = \mathbb{P}\Big(\sup_{t\in[0,1]} |B(t)| \le x\Big) = 1 - 2\sum_{k=1}^{\infty}(-1)^{k-1}e^{-2k^2x^2} = \frac{\sqrt{2\pi}}{x}\sum_{k=1}^{\infty} e^{-(2k-1)^2\pi^2/(8x^2)}. \tag{7.43}$$

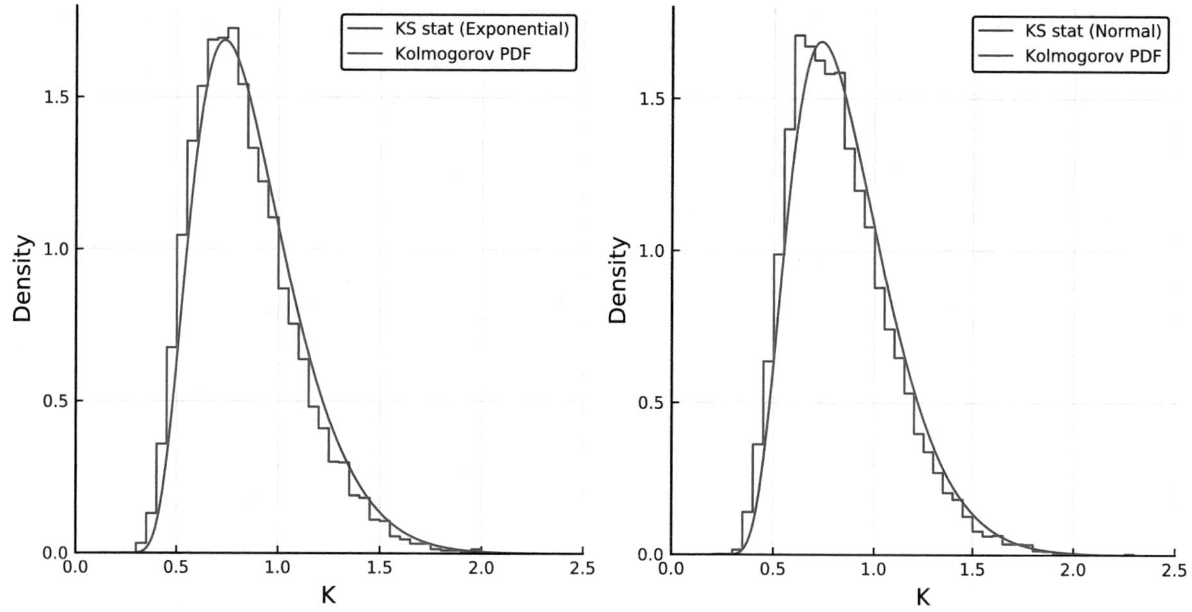

Figure 7.5: PDF of the Kolmogorov distribution, alongside histograms of K-S test statistics from normal and exponential populations.

This is sometimes called the *Kolmogorov distribution*. Thus, to obtain an approximate p-value for the Kolmogorov-Smirnov test using (7.36), we calculate

$$p = 1 - F_K(\sqrt{n}\tilde{S}). \tag{7.44}$$

Figure 7.5 generated by Listing 7.14 compares the PDF of the Kolmogorov distribution to the empirical distribution of $\tilde{S}$ scaled by $\sqrt{n}$ as on the left-hand side of (7.42). This is done for two different scenarios. First, data is sampled from an exponential distribution, and second, data is sampled from a normal distribution. As illustrated in the resulting Figure 7.5, the distributions of the Monte Carlo generated test statistics are in close agreement with the analytic PDF, regardless of what underlying distribution $F_0(x)$ the data comes from.

Listing 7.14: Comparisons of distributions of the K-S test statistic

```
1  using Distributions, StatsBase, HypothesisTests, Plots, Random; pyplot()
2  Random.seed!(0)
3
4  n = 25
5  N = 10^4
6  xGrid = -10:0.001:10
7  kGrid = 0:0.01:5
8  dist1, dist2 = Exponential(1), Normal()
9
10 function ksStat(dist)
11     data = rand(dist,n)
12     Fhat = ecdf(data)
13     sqrt(n)*maximum(abs.(Fhat.(xGrid) - cdf.(dist,xGrid)))
14 end
15
16 kStats1 = [ksStat(dist1) for _ in 1:N]
17 kStats2 = [ksStat(dist2) for _ in 1:N]
18
19 p1 = stephist(kStats1, bins=50,
20         c=:blue, label="KS stat (Exponential)", normed=true)
21 p1 = plot!(kGrid, pdf.(Kolmogorov(),kGrid),
22         c=:red, label="Kolmogorov PDF", xlabel="K", ylabel="Density")
23
24 p2 = stephist(kStats2, bins=50,
25         c=:blue, label="KS stat (Normal)", normed=true)
26 p2 = plot!(kGrid, pdf.(Kolmogorov(),kGrid),
27         c=:red, label="Kolmogorov PDF", xlabel="K", ylabel="Density")
28
29 plot(p1, p2, xlims=(0,2.5), ylims=(0,1.8), size=(800, 400))
```

In lines 4–8, we specify the sample size `n`, number of Monte Carlo repetitions `N`, grids for computation and plotting, and two distributions of the underlying population. In lines 10–14, the function `ksStat()` is created, which takes a distribution type as input, randomly samples `n` observations from it, calculates the ECDF of the data via the `ecdf()` function, and finally returns the left-hand side of (7.42) by calculating the K-S test statistic via (7.35), and multiplying this by `sqrt(n)`. Note that in line 12, the `ecdf()` function returns a cdf function type itself, which is stored as `Fhat`, and broadcasted over `xGrid` in line 13. In lines 16–17, a comprehension is used along with the `ksStat()` function to generate `N` K-S test statistics for each distribution of the population. The remainder of the code compares histograms of `kStats1` and `kStats2` against PDFs of the `Kolmogorov()` distribution.

Now that we have demonstrated that the distribution of the scaled Kolmogorov-Smirnov statistic is similar to the distribution of K as in (7.43), we demonstrate how the Kolmogorov-Smirnov statistic can be used to carry out a goodness-of-fit test. For this example, consider that a series of observations has been made from some unknown underlying gamma distribution with shape parameter 2 and mean 5. The question we then wish to ask is: *given the sample observations, is the underlying distribution exponential with the same mean?* The answer is false because an exponential distribution is a gamma distribution with a shape parameter of 1, not 2. However, with a finite number of observations, such as $n = 100$ in our case, we can only expect to give an approximate answer.

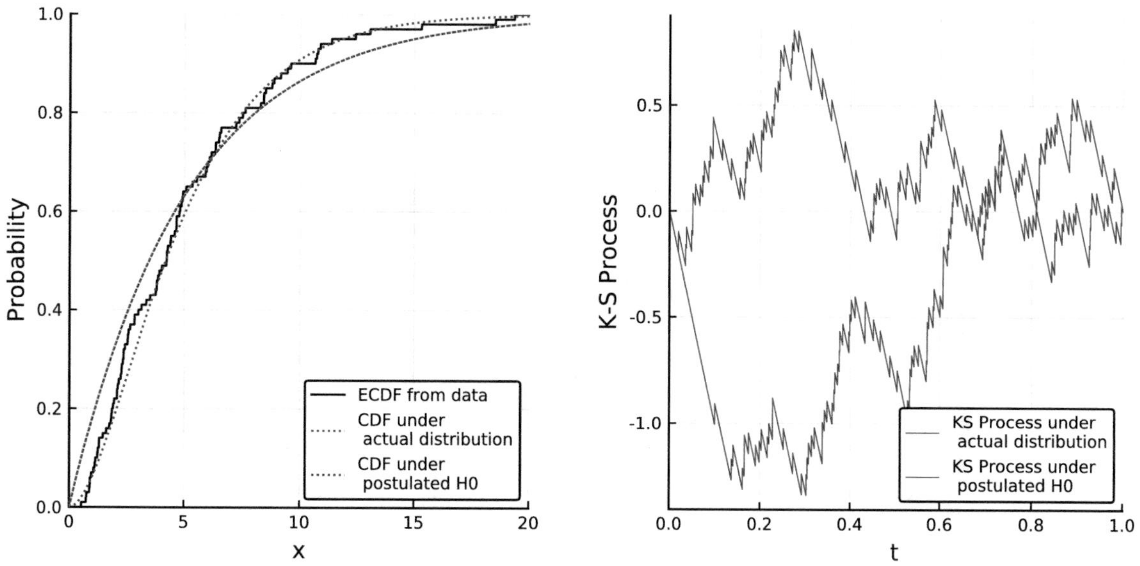

Figure 7.6: Left: CDFs and ECDF. Right: K-S processes scaled over $[0, 1]$.

To help illustrate the logic of the approach, see Figure 7.6 generated by Listing 7.15. The left plot presents the ECDF of the data plotted against the actual CDF (blue) as well as the postulated CDF (red). The ECDF follows the actual CDF quite closely but does not follow the postulated CDF well. Keep in mind that in a practical situation, we don't know the actual CDF. We only know the postulated CDF. Still we expect mild deviations under H_0, such that when composed with the CDF, behave approximately as a Brownian bridge (defined for $t \in [0, 1]$). In contrast, if irregular deviations appear we can conclude that H_0 does not hold. For this, look at the right-hand plot of Figure 7.6. The observed deviations (time stretched by the CDF) are in the red curve. Such a trajectory of a Brownian bridge is possible but not plausible. Instead, most trajectories will behave more like the blue curve.

Now the Kolmogorov-Smirnov statistic (7.35) is useful. Under H_0 (and for non-small n) it needs to follow a CDF as (7.43). Hence, this CDF can be used to compute the p-value using (7.44). Listing 7.15 computes the p-value by manually calculating a truncation of the series in (7.43), by using the `Kolmogorov()` distribution object, and by using `ApproximateOneSampleKSTest()` from the `HypothesisTests` package. As can be observed from the output, the resulting p-values are all in agreement at approximately 0.0545. Observe now that if $\alpha = 0.05$ then we fail to reject H_0 because there isn't sufficient evidence that the distribution deviates from an exponential distribution. With this example (using the same random number generation seed), if you were to increase the number of observations to $n = 200$, then the p-value changes to 0.0004, meriting rejection under any sensible significance level.

Listing 7.15: ECDF, actual and postulated CDF's, and their differences

```
using Random,Distributions,StatsBase,Plots,HypothesisTests,Measures; pyplot()
Random.seed!(3)

dist = Gamma(2, 2.5)
distH0 = Exponential(5)
n = 100
data = rand(dist,n)

Fhat = ecdf(data)
diffF(dist, x) = sqrt(n)*(Fhat(x) - cdf.(dist,x))
xGrid = 0:0.001:30
ksStat = maximum(abs.(diffF(distH0, xGrid)))

M = 10^5
KScdf(x) = sqrt(2pi)/x*sum([exp(-(2k-1)^2*pi^2 ./(8x.^2)) for k in 1:M])

println("p-value calculated via series: ",
        1-KScdf(ksStat))
println("p-value calculated via Kolmogorov distribution: ",
        1-cdf(Kolmogorov(),ksStat),"\n")

println(ApproximateOneSampleKSTest(data,distH0))

p1 = plot(xGrid, Fhat(xGrid),
        c=:black, lw=1, label="ECDF from data")
p1 = plot!(xGrid, cdf.(dist,xGrid),
        c=:blue, ls=:dot, label="CDF under \n actual distribution")
p1 = plot!(xGrid, cdf.(distH0,xGrid),
        c=:red, ls=:dot, label="CDF under \n postulated H0",
        xlims=(0,20), ylims=(0,1), xlabel = "x", ylabel = "Probability")

p2= plot(cdf.(dist,xGrid), diffF(dist, xGrid),lw=0.5,
        c=:blue,        label="KS Process under \n actual distribution")
p2 = plot!(cdf.(distH0,xGrid), diffF(distH0, xGrid), lw=0.5,
        c=:red, xlims=(0,1), label="KS Process under \n postulated H0",
    xlabel = "t", ylabel = "K-S Process")

plot(p1, p2, legend=:bottomright, size=(800, 400), margin = 5mm)
```

```
p-value calculated via series: 0.05473084786694438
p-value calculated via Kolmogorov distribution: 0.054730847866944266

Approximate one sample Kolmogorov-Smirnov test
----------------------------------------------
Population details:
    parameter of interest:   Supremum of CDF differences
    value under h_0:         0.0
    point estimate:          0.13421930779083405

Test summary:
    outcome with 95% confidence: fail to reject h_0
    two-sided p-value:           0.0545

Details:
    number of observations:   100
    KS-statistic:             1.3421930779083404
```

In lines 4 and 5, we set the actual underlying distribution (gamma) and postulated distribution (exponential), respectively. In line 6, we set the number of observations, `n`. The data is generated in line 7. The ECDF is created in line 9 and the process (7.40) is defined in line 9 via our function `diffF()` allowing different postulated (H_0) distributions. The Kolmogorov-Smirnov statistic is calculated in line 12. Line 15 implements our function `KScdf()` by truncating the series in (7.43) to `M`. We use it, as well as the `Kolmogorov()` distribution object to print out p-values in lines 17-20. Similarly, in line 21, we use `ApproximateOneSampleKSTest()` from `HypothesisTests()`. The remainder of the code creates Figure 7.6 where a point to note is the horizontal-axis values in lines 32 and 34 reflecting the composition in (7.41).

Testing for an Independent Sequence

One may also carry out some hypothesis tests for checking if a sequence of random variables is i.i.d. (independent and identically distributed). To illustrate one such test, we introduce the classic *Wald-Wolfowitz runs test*. Consider a sequence of data points $x_1, \ldots, x_n$ with sample mean $\overline{x}$. For simplicity, assume that no point equals the sample mean. Now transform the sequence to $y_1, \ldots, y_n$ via

$$y_i = x_i - \overline{x}.$$

We now consider the signs of y_i. For example, in a dataset with 20 observations, once considering the signs we may have a sequence such as

$$+-+-----++--+---++++,$$

indicating that the first is positive (greater than the mean), the second is negative (less than the mean), the third is positive, the fourth is negative, and this continues until the last four positive signs. Note that we assume no exact 0 for y_i and if such exist we can arbitrarily assign them to be either positive or negative. We then create the random variable, R, counting the number of *runs* in this sequence, where a run is a consecutive sequence of points having the same sign. In our example, the runs (visually separated by white space) are

$$+ \quad - \quad + \quad ----- \quad ++ \quad -- \quad + \quad --- \quad ++++.$$

Hence here $R = 9$. The essence of the Wald-Wolfowitz runs test is an approximation of the distribution of R under H_0. The null hypothesis is that the data is i.i.d. In that case, R can be shown to approximately follow a normal distribution with mean μ and variance σ^2, where

$$\mu = 2\frac{n_+ n_-}{n} + 1, \qquad \sigma^2 = \frac{(\mu-1)(\mu-2)}{n-1}. \tag{7.45}$$

Here n_+ is the number of positive values and n_- is the number of negative values. Note that n_+ and n_- are also random variables. Clearly $n_+ + n_- = n$, the total number of observations. In our example, $n = 20$, $n_+ = 11$, $n_- = 9$. You can use (7.45) to compute the mean and variance: $\mu = 10.9$ and $\sigma^2 = 4.64$. With such values at hand the test creates the p-value via

$$2\mathbb{P}\big(Z > \Big|\frac{R-\mu}{\sigma}\Big|\big),$$

where Z is a standard normal random variable. In this example, the p-value is 0.38 and hence we do not reject H_0 under any plausible α and don't conclude that there is any apparent violation of

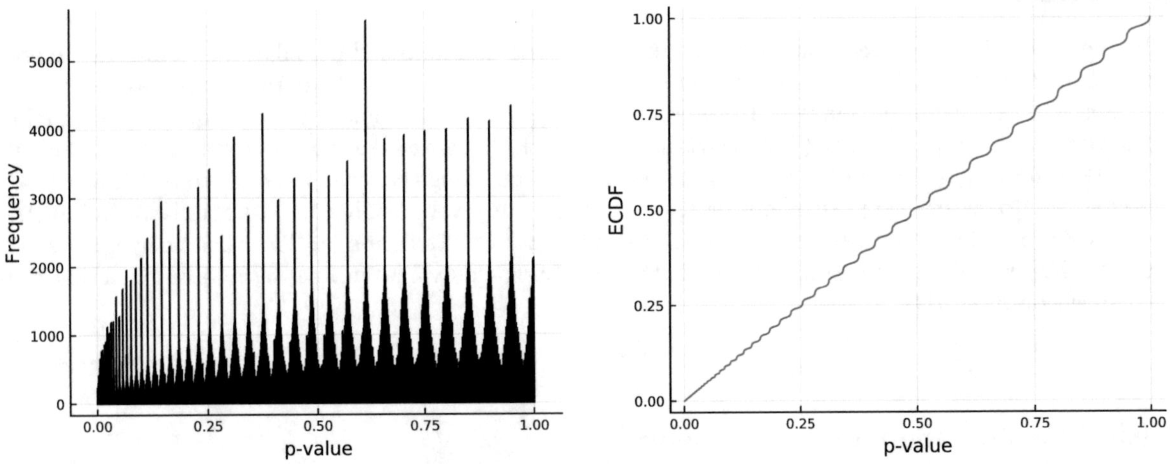

Figure 7.7: The distribution of the (approximate) p-value under H_0 for a Wald-Wolfowitz runs test.

model assumptions due to this test. However, if the p-value would have been significantly smaller we may have reason to suspect that some model assumptions are violated.

In Listing 7.16, we implement a Wald-Wolfowitz runs test for synthetic data. We generate 10^6 samples of i.i.d. standard normal sequences, each of length $n = 1,000$. Since this agrees with H_0 of the Wald-Wolfowitz runs test, we expect that the p-value follow an approximate uniform distribution (see Listing 7.19 of Chapter 7). To explore this, we plot a histogram of the p-value and its ECDF in Figure 7.7.

The distribution as appearing via the histogram is clearly not uniform. Spikes of high density appear and these are due to lattice effects (we are approximating a discrete random variable R, with a normal random variable). Nevertheless, the ECDF indicates that the distribution is almost uniform. The output of Listing 7.16 also presents the area to the left of the p-value for several specified α and illustrates that there is agreement. Hence, the normal approximation with parameters as in (7.45) appears to be very close.

Listing 7.16: The Wald-Wolfowitz runs test

```
1  using CSV, GLM, StatsBase, Random, Distributions, Plots, Measures; pyplot()
2  Random.seed!(0)
3
4  n, N = 10^3, 10^6
5
6  function waldWolfowitz(data)
7      n = length(data)
8      sgns = data .>= mean(data);
9      nPlus, nMinus = sum(sgns), n - sum(sgns)
10     wwMu = 2*nPlus*nMinus/n + 1
11     wwVar = (wwMu-1)*(wwMu-2)/(n-1)
12
13     R = 1
14     for i in 1:n-1
15         R += sgns[i] != sgns[i+1]
16     end
17
18     zStat = abs((R-wwMu)/sqrt(wwVar))
19     2*ccdf(Normal(),zStat)
20 end
21
22 experimentPvals = [waldWolfowitz(rand(Normal(),n)) for _ in 1:N]
23 for alpha in [ 0.001, 0.005, 0.01, 0.05, 0.1]
24     pva = sum(experimentPvals .< alpha)/N
25     println("For alpha = $(alpha), p-value area = $(pva)")
26 end
27
28 p1 = histogram(experimentPvals,bins = 5n, legend = false,
29     xlabel = "p-value", ylabel = "Frequency")
30
31 Fhat = ecdf(experimentPvals)
32
33 pGrid = 0:0.001:1
34 p2 = plot(pGrid,Fhat.(pGrid), legend =false, xlabel = "p-value", ylabel = "ECDF")
35 plot(p1, p2, size = (1000,400), margin = 5mm)
```

```
For alpha = 0.001, p-value area = 0.000855
For alpha = 0.005, p-value area = 0.005196
For alpha = 0.01, p-value area = 0.010739
For alpha = 0.05, p-value area = 0.05272
For alpha = 0.1, p-value area = 0.094269
```

In lines 6–20, we implement the `waldWolfowitz()` function. In line 8, we compare `data` to its mean and create a sequence of signs, `sgns`. Using the `sign()` function would have been possible, however by using >= we ensure to break ties with 0. We implement (7.45) in lines 9–11 and then lines 13–16 create the number of runs, `R`. We then calculate the p-value and return it in line 19. In line 22, we generate `N` p-values for the synthetic data for a sample size of `n`. These are then analyzed in lines 23–26 where we estimate the proportion of p-values less than a specified `alpha` and print these out comparing to `alpha`. The remainder of the code creates Figure 7.7. Note the use of `ecdf()` from `StatsBase`, similar to Listing 4.7 of Chapter 4.

7.5 More on Power

In this section, the concept of power is covered in greater depth together with related aspects of hypothesis testing such as the distribution of the p-value. Recall that as first introduced in Section 5.6 and summarized in Table 5.1, the statistical *power* of a hypothesis test is the probability of correctly rejecting H_0 as is given by $1 - \mathbb{P}(\text{Type II error})$. We now reinforce this idea through a concrete introductory example.

Consider a normal population with unknown parameters μ and σ, and say that we wish to conduct a one-sided hypothesis test on the population mean using the following hypothesis test setup:

$$H_0 : \mu = \mu_0 \qquad \text{and} \qquad H_1 : \mu > \mu_0. \tag{7.46}$$

Importantly, since power is the probability of a correct rejection, if the underlying (unknown) parameter μ varies greatly from the value under the null hypothesis μ_0, then the power of the test in this scenario is greater. Likewise, if the underlying parameter does not vary greatly from the value under the null hypothesis, the power of the test is lower. Similar effects occur due to the underlying (unknown) variance as well as the sample size. A lower variance implies higher power and larger sample sizes increase power. Also reducing (improving) α will decrease the power and vice versa.

In Listing 7.17, several different scenarios, labeled **A**, **B**, **C**, and **D**, are considered. For each, N test statistics are calculated via Monte Carlo simulation for N sample groups and the power of the test is estimated. First the underlying mean equals the mean under the null hypothesis and hence the power equals α. Then in each subsequent scenario, the parameters or sample size are changed in a way that power is increased. First in scenario **A**, the underlying mean is increased, then in scenario **B** it is increased further. Further in scenario **C** the sample size is increased, and finally in scenario **D** the standard deviation is decreased. As you can observe from the output of Listing 7.17 each of these incremental changes increases the power up to approximately 0.91 in case **D**. In practice, if keeping α constant, it is only the sample size that can be controlled, however understanding the effect of the other parameters on the power is important in deciding how large samples should be.

For some of these scenarios, Listing 7.17 employs kernel density estimation to plot the distribution of the test statistics. The resulting Figure 7.8 is similar to Figure 5.13; however, in this case, the focus is on power. The power under different scenarios is given by the area under each PDF to the right of the critical value boundary. Hence, the more "separation" that we can achieve from the distribution under H_0, the better. As a side point, note that the curves shown in Figure 7.8 could have alternatively been obtained analytically via the *non-central T-distribution*, a classical concept that we do not cover further.

Listing 7.17: Distributions under different hypotheses

```
using Random, Distributions, KernelDensity, Plots, LaTeXStrings; pyplot()
Random.seed!(1)

function tStat(mu0,mu,sig,n)
    sample = rand(Normal(mu,sig),n)
    xBar   = mean(sample)
    s      = std(sample)
    (xBar-mu0)/(s/sqrt(n))
end

mu0, mu1A, mu1B = 20, 22, 24
sig, n = 7, 5
N = 10^6
alpha = 0.05

dataH0  = [tStat(mu0,mu0,sig,n) for _ in 1:N]
dataH1A = [tStat(mu0,mu1A,sig,n) for _ in 1:N]
dataH1B = [tStat(mu0,mu1B,sig,n) for _ in 1:N]
dataH1C = [tStat(mu0,mu1B,sig,2*n) for _ in 1:N]
dataH1D = [tStat(mu0,mu1B,sig/2,2*n) for _ in 1:N]

tCrit = quantile(TDist(n-1),1-alpha)
estPwr(sample) = sum(sample .> tCrit)/N

println("Rejection boundary: ", tCrit)
println("Power under H0:  ", estPwr(dataH0))
println("Power under H1A: ", estPwr(dataH1A))
println("Power under H1B (mu's farther apart): ", estPwr(dataH1B))
println("Power under H1C (double sample size): ", estPwr(dataH1C))
println("Power under H1D (like H1C but std/2): ", estPwr(dataH1D))

kH0  = kde(dataH0)
kH1A = kde(dataH1A)
kH1D = kde(dataH1D)
xGrid = -10:0.1:15

plot(xGrid,pdf(kH0,xGrid),
        c=:blue, label="Distribution under H0")
plot!(xGrid,pdf(kH1A,xGrid),
        c=:red, label="Distribution under H1A")
plot!(xGrid,pdf(kH1D,xGrid),
        c=:green, label="Distribution under H1D")
plot!([tCrit,tCrit],[0,0.4],
        c=:black, ls=:dash, label="Critical value boundary",
        xlims=(-5,10), ylims=(0,0.4), xlabel=L"\Delta = \mu - \mu_0")
```

```
Rejection boundary: 2.131846786326649
Power under H0:  0.049598
Power under H1A: 0.134274
Power under H1B (mu's farther apart): 0.281904
Power under H1C (double sample size): 0.406385
Power under H1D (like H1C but std/2): 0.91554
```

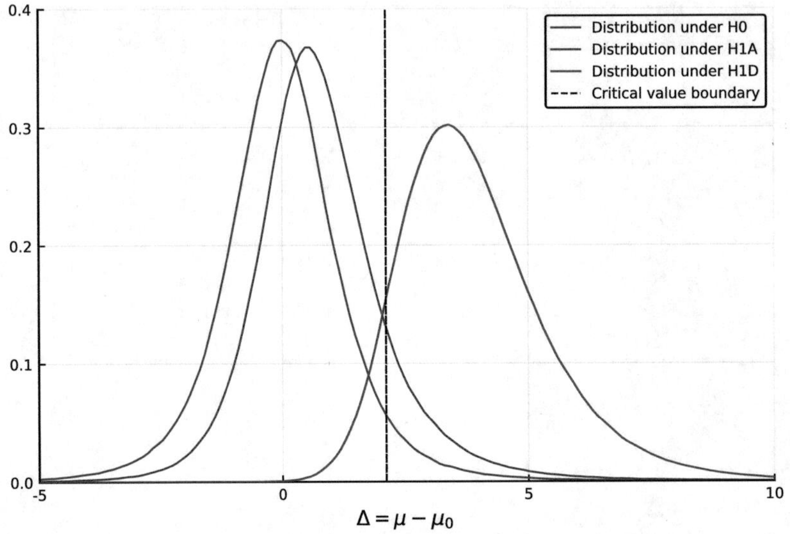

Figure 7.8: Numerically estimated distributions of the test statistic for various scenarios of values of the underlying parameter μ.

In lines 4–9, the function `tStat()` is defined, which returns the value of the test statistic for a randomly generated sample. Note here that `mu0` represents the value under the null hypothesis as used to calculate the test statistic while `mu` represents the value of the actual underlying mean, used to generate random samples. In line 11, we define different values of μ, matching H_0, scenario **A**, and scenario **B**. In line 12, the standard deviation and the number of observations are defined. In line 13, we define the number of Monte Carlo repetitions and in line 14 we define the significance level. In lines 16–20, the `tStat()` function is used along with a series of comprehensions to generate `N` test statistics for each of the scenarios as described above. In line 22, the critical value for the significance level `alpha` is calculated by using the `quantile()` function on a T-distribution `TDist()`, with `n-1` degrees of freedom. In line 23, the function `estPwr()` is defined, which takes an array of test statistics as input, and then approximates the corresponding power of the scenario as the proportion of statistics that exceeds `tCrit` calculated previously, i.e. the proportion of cases for which the null hypothesis was rejected. Note the use of the `.>` which returns an array of `true,false` values, which are then summed up and divided by `N`. Lines 25–30 print the output and estimate the power for each of the scenarios. Then lines 32–34 create `UnivariateKDE` objects representing kernel density estimates of the test statistics. These are then plotted in lines 37–42 using the `pdf()` function with a method used for kernel density estimates. Lines 43–45 plot the critical value.

Power Curves

From Listing 7.17, we can see that the statistical power of a hypothesis test can vary. It depends not only on the parameters of the test, such as the number of observations in the sample group n and the specified confidence level α, but also on the underlying parameter values μ and σ. Hence, a key aspect of *experimental design* involves determining the test parameters such that not only is the probability of a type I error controlled, but that the test is sufficiently powerful over a range of different scenarios. This is important, as in reality there are an infinite number of possibilities in H_1, any one of which could describe the underlying parameters. By designing a statistical test that has sufficient power, we aim to have confidence that if the underlying parameter deviates from the null

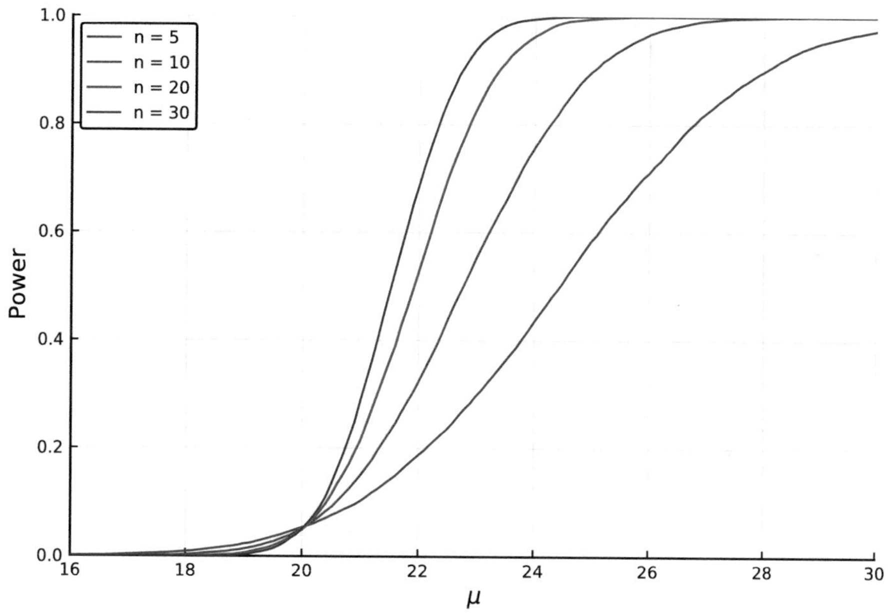

Figure 7.9: Power curves for the one-sided T-test with various sample sizes at $\alpha = 0.05$.

hypothesis, then this will be identified. Such planning is often aided by inspecting power curves.

A *power curve* is a plot of the power as a function of certain parameters. To illustrate this, we continue with the hypothesis formulation (7.46). Listing 7.18 estimates the power of a one-sided T-test. We estimate power of hypothesis test setup over a range of different values of μ, for various sample sizes of $n = 5, 10, 20, 30$. For each sample size, the power is estimated and the result is a power curve that can be plotted as shown in Figure 7.9.

Our focus is for $\mu > 20$. It can be seen that as the number of sample observations increases, the statistical power of the test increases. Similarly, the larger the μ is (compared to $\mu_0 = 20$) the higher the power. Observe also that at $\mu = \mu_0$ the power is $\alpha = 0.05$. An interesting subtle point to note is that where $\mu < 20$, the ordering of the curves is reversed. For example, one can see that in this region the scenario where $n = 30$ has less power than that for $n = 5$ due to the fact that the probability of rejecting the null hypothesis at all is lower. Another point to note is that the x-axis could be adjusted to represent the difference between the value of mu_0 under the null hypothesis, and the various possible values of μ as was done in Figure 7.8. Furthermore, one could make the axis scale invariant by dividing said difference by the standard deviation. Such curves are often seen in experimental design reference material.

With such a plot at hand, assume we are planning a costly (in the sample size) hypothesis test aiming to show that $\mu > 20$. Say that for sample size planning purposes, we have reason to believe that $\mu > 24$. Assume now that after fixing $\alpha = 0.05$ we wish to have power greater than 0.6. We then consider between the option of $n = 5$, $n = 10$, or $n = 20$ observations. Figure 7.8 illustrates that $n = 10$ is a sufficient number of observations. However, if we had plausible reason to believe that (under H_1) $\mu = 22$ then $n = 30$ observations are required to attain power > 0.6.

Listing 7.18: Power curves for different sample sizes

```
1  using Distributions, Plots, LaTeXStrings, Random; pyplot()
2
3  function tStat(mu0,mu,sig,n)
4      sample = rand(Normal(mu,sig),n)
5      xBar   = mean(sample)
6      s      = std(sample)
7      (xBar-mu0) / (s/sqrt(n))
8  end
9
10 function powerEstimate(mu0,mu1,sig,n,alpha,N)
11     Random.seed!(0)
12     sampleH1 = [tStat(mu0,mu1,sig,n) for _ in 1:N]
13     critVal  = quantile(TDist(n-1),1-alpha)
14     sum(sampleH1 .> critVal)/N
15 end
16
17 mu0 = 20
18 sig = 5
19 alpha = 0.05
20 N = 10^4
21 rangeMu1 = 16:0.1:30
22 nList = [5,10,20,30]
23
24 powerCurves = [powerEstimate.(mu0,rangeMu1,sig,n,alpha,N) for n in nList]
25
26 plot(rangeMu1,powerCurves[1],c=:blue, label="n = $(nList[1])")
27 plot!(rangeMu1,powerCurves[2],c=:red, label="n = $(nList[2])")
28 plot!(rangeMu1,powerCurves[3],c=:green, label="n = $(nList[3])")
29 plot!(rangeMu1,powerCurves[4],c=:purple, label="n = $(nList[4])",
30         xlabel= L"\mu", ylabel="Power",
31         xlims=(minimum(rangeMu1) ,maximum(rangeMu1)), ylims=(0,1))
```

In lines 3–8, the function `tStat()` is defined in an identical manner to Listing 7.17. In lines 10–15, the function `powerEstimate()` is created. It uses Monte Carlo to estimate the power of the one-sided hypothesis test (7.46), given the value under the null hypothesis `mu0`, and the actual parameter of the underlying process `mu1`. The other arguments of the function include the sample size `n`, the actual standard deviation `sig`, the significance level `alpha`, and the total number of Monte Carlo repetitions. Since we use common random numbers (see Section 10.6), in line 11 we fix the seed. In line 12, `tStat()` is used along with a comprehension to generate `N` test statistics from `N` independent sample groups. The test statistics are then stored as the array `sampleH1`. In line 13, the critical value for the given scenario of inputs is calculated in the same manner as in line 22 of Listing 7.17. In line 14, the proportion of test statistics greater than the critical value is calculated using the same approach as that of line 23 of Listing 7.17. In lines 17–22, the parameters of the problem are specified. The value under the null hypothesis `mu0`, the underlying variance of the unknown process `sig`, and the number of Monte Carlo repetitions `N`. The range over which the underlying mean of the process `mu1` will be calculated is specified as `rangeMu1`. A list of power curves depending on sample size are specified in `nList`. The actual simulation is carried out in line 24 where `powerCurves` is an array of arrays (one for each sample size in `nList`) with each entry being an array matching `powerEstimate()` broadcasted over `rangeMu1`. The remaining lines plot the curves.

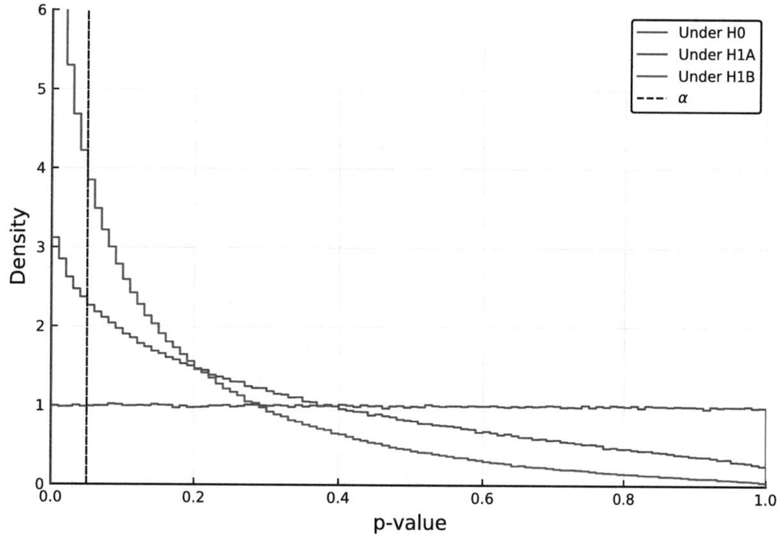

Figure 7.10: The distribution of the p-value under H_0 vs. the distribution of the p-value under H_1 with $\mu = 22$ (**A**) and H_1 with $\mu = 24$ (**B**).

Distribution of the p-value

Throughout this chapter, equations of the form $p = \mathbb{P}(S > u)$ were presented, where S is a random variable representing the test statistic, u is the observed test statistic, and p is the p-value of the observed test statistic. We now explore the distribution of the p-value.

To discuss the distribution of the p-value, denote the random variable of the p-value by P. Hence $P = 1 - F(S)$, where $F(\cdot)$ is the CDF of the test statistic under H_0. Note that P is just a transformation of the test statistic random variable S. Assume that S is continuous and assume that H_0 holds, hence $\mathbb{P}(S < u) = F(u)$. We now have,

$$\begin{aligned}\mathbb{P}(P > x) &= \mathbb{P}\big(1 - F(S) > x\big)\\ &= \mathbb{P}\big(F(S) < 1 - x\big)\\ &= \mathbb{P}\big(S < F^{-1}(1 - x)\big)\\ &= F(F^{-1}(1 - x))\\ &= 1 - x.\end{aligned}$$

Recall that for a uniform(0,1) random variable, the CCDF is $1 - x$ on $x \in [0, 1]$. Therefore under H_0, P is a uniform(0,1) random variable. This agrees with the fact that under H_0, the chance of rejecting H_0 is exactly α (this happens when $p < \alpha$).

If H_0 does not hold then $\mathbb{P}(S < u) \neq F(u)$ and the derivation above fails. In such a case, the distribution of the p-value is no longer uniform. In fact, in such a case, if the setting is such that the power of the test increases, then we expect the distribution of the p-value to be more concentrated around 0 than a uniform distribution.

In Listing 7.19, we revisit H_0, scenario **A**, and scenario **B** from Listing 7.17. For each case, we simulate Monte Carlo repetitions of the p-value, estimate the power, and plot the distribution in

Figure 7.10. Note that the power is the area to the left of α in the figure and as you can see from the output, it agrees with the output of Listing 7.17.

Listing 7.19: Distribution of the p-value

```
1  using Random, Distributions, KernelDensity, Plots, LaTeXStrings; pyplot()
2  Random.seed!(1)
3
4  function pval(mu0,mu,sig,n)
5      sample = rand(Normal(mu,sig),n)
6      xBar   = mean(sample)
7      s      = std(sample)
8      tStat  = (xBar-mu0) / (s/sqrt(n))
9      ccdf(TDist(n-1), tStat)
10 end
11
12 mu0, mu1A, mu1B  = 20, 22, 24
13 sig, n, N = 7, 5, 10^6
14
15 pValsH0 = [pval(mu0,mu0,sig,n) for _ in 1:N]
16 pValsH1A = [pval(mu0,mu1A,sig,n) for _ in 1:N]
17 pValsH1B = [pval(mu0,mu1B,sig,n) for _ in 1:N]
18
19 alpha = 0.05
20 estPwr(pVals) = sum(pVals .< alpha)/N
21
22 println("Power under H0:  ", estPwr(pValsH0))
23 println("Power under H1A: ", estPwr(pValsH1A))
24 println("Power under H1B (mu's farther apart): ", estPwr(pValsH1B))
25
26 stephist(pValsH0, bins=100,
27         normed=true, c=:blue, label="Under H0")
28 stephist!(pValsH1A, bins=100,
29         normed=true, c=:red, label="Under H1A")
30 stephist!(pValsH1B, bins=100,
31         normed=true, c=:green, label="Under H1B",
32     xlims=(0,1), ylims=(0,6), xlabel = "p-value", ylabel = "Density")
33
34 plot!([alpha,alpha],[0,6],
35         c=:black, ls=:dash, label=L"\alpha")
```

```
Power under H0:  0.049598
Power under H1A: 0.134274
Power under H1B (mu's farther apart): 0.281904
```

In lines 4 to 10, the function `pval()` is defined. It is similar to the `tStat()` function from Listings 7.17 and 7.18, but includes the extra line 9 which calculates the p-value from test statistic of line 8. Note the use of the `ccdf()` function. After defining the parameters in a way similar to the previous listings, we create arrays of simulated p-values in lines 15–17. These are then used to estimate the power in each case using our function `estPwr()` from line 20. The remainder of the code creates Figure 7.10.

Chapter 8

Linear Regression and Extensions

We now explore *regression analysis*, one of the most popular statistical techniques used in practice. We focus on *linear regression*. The key idea is to consider a *dependent variable* Y and see how it is affected by one or more *independent variables*, typically denoted by X. That is, regression analysis considers how X affects Y, and builds models for predicting future values of Y given future observed or postulated values of X. To build such models, we utilize collected (x, y) data and assume that future situations are resembled by the data at hand.

The variable Y is sometimes called the *response variable* or *dependent variable*, and similarly X is the *explanatory variable* or *independent variable*. Often there is a vector of explanatory variables, still denoted X. In the context of machine learning, we often call the elements of X or transformations of them, *features*. That is, X is the *feature vector* (or scalar) that is used to predict the value of Y.

When considering Y and X as random variables (with X possibly vector-valued), the phrase "*regression* of Y on X" signifies the *conditional expectation* of Y given an observed value of X, say $X = x$. That is, one may stipulate that at onset, both X and Y are random, and given some observed value of X, denoted by x, the regression is then given by

$$\hat{y} = \mathbb{E}[Y \mid X = x]. \tag{8.1}$$

Here, $\hat{y}$ is a *predictor* of the dependent variable Y, given an observation of the independent variable X. We can then say $\hat{y}(x)$ is the predicted value and use any input argument x for the prediction.

The simplest regression model for scalar X assumes that the regression function is affine (i.e. linear with an intercept term). That is,

$$\hat{y} = \mathbb{E}[Y \mid X = x] = \beta_0 + \beta_1 x.$$

Here, β_0 and β_1 represent the intercept and slope, respectively. In this case, a typical statistical model is to assume X is non-random and set,

$$Y = \beta_0 + \beta_1 X + \varepsilon,$$

where ε is considered as a noise term. In the basic case, ε is taken as a normally distributed random variable independent of everything else, with a variance that does not depend on X.

Y. Nazarathy and H. Klok, *Statistics with Julia*, Springer Series in the Data Sciences,
https://doi.org/10.1007/978-3-030-70901-3_8

A widely used method for finding estimates for β_0 and β_1 is *least squares estimation* or *least squares* for short. Given a series of observation tuples, $(x_1, y_1), \ldots, (x_n, y_n)$, which can be viewed as a "cloud of points", the least squares method finds the so-called "line of best fit", $\hat{y} = \hat{\beta}_0 + \hat{\beta}_1 x$, where $\hat{\beta}_0$ and $\hat{\beta}_1$ are estimates of β_0 and β_1 that are assumed to exist but are unknown. The estimates minimize a least squares cost, also known as a *loss function.*

The ideas of regression that we cover here span a broad spectrum of fields including engineering, statistics, data science, machine learning, and many aspects of artificial intelligence. Hence, in different domains, there is often a tradition to focus on different aspects of regression. For example, from an engineering or a machine learning perspective, one often focuses on the mechanics of least squares without giving much consideration to modeling assumptions and model interpretation. In such fields, the key is to create models based on *training data* with the aim making good predictions. As opposed to that, a classical statistical perspective applied to science (e.g. crop-science), often merits considering model assumptions in depth, so as to validate parameter interpretation and test for cause-and-effect relationships. Also, in machine learning, data science, and artificial intelligence, one may attempt to use advanced regression (or "regression like") models to create automated systems that predict well. Such usage of regression is covered partly towards the end of this chapter as we introduce GLM (*Generalized Linear Models*). Such a machine learning approach is also covered more broadly in the next chapter where we consider several methods from *supervised learning* dealing with both regression and classification problems.

As there are various approaches to considering regression, we try to strike a balance between them in the current presentation. For example, in Section 8.1, we focus on the many mechanistic aspects of least squares without considering any model assumptions. This is the basic engineering viewpoint. As opposed to that, Section 8.2 presents the basic linear regression model with one variable and its full statistical analysis including hypothesis testing, confidence intervals, and checking of model assumptions. Then subsequent sections progress into the rich world of vector-valued X, where we try to introduce the main concepts in a way that is useful both for traditional statisticians as well as machine learning professionals. In Section 8.3, we depart from simple linear regression and extend to vector-valued X. In Section 8.4 we show how extra features could be incorporated into the model when dealing with non-linear aspects, categorical variables, and interaction effects. Then in Section 8.5, we touch the complicated area of *model selection.* In Section 8.6, we discuss logistic regression and other generalized linear models without considering many statistical aspects, but rather taking a simple hands-on machine learning perspective. We then close with a brief taste of how to apply regression to time series in Section 8.7.

From a software perspective, the key tool used in this chapter is the Julia `GLM.jl` package, standing for "Generalized Linear Models". It is used for "linear models" but also allows for GLM. The reader should keep in mind that other Julia packages may also be suitable for regression including `Flux.jl`, `MultivariateStats.jl`, and `Lasso.jl`. The first two are used in the next chapter, while `Lasso.jl` is used briefly in this chapter in the context of model selection.

Since regression analysis is a broad subject one can spend much time exploring details. This chapter only presents key concepts and further exploration may be carried out by considering several additional textbooks. For an elementary introduction in the context of scientific applications, we recommend [B10]. Then a much more comprehensive classical treatment is in [M17]. More modern approaches are in [EH16] as well as [HTF01]. Finally, we also recommend Chapter 5 of [KBTV19].

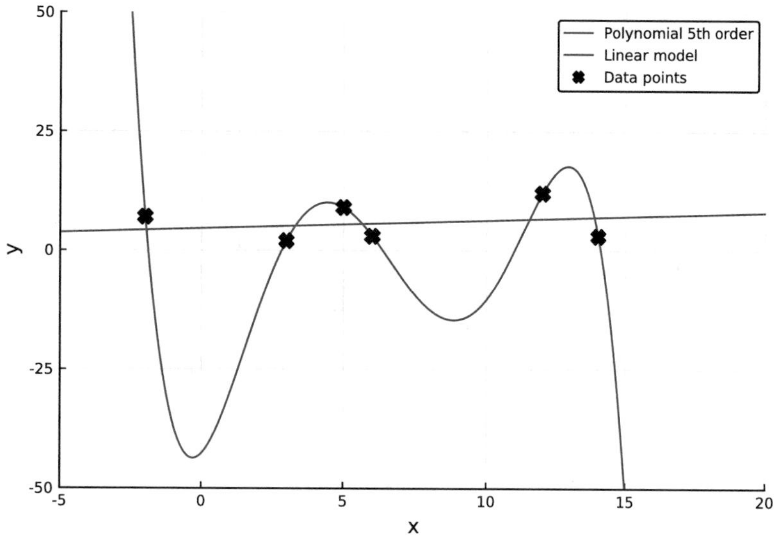

Figure 8.1: Cloud of point fitting via a fifth-degree polynomial, and a first-degree polynomial (linear model). Although the higher order polynomial fits the data perfectly, one could argue it is not better in practice

8.1 Clouds of Points and Least Squares

Consider a sequence of observations $(x_1, y_1), \ldots, (x_n, y_n)$, which when plotted on the Cartesian plane yield a *cloud of points*. If we assume that a functional relationship exists between x and y, such as $y = f(x)$, the first goal is to use these points to estimate the function $f(\cdot)$.

A classic non-statistical way of obtaining $f(\cdot)$ assumes that the observations exactly follow $y_i = f(x_i)$ for every i. This requires that there are no two distinct y values which share the same x value. One possible assumption is that $f(\cdot)$ is a polynomial of order $n-1$, where n is the number of points. Based on this assumption, *polynomial interpolation* can be carried out. This involves seeking the coefficients, $c_0, \ldots, c_{n-1}$ of the polynomial,

$$f(x) = c_{n-1}x^{n-1} + \ldots + c_2x^2 + c_1x + c_0.$$

The coefficients can be found by constructing a *Vandermonde matrix* as shown in the left hand side of (8.2), and then solving for the coefficients $(c_0, \ldots, c_{n-1})$ using,

$$\begin{bmatrix} 1 & x_1 & x_1^2 & \cdots & x_1^{n-1} \\ 1 & x_2 & x_2^2 & \cdots & x_2^{n-1} \\ 1 & x_3 & x_3^2 & \cdots & x_3^{n-1} \\ \vdots & \vdots & \vdots & \ddots & \vdots \\ 1 & x_n & x_n^2 & \cdots & x_n^{n-1} \end{bmatrix} \begin{bmatrix} c_0 \\ c_1 \\ c_2 \\ \vdots \\ c_{n-1} \end{bmatrix} = \begin{bmatrix} y_1 \\ y_2 \\ y_3 \\ \vdots \\ y_n \end{bmatrix}. \tag{8.2}$$

This is a linear system of equations with n equations and n unknowns.

As an example consider the 6 data points shown in Figure 8.1. In this figure, a fifth-degree polynomial is fit to these points by solving (8.2). This polynomial perfectly fits the data points. However, one can argue that a linear model may be better because it will be able to better accommodate future observations that will surely deviate from the-fifth degree polynomial. By requiring every point to agree with $f(\cdot)$ exactly, the approach often results in an "over-fit" model. This phenomenon, known as *over fitting*, is common throughout data analysis. It is often possible to find a model that describes the observed data exactly, but when new observations are made performs poorly. Over-fit models are often too complicated for the scenario at hand.

We create Figure 8.1 in Listing 8.1 where polynomial interpolation is carried out by constructing a Vandermonde matrix which is then used to solve for the coefficients c_i. The resulting polynomial fit to the data (with coefficients rounded to two significant digits) is,

$$y = -0.01x^5 + 0.26x^4 - 2.73x^3 + 8.96x^2 + 6.2x - 42.72.$$

This polynomial is then plotted against the line,

$$y = 4.58 + 0.17x$$

and displayed in Figure 8.1. The parameters of this line were obtained via the least squares method, which is described in the next subsection. One can see that although the polynomial fits the data exactly, it is much more complicated than the line and appears to be overfitting the data. That is, if an additional seventh observation was recorded then the straight line may be a far better predictor.

Listing 8.1: Polynomial interpolation vs. a line

```
1   using Plots; pyplot()
2
3   xVals = [-2,3,5,6,12,14]
4   yVals = [7,2,9,3,12,3]
5   n = length(xVals)
6
7   V = [xVals[i+1]^(j) for i in 0:n-1, j in 0:n-1]
8   c = V \ yVals
9   xGrid = -5:0.01:20
10  f1(x) = c'*[x^i for i in 0:n-1]
11
12  beta0, beta1 = 4.58, 0.17
13  f2(x) = beta0 + beta1*x
14
15  plot(xGrid,f1.(xGrid), c=:blue, label="Polynomial 5th order")
16  plot!(xGrid,f2.(xGrid),c=:red, label="Linear model")
17  scatter!(xVals,yVals,
18          c=:black, shape=:xcross, ms=8,
19          label="Data points", xlims=(-5,20), ylims=(-50,50),
20          xlabel = "x", ylabel = "y")
```

In line 7, the matrix `V` is defined, which represents the Vandermonde matrix as shown in (8.2). In line 8, the \ operator is used to solve the system of equations shown in (8.2), returning the coefficients as an array, which is then stored as `c`. In line 10, the function `f1()` is defined, which uses the inner product by multiplying `c'` with an array of monomials, and describes our polynomial of order `n-1`. Line 13 the function `f2()` is defined, which describes our linear model. Note the use of hard coded coefficients here. Note the use of mapping `f1()` and `f2()` over `xGrid` via the broadcast operator "." in lines 15 and 16.

Fitting a Line Through a Cloud of Points

A line of the form $y = f(x)$, with the function $f(x) = \beta_0 + \beta_1 x$, may be a sensible model. However, it typically cannot be interpolated. It is obvious that unless the points lie exactly on a line, the desire of $y_i = f(x_i)$ will not be satisfied for most of the points. The question then arises of how to best select β_0 and β_1? That is, how to fit a line through a cloud of points?

The typical approach is to select β_0 and β_1 such that the deviations between y_i and $\hat{y}_i$ are minimized, where $\hat{y}_i$ is the model prediction for the given x_i. That is,

$$\hat{y}_i = \beta_0 + \beta_1 x_i.$$

There is no universal way for measuring such deviations, instead there are several different measures. Here, the two most common measures, commonly known as *loss functions*, are presented. The L_2 *norm* (or *Euclidean norm*) based loss function, and the L_1 *norm* based loss function, which are defined, respectively, as

$$L^{(2)}(\beta_0, \beta_1) := \sum_{i=1}^{n} (\hat{y}_i - y_i)^2 \qquad \text{and} \qquad L^{(1)}(\beta_0, \beta_1) := \sum_{i=1}^{n} \big|\hat{y}_i - y_i\big|. \tag{8.3}$$

Both of these are based on the elements $(e_1, \dots, e_n)$, where $e_i := \hat{y}_i - y_i$, and these are also known as the *errors* or *residuals*. The first is related to the L_2 norm of these values and the second is related to the L_1 norm of these values. In general, given a vector $w = (w_1, \dots, w_n)$, the L_2 norm, often denoted by $||w||_2$ is given by

$$||w||_2 = \sqrt{\sum_{i=1}^{n} w_i^2}. \tag{8.4}$$

Note that if no ambiguity is present, the notation $||w||$ is used instead of $||w||_2$. The L_1 norm of w is given via

$$||w||_1 = \sum_{i=1}^{n} |w_i|. \tag{8.5}$$

Both of these are special cases of the L_q norm where $q \geq 1$. This norm (sometimes denoted L_p) is given by

$$||w||_q = \Big(\sum_{i=1}^{n} |w_i|^q\Big)^{1/q}.$$

Going back to (8.3) we see that $L^{(1)}$ is exactly the L_1 norm of $(e_1, \dots, e_n)$ and $L^{(2)}$ is the square of the L_2 norm of $(e_1, \dots, e_n)$. If $L^{(1)}$ or $L^{(2)}$ are zero, then $y_i = f(x_i)$ for all i exactly. Further, having any of these norms as close to zero as possible ensures a "close fit" between the points and the fitted line $f(x) = \beta_0 + \beta_1 x$. Hence, estimates of these coefficients can be obtained via the optimization (minimization) problem

$$\min_{\beta_0, \beta_1} L^{(\ell)}(\beta_0, \beta_1), \tag{8.6}$$

where ℓ is either 1 or 2. Thus, fitting a line through a cloud of points first requires choosing $\ell = 1$ or $\ell = 2$ and then finding β_0 and β_1 so as to optimize (8.6).

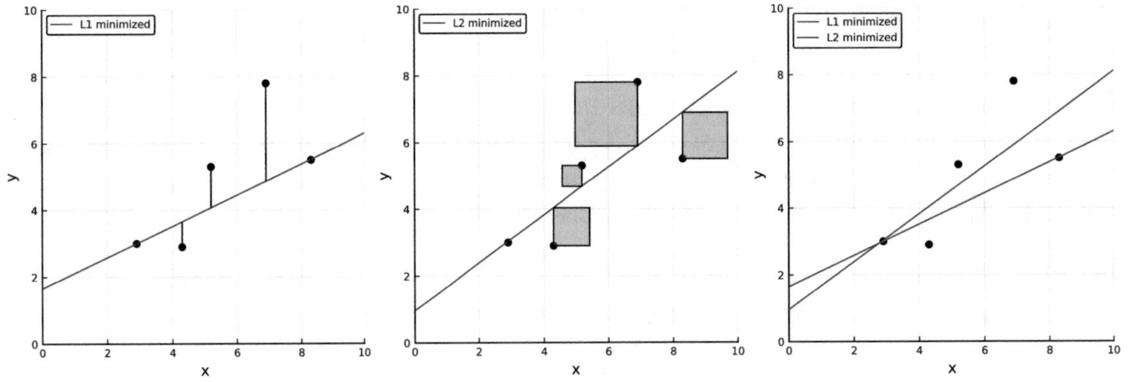

Figure 8.2: The blue line minimizes $L^{(1)}$ (the sum of the lengths of the black lines) and the red line minimizes $L^{(2)}$ (the sum of the areas of the squares). .

In practice, the most common and simplest method is to focus on $\ell = 2$. This is due to the analytic tractability of the L_2 norm and the minimization of $L^{(2)}$. Minimization of $L^{(2)}$ is presented via least squares later in this section. However, at this point, both $\ell = 1$ and $\ell = 2$ are presented, and we carry out the optimization for (8.6) via naive Monte Carlo guessing. This is to understand the differences between $\ell = 1$ and $\ell = 2$ qualitatively. The remainder of the chapter then focuses almost exclusively on $\ell = 2$, with $\ell = 1$ and other loss measures left to further reading.

Figure 8.2 presents $\ell = 1$ and $\ell = 2$ minimization for a small dataset with 5 points. The right most plot presents the lines for both objectives (minimized loss functions) simultaneously. As can be observed, the lines are significantly different for the same dataset. This is because each objective aims at minimizing something different. The $\ell = 1$ case, presented also in the left most plot, aims to minimize the sum of lengths from each point to the line. The $\ell = 2$ case, also presented in the middle plot, aims to minimize the total *sum of squares.*

The $\ell = 1$ case is generally more robust to effects from extreme or "far away" observations. As opposed to that the $\ell = 2$ case is generally more sensitive to such outliers. This is evident in Figure 8.2 by considering the effect of the point $(6.9, 7.8)$, i.e. the point with highest y value. That point 'pulls' the red $\ell = 2$ line much higher than it pulls the $\ell = 1$ line. This is because of the quadratic loss of $\ell = 2$ as opposed to the linear loss of $\ell = 1$. Hence, from this perspective, one can argue that $\ell = 1$ minimization is superior. However, $\ell = 2$ is very natural from a linear algebraic point of view and has simple and beautiful solutions via least squares, covered in the next subsection and used in most of this chapter.

The figure is generated by Listing 8.2 where uniform random values over the grid $[0, 5] \times [0, 5]$ are trialled for β_0 and β_1. For each pair of values, the $L^{(1)}$ and $L^{(2)}$ costs (loss function values) are compared to their previous values, and, if the costs are lower, then the corresponding values for β_0 and β_1 adopted. By repeating this process N times, for large N, we aim to obtain coefficient values that closely approximate those that minimize $L^{(1)}$ and $L^{(2)}$. This method is by no means an efficient method. We simply present it for pedagogical purposes. Note that for clarity in this example, β_0 and β_1 are denoted in the code via `alpha` and `beta`, respectively. Also note that the values that minimize (8.6) also minimize monotonic transformations of those objectives (such as square root over positive numbers). Hence, we can compute the L_2 norm (line 19) instead of the sum of squares $L^{(2)}$ and still find the minimizer.

Listing 8.2: L1 and L2 norm minimization by Monte Carlo guessing

```
using DataFrames, Distributions, Random, LinearAlgebra, CSV, Plots; pyplot()
Random.seed!(0)

data = CSV.read("../data/L1L2data.csv")
xVals, yVals, n = data.X, data.Y, size(data)[1]

N = 10^6
alphaMin, alphaMax, betaMin, betaMax = 0, 5, 0, 5

alpha1, beta1, alpha2, beta2, bestL1Cost, bestL2Cost = 0.0,0.0,0.0,0.0,Inf,Inf
for _ in 1:N
    rAlpha,rBeta=rand(Uniform(alphaMin,alphaMax)),rand(Uniform(betaMin,betaMax))
    L1Cost = norm(rAlpha .+ rBeta*xVals - yVals,1)
    if L1Cost < bestL1Cost
        global alpha1 = rAlpha
        global beta1 = rBeta
        global bestL1Cost = L1Cost
    end
    L2Cost = norm(rAlpha .+ rBeta*xVals  - yVals)
    if L2Cost < bestL2Cost
        global alpha2 = rAlpha
        global beta2 = rBeta
        global bestL2Cost = L2Cost
    end
end

println("L1 line: $(round(alpha1,digits = 2)) + $(round(beta1,digits = 2))x")
println("L2 line: $(round(alpha2,digits = 2)) + $(round(beta2,digits = 2))x")

d = yVals - (alpha2 .+ beta2*xVals)
rectangle(x, y, d) = Shape(x .- [0,d,d,0,0], y .- [0,0,d,d,0])

p1 = scatter(xVals,yVals, c=:black, ms=5, label="")
p1 = plot!([0,10],[alpha1, alpha1 .+ beta1*10], c=:blue,label="L1 minimized")
for i in 1:n
    x,y = xVals[i],yVals[i]
    p1 = plot!([x, x], [y, alpha1 .+ beta1*x],color="black", label="")
end

p2 = scatter(xVals,yVals, c=:black, ms=5, label="")
p2 = plot!([0,10],[alpha2, alpha2 .+ beta2*10],c=:red,label="L2 minimized")
for i in 1:n
    x,y = xVals[i],yVals[i]
        p2 = plot!(rectangle(x,y,d[i]), fc=:gray, fa=0.5, label="")
end

p3 = scatter(xVals,yVals, c=:black, ms=5, label="")
p3 = plot!([0,10],[alpha1, alpha1 .+ beta1*10], c=:blue, label="L1 minimized")
p3 = plot!([0,10],[alpha2, alpha2 .+ beta2*10], c=:red, label="L2 minimized")

plot(p1, p2, p3, layout = (1,3),
        ratio=:equal, xlims=(0,10), ylims=(0,10),
        legend=:topleft, size=(1200, 400),
        xlabel = "x", ylabel = "y")
```

```
L1 line: 1.65 + 0.47x
L2 line: 0.96 + 0.72x
```

Lines 4–5 set up the data in the arrays `xVals`, `yVals`, each of size `n`. Lines 7–8 set up the randomized optimization parameters where `N` is the number of repetitions and the other variables define the bounding box for search. Lines 10–25 constitute the optimization. In each iteration of the loop, `bestL1Cost` and `bestL2Cost` are adjusted if lower costs are found. In that case, the optimizers are also set (lines 15–16 for $\ell = 1$ and lines 21–22 for $\ell = 2$). Note the use of `norm()` in line 13 with the second parameter equalling `1` indicating the L_1 norm. Similarly, `norm()` in line 19 without the second parameter has the default behavior of the L_2 norm. The best parameters found are printed by lines 27–28. The remainder of the code creates Figure 8.2. Note the use of `Shape()` in line 31 for creating objects used for the center plot. We use this constructor to create our own function, `rectangle()`. It is then used in line 44 for each data point.

Least Squares

Having explored the fact that there are multiple ways to fit a line through a cloud of points, we now focus on the most common and mathematically simple way. This is the method of *least squares.* It involves finding the values of β_0 and β_1 that minimize $L^{(2)}$. Observe that the loss function $L^{(2)}$ can be written in several alternative ways, such as

$$L^{(2)} = \sum_{i=1}^{n} e_i^2 = \sum_{i=1}^{n} (y_i - \hat{y}_i)^2 = \sum_{i=1}^{n} (y_i - \beta_0 - \beta_1 x_i)^2 = ||y - A\beta||^2. \tag{8.7}$$

In the last representation, we use the L_2 norm (denoted here simply via $||\cdot||$) and square it. This uses the fact that for $e = (e_1, \ldots, e_n)$,

$$||e||^2 = \sum_{i=1}^{n} e_i^2.$$

In our case, $e = y - A\beta$ is determined via the (column) vector of coefficients $\beta = (\beta_0, \beta_1)$, the (column) vector of y observations $y = (y_1, y_2, \ldots, y_n)$, and the *design matrix* $A \in \mathbb{R}^{n\times 2}$. The design matrix in this case is

$$A = \begin{bmatrix} 1 & x_1 \\ 1 & x_2 \\ \vdots & \vdots \\ 1 & x_n \end{bmatrix}. \tag{8.8}$$

Later in Section 8.3 when we extend to $p \geq 2$ variables, the vector of coefficients becomes $\beta = (\beta_0, \beta_1, \ldots, \beta_p)$ and the $n \times (p+1)$ design matrix is extended as well (see equation (8.40) below).

In general, once we see that the loss function can be represented via $||y - A\beta||^2$, we may also consider *arbitrary least squares problems*, not directly related to fitting a line through a cloud of points. For this, just consider any $n \times (p+1)$ matrix A and y as any n-vector. There is no need to stick to the form (8.8) (or the more general (8.40)) for A.

The goal is now to minimize $||y - A\beta||^2$ by finding the best vector $(\beta_0, \beta_1, \ldots, \beta_p)$. This can be thought of as an attempt to "solve" the system of equations,

$$A\beta = y. \tag{8.9}$$

If the vector y happens to be in the column space (linear combinations of columns of A) then a solution β to (8.9) exists. However if not, we might as well try to minimize the (normed) difference between the left hand side and the right hand side of (8.9). Hence we seek minimization of $||y - A\beta||$ or equivalently of $||y - A\beta||^2$.

We typically assume that A is a "skinny" matrix meaning that $p + 1 \leq n$. It turns out that the theory of least squares is simplest when A is a *full rank* matrix, i.e. has rank $p + 1$ and hence all the columns are linearly independent. In this case, the *Grahm matrix*, $A^\top A$, is *non-singular* and the *Moore-Penrose pseudo-inverse* exists. That important matrix is defined as follows:

$$A^\dagger := (A^\top A)^{-1} A^\top. \tag{8.10}$$

Then, it follows from the theory of linear algebra that

$$\hat{\beta} = A^\dagger y, \tag{8.11}$$

minimizes $L^{(2)}$. Hence, the simple and neat matrix formulae (8.10) and (8.11) elegantly solve the least squares problem. We now present a few alternative linear algebraic forms. These may be skipped on a first reading and you may also consult [BL18] and [S18] for the linear algebra background, extensions, and derivations of these results.

An alternative representation of least squares can be shown by considering the *QR factorization* of A, and denoting $A = QR$, where Q is a matrix of orthonormal columns and R is an upper triangular matrix. Again, consult [BL18] or similar texts for details. In this case, it is easy to see that

$$A^\dagger = R^{-1} Q^\top. \tag{8.12}$$

Further, even if A is not full rank, we may compute the pseudo-inverse by considering the *singular value decomposition* (SVD) of A. Here, $A = U\Sigma V^\top$, where U is a $n \times n$ orthogonal matrix, Σ is an $n \times (p+1)$ matrix with non-zero elements only on the diagonal, and V is a $(p+1) \times (p+1)$ orthogonal matrix. Consult [S18] or similar texts for details. In such a case

$$A^\dagger = V\Sigma^+ U^\top, \tag{8.13}$$

where Σ^+ is contains the reciprocals of the non-zero (diagonal) elements of Σ and has 0 values elsewhere. The usefulness of (8.13) as opposed to the representations (8.12) and (8.10) is that it does not require A to be full rank.

Practically, for solving least squares, it is easiest to make use of the powerful Julia *backslash* operator, "\". This notation, popularized by precursor languages such as MATLAB, treats $Ab = y$ as a general system of equations and allows us to write `b = A \y` as the "solution" for b even if no solution exists. If A happens to be square and non-singular then analytically (however, not always numerically), it is equivalent to the Julia expression `b = inv(A)*y`. However, from a numerical and performance perspective, the use of backslash is generally preferred as it calls upon dedicated routines from LAPACK. This is the linear algebra package, initially bundled into MATLAB, but also employed by Julia and a variety of other scientific computing systems. More importantly for our case, when A is skinny and full rank, evaluation of `b = A \y`, produces the least squares solution. That is, it sets, $b = A^\dagger y$ in a numerically efficient manner.

There are many ways to derive the optimal $\hat{\beta}$ of (8.11) based on $A^\dagger$. A straightforward approach that also sheds light on the *gradient descent* method is to consider the representation,

$$L^{(2)} = \sum_{i=1}^{n} \Big(\sum_{j=1}^{p+1} A_{ij}\beta_j - y_i \Big)^2.$$

By treating $L^{(2)}$ as a function of $\beta_1, \ldots, \beta_{p+1}$, we can evaluate its gradient, by calculating the derivative with respect to each β_k for $k = 1, \ldots, p+1$, as follows:

$$\begin{aligned}
\frac{\partial L^{(2)}}{\partial \beta_k} &= 2\sum_{i=1}^{n} A_{ik}\Big(\sum_{j=1}^{p+1} A_{ij}\beta_j - y_i \Big) \\
&= 2\sum_{i=1}^{n} (A^\top)_{ki}(A\beta - y)_i \\
&= \Big(2A^\top (A\beta - y) \Big)_k .
\end{aligned}$$

Putting these components together we see that the gradient is,

$$\nabla L^{(2)} = 2A^\top (A\beta - y). \tag{8.14}$$

This $p+1$-dimensional vector has A and y fixed and for each $\beta = (\beta_1, \ldots, \beta_{p+1})$ points in the direction of steepest ascent of $L^{(2)}(\beta)$. One use of (8.14) is for first-order optimization conditions. Equating it to zero, we obtain the so-called *normal equations*,

$$A^\top A\beta = A^\top y. \tag{8.15}$$

It can be shown that if A is full rank then there is a unique solution to (8.15) and this yields the least squares solution to the optimization problem. Further if A is not full rank, then there are many solutions to (8.15), however, the solution with minimal norm $||\beta||$ is the least squares solution given by $A^\dagger y$ with $A^\dagger$ given by (8.13) based on SVD. Hence, in any case, the normal equations are central to least squares.

The gradient expression (8.14) is not only needed for obtaining the normal equations (8.15). It also allows finding the least squares solution via *gradient descent.* This algorithm running over iterations indexed by $t = 0, 1, 2, \ldots$, works by taking small steps in the direction opposite to the gradient. That is, it traverses "downhill" each time trying to go in the steepest direction down. Gradient descent is central to machine learning and is covered in slightly greater depths in the next chapter. However, we now briefly describe how it can be used for least squares.

In its simplest form, steps are determined by a fixed $\eta > 0$ called the *learning rate.* After some arbitrary initialization with vector $\beta(0)$, the next vector $\beta(1)$ is obtained via.

$$\beta(t+1) = \beta(t) - \eta \nabla L^{(2)}\big(\beta(t)\big). \tag{8.16}$$

The algorithm then repeats (8.16) until $\beta(t+1)$ and $\beta(t)$ are close or until some similar stopping criteria is met. With some fixed $\varepsilon > 0$, we can specify the stopping criteria as

$$||\beta(t) - \beta(t+1)|| < \varepsilon, \tag{8.17}$$

or some similar threshold. The final $\beta(t+1)$ is used as the least squares estimate and if η and ε are well chosen then it often closely approximates the optimal β. Gradient descent is very popular for large least squares problems such as very large p in the order of thousands or more. This is common in machine learning, however, in classic applications of statistics it isn't often used.

We now return to $p = 2$ and the specific A of (8.8) used to fit a line through a cloud of points on the plane. Note that, instead of denoting β as (β_1, β_2), we use the more common notation $\beta = (\beta_0, \beta_1)$, where we also use $\hat{\beta}_i$ to indicate an estimator for β_i. In this case, the normal equations of (8.15) can be shown to read as

$$n\hat{\beta}_0 + \hat{\beta}_1 \sum_{i=1}^{n} x_i = \sum_{i=1}^{n} y_i,$$
$$\hat{\beta}_0 \sum_{i=1}^{n} x_i + \hat{\beta}_1 \sum_{i=1}^{n} x_i^2 = \sum_{i=1}^{n} y_i x_i.$$

The solution to these equations is the least squares estimators $\hat{\beta}_0$ and $\hat{\beta}_1$. Using the sample means, $\overline{x}$ and $\overline{y}$ the estimators can be represented as

$$\hat{\beta}_0 = \overline{y} - \hat{\beta}_1 \overline{x}, \qquad \hat{\beta}_1 = \frac{n \sum_{i=1}^{n} y_i x_i - \left(\sum_{i=1}^{n} y_i\right)\left(\sum_{i=1}^{n} x_i\right)}{n \sum_{i=1}^{n} x_i^2 - \left(\sum_{i=1}^{n} x_i\right)^2}. \tag{8.18}$$

In a different format, the following quantities are also commonly used:

$$S_{xx} = \sum_{i=1}^{n} (x_i - \overline{x})^2, \qquad S_{xy} = \sum_{i=1}^{n} (y_i - \overline{y})(x_i - \overline{x}). \tag{8.19}$$

These yield an alternative formula for $\hat{\beta}_1$,

$$\hat{\beta}_1 = \frac{S_{xy}}{S_{xx}}. \tag{8.20}$$

Finally, one may also use the sample correlation, and sample standard deviations to obtain $\hat{\beta}_1$,

$$\hat{\beta}_1 = \text{corr(x,y)} \frac{\text{std}(y)}{\text{std}(x)}. \tag{8.21}$$

Hence, in summary, we see that there are many alternative ways of solving the least squares problem. In Listing 8.2, each of the representations covered above is used to obtain the least squares estimate for the same dataset used in Listing 8.2. The purpose is to illustrate that a variety of alternative methods, representations. and commands can be used to solve least squares. In total, 11 different methods are used, labeled `A` to `K`, and each one obtains the same estimates for β_0 and β_1 from the data. Approaches `A` and `B` use the formulas above for the case of $p = 2$ (simple linear regression). Approaches `C`, `D`, `E`, `F`, `G`, and `H` work with either the normal equations, (8.15), or the pseudo inverse, $A^{\dagger}$. Approach `I` executes a gradient descent algorithm (see code comments below). Finally, `J` and `K` call upon the `GLM` statistical package, which is covered in more detail in Section 8.2. From the output, it can be seen that all approaches yield the same estimates.

Listing 8.3: Computing least squares estimates

```
using DataFrames, GLM, Statistics, LinearAlgebra, CSV
data = CSV.read("../data/L1L2data.csv")
xVals, yVals = data[:,1], data[:,2]
n = length(xVals)
A = [ones(n) xVals]

# Approach A
xBar, yBar = mean(xVals),mean(yVals)
sXX, sXY = ones(n)'*(xVals.-xBar).^2 , dot(xVals.-xBar,yVals.-yBar)
b1A = sXY/sXX
b0A = yBar - b1A*xBar

# Approach B
b1B = cor(xVals,yVals)*(std(yVals)/std(xVals))
b0B = yBar - b1B*xBar

# Approach C
b0C, b1C = A'A \ A'yVals

# Approach D
Adag = inv(A'*A)*A'
b0D, b1D = Adag*yVals

# Approach E
b0E, b1E = pinv(A)*yVals

# Approach F
b0F, b1F = A\yVals

# Approach G
F = qr(A)
Q, R = F.Q, F.R
b0G, b1G = (inv(R)*Q')*yVals

# Approach H
F = svd(A)
V, Sp, Us = F.V, Diagonal(1 ./ F.S), F.U'
b0H, b1H = (V*Sp*Us)*yVals

# Approach I
eta, eps = 0.002, 10^-6.
b, bPrev = [0,0], [1,1]
while norm(bPrev-b) >= eps
    global bPrev = b
    global b = b - eta*2*A'*(A*b - yVals)
end
b0I, b1I = b[1], b[2]

# Approach J
modelJ = lm(@formula(Y ~ X), data)
b0J, b1J = coef(modelJ)

# Approach K
modelK = glm(@formula(Y ~ X), data, Normal())
b0K, b1K = coef(modelK)
println(round.([b0A,b0B,b0C,b0D,b0E,b0F,b0G,b0H,b0I,b0J,b0K],digits=3))
println(round.([b1A,b1B,b1C,b1D,b1E,b1F,b1G,b1H,b1I,b1J,b1K],digits=3))
```

```
[0.945, 0.945, 0.945, 0.945, 0.945, 0.945, 0.945, 0.945, 0.944, 0.945, 0.945]
[0.716, 0.716, 0.716, 0.716, 0.716, 0.716, 0.716, 0.716, 0.717, 0.716, 0.716]
```

Observe that in line 5, we construct the design matrix `A` according to (8.8). Lines 8–11 implement (8.18) using (8.20). In line 9, we use an inner product with `ones(n)` for the first element, `sXX`. Then, for the second element, we use the `dot()` function which takes the inner product of both its arguments. This is simply to see two different ways of executing inner products. Lines 14–15 implement (8.21). This uses the `cor()` and `std()` functions from `Statistics`. Line 18 is a direct solution of the normal equations (8.15). Here, we use the backslash operator to solve the equations. The expression `A'A \ A'yVals` is an array with the solution. But we transform it into the individual elements, `b0C` and `b1C`. Lines 21 and 22, do the same thing, by finding $A^\dagger$, denoted `Adag` in line 21. Then it applied (as a linear transformation) to `yVals` in line 22. Line 25 shows the use of the `pinv()` function that computes $A^\dagger$ directly. Line 28 computes $\hat{\beta}$ by using the backslash "\" operator. This delegates the exact numerical aspect to Julia, as opposed to forcing it directly as in the previous lines. It is generally the preferred method. Lines 31–33 use QR-factorization. In line 31, the object `F` is assigned the result of `qr()` from `LinearAlgebra`. Then the specific `Q` and `R` matrices are obtained from that object via `F.Q` and `F.R` respectively. Line 33 then implements (8.12), representing $A^\dagger$ via the code `inv(R)*Q'`. Lines 36–38 use the SVD formula (8.13). First `svd()` from `LinearAlgebra` is used in line 36. Then the specific matrices are pull from the decomposition in line 37. Observe the use of `Diagonal()` to create a diagonal matrix with the reciprocals of the ordered singular values, `F.S`. Lines 41–47 implement a gradient descent. The learning rate η and the stopping threshold ε are set in line 41. The initial guess is in line 42. The iteration then proceeds until (8.17) is satisfied. Lines 50–51 and 54–55 use the `lm()` and the `glm()` functions from the `GLM` package. In both cases, the result is in a model object, denoted `modelJ` and `modelK` in the code. Then the `coef()` function retrieves the estimates from the model objects. We elaborate more on `GLM` in the next section.

8.2 Linear Regression with One Variable

Having explored the notion of the line of best fit and least squares in the previous section, we now move onto the most basic statistical application: *linear regression with one variable* also known as *simple linear regression*. This is the case where we assume the following relationship between the random variables Y_i and X_i:

$$Y_i = \beta_0 + \beta_1 X_i + \varepsilon_i. \tag{8.22}$$

Here, ε_i is normally distributed with zero mean and variance σ^2 and is assumed independent across observation indexes i. By assuming such a specific form of the error term, ε, we are able to say more about the estimates of β_0 and β_1 than we did in the previous section. While least squares procedures gave us a solid way to obtain $\hat{\beta}_0$ and $\hat{\beta}_1$, by themselves, such procedures don't give any information about the reliability of the estimates. Hence, by assuming a model such as (8.22), we can go further. Specifically, we can carry out statistical inference for the unknown parameters, β_0, β_1, and σ^2. This includes confidence intervals and hypothesis tests.

In this *linear regression* context, while we denote Y_i and X_i as random variables in (8.22), we assume that the Y values are random and the X values are observed and not random. We then obtain estimates of β_0, β_1 and σ^2 given the sample, $(x_1, y_1), \dots, (x_n, y_n)$. The x_i coordinates, assumed deterministic, are sometimes called the *design*. The y_i coordinates are assumed as observations following the relationship in (8.22) given that $X_i = x_i$. Observe that once conditioning on $X_i = x_i$, each Y_i is normally distributed with mean $\beta_0 + \beta_1 x_i$ and variance σ^2.

The standard statistical way to estimate β_0, β_1 and σ^2 is to use maximum likelihood estimation, see Section 5.4. Since ε_i are independent and assumed normally distributed, the likelihood function is

$$L(\beta_0, \beta_1, \sigma^2 \mid y, x) = \prod_{i=1}^{n} \frac{1}{\sqrt{2\pi\sigma^2}} e^{-\frac{(y_i-\beta_0-\beta_1 x_i)^2}{2\sigma^2}}.$$

Now the log-likelihood function evaluates to

$$\ell(\beta_0, \beta_1, \sigma^2 \mid y, x) = -\frac{n}{2}\log 2\pi - \frac{n}{2}\log\sigma^2 - \frac{1}{2\sigma^2}\sum_{i=1}^{n}(y_i - \beta_0 - \beta_1 x_i)^2. \tag{8.23}$$

As it is a function of β_0, β_1, and σ^2, MLE dictates that we maximize it jointly over these three unknown parameters. It can then be shown to be maximized by the same least squares estimates presented in the previous section. That is, imposing a normality assumption on the residuals as in (8.22) remarkably implies that the MLE estimate of the parameters is the least squares estimate given for example via formula, (8.18). Note that if ε_i from (8.22) are not normally distributed, or not independent of any other ε_j, or not with constant variance, then the MLE for β_0 and β_1 is no longer the least squares estimate.

Further, the MLE maximizer for σ^2 is

$$\hat{\sigma}^2 = \frac{1}{n}\sum_{i=1}^{n}(y_i - \hat{\beta}_0 - \hat{\beta}_1 x_i)^2. \tag{8.24}$$

With estimators such as $\hat{\beta}_0$, $\hat{\beta}_1$,and $\hat{\sigma}^2$ at hand, we can now carry out statistical inference for the regression model. The least squares estimators $\hat{\beta}_0$ and $\hat{\beta}_1$ are unbiased estimators for β_0 and β_1, respectively. However it turns out that $\hat{\sigma}^2$ is a slightly biased estimator of σ^2. An unbiased estimator, denoted MSE (for Mean Square Error) is

$$\text{MSE} = \frac{SS_{\text{residuals}}}{n-2}, \qquad \text{with} \qquad SS_{\text{residuals}} = \sum_{i=1}^{n}(y_i - \hat{\beta}_0 - \hat{\beta}_1 x_i)^2. \tag{8.25}$$

The denominator, $n-2$ is called the *degrees of freedom*. It is also sensible to consider $SS_{\text{residuals}}$ in comparison to $SS_{\text{total}} = \sum_{i=1}^{n}(y_i - \bar{y})^2$. The former measures variation of residuals around the regression line and the latter measures total variation of the dependent variable. Sums of squares decompositions hold for linear regression as they do for ANOVA (see Section 7.3). This also motivates computing the quantity "*R squared*", some times called the *coefficient of determination* and defined via

$$R^2 = 1 - \frac{SS_{\text{residuals}}}{SS_{\text{total}}} \in [0, 1]. \tag{8.26}$$

If R^2 is close to 1 then it implies that the residual variation is low, whereas, if R^2 is close to 0, it implies that most of the total variation is due to residuals. Considering R^2 as an index for the tightness of the fit is common practice since it describes the proportion of the variance of Y that is predictable via X. Hence, in summary, when executing linear regression we are presented with numerical values for $\hat{\beta}_0$, $\hat{\beta}_1$, MSE, and R^2.

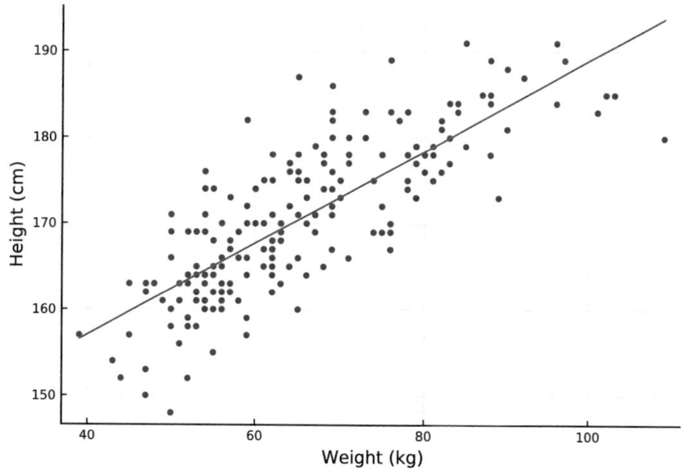

Figure 8.3: A scatter plot of Height vs. Weight with a line of best fit obtained via linear regression.

Using the `GLM` Package

The main Julia package that we use for carrying out linear regression is `GLM.jl` which stands for *Generalized Linear Models.* We describe the "generalized" notion in Section 8.6 and for now use the package for nothing more than the statistical inference of the model (8.22). In fact, we have already briefly used this package in Listing 8.3, lines 50-55, and in the ANOVA example of Section 7.3.

To illustrate usage, in Listing 8.4, we consider the `weightHeight.csv` dataset relating weights and heights of individuals. We carry out linear regression by invoking several alternative functions from `GLM`. These include `lm()` and the analogous use of `fit()` with a `LinearModel` argument. An alternative is to use `glm()` or the analogous use of `fit()` with a `GeneralizedLinearModel` argument together with `Normal()` and `IdentityLink()`. In all cases, we specify a formula via

```
@formula(Height ~ Weight).
```

Such a Julia formula macros (marked by `@`) is implicitly supplied by the package `StatsModels`. The formula indicates that we are seeking a model where `Height`, the dependent Y variable, is represented in terms of `Weight`, the independent X variable. These are the names of the variables (columns) in the dataset. An alternative valid macro is

```
@formula(Height ~ Weight + 1).
```

Here, the `+1` explicitly indicates to add an intercept term to the regression. However, by default it is not needed and is already assumed. If however you wish to carry out the regression without the intercept term then use `-1`.

As can be observed from the output of Listing 8.4, there are only very minor differences when using `lm()` vs. `glm()` (alternatively `fit()` with `LinearModel` vs. `GeneralizedLinearModel`). These have to do with the interpretation of the distribution of the test statistic for checking if $\hat{\beta}_i$ is significantly different from 0. Essentially, when using `lm()` a T-distribution is used and when using `glm()` an normal distribution is used. This slightly affects the p-value and confidence bands for the parameter estimates, but not more. We elaborate on this below.

The outcome of `lm()`, `glm()`, or `fit()` is a model object that can then be used in various ways. A summary of the model can be printed using `println()` or similar. Further, functions such as `deviance()`, `stderror()`, `dof_residual()`, `vcov()`, and `r2()` and most importantly `coef()` can be applied to the model to obtain results from the regression. We use these functions in Listing 8.4. The listing also produces Figure 8.3. Notice that in this listing `lm1` and `lm2` are identical. Similarly, `glm1` and `glm2` are identical. We simply chose to present the `fit()` counterparts so that you see alternative syntax to apply `lm()` and `glm()`.

The key components of the output tables are in the `Estimate` collumn. These are $\hat{\beta}_0$ and $\hat{\beta}_1$ marked via `(Intercept)` and `Weight`, respectively. Interpretation of other important components including `Pr(>|z|)`, the p-value and the confidence bands is described in the examples that follow.

Listing 8.4: Simple linear regression with GLM

```
using DataFrames, GLM, Statistics, CSV, Plots; pyplot()

data = CSV.read("../data/weightHeight.csv")

lm1 = lm(@formula(Height ~ Weight), data)
lm2 = fit(LinearModel,@formula(Height ~ Weight), data)

glm1 = glm(@formula(Height ~ Weight), data, Normal(), IdentityLink())
glm2 = fit(GeneralizedLinearModel,@formula(Height ~ Weight), data, Normal(),
        IdentityLink())

println("***Output of LM Model:")
println(lm1)
println("\n***Output of GLM Model:")
println(glm1)

pred(x) = coef(lm1)'*[1, x]

println("\n***Individual methods applied to model output:")
println("Deviance: ",deviance(lm1))
println("Standard error: ",stderror(lm1))
println("Degrees of freedom: ",dof_residual(lm1))
println("Covariance matrix: ",vcov(lm1))

yVals = data.Height
SStotal = sum((yVals .- mean(yVals)).^2)

println("R squared (calculated in two ways):",r2(lm1),
        ",\t", 1 - deviance(lm1)/SStotal)

println("MSE (calculated in two ways: ",deviance(lm1)/dof_residual(lm1),
        ",\t",sum((pred.(data.Weight) - data.Height).^2)/(size(data)[1] - 2))

xlims = [minimum(data.Weight), maximum(data.Weight)]
scatter(data.Weight, data.Height, c=:blue, msw=0)
plot!(xlims, pred.(xlims),
        c=:red, xlims=(xlims),
        xlabel="Weight (kg)", ylabel="Height (cm)", legend=:none)
```

```
***Output of LM Model:
Height ~ 1 + Weight
Coefficients:
------------------------------------------------------------------------------
              Estimate  Std. Error  t value  Pr(>|t|)  Lower 95%   Upper 95%
------------------------------------------------------------------------------
(Intercept)  135.793     1.95553    69.4404    <1e-99  131.937     139.649
Weight         0.532299  0.0293556  18.1328    <1e-43    0.474409    0.590188
------------------------------------------------------------------------------

***Output of GLM Model:
Height ~ 1 + Weight
Coefficients:
------------------------------------------------------------------------------
              Estimate  Std. Error  z value  Pr(>|z|)  Lower 95%   Upper 95%
------------------------------------------------------------------------------
(Intercept)  135.793     1.95553    69.4404    <1e-99  131.96      139.626
Weight         0.532299  0.0293556  18.1328    <1e-72    0.474763    0.589835
------------------------------------------------------------------------------

***Individual methods applied to model output:
Deviance: 5854.057142765537
Standard error: [1.9555309971174064, 0.029355593998568814]
Degrees of freedom: 198.0
Covariance matrix: [3.8241014806 -0.05628525996; -0.05628525996 0.0008617508990]
R squared (calculated in two ways):0.6241443400847023,  0.6241443400847023
MSE (calculated in two ways: 29.56594516548251, 29.56594516548251
```

In line 3, we read the dataset. It is comprised of `Height` and `Weight`. In line 5-10, we create alternative models using package `GLM` as described above. In practice, you would only use one of these four alternative ways for creating the model. Lines 12–15 print the `lm()`- and `glm()`-based models. In both cases, the formula of the model is printed followed by a table that lists the coefficient estimates, followed by standard errors, test statistics (t-value or z-value for `lm()` or `glm()`, respectively) and then p-values. We explain the meaning of these tables in the sequel. In line 17, we create the `pred()` function which uses the model to predict $\hat{y}$ for a given x. It does this by taking the inner product of the coefficient vector obtained via `coef()` and the vector `[1, x]`. Lines 19–32 present a variety of descriptors associated with the model. The function `deviance()` yields $SS_{\text{residuals}}$. The function `stderror()` yields standard error for the coefficient estimates (these are in agreement with the values in the tables). The function `dof_residual()` yields 198. This is the number of observations, 200 minus 2 as per the numerator of (8.25). The function `vcov()` yields the covariance matrix of the estimators as discussed in the sequel. We then present R^2, both using the function `r2()` and via a manual calculation. We also present the MSE both using `deviance()` and via a manual calculation. Lines 34–38 produce Figure 8.3 using our function `pred()`.

The Distribution of the Estimators

As the least squares estimators, $\hat{\beta}_0$ and $\hat{\beta}_1$, for the model (8.22) are random variables, we may compute their distribution. For this recall (8.11) and notice that the vector $\hat{\beta}$ is obtained via $\hat{\beta} = (A^\top A)^{-1}A^\top Y$. Combine this with (8.22) which by recalling (8.8) can be written in matrix form via

$$Y = A\beta + \varepsilon.$$

This yields,

$$\hat{\beta} = (A^\top A)^{-1}A^\top(A\beta + \varepsilon) = \beta + (A^\top A)^{-1}A^\top\varepsilon.$$

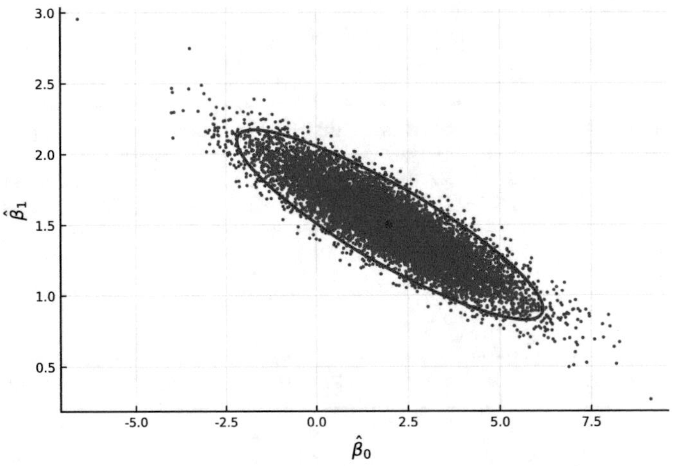

Figure 8.4: An illustration of the distribution of the estimators $\hat{\beta}_0$ and $\hat{\beta}_1$ with approximatly 95% of the simulated points colored green.

Hence, $\hat{\beta}$ is a Gaussian random vector because it is an affine transformation of ε, similar to (3.29) in Chapter 3. We can now investigate its mean vector and covariance matrix and by this know the mean and variance of the individual components as well as their covariance.

Since ε is a zero mean Gaussian random vector, we have that $\mathbb{E}[\hat{\beta}] = \beta$ or written element wise in the case of simple linear regression

$$\mathbb{E}[\hat{\beta}_0] = \beta_0, \qquad \mathbb{E}[\hat{\beta}_1] = \beta_1. \tag{8.27}$$

We thus see that the estimator $\hat{\beta}$ is unbiased. Further, to compute the covariance matrix of the estimators consider $\hat{\beta} - \beta = (A^\top A)^{-1}A^\top \varepsilon$, and take the self outer product $(\hat{\beta} - \beta)(\hat{\beta} - \beta)^\top$ to get the matrix $(A^\top A)^{-1}A^\top \varepsilon\varepsilon^\top A(A^\top A)^{-1}$. The expectation of $\varepsilon\varepsilon^\top$ is $\sigma^2 I$ and hence the expectation of $(\hat{\beta} - \beta)(\hat{\beta} - \beta)^\top$ reduces to $\sigma^2(A^\top A)^{-1}$. Hence, in summary,

$$\text{Cov}(\hat{\beta}) = \sigma^2(A^\top A)^{-1}.$$

For the case of simple linear regression, $A^\top A$ is a 2×2 matrix with an explicit inverse. It isn't hard to obtain,

$$\text{Cov}\left(\begin{bmatrix} \hat{\beta}_0 \\ \hat{\beta}_1 \end{bmatrix}\right) = \begin{bmatrix} \frac{\sigma^2 \sum_{i=1}^n x_i^2}{nS_{xx}} & -\frac{\bar{x}\sigma^2}{S_{xx}} \\ -\frac{\bar{x}\sigma^2}{S_{xx}} & \frac{\sigma^2}{S_{xx}} \end{bmatrix}, \tag{8.28}$$

with S_{xx} as defined in (8.19). These distributional properties of the elements of $\hat{\beta}$ are central to the hypothesis tests and confidence intervals that we present in the subsections that follow. For now, we illustrate the distribution of the random vector $\hat{\beta}$ in Listing 8.5 where we generate synthetic data according to (8.22) with $n = 10$ observations. We repeat this 10^4 times, each time obtaining new estimates $\hat{\beta}_0$ and $\hat{\beta}_1$. A point cloud showing the distribution of the estimators is plotted in Figure 8.4, where a 95% *confidence ellipse* is sketched. From the output, it is evident that approximately 95% of the points fall in this ellipse.

Listing 8.5: The distribution of the regression estimators

```
using DataFrames, GLM, Distributions, LinearAlgebra, Random
using Plots, LaTeXStrings;pyplot()
Random.seed!(0)

beta0, beta1 = 2.0, 1.5
sigma = 2.5
n, N = 10, 10^4
alpha = 0.05

function coefEst()
    xVals = collect(1:n)
    yVals = beta0 .+ beta1*xVals + rand(Normal(0,sigma),n)
    data = DataFrame([xVals,yVals],[:X,:Y])
    model = lm(@formula(Y ~ X), data)
    coef(model)
end

ests = [coefEst() for _ in 1:N]

xBar = mean(1:n)
sXX = sum([(x - xBar)^2 for x in 1:n])
sx2 = sum([x^2 for x in 1:n])
var0 = sigma^2 * sx2/(n*sXX)
var1 = sigma^2/sXX
cv = -sigma^2*xBar/sXX

mu = [beta0, beta1]
Sigma = [var0 cv; cv var1]

A = cholesky(Sigma).L
Ai = inv(A)

r = quantile(Rayleigh(),1-alpha)
isInEllipse(x) = norm(Ai*(x-mu)) <= r
estIn = isInEllipse.(ests)

println("Proportion of points inside ellipse: ", sum(estIn)/N)

scatter(first.(ests[estIn]),last.(ests[estIn]),c=:green, ms=2, msw=0)
scatter!(first.(ests[.!estIn]),last.(ests[.!estIn]),c=:blue, ms=2, msw=0)

ellipsePts = [r*A*[cos(t),sin(t)] + mu  for t in 0:0.01:2pi]
scatter!([beta0],[beta1],c=:red, ms=5, msw = 0)
plot!(first.(ellipsePts),last.(ellipsePts),
        c=:red, lw=2, legend=:none,
        xlabel=L"\hat{\beta}_0", ylabel=L"\hat{\beta}_1")
```

```
Proportion of points inside ellipse: 0.9471
```

In lines 5-8, we set the parameters of the simulation. The assumed values are $\beta_0 = 2.0$, $\beta_1 = 1.5$, and $\sigma^2 = 2.5^2$. The sample size is at $n = 10$ and the number of simulated samples is 10^4. The variable `alpha` is used to control the dimensions of the ellipse such that `1-alpha` of the points are expected to fall within it. The function `coefEst()` in lines 10-16 first creates a sequence $x_1, \ldots, x_n$ on $1, 2, \ldots, n$. Then in line 12, $y_1, \ldots, y_n$ is generated according to (8.22). These values are then set in a `DataFrame`, and a linear model is created. The return value in line 15 is the two-dimensional vector `coef(model)`. In line 18. we create an array, `ests`, of coefficients values, repeating least squares estimation `N` times. These are the points plotted in Figure 8.4. Each element of `ests` is an array of length 2. Lines 20–28 construct the mean vector `mu` and covariance matrix `Sigma` according to (8.27) and (8.28), respectively. Lines 30–31 find the matrix `A` and its inverse `Ai` used for plotting the ellipse and checking if a point is inside the ellipse. See similar use of the Cholesky factorization in Listing 3.32 of Chapter 3. The variable `r` in line 33 uses the fact that the radius of a standard bi-variate normal distribution follows a Rayleigh distribution. Then the function in line 34 gets a point `x` and returns `true` if it falls within the ellipse, and `false` otherwise. This is used to create the array of booleans `estIn`. The proportion of points in the ellipse is printed in line 37. The remainder of the code creates the figure. Observe the use of `.!estIn` in line 40 to negate all the elements of `estIn`. Also observe the use of the transformation '`r*A*`' on the unit circle in line 42. After shifting using '`+ mu`' it creates the ellipse which is displayed in lines 44–46.

Statistical Inference for Simple Linear Regression

Now that we understand the distribution of $\hat{\beta}_0$ and $\hat{\beta}_1$, we can make use of it for confidence intervals and hypothesis tests associated with the regression line. The main hypothesis test for simple linear regression is for the hypothesis:

$$H_0 : \beta_1 = 0, \qquad \text{vs.} \qquad H_1 : \beta_1 \neq 0. \tag{8.29}$$

Here, H_0 implies that X has no effect on Y, whereas H_1 implies that there is an effect. A similar, although less popular, hypothesis test may also be carried out for the intercept β_0:

$$H_0 : \beta_0 = 0, \qquad H_1 : \beta_0 \neq 0. \tag{8.30}$$

In both hypothesis tests (8.29) and (8.30), we use a test statistic of the form

$$T_i = \frac{\hat{\beta}_i}{S_{\hat{\beta}_i}},$$

where $i = 1$ for (8.29) and $i = 0$ for (8.30). In each case, the estimate of the standard error for $\hat{\beta}_i$ differs. For $i = 1$, we have,

$$S_{\hat{\beta}_1} = \sqrt{\frac{\text{MSE}}{S_{xx}}}, \tag{8.31}$$

whereas for the intercept case, $i = 0$, we have,

$$S_{\hat{\beta}_0} = \sqrt{\text{MSE}\left(\frac{1}{n} + \frac{\overline{x}^2}{S_{xx}}\right)}.$$

Both cases, make use of the MSE as defined in (8.25). It turns out that in both cases, under H_0, the test statistic, T_i is distributed according to a T-distribution with $n-2$ degrees of freedom. This can

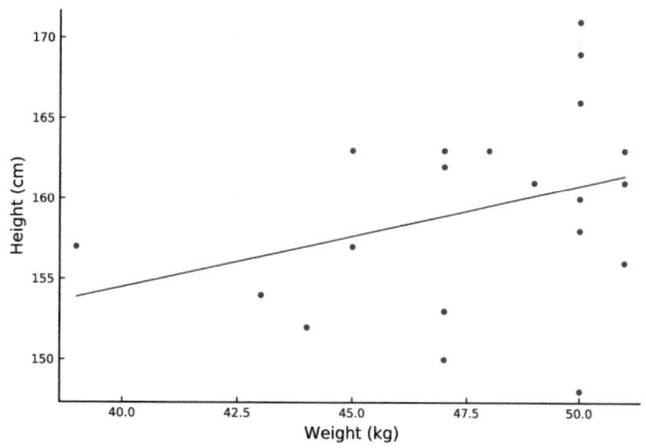

Figure 8.5: A scatter plot of Height vs. Weight with a line of best fit obtained via linear regression. While the line appears to be sloping up, the p-value is actually 0.1546, hence, there isn't evidence of an effect of weight on height.

now be used to test the hypothesis (8.29) or (8.30) in a similar manner to that presented in Chapter 7. Note that it is also possible to adapt the hypothesis to test for $H_0 : \beta_i = \delta$, $H_1 : \beta_i \neq \delta$, for any desired δ or to create one-sided hypothesis tests. However in practice, the test (8.29) is most useful because it answers the scientific question of X having an effect on Y or not. Using the T-distribution and the standard error (8.31), we can also obtain confidence intervals for $\hat{\beta}_0$ and $\hat{\beta}_1$.

In Figure 8.5, we present a scatter plot of height as a function of weight together with a fitted regression line. This is from the `weightHeight.csv` dataset where we purposefully only use the 20 observations with the lowest weight. For these observations, it may appear visually that weight affects height because the line has a non-zero slope. However, is it statistically significant?

We carry out this hypothesis test in Listing 8.6 where we also generate Figure 8.5. In this case, we are not able to reject H_0 of (8.29) under any reasonable confidence level. This is due to the p-value of 0.1546 resulting from $\hat{\beta}_1 = 0.628733$ and a standard error (8.31) of 0.423107. The listing calculates the p-value and confidence interval for $\hat{\beta}_0$, manually in addition to using `GLM` output.

Listing 8.6: Statistical inference for simple linear regression

```
using CSV, GLM, Distributions, Plots; pyplot()

data = CSV.read("../data/weightHeight.csv")
df = sort(data, :Weight)[1:20,:]
model = lm(@formula(Height ~ Weight), df)
pred(x) = coef(model)'*[1, x]
tStat = coef(model)[2]/stderror(model)[2]
n = size(df)[1]
pVal = 2*ccdf(TDist(n-2),tStat)
println("Manual Pval: ", pVal)
println("Manual Confidence Interval: ",
    (coef(model)[2] - quantile(TDist(n-2), 0.975)*stderror(model)[2],
     coef(model)[2] + quantile(TDist(n-2), 0.975)*stderror(model)[2]))
println(model)

scatter(df.Weight, df.Height,c=:blue, msw=0)
xlims = [minimum(df.Weight), maximum(df.Weight)]
plot!(xlims, pred.(xlims),
    c=:red, legend=:none, xlabel = "Weight (kg)", ylabel = "Height (cm)")
```

```
Manual Pval: 0.15458691273390412
Manual Confidence Interval: (-0.2601822581996819, 1.5176487875509448)
Height ~ 1 + Weight
Coefficients:
--------------------------------------------------------------------------
             Estimate  Std. Error  t value  Pr(>|t|)  Lower 95%  Upper 95%
--------------------------------------------------------------------------
(Intercept) 129.359     20.2252    6.39594    <1e-5   86.8678    171.851
Weight        0.628733   0.423107  1.48599   0.1546  -0.260182     1.51765
--------------------------------------------------------------------------
```

In line 3, we read the dataset and then sort the data frame according to `:Weight` in line 4. We then sort by `:Weight` and keep the first 20 entries. Line 5 creates `model` and line 7 defines the function `pred()` similar to previous examples. In line 7, we calculate the T-statistic using the `coef()` and `stderror()` functions applied to `model`. The p-value is calculated in line 9. It is printed on line 10 and you may verify that the result agrees with the output of `model` printed on line 14. Lines 11–13 calculate and print a confidence interval for $\hat{\beta}_1$, also in agreement with the model output. Note that if you attempt to use `glm()` in line 6 together with `Normal()` and `IdentityLink()` as in Listing 8.4 then the p-value and confidence interval will differ because a normal distribution is used instead of a T-distribution. Hence, `lm()` should be used (for such small samples) as opposed to `glm()`. The remainder of the code creates Figure 8.5.

Confidence Bands and Prediction Bands

After collecting data, having a predicted model, $y = \hat{\beta}_0 + \hat{\beta}_1 x$ is useful for determining a prediction $\hat{y}(x^*)$ for every independent variable value x^*. For example, with the `weightHeight.csv` data used in Listing 8.4, based on $n = 200$ observations, we approximately have $\hat{\beta}_0 = 135.8$ and $\hat{\beta}_1 = 0.53$. Hence, if we consider an individual weighing 87 kg then based on this model, their predicted height is

$$\hat{y}(87) = 135.8 + 0.53 \times 87 = 181.9\,\text{cm}.$$

Having such a prediction is useful and is the core purpose of the model. However, we would also like to obtain uncertainty bounds around $\hat{y}(87)$. Further, if instead of just $x^* = 87$ we would use x^* over some interval, then we would like to obtain *uncertainty bands.*

For this, we need to differentiate between two possible meanings of $\hat{y}(87) = 181.9$ or any other $\hat{y}(x^*)$. One meaning is that according to the model, 181.9cm is the expected height of individuals with a weight of 87kg. Another meaning is that 181.9cm is the predicted height of an arbitrary individual with a weight of 87kg. In both, the expected value and predicted value cases, our best possible estimate is $\hat{y}(x^*)$. However, when we consider uncertainty bounds (or bands) the width differs depending on if we consider the expected value or the predicted value. This is similar to the difference between confidence intervals and prediction intervals presented in Chapter 6.

When considering expected values, the formula for *confidence bands* is:

$$\hat{y}(x^*) \pm t_{1-\frac{\alpha}{2},n-2} \sqrt{\text{MSE} \times \Big(\frac{1}{n} + \frac{(x^* - \bar{x})^2}{S_{xx}}\Big)}. \tag{8.32}$$

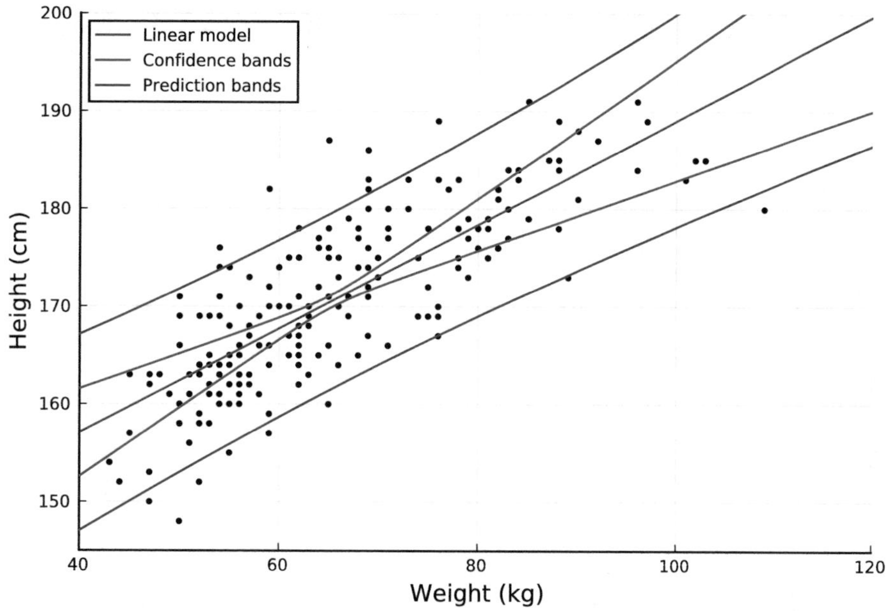

Figure 8.6: A scatter plot of height vs. weight with confidence bands, predication bands, and line of best fit.

Further, when considering predicted values, the formula for *prediction bands* is

$$\hat{y}(x^*) \pm t_{1-\frac{\alpha}{2},n-2}\sqrt{\text{MSE} \times \Big(1 + \frac{1}{n} + \frac{(x^* - \bar{x})^2}{S_{xx}}\Big)}. \tag{8.33}$$

In both cases, $1 - \alpha$ is the confidence level and quantiles of a T-distribution with $n - 2$ degrees of freedom are used. However, in the prediction interval case (8.33), there is an additional 1 term not appearing in (8.32).

We illustrate these bands in Figure 8.6 generated by Listing 8.7. As can be observed from the figure, prediction bands are wider than confidence bands. If you were wishing to use the model to predict the height of a specific individual based on their weight you would use the blue prediction bands for uncertainty quantification. However, if you wanted to get a feel for possible models that could have resulted, you would use the green confidence bands.

Listing 8.7: Confidence and prediction bands

```
1  using CSV, GLM, Distributions, Plots; pyplot()
2  data = CSV.read("../data/weightHeight.csv")
3  n = size(data)[1]
4  model = fit(LinearModel, @formula(Height ~ Weight), data)
5
6  alpha = 0.1
7  tVal = quantile(TDist(n-2),1-alpha/2)
8
9  xbar = mean(data.Weight)
10 Sxx = std(data.Weight)*(n-1)
11 MSE = deviance(model)/(n-2)
12 pred(x) = coef(model)'*[1, x]
13
14 interval(x,sign,prediction = 0) = sign(pred(x),
15                         tVal * sqrt(MSE*(prediction+1/n+(x-xbar)^2/Sxx)) )
16
17 xGrid = 40:1:140
18 scatter(data.Weight, data.Height, c=:black, ms=2, label="")
19 plot!(xGrid,pred.(xGrid), c=:red, label="Linear model")
20 plot!(xGrid,interval.(xGrid,+),c=:green, label="Confidence bands")
21 plot!(xGrid,interval.(xGrid,-),c=:green, label="")
22 plot!(xGrid,interval.(xGrid,+,1),c=:blue,label="Prediction bands")
23 plot!(xGrid,interval.(xGrid,-,1),
24         c=:blue, label="", xlims=(40, 120), ylims=(145, 200), legend=:topleft,
25         xlabel = "Weight (kg)", ylabel = "Height (cm)")
```

In lines 2–4, we read the dataset and fit the model as in previous examples. We also set the number of observations, n. In lines 6–7, we set the significance level, alpha and the corresponding T-value. In lines 9–12, we calculate summary statistics, $\overline{x}$, S_{xx} and MSE. In line 12, we define the function pred(), determining $\hat{y}(x)$. In lines 14–15, we define the function interval() which is the main focus of this example. It is designed to implement both the upper bound and lower bound in (8.32) and (8.33). The argument x is x^*. The argument sign can literally be "+" or "-". The argument prediction has a default value of 0 indicating this is a confidence band as in (8.32). However, if the value is set to 1, then equation (8.33) is obtained. The remainder of the code creates Figure 8.6. Observe the use of our interval() function applied via the broadcast "." operator to xGrid. The second argument, + or - are actually the functions plus and minus, respectively.

Checking Model Assumptions

As explored so far, using a statistical model as in (8.22) allows us to make a variety of conclusions about the population. These include prediction, hypothesis testing, and confidence bands as presented above. However, the validity of these techniques relies on several *model assumptions*. These include:

Assumption I: A linear relationship between variables. More specifically $\mathbb{E}[Y \mid X = x]$ is a linear (affine) function in x.

Assumption II: Normally distributed errors around the regression line.

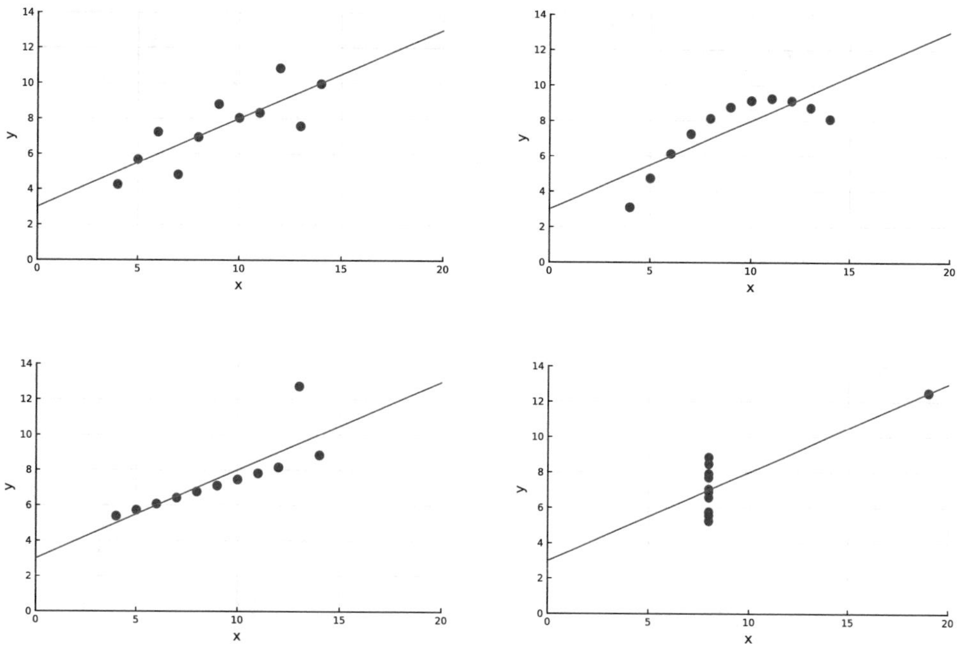

Figure 8.7: Plot of Anscombe's quartet. Although each data block has nearly identical descriptive statistics, and each results in almost identical linear models, it can be seen that the underlying data blocks are very different.

Assumption III: Equal variance for the errors around the regression line.

Assumption IV: Mutually independent errors.

Assumption V: No outliers, i.e. the previous assumptions hold for the full dataset.

While a least squares line passing through a cloud of points can always be found, if any of the assumptions I–V do not hold, then the validity of the statistical results is questionable. As a basic illustration, let us explore how things can go wrong with a classic adversarial example called *Anscombe's quartet.* This is a collection of four data blocks, each with observations of the form $(x_1, y_1), \dots, (x_n, y_n)$. Anscombe's quartet is useful in highlighting the dangers of applying a wrong model to the data. Although each of its four data blocks has identical estimates for the regression line as well as for the coefficient of determination R^2, the nature of each underlying data block is vastly different. This becomes obvious once the data blocks are visualized as in Figure 8.7. All blocks with the exception of the top left block appear to have some model assumptions violated.

Figure 8.7 is created via Listing 8.8 where the Anscombe's quartet data is loaded from the `RDatasets` package. One point arising from the output is that R^2 cannot be used to determine if model assumptions are satisfied or not. That is, the value of R^2 from (8.26) does not give an indication about the validity of the model assumptions, it only gives an indication of the tightness of the data around the prediction line. We mention this because sometimes users of linear regression are tempted to make this mistake with R^2. Other, more valid ways for checking model assumptions are described below.

Listing 8.8: The Anscombe quartet datasets

```
1  using RDatasets, DataFrames, GLM, Plots, Measures; pyplot()
2
3  df = dataset("datasets", "anscombe")
4
5  model1 = lm(@formula(Y1 ~ X1), df)
6  model2 = lm(@formula(Y2 ~ X2), df)
7  model3 = lm(@formula(Y3 ~ X3), df)
8  model4 = lm(@formula(Y4 ~ X4), df)
9
10 println("Model 1. Coefficients: ", coef(model1),"\t R squared: ",r2(model1))
11 println("Model 2. Coefficients: ", coef(model2),"\t R squared: ",r2(model2))
12 println("Model 3. Coefficients: ", coef(model3),"\t R squared: ",r2(model3))
13 println("Model 4. Coefficients: ", coef(model4),"\t R squared: ",r2(model4))
14
15 yHat(model, X) = coef(model)' * [ 1 , X ]
16 xlims = [0, 20]
17
18 p1 = scatter(df.X1, df.Y1, c=:blue, msw=0, ms=8)
19 p1 = plot!(xlims, [yHat(model1, x) for x in xlims], c=:red, xlims=(xlims))
20
21 p2 = scatter(df.X2, df.Y2, c=:blue, msw=0, ms=8)
22 p2 = plot!(xlims, [yHat(model2, x) for x in xlims], c=:red, xlims=(xlims))
23
24 p3 = scatter(df.X3, df.Y3, c=:blue, msw=0, ms=8)
25 p3 = plot!(xlims, [yHat(model3, x) for x in xlims], c=:red, xlims=(xlims))
26
27 p4 = scatter(df.X4, df.Y4, c=:blue, msw=0, ms=8)
28 p4 = plot!(xlims, [yHat(model4, x) for x in xlims], c=:red, msw=0, xlims=(xlims))
29
30 plot(p1, p2, p3, p4, layout = (2,2), xlims=(0,20), ylims=(0,14),
31         legend=:none, xlabel = "x", ylabel = "y",
32         size=(1200, 800), margin = 10mm)
```

```
Model 1. Coefficients: [3.00009, 0.500091]      R squared: 0.6665424595087749
Model 2. Coefficients: [3.00091, 0.5]           R squared: 0.6662420337274844
Model 3. Coefficients: [3.00245, 0.499727]      R squared: 0.6663240410665592
Model 4. Coefficients: [3.00173, 0.499909]      R squared: 0.6667072568984651
```

In line 3, Anscombe's quartet dataset is loaded from the `RDatasets` package via the `dataset()` function. This function takes two arguments, the name of the data package in `RDatasets` containing Anscombe's quartet (`"datasets"`), and the name of the dataset (`"Anscombe"`). In lines 5–8, a linear model for each of the four datasets is created. Note that `df` has 8 columns in total, with the individual four datasets comprised of x-y pairs (e.g. `X1`, `Y1`). Note that each model is a simple linear model of the form (8.22). In lines 10–13, we print the $\hat{\beta}_0$, $\hat{\beta}_1$, and R^2 for each model. In line 15, the function `yHat()` is created, which takes a model type as input, and an x value. This function is then used to plot the regression line from each model for each data block in lines 19, 22, 25, and 28.

In general, most accepted techniques for checking model assumptions involve considering the *residuals* $\{e_i\}$. These are constructed by estimating $\hat{y}_i$ for each value of x_i and then setting,

$$e_i = y_i - \hat{y}_i, \qquad \text{for} \qquad i = 1, \dots, n. \tag{8.34}$$

The residuals were already presented in (8.7). While least squares minimizes the sum of squares, the sum of residuals is always zero. This can be shown using the normal equations. Hence, the arithmetic mean of the residuals is also 0. Analysis of model assumptions then amounts to analyzing the way in which the residuals fluctuate around 0. This is often done via a *residual plot*, where all the residuals for a dataset are plotted. Such a visualization then allows us to see if there is any strong indication that assumptions I, II, III, IV, or V don't hold.

Sometimes we progress beyond a residual plot and use some of the model insight to further analyze the residuals. According to the model assumptions, the random error terms $\{\varepsilon_i\}$ in (8.22) should be a sequence of i.i.d. normally distributed random variables. However, the measured residuals $\{e_i\}$ are not generally independent and identically distributed even when the model assumptions hold. To see this, consider how we obtain the vector $\hat{y}$ via $\hat{y} = A\hat{\beta} = AA^{\dagger}y$. Here, it is useful to define the *projection matrix*

$$H = AA^{\dagger} = A(A^{\top}A)^{-1}A^{\top},$$

and thus $\hat{y} = Hy$. Note that H is also sometimes referred to as the *hat matrix.* Now considering the elements $\{e_i\}$ in the residual vector e, we have

$$e = (I - H)y. \tag{8.35}$$

The representation in (8.35) is useful because we can treat y as a random vector via (8.22). This means that each entry in y is normally distributed with variance σ^2 and the entries are independent. Hence, the covariance matrix of the residuals is $\sigma^2(I - H)$. Also, the mean vector of the residuals is the 0 vector because the least squares estimates are unbiased. Importantly, $I - H$ is generally not a diagonal matrix and hence residuals are dependent. Further, the diagonal entries of H (or $I - H$) are generally not constant and the variance of the i'th residual is $\sigma^2(1 - H_{ii})$, where H_{ii} is the i'th diagonal entry from the projection matrix.

The entry, H_{ii}, is sometimes called the i'th *leverage* and it can be shown that $H_{ii} \in [0,1]$. For simple linear regression ($p = 1$ as in (8.22)), we have that

$$H_{ii} = \frac{1}{n} + \frac{(x_i - \overline{x})^2}{\sum_{j=1}^{n}(x_j - \overline{x})^2}, \tag{8.36}$$

and observing this formula indicates the meaning of the term "leverage". For $p > 1$, as is covered in the next section, the simple (8.36) does not hold, still the meaning of "leverage" carries through.

With such insight about the structure of the residuals, there are now several types of methods for checking model assumptions. One approach is based on the set of *studentized residuals* defined via

$$t_i = \frac{e_i}{\sqrt{\text{MSE}(1 - H_{ii})}} \qquad \text{for} \qquad i = 1, \ldots, n. \tag{8.37}$$

Here, MSE estimates the variance of ε_i and is taken from (8.25). An alternative version exists where instead of MSE, for each i, we use an estimate of σ^2 that is based on all the data with the exception of the i'th observation. This case, sometimes called *externally studentized*, implies that each of the t_i is exactly T-distributed. However, there isn't much harm to stick with (8.37), sometimes called *internally studentized.* In both cases, if the model assumptions hold, the studentized residuals approximately follow a standard normal distribution for non-small n. Nevertheless, they are generally not independent. We note that there is some confusion of terminology in the literature. Some sources

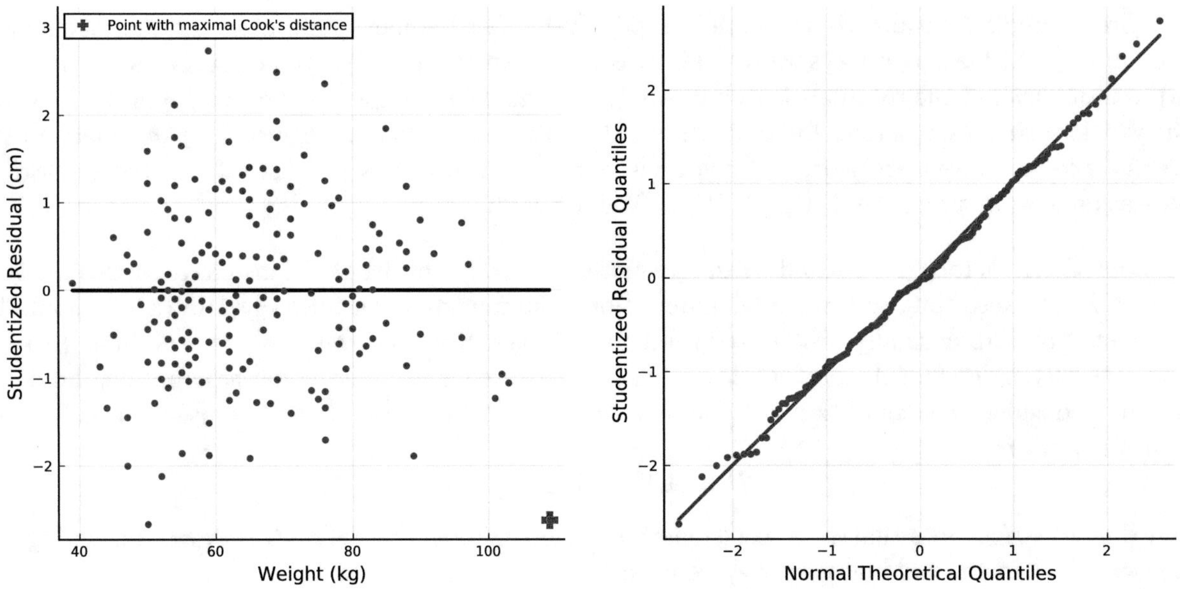

Figure 8.8: Left: the studentized residuals of the data. Right: a normal probability plot of the studentized residuals.

refer to (8.37) as *standardized residuals* and only to the externally studentized case as "studentized residuals".

With the set of studentized residuals, $\{t_i\}$ available, it is then common practice to plot these residuals together with a normal probability plot (see Listing 4.18 in Chapter 4). Such plots, as in Figure 8.8 can then be visually inspected for deviations from assumptions I–V. In the case of this figure, there does not appear to be cause for general alarm regarding deviation from model assumptions. This figure is created by Listing 8.9 where we use the `weightHeight.csv` data to which a regression line was fit in Figure 8.3.

A related type of analysis that can be carried out is *influence diagnosis* where each data point is examined for its influence on the data set. This allows us to find if there are certain points that disproportionally leverage the data, hinting at the fact that these points are systematic *outliers*, or that there is a general violation of model assumptions. One such common method is called *Cook's distance* which can be calculated for each point. We don't present the details beyond a working formula for Cook's distance which is

$$D_i = \frac{t_i^2}{\tau} \frac{H_{ii}}{(1 - H_{ii})}. \tag{8.38}$$

Here, τ is the rank of the matrix H which can also be found via the trace (the sum of the diagonal elements). As a rough rule of thumb, a data point with $D_i > 1$ may be considered as *influential* and should be further investigated. In the case of Listing 8.9, we see that the point with the highest Cook's distance has $D_i \approx 0.22$, which isn't an alarming value. Still, we plot this point in Figure 8.8.

Listing 8.9: Plotting the studentized residuals

```
1  using CSV, GLM, LinearAlgebra, StatsPlots, Plots, Measures; pyplot()
2
3  df = CSV.read("../data/weightHeight.csv")
4  n = size(df)[1]
5
6  model = lm(@formula(Height ~ Weight), df)
7  MSE = deviance(model)/dof_residual(model)
8  pred(x) = coef(model)'*[1, x]
9
10 A = [ones(n) df.Weight]
11 H = A*pinv(A)
12 residuals = (I-H)*df.Height
13 studentizedResiduals = residuals ./ (sqrt.(MSE*(1 .- diag(H))))
14
15 tau = rank(H)
16 println("tau = ", tau, " or ", sum(diag(H)), " or ", tr(H))
17
18 cookDistances = (studentizedResiduals.^2/tau) .* diag(H) ./ (1 .- diag(H))
19 maxCook, indexMaxCook = findmax(cookDistances)
20 println("Maximal Cook's distance = ", maxCook)
21
22 p1 = plot([minimum(df.Weight),maximum(df.Weight)],[0,0],
23                 lw = 2, c=:black, label=:none)
24      scatter!([df.Weight[indexMaxCook]],[studentizedResiduals[indexMaxCook]],
25                 c=:red, ms = 10, msw = 0, shape =:cross,
26                 label = "Point with maximal Cook's distance",legend=:topleft)
27
28     scatter!(df.Weight, studentizedResiduals, xlabel = "Weight (kg)",
29              ylabel = "Studentized Residual (cm)",c=:blue,msw=0,label=:none)
30
31 p2 = qqnorm(studentizedResiduals, msw=0, lw=2, c =[:red :blue],legend = false,
32         xlabel="Normal Theoretical Quantiles",
33         ylabel="Studentized Residual Quantiles")
34
35 plot(p1,p2,size=(1000,500),margin = 5mm)
```

```
tau = 2 or 2.0000000000000004 or 1.9999999999999998
Maximal Cook's distance = 0.2216877435476448
```

In line 6, we fit the model similar to previous examples. In line 7, we compute `MSE` similar to Listing 8.4. Lines 10–13 compute the studentized residuals using (8.37) and the fact that `pinv()` implements the pseudo-inverse $A^{\dagger}$. Note the use of `diag()` in line 13 to extract the diagonal of the matrix and the many applications of element wise operations with a dot. In line 15, we obtain the rank of H using `rank()` from `LinearAlgebra`. An alternative is to compute the sum of the diagonal elements directly, also known as the *trace* of the matrix. We do this via `tr()`. In line 18, we find Cook's distance of every point via (8.38). Then in line 19, we find the point with the maximal Cook's distance. The remainder of the code creates Figure 8.8 where in lines 31–33 we use `qqnorm()` from `StatsPlots` for the normal probability plot (see also Listing 4.18).

8.3 Multiple Linear Regression

Having explored many aspects of linear regression with one variable, we now consider situations involving more than one variable. This is sometimes called *multiple regression*. We generalize model (8.22) by extending the case of a single independent variable to p independent variables:

$$Y_i = \beta_0 + \beta_1 X_{i1} + \beta_2 X_{i2} + \ldots + \beta_p X_{ip} + \varepsilon_i. \tag{8.39}$$

Here, ε_i is still a single normally distributed random variable for every i and it is now added to a linear combination of p explanatory variables and an intercept term. The same model assumptions seen for $p = 1$ are assumed to hold, now extended to multiple dimensions.

The data for this model involves n tuples of the form: $(x_{11}, \ldots, x_{1p}, y_1), \ldots, (x_{n1}, \ldots, x_{np}, y_n)$, where $n > p$. Each such tuple is an observation with a dependent variable y_i and independent variables $x_{i1}, \ldots, x_{ip}$. In this case, the least squares estimation minimizing (8.7) carries over in a straightforward manner as was already alluded to in Section 8.1. This is now with an $n \times (p+1)$ design matrix A, generalizing (8.8) to

$$A = \begin{bmatrix} 1 & x_{11} & \ldots & x_{1p} \\ 1 & x_{21} & \ldots & x_{2p} \\ \vdots & \vdots & & \vdots \\ 1 & x_{n1} & \ldots & x_{np} \end{bmatrix}. \tag{8.40}$$

In addition to the least squares results of Section 8.1, many of the statistical analysis results of Section 8.2 carry over from the case of $p = 2$ to general p. The formulas for β_0 and β_1 of (8.18), (8.20), (8.21), and (8.28) are specific for $p = 2$, however most other formulas in the chapter are for arbitrary p. In certain cases, formulas need to be generalized, replacing $n - 2$ with $n - p$ or similar, however, these details are not critical to our exploration and can be found in texts such as [M17].

We first consider a simple example and then explore an issue found in multiple linear regression that is not exhibited in simple linear regression, namely collinearity. The focus in this section is on the case where all the variables are continuous in nature. In the next section, we describe how to deal with categorical variables and other aspects that arise in regression modeling with multiple variables.

A Multiple Linear Regression Example

We use the `cpus` dataset from the `RDatasets` `MASS` package. This dataset is a record of the relative performance measured of several old CPUs (relative to an IBM 370/158-3). Each CPU observation comes with respective attributes, such as clock speed, cache size, etc.

In Listing 8.10, linear regression is used to fit a model that predicts the performance of a CPU based on several characteristics. In creating this regression model, we assume that a linear relationship can hold for the "CPU frequency" but not for its reciprocal, the "clock cycle speed". For this, we created a new variable, `Freq` by transforming the original variable `CycT`. Such *variable transformations* are often common in regression analysis as one often seeks to transform the data so that a linear model like (8.39) is applicable. To consider if and when to carry out such transformations, a first step is to visualize the data as in scatter plot matrix of Listing 4.24. For brevity, we omit these considerations in this example and assume that the need for the transformation is predetermined.

As you may see from the output of the listing, there are now $p+1$ parameter estimates (5 in our case). Each estimate has its own standard error, T-value, a resulting p-value, and a confidence interval, similar to the simple linear regression case (see, for example, the output of Listing 8.6). Such a display of the variables often allows us to get an immediate feel for the quality of the regression model. For example in our case, we see that all variables exhibit extremely low p-values with the exception of `Freq`. Nevertheless, with a p-value of around 8%, that variable is still potentially meaningful for the model. Hence, initially, this may give an indication that all variables are needed.

Indeed, the p-value column can be used as a first guide for considering which variables are meaningful for the regression. However, since multiple tests are being carried out in parallel, the meaning of the p-value is not as straightforward as if only a single comparison is carried out. Further, the p-values of different variables influence each other. For example, if you remove or add some variables, then generally, all p-values change. We discuss such aspects of *model selection* in more detail in Section 8.5.

For now, let us use multiple regression for prediction. With such a five-parameter model at hand, we compare two hypothetical computers, `A` and `B`. Computer `A` has a smaller cache but is 100 times faster in terms of frequency. The predicted values for the two computers are presented and we see that computer `B` is expected to have slightly better performance according to this model.

Listing 8.10: Multiple linear regression

```julia
using RDatasets, GLM, Statistics

df = dataset("MASS", "cpus")
df.Freq = map( x->10^9/x , df.CycT)

model = lm(@formula(Perf ~ MMax + Cach + ChMax + Freq), df)
pred(x) = round(coef(model)'*vcat(1,x),digits = 3)

println("n = ", size(df)[1])
println("(Avg,Std) of observed performance: ", (mean(df.Perf),std(df.Perf)))
println(model)
println("Estimated performance for computer A: ", pred([32000, 32, 32, 4*10^7]))
println("Estimated performance for computer B: ", pred([32000, 16, 32, 6*10^7]))
```

```
n = 209
(Avg,Std) of observed performance: (105.61722488038278, 160.83058719907774)

Perf ~ 1 + MMax + Cach + ChMax + Freq

Coefficients:
---------------------------------------------------------------------------------
                 Estimate    Std. Error   t value  Pr(>|t|)     Lower 95%    Upper 95%
---------------------------------------------------------------------------------
(Intercept)  -46.5763      7.62382      -6.10931    <1e-8    -61.6078      -31.5447
MMax           0.00841457  0.000639214  13.1639     <1e-28     0.00715426    0.00967489
Cach           0.872508    0.152825      5.70919    <1e-7      0.571189      1.17383
ChMax          0.96736     0.234847      4.11911    <1e-4      0.504321      1.4304
Freq           9.74951e-7  5.5502e-7     1.75661    0.0805    -1.1936e-7     2.06926e-6
---------------------------------------------------------------------------------
Estimated performance for computer A: 320.564
Estimated performance for computer B: 326.103
```

In line 3, the `"cpus"` dataset from the `"MASS"` RDatasets package is stored as the data frame `df`. In line 4, the "cycle time" `CycT` is transformed into the "frequency" (`Freq`) via the `map()` function. Since the cycle time is in nanoseconds, one can calculate the cycles per second via `10^9/x`. This is the anonymous function presented as the first argument to `map()`. In line 6 the model is created, where the response variable `Perf` is the "performance", and the explanatory variables are: `MMax`-the maximum main memory in kilobytes (KB), `Cach`-the cache size in KB, `ChMax`-the maximum number of channels, and `Freq`-the frequency in cycles per second. Note the use of `@formula()` for elegantly specifying such a model. In line 7, we created our `pred()` function where the inner product of $\hat{\beta}$ represented via `coef(model)` is taken with the vector `vcat(1,x)`. The remainder of the code presents the output with predictions of computer `A` and computer `B` calculated and printed in lines 12–13.

Collinearity

When conducting multiple linear regression it is possible for some subset of the explanatory variables, $X_1, \ldots, X_p$, to be statistically dependent. This situation is called *collinearity* or *multicollinearity* and it violates the model assumptions and can thus cause problems in model interpretation.

In extreme situations, collinearity may be due to full redundancy of the data or multiple readings perhaps using different scales. Imagine, for example, that temperature readings are present both in the Centigrade scale as one variable and the Fahrenheit scale as another variable. In other situations, collinearity is present due to inherent statistical relationships between variables. For example, assume we are trying to predict the salary of individuals, Y. For this, we consider the age of individuals, X_1 and the years of experience, X_2. In this case, assuming there are no career interruptions, we have

$$X_2 = X_1 - D,$$

where D, a *latent variable*, is the age of the individual at which she started employment. This immediately renders X_1 and X_2 to be statistically dependent random variables. Such a dependence may be very strong if the variability of D isn't large. If D happens to be a deterministic fixed value then there is even perfect collinearity between X_1 and X_2.

Such a case of perfect collinearity renders the design matrix A, (8.40) to have less than $p+1$ independent columns (the matrix is *less than full rank*). In such a case, the matrix $A^\top A$ is not invertible and the pseudo-inverse cannot be computed as in (8.10) or (8.12). Still there are ways around this problem via the singular value decomposition as in (8.13), or other means such as ridge regression (presented in Section 9.3 of the next chapter), or LASSO presented in Section 8.5. However, in many cases collinearity isn't perfect, and hence pure algebraic problems don't exist. Nevertheless, numerical and statistical problems can still persist.

A consequence of collinearity, even if not perfect, is a breakdown of the model assumptions of the linear regression model. This typically does not affect the least squares estimates, however, it does imply that the model is extremely sensitive to perturbations of the data. It also means that p-values and other statistical summaries from the model are distorted.

There are several alternative ways for detecting the presence of collinearity. A first basic approach is to consider the correlation between different explanatory variables and if it is close to 0 then to assume that no collinearity exists, and otherwise, to consider elimination of one of the variables that exhibit high correlation with another variable. This method may work well if there is a small number of variables, however, when there are multiple variables, small correlations between pairs of variables may hide collinearity that is based on the interplay of multiple variables.

An alternative is to consider a method related to R^2 called the *Variance Inflation Factor (VIF)*. The VIF can be calculated for each explanatory variable in (8.39) by first regressing the variable against the others. For example, VIF_1 is calculated with the aid of the regression model,

$$X_{i1} = \tilde{\beta}_0 + \tilde{\beta}_2 X_{i2} + \ldots + \tilde{\beta}_p X_{ip} + \varepsilon_i.$$

Similarly, for any other explanatory variable j, VIF_j is calculated with the aid of the regression model,

$$X_{ij} = \tilde{\beta}_0 + \tilde{\beta}_1 X_{i1} + \ldots + \tilde{\beta}_{j-1} X_{i,j-1} + \tilde{\beta}_{j+1} X_{i,j+1} + \ldots + \tilde{\beta}_p X_{ip} + \varepsilon_i.$$

In each case, we compute the coefficient of determination of the model and denote it R_j^2 when variable j on the left hand side of the model. Then we set

$$\text{VIF}_j = \frac{1}{1 - R_j^2}, \tag{8.41}$$

and the square root of this value indicates the factor by which the standard error increases compared to if variable j had no correlation with any other variable.

Now by computing $\text{VIF}_1, \ldots, \text{VIF}_p$ we may see which variable causes the highest variance inflation and if that value is greater than some threshold (sometimes taken to be at around 10), then we can remove the variable with the maximal VIF. The process can then continue recursively. This method then presents one well-defined approach for detecting and handling collinearity.

In Listing 8.11, we present an artificial example with X_1, X_2 and X_3, where

$$X_3 = X_1 + 2X_2 + \text{noise}. \tag{8.42}$$

This way when the variance of the noise is low, there is collinearity. However, when the variance of the noise is high, the level of collinearity is not high. This type of example allows us to modify the variance of the noise and see the effect on the correlation between variables and on the VIF. Line 13 in Listing 8.11 implements (8.42), where η is the standard deviation of the noise. Observe from the output that when $\eta = 200$, VIF_3 is low and so are the correlations. As opposed to that, as η decreases towards 0 the correlations increase and so does VIF_3. In practice, if we were observing this system in a situation similar to $\eta = 50$ we may retain all variables, however, if we were in a situation similar to $\eta = 10$ we would remove X_3 due to the high VIF.

The example goes all the way to $\eta = 0$ in which case X_3 is a perfect linear combination of X_1 and X_2. In this case, the model fitting in line 25 fails and throws an *exception*. We catch this exception in lines 27–28 and can potentially handle it in line 39.

Listing 8.11: Exploring collinearity

```
using Distributions, GLM, DataFrames, Random, LinearAlgebra
Random.seed!(0)

n = 100
beta0, beta1, beta2, beta3 = 10, 30, 60, 90
sig, sigX = 25, 5
etaVals = [200.0, 100.0 , 50.0, 10.0, 1.0, 0.1, 0.01, 0.0]

function createDataFrame(eta)
    Random.seed!(1)
    x1 = round.(collect(1:n) + sigX*rand(Normal(),n),digits = 5)
    x2 = round.(collect(1:n) + sigX*rand(Normal(),n),digits = 5)
    x3 = round.(x1 +2*x2 + eta*rand(Normal(),n),digits = 5)
    y = beta0 .+ beta1*x1 + beta2*x2 + beta3*x3 + sig*rand(Normal(),n)
    return DataFrame(Y = y, X1 = x1, X2 = x2, X3 = x3)
end

VIF3() = 1/(1-r2(lm(@formula(X3 ~ X1 + X2),df)))

for eta in etaVals
    print("eta = $(eta): ")
    df = createDataFrame(eta)
    glmOK = true
    try
        global model = lm(@formula(Y ~ X1 + X2 + X3),df)
    catch err
        println("Exception with GLM: ", err)
        glmOK = false
    end

    if glmOK
        covMat = vcov(model)
        sigVec = sqrt.(diag(covMat))
        corrmat = round.(covMat ./ (sigVec*sigVec'),digits=5)
        println("Corr(X1,X3) = ", corrmat[2,4],
                ",\t Corr(X2,X3) = ",corrmat[3,4],
                ",\t VIF3 = ", round.(VIF3(),digits=2) )
    else
         println("\t In this case we may use SVD or ridge regression if needed.")
    end
end
```

```
eta = 200.0: Corr(X1,X3) = -0.18096,     Corr(X2,X3) = 0.02747,        VIF3 = 1.51
eta = 100.0: Corr(X1,X3) = -0.21046,     Corr(X2,X3) = -0.03728,       VIF3 = 2.31
eta = 50.0: Corr(X1,X3) = -0.26771,      Corr(X2,X3) = -0.1646,        VIF3 = 5.05
eta = 10.0: Corr(X1,X3) = -0.61425,      Corr(X2,X3) = -0.76913,       VIF3 = 81.39
eta = 1.0: Corr(X1,X3) = -0.98805,       Corr(X2,X3) = -0.99699,       VIF3 = 7611.34
eta = 0.1: Corr(X1,X3) = -0.99987,       Corr(X2,X3) = -0.99997,       VIF3 = 756822.99
eta = 0.01: Corr(X1,X3) = -1.0,          Corr(X2,X3) = -1.0,           VIF3 = 7.56e7
eta = 0.0: Exception with GLM: PosDefException(4)
         In this case we may use SVD or ridge regression if needed.
```

In lines 4–7, we define the basic parameters of this experiment: `n` is the number of observations; the `beta` variables are the actual β_i values used to generate the data; `sig` is the standard deviation of the error term; `sigX` determines variability of the x-values; and `etaVals` determines a range of values for the variability of the noise in (8.42). In lines 9–16, we define the function `createDataFrame()`. It creates data based on `eta`, which determines the variability of the noise in (8.42). In line 18, we define `VIF3()` for calculating the variance influence factor of X_3. Lines 20–41 iterate over `etaVals`, each time trying to fit a linear model in line 25. Here, we use Julia's `try-catch` mechanism to catch an exception in case `lm()` throws an exception. This happens in the case of `eta = 0.0` in which case the matrix $A^\top A$ is singular. Then in lines 32–37, we output covariance values and GLM output when an exception isn't thrown. If an exception is thrown we execute line 39. See Listing 9.10 in Chpater 9 for ridge regression.

8.4 Model Adaptations

In certain situations, directly using the variables $X_{i1}, \ldots, X_{ip}$ as in the model (8.39) is not sensible. This may be the case when the relationship between X and Y is not linear, when some of the elements of X are discrete categorical variables, or when elements of X interact such that the value of one element of X influences the effect that other elements of X have on Y. Luckily, linear regression is still useful for many of these situations once certain *model transformations* are applied.

We have already carried out a simple model transformation in Listing 8.10, where we took the variable `CycT` and created the variable `Freq` via the reciprocal of `CycT`. We now extend this idea by illustrating common model transformations. In general, we may transform the p variables of (8.39) to ℓ variables (or features) by specifying ℓ functions

$$f_1(x_1, \ldots, x_p), \ldots\ldots, f_\ell(x_1, \ldots, x_p). \tag{8.43}$$

We then define the transformed variables, $W_{ij} = f_j(X_{i1}, \ldots, X_{ip})$ for $j = 1, \ldots, \ell$ and $i = 1, \ldots, n$. The regression model then becomes

$$Y_i = \beta_0 + \beta_1 W_{i1} + \beta_2 W_{i2} + \ldots + \beta_\ell W_{i\ell} + \varepsilon_i. \tag{8.44}$$

This model has $\ell + 1$ parameters that are to be estimated, in comparison to $p + 1$ in the original number. We can have that $\ell > p$, $\ell = p$, or $\ell < p$. In any case, such transformations can go a long way in extending the power of linear regression. We now illustrate the basic cases.

Fitting to a Polynomial

Observe Figure 8.9 where the data points clearly don't fall on a straight line. In such a case, we may wish to fit a parabola via the model

$$Y_i = \beta_0 + \beta_1 X_i + \beta_2 X_i^2 + \varepsilon. \tag{8.45}$$

This can be achieved by setting $f_1(x) = x$ and $f_2(x) = x^2$. To see this, consider (8.43) and (8.44) with $p = 1$ and $\ell = 2$.. Other higher order polynomials are also possible by extending this idea. Similarly, any functional form may be chosen for the transformation function $f_j(\cdot)$, all with a purpose of describing the functional relationship between X and Y properly. Once such a transformation has

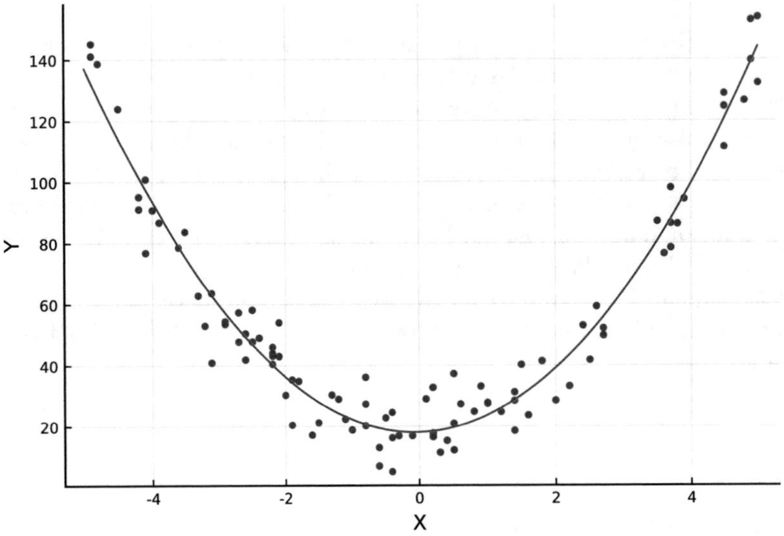

Figure 8.9: Plot of the linear regression polynomial model, alongside the original data points.

taken place, a linear regression model can be fit to the transformed data. In this case, it yields parameter estimates for β_0, β_1, and β_2.

In Listing 8.12, we first load the original data, then append an additional column `X2` to our data frame. The values in this appended column `X2` take on the square of the values stored in column `X` (this is the transformation based on $f_2(x)$). Then the `lm()` function is used to fit a linear model and the resulting solution is plotted along with the original data points in Figure 8.9. It can be seen from the resulting output that the predicted model is polynomial in nature and is a good fit for the data. Also, the p-values in the output can still be used as a guideline for model fit. Since they are all small, it indicates that both the original variable X and the new variable X^2 are useful for describing Y. The listing also shows how to use the `GLM` package to directly specify a quadratic relationship. This is via `@formula(Y ~ X*X)`.

Listing 8.12: Linear regression of a polynomial model

```
using CSV, GLM, Plots; pyplot()

data = CSV.read("../data/polynomialData.csv")
data.X2 = abs2.(data.X)
model1 = lm(@formula(Y ~ X + X2), data)
model2 = lm(@formula(Y ~ X*X), data)
println(model1)
println(model2)

yHat(x) = coef(model1)[1] + coef(model1)[2]*x + coef(model1)[3]*x^2

xGrid = -5:0.1:5
scatter(data.X, data.Y, c=:blue, msw=0)
plot!(xGrid, yHat.(xGrid),
    xlabel="X", ylabel="Y", c=:red, legend=:none)
```

```
Y ~ 1 + X + X2
Coefficients:
-----------------------------------------------------------------------------
              Estimate   Std. Error   t value   Pr(>|t|)  Lower 95%   Upper 95%
-----------------------------------------------------------------------------
(Intercept)  18.0946     1.16533     15.5274     <1e-27   15.7817     20.4074
X             0.705362   0.298831     2.36041    0.0203    0.112266    1.29846
X2            4.90144    0.108245    45.281      <1e-66    4.6866      5.11628
-----------------------------------------------------------------------------

Y ~ 1 + X + X & X
Coefficients:
-----------------------------------------------------------------------------
              Estimate   Std. Error   t value   Pr(>|t|)  Lower 95%   Upper 95%
-----------------------------------------------------------------------------
(Intercept)  18.0946     1.16533     15.5274     <1e-27   15.7817     20.4074
X             0.705362   0.298831     2.36041    0.0203    0.112266    1.29846
X & X         4.90144    0.108245    45.281      <1e-66    4.6866      5.11628
-----------------------------------------------------------------------------
```

In line 3, the data from the CSV file is loaded and stored as the data frame `data`. In line 4, a second column, `data.X2`, is appended to the data frame `data`. The values are defined as the squares of the values in column `X` through the use of the `abs2()` and `'.'`. In line 5, the linear model (8.45) is implemented via the `@formula` macro, and this model is then fit via `lm()` and stored as `model1`. In line 6, we create `model2` this time explicitly stating a quadratic relationship via `X*X`. In line 8, the function `yHat(x)` is created which makes model predictions for a given value of `x` based on the coefficients of `model1`. The coefficients of `model2` are the identical. The remainder of the code creates Figure 8.9.

Categorical Variables

A very common model adaptation deals with categorical variables. As discussed previously in Section 4.1, while a numerical variable such as temperature, length, or mass can take on any value in a continuous domain, a *categorical variable* is a variable that only takes on discrete values from a finite set. In many cases, the possible discrete values, *groups*, or *levels* of a categorical variable are not ordered in nature and the variable is called a *nominal* variable. An example is color, where only specific colors such as "red", "green", "yellow", and "blue" are specified. In other cases, the categorical variable is *ordinal*, in which case there is a clear ordering between the levels. In principle, this allows us to convert the variable into a numerical variable for the regression. For example if the variable is a course grade and has levels "F", "D", "C", "B", or "A" then it may converted into a numerical variable with corresponding values such as 55, 65, 75, 85, and 95. In such a case, regression may be executed by considering the variable with the numerical levels. However, one must keep in mind that when there are multiple variables at play, the manner in which numerical values are assigned to an ordinal categorical variable, affects the results. Hence, this method of dealing with (ordinal) categorical variables is questionable.

Instead, there is a more common approach for dealing with categorical variables which works for both ordinal and nominal variables and ignores the value or meaning of the levels. The idea is to use *indicator variables*, also called *dummy variables* in this context. Assume that a categorical variable

$X_{i,c}$ has L levels. We then assign the levels to the integers $1, 2, \ldots, L$ in some arbitrary manner where we call the L'th level the *base level.* We then create $L-1$ indicator variables $\mathbf{1}_{i,c,1}, \mathbf{1}_{i,c,2}, \ldots, \mathbf{1}_{i,c,L-1}$, where

$$\mathbf{1}_{i,c,k} = \begin{cases} 1 & \text{if } X_{i,c} \text{ is at level } k, \\ 0 & \text{otherwise.} \end{cases} \qquad \text{for levels} \qquad k = 1, \ldots, L-1.$$

That is, for each categorical variable $X_{i,c}$, we create a vector $(\mathbf{1}_{i,c,1}, \mathbf{1}_{i,c,2}, \ldots, \mathbf{1}_{i,c,L-1})$ of length $L-1$ that is either all 0's if $X_{i,c,k} = L$ (the base level) or has a single 1 at the coordinate matching the value of $X_{i,c}$ when the level is not the base level. This is very similar to *one hot encoding* used in the next chapter as is common in machine learning, however, it differs in that the length of the vector is $L-1$ and not L. Note that if you were to use L dummy variables instead of $L-1$ then the design matrix, A, would not be full rank, hence we employ $L-1$ variables. Also note that in the case of a categorical variable with only two levels, if we encode $X_{i,c}$ as a *binary variable* from the outset then no further encoding is needed since $\mathbf{1}_{i,c,1} = X_{i,c}$.

Now with these dummy variables in place, the regression model is set up as follows. Say, for example, we have a model with three numerical variable $X_{i,1}$, $X_{i,2}$, and $X_{i,3}$ as well as two categorical variables $X_{i,4}$ and $X_{i,5}$ with corresponding number of levels L_4 and L_5, respectively. Then the regression model is

$$Y_i = \beta_0 + \beta_1 X_{i,1} + \beta_2 X_{i,2} + \beta_3 X_{i3} + \sum_{k=1}^{L_4-1} \beta_{4,k} \mathbf{1}_{i,4,k} + \sum_{k=1}^{L_5-1} \beta_{5,k} \mathbf{1}_{i,5,k} + \varepsilon_i. \tag{8.46}$$

Here, the coefficients $\beta_{4,1}, \ldots, \beta_{4,L_4-1}$, determine the effect of the various levels of $X_{i,4}$ on Y_i, relative to the base level. Similarly $\beta_{5,1}, \ldots, \beta_{5,L_5-1}$ are the level coefficients for X_{i5}.

Similar to (8.46) any other combination of numerical and categorical variables is possible. As an example, in Listing 8.13, we consider a simpler case based on the `weightHeight.csv` data. Here, Y_i is the height of the person, $X_{i,1}$ is their weight, and $X_{i,2}$ is the categorical variable indicating their gender. The dataset only includes values of `M` and `F` for male and female, but we purposefully modify a few arbitrary entries to have other genders which we simply label as `O1` and `O2`. This allows us to consider a case with $L = 4$. Then when we use the `GLM` package to encode the categorical variable where we use the `contrasts` option, we explicitly state the base gender as `F`. We also explicitly encode the levels as $k = 1$ for `M`, $k = 2$ for `O1`, and $k = 3$ for `O2`. See line 11 in Listing 8.13.

The associated model is then

$$Y_i = \beta_0 + \beta_1 X_{i1} + \beta_{21} \mathbf{1}_{i,2,1} + \beta_{22} \mathbf{1}_{i,2,2} + \beta_{23} \mathbf{1}_{i,2,3} + \varepsilon_i, \tag{8.47}$$

where the coefficients have the following interpretation. The intercept β_0 is as in previous models, however it now corresponds to gender `F` as this is the base level of the categorical variable. Similarly, $\beta_0+\beta_1 x$ is the predicted height for an individual of weight x that is of gender `F`. Then any modification to the gender is expected to modify the predicted height with a value of β_{21} for gender `M`, β_{22} for gender `O1`, and β_{23} for gender `O2`.

As apparent from the output of Listing 8.13, the coefficients β_0, β_1, and β_{21} (for males) have significant p-values, however, the coefficients for the genders `O1` and `O2` have non-significant p-values. This is expected because we randomly assigned these genders to the data. The listing also creates Figure 8.10 which presents the fit of the models for males and females, plotting other genders as green points. Observe that as dictated by the model (8.47), the slope of both lines is identical and hence they are parallel. That is, the effect of gender is only a shift of the line. The lines would generally not be parallel if there were interaction terms, a topic which we cover next.

Listing 8.13: Regression with categorical variables - no interaction effects

```
1  using CSV, GLM, Plots, Random; pyplot()
2  Random.seed!(0)
3
4  df = CSV.read("../data/weightHeight.csv", copycols = true)
5  n = size(df)[1]
6  df[shuffle(1:n),:] = df
7  df[[10,40,60,130,140,175,190,200],:Sex] .= "O1"
8  df[[9,44,63,132,138,172,192,199],:Sex] .= "O2"
9
10 model = lm(@formula(Height ~ Weight + Sex), df,
11         contrasts=Dict(:Sex=>DummyCoding(base="F",levels=["M","O1","O2","F"])))
12 b0, b1, b2, b3, b4  = coef(model)
13 pred(weight,sex) = b0+b1*weight+b2*(sex=="M")+b3*(sex=="O1")+b3*(sex=="O2")
14 println(model)
15
16 males = df[df.Sex .== "M",:]
17 females = df[df.Sex .== "F",:]
18 other = df[(df.Sex .!= "M") .& (df.Sex .!= "F"),:]
19
20 xlim = [minimum(df.Weight), maximum(df.Weight)]
21 scatter(males.Weight, males.Height, c=:blue, msw=0, label="Males")
22 plot!(xlim, pred.(xlim,"M"), c=:blue, label="Male model")
23
24 scatter!(females.Weight, females.Height, c=:red, msw=0, label="Females")
25 plot!(xlim, pred.(xlim,"F"),
26         c=:red, label="Female model", xlims=(xlim),
27         xlabel="Weight (kg)", ylabel="Height (cm)", legend=:topleft)
28
29 scatter!(other.Weight, other.Height, c=:green, msw=0, label="Other")
```

```
Height ~ 1 + Weight + Sex
------------------------------------------------------------------------------
               Estimate  Std. Error   t value  Pr(>|t|)   Lower 95%   Upper 95%
------------------------------------------------------------------------------
(Intercept)  142.194       2.17705    65.3152    <1e-99  137.901     146.488
Weight         0.395869    0.037162   10.6525    <1e-20    0.322578    0.469161
Sex: M         5.61479     1.01905     5.50982   <1e-6     3.60501     7.62457
Sex: O1        3.12116     1.88874     1.65251   0.1000   -0.603814    6.84614
Sex: O2        2.0663      1.92299     1.07452   0.2839   -1.72624     5.85884
------------------------------------------------------------------------------
```

We read the data in line 4 and shuffle the observations in line 6. In lines 7–8, we modify some selected indices to have genders other than "male" and "female". We denote these `"O1"` and `"O2"`. The model is specified in lines 10–11 where since `:Sex` is a categorical variable, we also specify the coding using the `contrasts` option of `lm()`. Without this option, `:Sex` is still treated as categorical but will have some arbitrary coding. Here, we use `DummyCoding()` to specify that the base level is `"F"` and hence won't have a regression coefficient associated with it. In line 12, we set the variables `b0`, `b1`,...,`b4` as the regression coefficients. These allow us to create the `pred()` function in line 13. In line 14, we print the model. Line 16–18 are used to create the data for plotting. Note the use of the element-wise and, `.&` in line 18. The remainder of the code creates Figure 8.10.

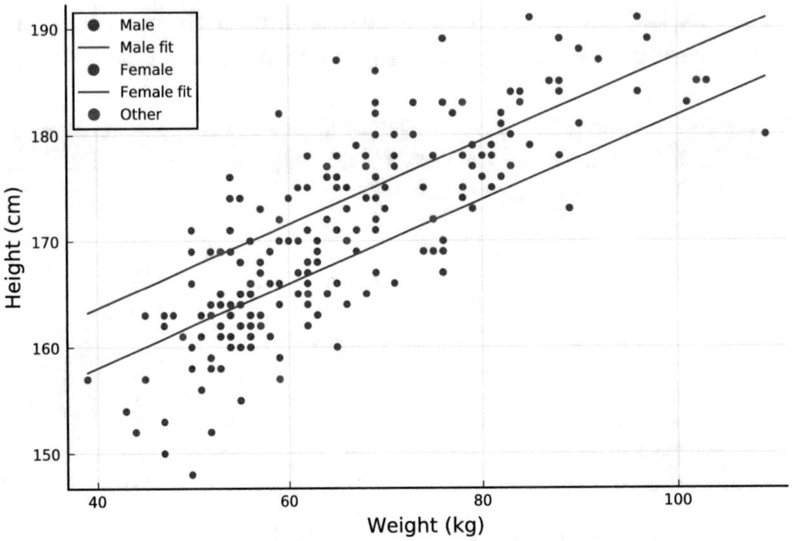

Figure 8.10: Linear model with categorical (sex) variable plotted for the two statistically significant levels of the variable, male and female. Other genders are plotted as green points.

Interactions

As can be observed from model (8.47) and Figure 8.10, the effect of the continuous variable weight on height is the same for all genders. Further, as may be read from the output of Listing 8.13, each additional kilogram implies approximately 0.396 centimeters of height and this holds identically for males, females, or members of the other genders. Similarly, in a model such as (8.46), the effect of the continuous terms X_1, X_2, and X_3 on Y is not altered by the values of the categorical variables. Such a setting is said to have *no interaction effects.*

In many cases, modeling with no interaction effects is sensible as it keeps the number of parameters in the model low. However, in other cases, one may expect that the level of categorical variables will induce a different relationship between the continuous variable and the response variable. For the weight and height example, we now assume that there are $L = 3$ genders, `F`, `M`, and `O`. In this case, one possible model with *interaction effects* is

$$Y_i = \beta_0 + \beta_1 X_{i1} + \big(\beta_0^{21} + \beta_1^{21} X_{i1}\big)\mathbf{1}_{i,2,1} + \big(\beta_0^{22} + \beta_1^{22} X_{i1}\big)\mathbf{1}_{i,2,2} + \varepsilon_i. \tag{8.48}$$

Compare model (8.48) with (8.47). One difference is that in (8.47) there are $L = 4$ genders while now we arbitrarily reduced the number to $L = 3$. But more importantly, observe that in (8.48) the continuous variable X_{i1} has potentially a different slope for each level of the categorical gender variable. The slope β_1 is the slope at the base level, `F`. The slope β_1^{21} is the slope when the gender is of level 1, `M`. Finally, the slope β_1^{22} is the slope when the gender is of level 2, `O`. We thus say that there is an interaction between gender and the effect of weight on height. Note that an alternative representation of (8.48) is

$$Y_i = \beta_0 + \beta_0^{21}\mathbf{1}_{i,2,1} + \beta_0^{22}\mathbf{1}_{i,2,2} + \big(\beta_1 + \beta_1^{21}\mathbf{1}_{i,2,1} + \beta_1^{22}\mathbf{1}_{i,2,2}\big)X_{i1} + \varepsilon_i.$$

Listing 8.14 deals with such a model. In Julia, as specified by the `@formula` macro from

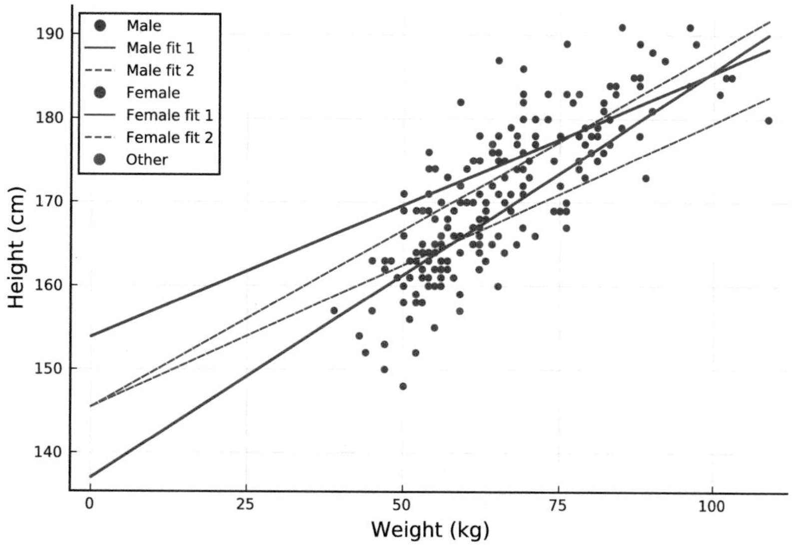

Figure 8.11: Linear model with interaction effects. Two types of models.

StatsModels.jl, interaction as in (8.48) is achieved by using `*`. See line 10 in Listing 8.14 where we specify `Weight * Sex` in the formula. This creates a model of the form

```
Height ~ 1 + Weight + Sex + Weight & Sex
```

as apparent when considering the output of Listing 8.14. An alternative is to specify `Weight & Sex` as in line 11. Here, the model is created without the individual `Sex` term:

```
Height ~ 1 + Weight + Weight & Sex
```

This corresponds to

$$Y_i = \beta_0 + \big(\beta_1 + \beta_1^{21}\mathbf{1}_{i,2,1} + \beta_1^{22}\mathbf{1}_{i,2,2}\big)X_{i1} + \varepsilon_i,$$

and implies that the y-intercept term is shared for all genders. Figure 8.11 plots predictions for each model variant. In both models the male slope and female slope is different, however, for the second model, males and females share the same intercept.

An alternative means of specifying the second model is `Weight + Weight & Sex` as appearing line 12. This creates the model,

```
Height ~ 1 + Weight + Weight & Sex
```

which when considering the output of Listing 8.14, may initially appear to differ from the second model. However, observe that the parameter estimates for both models are equivalent by inspecting the output generated by lines 21 and 22 of the listing.

When considering examples with multiple variables many more variants are possible. However, in practice, the inclusion of interaction terms is recommended only if there is a strong scientific belief that such effects are plausible. Further in the case of two continuous variables, say X_1 and X_2, including a multiplicative effect via X_1X_2 is also sometimes called an interaction effect. We don't present such an example here.

Listing 8.14: Regression with categorical variables - with interaction effects

```
using CSV, GLM, Plots, Random; pyplot()
Random.seed!(0)

df = CSV.read("../data/weightHeight.csv", copycols = true)
n = size(df)[1]
df[shuffle(1:n),:] = df
df[[10,40,60,130,140,175,190,200],:Sex] .= "O"

conts = Dict(:Sex=>DummyCoding(base="F",levels=["F","M","O"]))
model1 = lm(@formula(Height ~ Weight * Sex), df, contrasts=conts)
model2 = lm(@formula(Height ~ Weight & Sex), df, contrasts=conts)
model3 = lm(@formula(Height ~ Weight + Weight & Sex), df, contrasts=conts)

a0, a1, a2, a3, a4, a5 = coef(model1)
b0, b1, b2, b3 = coef(model2)
c0, c1, c2, c3 = coef(model3)

println(model1)
println(model2)
println(model3)
println("Model2 and Model3 equivalence: ",
    round.((b0 - c0, b1 - c1, b2 - (c1+c2), b3 - (c1+c3)),digits=5))

pred1(weight,sex) =  a0 + a1 * weight +
                          (sex=="M")*(a2+a4*weight)+
                          (sex=="O")*(a3+a5*weight)
pred2(weight,sex) =  b0 + weight*(b1*(sex=="F") + b2*(sex=="M") + b3*(sex=="O"))

males,females,other=df[df.Sex .=="M",:],df[df.Sex .=="F",:],df[df.Sex .=="O",:]

xlim = [0, maximum(df.Weight)]
scatter(males.Weight, males.Height, c=:blue, msw=0, label="Male")
plot!(xlim, pred1.(xlim,"M"), c=:blue, lw = 1.5, label="Male fit 1")
plot!(xlim, pred2.(xlim,"M"), c=:blue, linestyle=:dash, label="Male fit 2")
scatter!(females.Weight, females.Height, c=:red, msw=0, label="Female")
plot!(xlim, pred1.(xlim,"F"), c=:red, lw=1.5,label="Female fit 1",xlims=(xlim))
plot!(xlim, pred2.(xlim,"F"), c=:red, linestyle=:dash, label="Female fit 2")
scatter!(other.Weight, other.Height, c=:green, msw=0, label="Other",
        xlabel="Weight (kg)", ylabel="Height (cm)", legend=:topleft)
```

```
Height ~ 1 + Weight + Sex + Weight & Sex
-------------------------------------------------------------------------------------
                    Estimate  Std. Error     t value  Pr(>|t|)    Lower 95%    Upper 95%
-------------------------------------------------------------------------------------
(Intercept)       137.014      3.80985    35.9631       <1e-87  129.5         144.528
Weight              0.486337   0.0661584   7.35109      <1e-11    0.355855      0.616819
Sex: M             16.8427     5.23404     3.21791      0.0015    6.51974      27.1656
Sex: O             -0.30733   12.1393     -0.0253169    0.9798  -24.2494       23.6347
Weight & Sex: M    -0.170169   0.0810652  -2.09916      0.0371   -0.330051     -0.0102863
Weight & Sex: O     0.0435188  0.189214    0.229998     0.8183   -0.329661      0.416699
-------------------------------------------------------------------------------------

Height ~ 1 + Weight & Sex
---------------------------------------------------------------------------------
                    Estimate  Std. Error   t value  Pr(>|t|)   Lower 95%   Upper 95%
---------------------------------------------------------------------------------
(Intercept)       145.487     2.60535     55.8416     <1e-99  140.349     150.625
Weight & Sex: F     0.340394  0.04569      7.45007    <1e-11    0.250287    0.430501
Weight & Sex: M     0.424167  0.0343884   12.3346     <1e-25    0.356348    0.491986
Weight & Sex: O     0.396413  0.0483823    8.19336    <1e-13    0.300997    0.49183
---------------------------------------------------------------------------------

Height ~ 1 + Weight + Weight & Sex
-----------------------------------------------------------------------------------
                    Estimate  Std. Error   t value  Pr(>|t|)    Lower 95%   Upper 95%
-----------------------------------------------------------------------------------
(Intercept)       145.487      2.60535    55.8416     <1e-99  140.349       150.625
Weight              0.340394   0.04569     7.45007    <1e-11    0.250287      0.430501
Weight & Sex: M     0.0837732  0.0159008   5.2685     <1e-6     0.0524147     0.115132
Weight & Sex: O     0.0560195  0.0295724   1.89432    0.0597   -0.0023014     0.11434
-----------------------------------------------------------------------------------
Model2 and Model3 equivalence: (-0.0, 0.0, 0.0, 0.0)
```

This code listing is similar to Listing 8.13 with some minor differences. Here we modify the data in line 7 to have an additional `"O"` sex type (as opposed to two additional types in Listing 8.13). Then in lines 10–12, we use three different model specifications, all featuring interactions. The `Weight * Sex` formula in `model1` induces both constant terms and specific slope terms for each categorical level. As opposed to that, the `Weight & Sex` formula in line 11 includes only specific slope terms for each categorical level, but the levels share the same constant term. Then the `Weight + Weight & Sex` formula in line 12 implies effectively the same model only that the coefficients need to be added up. We show the equivalence between `model2` and `model3` by printing out the difference in the coefficient interpretations in lines 21–22. We use the `model1` coefficients in `pred1()` function in lines 24–26. We then use the `model2` coefficients in `pred2()` function in line 27. The remainder of the code creates Figure 8.11.

Simpson's Paradox

Having covered categorical variables and interaction effects, we now investigate the so-called *Simpsons's paradox*, or *Yule Simpson effect*. This paradox is the observation that a trend present in the data can disappear or reverse when the data is divided into subgroups. Although simple in intuition, it is an important concept to remember. One must always be careful when constructing a model, as there may be another hidden factor within the data that may significantly change the results and conclusions of our analysis.

Consider Figure 8.12 based on the `IQalc.csv` dataset. The horizontal axis represents measurements of individuals' IQs and the vertical axis is a measurement of their weekly alcohol consumption via an "alcohol consumption metric". In the left hand plot, data for all 600 individuals is presented, and it appears that there is a positive relationship between IQ and the alcohol consumption metric. However, when partitioning these individuals into three subsets based on another attribute, a different picture appears. In the right hand plot, it appears that the effect of IQ on the alcohol metric is negative for each of the groups. Note that while we don't print this, the p-value for the effect of IQ is less than 0.05 for each of the regression lines.

In Listing 8.15, we create Figure 8.12 in a straightforward manner, similar to the previous code examples.

Listing 8.15: Simpson's paradox

```
1   using CSV, GLM, Plots; pyplot()
2
3   df = CSV.read("../data/IQalc.csv")
4   groupA = df[df.Group .== "A", :]
5   groupB = df[df.Group .== "B", :]
6   groupC = df[df.Group .== "C", :]
7
8   model  = fit(LinearModel, @formula(AlcConsumption ~ IQ), df)
9   modelA = fit(LinearModel, @formula(AlcConsumption ~ IQ), groupA)
10  modelB = fit(LinearModel, @formula(AlcConsumption ~ IQ), groupB)
11  modelC = fit(LinearModel, @formula(AlcConsumption ~ IQ), groupC)
12
13  pred(x)  = coef(model)'  * [1, x]
14  predA(x) = coef(modelA)' * [1, x]
15  predB(x) = coef(modelB)' * [1, x]
16  predC(x) = coef(modelC)' * [1, x]
17
18  xlims = collect(extrema(df.IQ))
19
20  p1 = scatter(df.IQ, df.AlcConsumption, c=:black, msw=0, ma=0.2, label="")
21       plot!(xlims, pred.(xlims), c=:black, label="All data")
22
23  p2 = scatter(groupA.IQ, groupA.AlcConsumption, c=:blue, msw=0, ma=0.2, label="")
24       scatter!(groupB.IQ, groupB.AlcConsumption, c=:red, msw=0, ma=0.2, label="")
25       scatter!(groupC.IQ, groupC.AlcConsumption, c=:green,msw=0, ma=0.2, label="")
26       plot!(xlims, predA.(xlims), c=:blue, label="Group A")
27       plot!(xlims, predB.(xlims), c=:red, label="Group B")
28       plot!(xlims, predC.(xlims), c=:green, label="Group C")
29
30  plot(p1, p2, xlims=(xlims), ylims=(0,1),
31          xlabel="IQ", ylabel="Alcohol Metric", size=(800,400))
```

In line 3, the data is loaded and, in lines 4–6, we create three data frames, one for each group. In lines 8–11, we create four models where `model` includes all the data, while `modelA`, `modelB`, and `modelC`, each include data for a specific group. The predictor functions are then created in lines 13–16 The remainder of the code creates Figure 8.12.

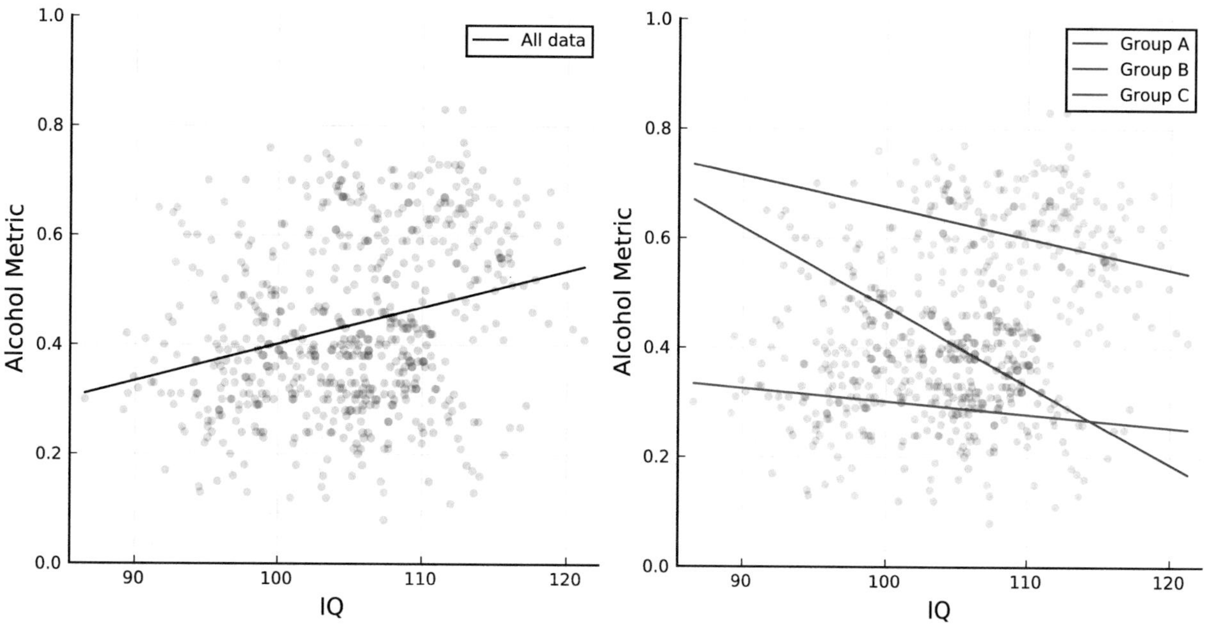

Figure 8.12: An illustration of Simpson's paradox. The trend in the data reverses when the additional variable (group) is taken into account.

8.5 Model Selection

The act of executing *model selection* involves selecting the best model from a set of possible models. As the previous Simpson's paradox example illustrated, different models can yield wildly different results and interpretations. Hence, selecting a model that represents reality well is important. Although this can involve aspects of experimental design, here we focus purely on the task of choosing the best model from a given dataset. The main idea is to select the variables (features) that describe reality well and in a manner supported by the data. Importantly, this includes the selection of transformations of the variables, as presented in the previous section, including interactions, polynomial transformations, and other forms.

Since many possible models exist, we typically evaluate a model not just based on its accuracy but also based on the model complexity. That is, if two models have roughly the same statistical properties, then the simpler model is typically preferable. This is because in such cases it is generally more likely that the simpler model would *generalize* well. An over-complicated model may lead to issues such as over-fitting which has the consequence that slight modifications to the data, or new data, would not be well described. See Figure 8.1 for a classic, yet exaggerated, illustration of such a tradeoff.

There are many methods in statistics and machine learning for model selection, none of which are accepted as the universally ideal method. In fact, many researchers still believe that the best way to carry out the model selection (even for linear models) is an open problem. Nevertheless, in practice, models need to be chosen and hence several popular model selection methods are often employed. In this section, we only briefly illustrate two such methods, *stepwise regression* (*backward elimination*), and the more modern *LASSO*-based method.

In addition to these methods, there are many other methods that we don't cover here. These include usage of the *Akaike Information Criterion (AIC)* which works well for almost any maximum likelihood-based estimation. Further, there is the related *Bayesian Information Criterion (BIC)*, computation of the *Predicted Residual Error Sum of Squares (PRESS)*, usage of *Adjusted* R^2, and other tools based on prediction evaluation such as *cross validation*. For example, in Listing 8.19, we compare different models based on their performance on a *test set*.

Classical Approach for Model Selection: Stepwise Regression

The classic approach for model selection in linear regression is called *stepwise regression*. The idea is to incrementally expand, or collapse the model, each time evaluating p-values for making the next choice of which variable to add or remove. The expanding method is called *forward selection* and is based on adding variables to the model one by one. The collapsing approach is called *backward elimination* and is based on removing variables one by one. Both of these methods are considered outdated and in many ways inferior to the LASSO method that we present below. Nevertheless, many classic studies are based on these methods and hence understanding how they operate may be useful. Here, we only focus on backward elimination.

As an algorithm, backward elimination begins with the full set of variables (features). This may also include any potential variable transformations employed. Then a regression model is fit and the p-value associated with each variable is examined and compared to a predefined threshold p^* such as $p^* = 0.05$. If all p-values are less than p^* then the full model is retained. Otherwise, the algorithm dictates to remove the variable with the highest p-value. Then a model is fit again to the reduced set of variables, and the process repeats. If all variables have p-values less than or equal to p^* the algorithm terminates, otherwise, the highest p-value variable is removed, and so on.

Note that during these iterations, the p-values are generally modified as the reduction of variables in the model affects the p-values of all other variables. A plus of backward elimination is that it is simple to execute, as we do directly in Listing 8.16. A drawback is that there is no theoretical foundation justifying why such a *greedy* algorithm works well.

In Listing 8.16, we carry out backward elimination on a starting model similar to the model of Listing 8.10. However, we purposefully add two variables, `Junk1` and `Junk2` that are independent of the response variable `Perf`, and hence should ideally be knocked out via backward elimination. Indeed, we see that in the first iteration `Junk2` is removed, then `ChMax`, then `Junk1`, and then `Freq`. This leaves us with a model with two independent variables `MMax` and `Cach`. The listing makes use of `StatsModels.jl`, a package that is implicitly included via `GLM` in most code examples. Typically, we would use the `@formula` macro to specify a model, however here the formula is constructed programmatically in line 16 using `term()` from `StatsModels.jl`.

An additional scenario in backward elimination that we do not cover here deals with non-binary categorical variables. In such cases, as described in the section above, each categorical variable induces several dummy variables in the model. This creates a potential problem for backward elimination because all dummy variables should be eliminated together, or retained. One way to circumvent this is via the *partial F-test*, a concept that we do not cover further, but was used in Listing 7.10, using the `ftest()` function. We omit further details.

Listing 8.16: Backward elimination in linear models

```
using StatsModels, RDatasets, DataFrames, GLM, Random
Random.seed!(0)

n = 30
df = dataset("MASS", "cpus")[1:n,:]
df.Freq = map( x->10^9/x , df.CycT)
df = df[:, [:Perf, :MMax, :Cach, :ChMax, :Freq]]
df.Junk1 = rand(n)
df.Junk2 = rand(n)

function stepReg(df, reVar, pThresh)
    predVars = setdiff(propertynames(df), [reVar])
    numVars = length(predVars)
    model = nothing
    while numVars > 0
        fm = term(reVar) ~ term.((1, predVars...))
        model = lm(fm, df)
        pVals = coeftable(model).cols[4][2:end]
        println("Variables: ", predVars)
        println("P-values = ", round.(pVals,digits = 3))
        pVal, knockout = findmax(pVals)
        pVal < pThresh && break
        println("\tRemoving the variable ", predVars[knockout],
                " with p-value = ", round(pVal,digits=3))
        deleteat!(predVars,knockout)
        numVars = length(predVars)
    end
    model
end

model = stepReg(df, :Perf, 0.05)
println(model)
```

```
Variables: [:MMax, :Cach, :ChMax, :Freq, :Junk1, :Junk2]
P-values = [0.0, 0.063, 0.695, 0.344, 0.449, 0.901]
        Removing the variable Junk2 with p-value = 0.901
Variables: [:MMax, :Cach, :ChMax, :Freq, :Junk1]
P-values = [0.0, 0.053, 0.654, 0.323, 0.433]
        Removing the variable ChMax with p-value = 0.654
Variables: [:MMax, :Cach, :Freq, :Junk1]
P-values = [0.0, 0.023, 0.332, 0.474]
        Removing the variable Junk1 with p-value = 0.474
Variables: [:MMax, :Cach, :Freq]
P-values = [0.0, 0.008, 0.211]
        Removing the variable Freq with p-value = 0.211
Variables: [:MMax, :Cach]
P-values = [0.0, 0.011]

Perf ~ 1 + MMax + Cach
--------------------------------------------------------------------------------
               Estimate   Std. Error   t value  Pr(>|t|)    Lower 95%  Upper 95%
--------------------------------------------------------------------------------
(Intercept)  -39.7959    24.1264      -1.64948    0.1106  -89.2991     9.70734
MMax           0.0113988  0.00108912  10.4661     <1e-10    0.00916411 0.0136335
Cach           0.904057   0.332653     2.71772    0.0113    0.22151    1.5866
--------------------------------------------------------------------------------
```

In lines 4-9, we set up the data frame. It is based on a small number of observations, `n`. We modify the original variables in lines 6–7 and then add two additional variables based on random elements, `Junk1`, and `Junk2`. The function `stepReg()` defined in lines 11–29 executes backward elimination on the data frame `df`, with response variable `reVar`, and a p-value threshold `pThresh`. In line 12, `predVars` is the array of symbols of the independent variables after `setdiff()` is used to remove the response variable from the `propertynames()` of all variables in the data frame. In line 13, we initialize the number of variables `numVars`. In line 14, the model is initialized as an empty object using `nothing`. The loop in lines 15–27 is executed as along as there are variables or until the maximum of the p-values is less than `pThresh` (line 21). In line 21, we programmatically construct a formula, `fm`. We use here the `term()` function from `StatsModels.jl`. In line 17, we execute a linear model fit for this formula. In line 18, we read the p-values by applying `coeftable()` to the model. We take indices `2:end` to avoid the intercept. After printing out information about the current iteration in lines 19–20, we find the maximal p-value and its index in line 21. Then in line 22, we `break` if needed. This uses the short circuit `&&` evaluation. Otherwise, we remove the variable in line 25 using `deleteat!()` and the process repeats. We use our function `stepReg()` in line 31 to find a reduced model which we print in line 32.

Model Selection via LASSO

A completely different approach for model selection is via the *least absolute shrinkage and selection operator* method known as *LASSO* for short. This method implements a concept known as *regularization*, further discussed in Section 9.3. As opposed to stepwise variable selection algorithms that deal directly with elimination of variables, LASSO eliminates variables as a by-product of regularization. The basic minimization objective of regression $||A\beta - y||_2^2$ is augmented with an additional *penalty* or *regularization* term as follows:

$$\min_{\beta \in \mathbb{R}^{p+1}} ||A\beta - y||_2^2 + \lambda ||\beta||_1. \tag{8.49}$$

Recall that $|| \cdot ||_1$ and $|| \cdot ||_2$ are the L_1 and L_2 norms appearing in (8.5) and (8.4), respectively. Here, $\lambda \geq 0$ and the L_1-based term $\lambda||\beta||_1$ induces a penalty for high values of β. The minimization problem (8.49) is equivalent to minimization of the basic regression objective $||A\beta - y||_2^2$ together with an additional constraint of $||\beta||_1 \leq t$. Further, the theory of optimization implies that there is mapping between λ and t such that if $\lambda = 0$ then $t = \infty$ and as λ grows t decreases. This decrease is known as *shrinkage* and implies that for increasing values of λ the constraint on $||\beta||_1$ becomes more and more stringent. Then a consequence of the L_1 norm is that as λ grows, more and more elements of β are zeroed out and are thus "eliminated" from the model.

The LASSO objective in its own right is useful as a model beyond model selection, however, here we present a simple example of how to use it for model selection. Assume that we wish to keep only three variables in the model. In that case, we can fit a LASSO model as (8.49) for an increasing collection of λ values and seek the value λ^* such that, for $\lambda < \lambda^*$, there are more than 3 variables in the model and, at λ^*, there are 3 variables. We then read off the coefficients β and keep the variables that have non-zero β values, while eliminating those that have zero values.

This process is demonstrated in Listing 8.17 which uses the `Lasso` package and also creates Figure 8.13. As apparent from the output, $\lambda^* = 54.2$ is the minimal λ value in which the number

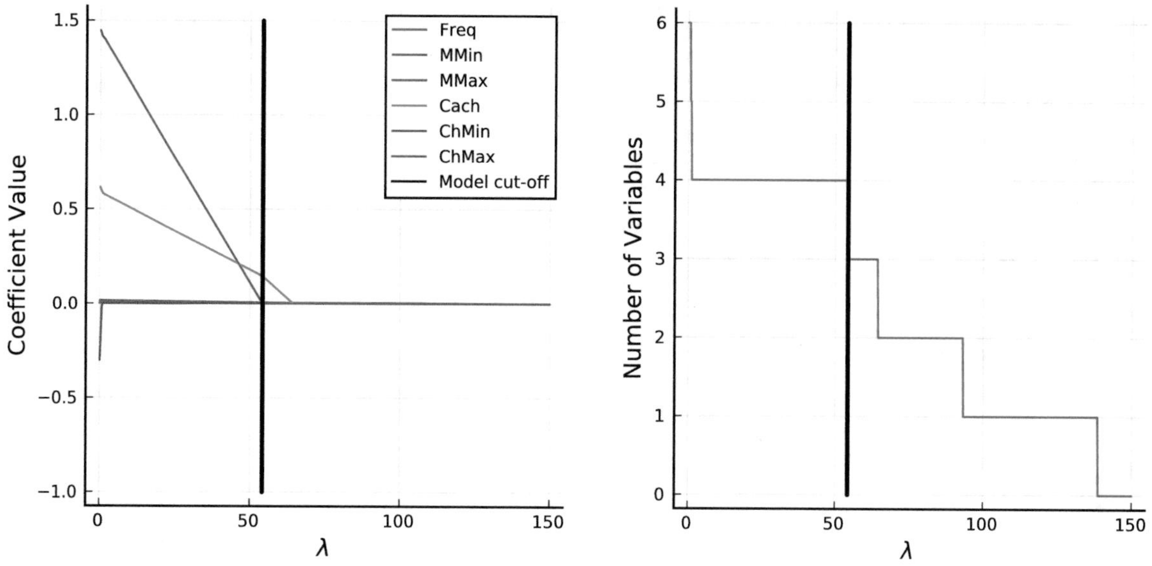

Figure 8.13: Model selection using LASSO.

of variables is 3. The left hand side of Figure 8.13 is a *trace plot* which shows how the values of the coefficients shrink when λ increases. Observe that the trace plot has piecewise breakpoints. The right hand plot is less standard and presents the number of (non-zero) variable coefficients in the model as a function of λ. Another common plot that we don't present here is a plot of the loss function as a function of λ. With such a plot, it is often common to search for a *knee point* when using LASSO for model selection. Such a point is the point at which increasing λ (removing more variables) increases the loss significantly, but up to it increases of λ do not cause heavy increases in the loss. Choosing such a point is often a subjective matter.

As with stepwise regression, when using LASSO with categorical variables more care is needed. Here, there is also the potential problem that non-binary categorical variables are related to several dummy variables, and then certain dummy variables may be removed from the model, while others are not. For this, there is an algorithm called *Group LASSO* which we do not discuss further. Note also that there are alternative formulas to LASSO, where the intercept term, β_0, is not included in the penalty term of (8.49). Such a formulation implies that the intercept β_0 will not be knocked out of the model. Further, in the same manner, it is possible to retain other coefficients in the model by omitting them from the $\lambda||\beta||_1$ term.

Finally, we mention that unlike regression (as well as ridge regression covered in the next chapter), LASSO is not explicitly solved but rather always requires iterative algorithms. The `Lasso` package implements one such algorithm called *glmnet coordinate descent.* Another common algorithm that can be adapted to work for LASSO is *Least Angle Regression (LARS).*

Listing 8.17: Using LASSO for model selection

```
1  using RDatasets, DataFrames, Lasso, LaTeXStrings, Plots, Measures; pyplot()
2
3  df = dataset("MASS", "cpus")
4  df.Freq = map( x->10^9/x , df.CycT)
5  df = df[:, [:Perf, :Freq, :MMin, :MMax, :Cach, :ChMin, :ChMax]]
6  X = [df.Freq df.MMin df.MMax df.Cach df.ChMin df.ChMax]
7  Y = df.Perf
8
9  targetNumVars = 3
10
11 lambdaStep = 0.2
12 lamGrid = collect(0:lambdaStep:150)
13 lassoFit = fit(LassoPath,X, Y, λ = lamGrid);
14 dd = Array(lassoFit.coefs)'
15 nV = sum(dd .!= 0.0 ,dims=2)
16
17 goodLambda = lamGrid[findfirst((n)->n==targetNumVars,nV)]
18 newFit = fit(LassoPath,X, Y, λ = [goodLambda - lambdaStep, goodLambda])
19 println(newFit)
20 println("Coefficients: ", Array(newFit.coefs)'[2,:])
21
22 p1 = plot(lassoFit.λ, dd, label = ["Freq" "MMin" "MMax" "Cach" "ChMin" "ChMax"],
23     ylabel = "Coefficient Value")
24 plot!([goodLambda,goodLambda],[-1,1.5],c=:black, lw=2, label = "Model cut-off")
25
26 p2 = plot(lassoFit.λ,nV, ylabel = "Number of Variables",legend = false)
27 plot!([goodLambda,goodLambda],[0,6],c=:black, lw=2, label = "Model cut-off")
28
29 plot(p1,p2,xlabel= L"\lambda", margin = 5mm, size = (800,400))
```

```
LassoPath (2) solutions for 7 predictors in 73 iterations):
---------------------------
          λ   pct_dev  ncoefs
---------------------------
[1]   54.0  0.673401       4
[2]   54.2  0.672031       3
---------------------------
Coefficients: [0.0, 0.005294166576, 0.0056046537521, 0.14520251523, 0.0, 0.0]
```

Lines 3–5 are similar to the previous listing creating `df` with the right variables. With then create the matrix `X` and the array `Y` to be used by for LASSO. In line 9, `targetNumVars` is set. This is the number of variables that will remain in the model after elimination. Lines 11–12 set `lamGrid` which is the range of λ values over which to execute LASSO. The actual fitting is in line 13 with `fit(LassoPath,...)`. The raw coefficients result of the LASSO fit is in `lassoFit.coefs`. We then convert this to an array using `Array()` and a transpose operator to arrive at `dd`. The elements of this array are then compared to `0.0` and summed on the second dimension to arrive at `nV` in line 15. This array specifies the number of non-zero variables for each step in the lasso path. Then in line 17, we set `goodLambda` as the first value where we have the target number of variables. Note the use of `findfirst()`. Then in line 18, we execute another LASSO fit, called `newFit` with only two values of λ, the `goodLambda`, and one value smaller than that. When we print the results in line 19, we see that indeed, for the smaller λ, there are 4 variables, while for `goodLambda`, there are the desired number of variables. We print the coefficients of the variables in line 20 and this gives an indication of which variables were selected. The remainder of the code creates Figure 8.13.

8.6 Logistic Regression and the Generalized Linear Model

Up to now, we dealt with statistical models of the form $Y = \beta^\top X + \varepsilon$ (where we consider the first element of the vector X to be 1 allowing for a β_0 intercept term). These models are linear because the conditional expectation of Y given X is linear/affine. Still, we were able to transform elements of X, as in the polynomial example (8.45) to accommodate some non-linear relationships as long as transformed values of X interact linearly. Since X is considered non-random in the regression, such transformations allowed us to stay in the realm of linear regression and least squares. However, what if we wanted to transform Y? This is where *Generalized Linear Models (GLM)* come into the picture.

For a GLM, we choose a one-to-one real function $g(\cdot)$ and call it the *link function*. We then set,

$$g(Y) = \beta^\top X + \varepsilon, \qquad \text{or} \qquad Y = g^{-1}(\beta^\top X + \varepsilon),$$

where $g^{-1}(\cdot)$ is the *inverse link function*. Now remembering from (8.1) that in regression we wish to consider the conditional expectation of Y given $X = x$, we have

$$\hat{y}(x) = \mathbb{E}[g^{-1}(\beta^\top x + \varepsilon)].$$

The random component in this expectation is ε and for any distribution of ε and every link function, there is some expected value function $\hat{y}(x)$. In this linear regression case, the link function is just the identity function $g(y) = y$ in which case $\hat{y}(x) = \beta^\top x$. This is because expectation is linear and ε has zero mean. However, in the generalized linear model, the expected value is generally more complicated. In the $g(y) = y$ case and assuming ε is normally distributed, $\hat{\beta}$ found via least squares was also the MLE. This was discussed briefly when we presented the log-likelihood (8.23). However, in the GLM case, least squares does not generally yield the MLE. Instead, we need to use other numerical methods for estimating the parameters.

It turns out that when ε follows a distribution from an *exponential family*, there is a corresponding suitable link function, $g(\cdot)$ that allows finding the MLE of β using efficient algorithms. The exponential family of distributions is actually a "clan" of distributions that encompasses the normal distribution, the exponential distribution, the Poisson distribution, and many more common distributions. We don't discuss it further in the book, however, its relevance to GLM is interesting. The practical point is that there are distribution-link function pairs that allow for simple modeling and efficient procedures for finding the MLE $\hat{\beta}$.

When we fit a GLM model to data, Julia's `GLM` package, searches for the MLE. It then uses asymptotic properties of MLE for yielding standard errors and p-values associated with each $\hat{\beta}_i$. This is where using `glm()` slightly differs from `lm()` which uses the T-distribution for p-values (see Listing 8.4). In contrast, `glm()` just uses the normal distribution for inference.

We first focus on one of the most common GLM models, *logistic regression*. We then present an additional example, using GLM for prediction. Keep in mind that logistic regression is also revisited in Section 9.2 in the context of supervised learning classification. In fact, in machine learning, it has become one of the cornerstone algorithms for classification and is often the gateway to more advanced deep learning models.

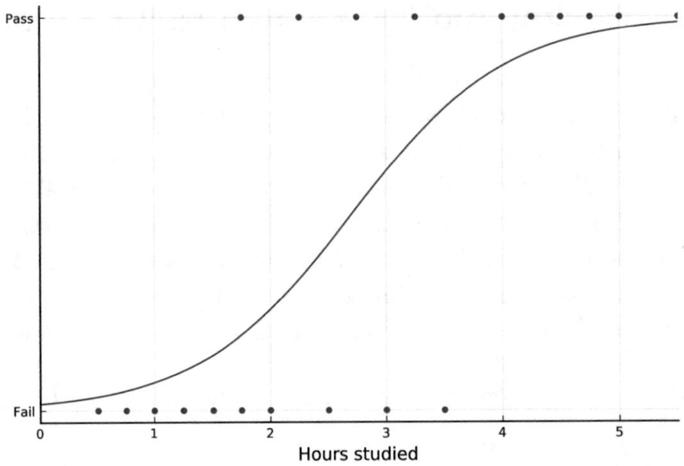

Figure 8.14: Logistic regression for exam performance.

Logistic Regression

Sometimes a dataset may consist of predictor variables which are continuous, but where the recorded outcomes are binary in nature, such as heads/tails, success/failure, or 0/1. In these cases, it makes no sense to use a linear model. Instead, *logistic regression* can be used to calculate the parameters of a *logistic model.*

In a logistic model, the outcomes are represented by the indicator function, with positive (i.e. successful) outcomes considered 1, and negative outcomes considered 0. The logistic model then considers that the *log of the odds* is a linear combination of the predictors, as shown below,

$$\log\left(\frac{Y}{1-Y}\right) = \beta_0 + \beta_1 X_1 + \ldots + \beta_p X_p + \varepsilon \qquad \text{where} \qquad Y = \mathbb{P}(\text{success}). \tag{8.50}$$

Note that the left hand side of this equation is also known as the *logit function.* Considering (8.50) as a GLM, we have that

$$g(y) = \log\left(\frac{y}{1-y}\right) \qquad \text{and} \qquad g^{-1}(u) = \frac{1}{1+e^{-u}} = \frac{e^u}{1+e^u}.$$

The inverse of the logit function is called the *sigmoid function* also known as a *logistic function.* One suitable distribution for ε is a Bernoulli distribution. In this case, we have that $\hat{y}(x) = g^{-1}(\beta x)$

In the case of a single predictor variable, this can be re-arranged as

$$\hat{y}(x) = \frac{1}{1+e^{-(\beta_0+\beta_1 x)}}. \tag{8.51}$$

In Listing 8.18 below, an example is presented based on the results of an exam for a group of students, with only one predictor variable. In this example, the result for each student has been recorded as either pass (1), or fail (0), along with the number of hours each student studied for the exam. In this example, the GLM package is used to perform logistic regression. The resulting model is plotted in Figure 8.14, and the associated model and coefficients are printed below.

Listing 8.18: Logistic regression

```
1  using GLM, DataFrames, Distributions, Plots, CSV; pyplot()
2
3  data = CSV.read("../data/examData.csv")
4  model = glm(@formula(Pass ~ Hours), data, Binomial(), LogitLink())
5  println(model)
6
7  b0, b1 = coef(model)
8  pred(x) = 1/(1+exp(-(b0 + b1*x)))
9
10 xGrid = 0:0.1:maximum(data.Hours)
11 scatter(data.Hours, data.Pass, c=:blue, msw=0)
12 plot!(xGrid, pred.(xGrid),
13         c=:red, xlabel="Hours studied", legend=:none,
14         xlims=(0, maximum(data.Hours)), yticks=([0:1;], ["Fail", "Pass"]))
```

```
Pass ~ 1 + Hours

Coefficients:
---------------------------------------------------------------------------
              Estimate  Std. Error   z value  Pr(>|z|)  Lower 95%  Upper 95%
---------------------------------------------------------------------------
(Intercept)  -4.07771    1.76098   -2.31559    0.0206  -7.52918   -0.626248
Hours         1.50465    0.628716   2.3932     0.0167   0.272384   2.73691
---------------------------------------------------------------------------
```

In line 3, the data is loaded into the data frame `data`. In line 4, the `glm()` function is used to create a logistic model. The `@formula` macro is used to create a formula as the first argument, where the predictor variable is the `Hours` column, and the response is the `Pass` column. The data frame `data` is specified as the second argument. The third argument is the distribution of the error, and the fourth argument is the type of link function used, specified here as `LogitLink()`. In line 7, the coefficients of the model are stored as `b0` and `b1`. In line 8, we define the prediction function `pred()` where we manually implement the sigmoid function of $\beta_0 + \beta_1 x$ as in (8.51). The remainder of the code creates Figure 8.14.

Other GLM Examples

Generalized linear models go well beyond linear regression and logistic regression. For example, another common model is *Poisson regression* where the link function is $\log(\cdot)$ and the distribution of ε is Poisson. This type of model is often used for *counts data*. Other common examples exist as well, sometimes used for a specific application and at other times used simply to broaden the family of statistical models available.

Sometimes, for a given dataset one may wish to try multiple GLM models to see which one performs best. One naive method for such model selection is to split the data into a *training set* and *testing set*. In fact, this is the most common machine learning approach, employed in *supervised learning* which we cover in the next chapter. We now illustrate such a basic approach for selecting a GLM model.

In Listing 8.19, we consider the `cpus` dataset and use 80% of the data for training and 20% for testing (or validation). We fit a linear model, a Poisson regression model, and a GLM model with a gamma error distribution and an inverse link function, $1/x$. All three are common. We then compute the loss for each model to see that in this case the second model, Poisson regression appears to work best. This is only an elementary example of GLM, aiming to show how to use different models with the `glm()` function.

Listing 8.19: Exploring generalized linear models

```
using GLM, RDatasets, DataFrames, Distributions, Random, LinearAlgebra
Random.seed!(0)

df = dataset("MASS", "cpus")
n = size(df)[1]
df = df[shuffle(1:n),:]

pTest = 0.2
lastTindex = Int(floor(n*(1-pTest)))
numTest = n - lastTindex

train = df[1:lastTindex,:]
test = df[lastTindex+1:n,:]

form = @formula(Perf~CycT+MMin+MMax+Cach+ChMin+ChMax)
model1 = glm(form, train, Normal(),  IdentityLink())
model2 = glm(form, train, Poisson(), LogLink())
model3 = glm(form, train, Gamma(),  InverseLink())

invIdenityLink(x) = x
invLogLink(x) = exp(x)
invInverseLink(x) = 1/x

A = [ones(numTest) test.CycT test.MMin test.MMax test.Cach test.ChMin test.ChMax]
pred1 = invIdenityLink.(A*coef(model1))
pred2 = invLogLink.(A*coef(model2))
pred3 = invInverseLink.(A*coef(model3))

actual = test.Perf
lossModel1 = norm(pred1 - actual)
lossModel2 = norm(pred2 - actual)
lossModel3 = norm(pred3 - actual)

println("Model 1: ", coef(model1))
println("Model 2: ", coef(model2))
println("Model 3: ", coef(model3))
println("\nLoss of models 1,2,3: ",(lossModel1 ,lossModel2, lossModel3))
```

```
Model 1: [-60.3184, 0.055669, 0.0175752, 0.00435287, 0.907027, -1.76895, 2.36445]
Model 2: [3.91799, -0.00161407, 9.77454e-6, 3.1481e-5, 0.00577684, 0.00519083, 0.00156646]
Model 3: [0.00993825, 6.12058e-5, 1.42618e-7, -1.5675e-7, -2.71825e-5, -7.61944e-5, 1.2299e-5]

Loss of models 1,2,3: (558.944925722054, 360.11867318943433, 577.4165274822029)
```

In lines 4–6, we setup the data frame based on the `cpus` dataset and randomly shuffle the rows in line 6. In lines 8–10, we determine the indices of the training set and test set. The training set data frame, `train`, is then determined in line 12. The test set data frame, `test` is determined in line 13. Line 15 sets the `formula` to be used for GLM. The dependent variable is `Perf` and the independent variables are `CycT`, `MMin`, `MMax`, `Cach`, `ChMin`, and `ChMax`. Lines 16–18 create three `glm` models: `model1` is a standard linear model; `model2` has a `LogLink()` link function with `Poisson()` error; and `model3` has an `InverseLink()` link function with `Gamma()` error. In lines 20–22, we define the inverse link functions for each of those models. The design matrix for the test data, `test` is constructed in line 24. We then use it to create the predictions of the test data in lines 25–27. The in lines 30–32, we evaluate the performance of these three models by computing the loss. The results are printed in lines 34–37.

8.7 A Taste of Time Series and Forecasting

So far in this chapter, we dealt with models that relate the independent variables X and the dependent variable Y. We now focus on a special case in which X represents time and Y is some measured value. Data that represents this situation is called a *time-series*. We have already dealt with basic visualizations of time series in Listing 4.19 and Listing 4.20 of Chapter 4. We now present additional examples related to *forecasting* using linear regression and autocorrelation analysis.

The field of time series is well developed and includes a variety of *time-series models* for prediction and forecasting. Many models are based on stochastic processes, a concept that we visit in Chapter 10. Nevertheless, in this book, we don't explicitly deal with common time-series models such as *Autoregressive Moving Average Models* (ARMA) or similar models. For learning about such models, we suggest [SS17]. An exception is in Listing 10.13 of Chapter 10 where we deal with Kalman filtering. However, as a general rule, the broad area of time-series modeling is not covered in this book. Nevertheless, the purpose of this section is to present the reader with a taste of time-series forecasting, especially using basic tools of linear regression.

We deal with the dataset `oneOnEpsilonBlogs.csv` which records visits to the blog posts available via the website https://oneonepsilon.com/. This is a mathematics education website. The dataset was created based on Google Analytics data of the website and records the number of daily site visits during the period of June 15, 2018–May 31, 2020. This is a period spanning 717 days. The typical goal one has with such a time series is prediction and forecasting: Given historical measurements and current trends, what is the future prediction of daily visits.

In Listing 8.20, we carry out exploratory data analysis of this dataset, creating Figure 8.15. This analysis involves transformations of the time series, some of which are done via the `TimeSeries` package which supplies `TimeArray` types for recording time series. The top left plot presents the original time series in blue together with a 7 day moving average in red. The average is calculated using formula (4.10) from Chapter 4, with $L = 7$. The top right plot "zooms in" on the days during the year 2020.

The next two plots (middle row of Figure 8.15) present the difference between the actual time series and the weekly average. Such a transformation, sometimes called *demeaning*, is common practice in time series for the purpose of removing trends. As is apparent from the right hand plot

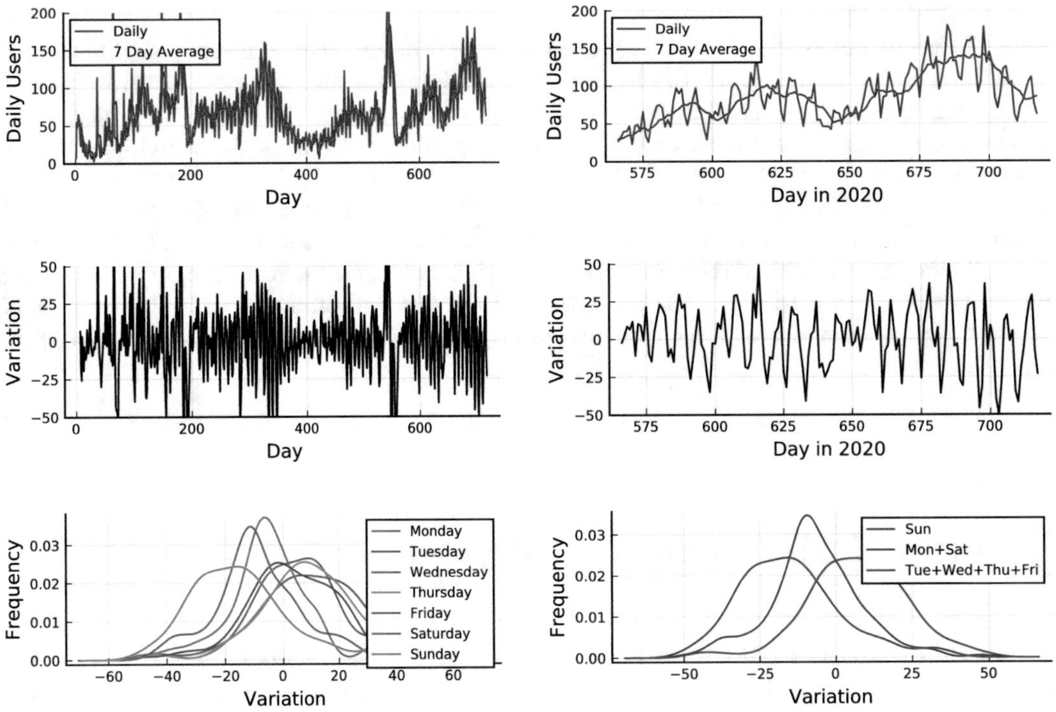

Figure 8.15: Exploratory data analysis of time-series data.

(focusing on the days in 2020), there is a weekly trend for website visits. This is not unexpected because educational blog posts tend to be visited more during the week than on weekends.

The analysis appearing in Figure 8.15 continues with the purpose of understanding the weekly variation. In the bottom left plot, smoothed histograms of the weekly variations are plotted separately for each day of the week. By looking at the distributions in the bottom left plot, it is apparent that the days can be roughly clustered into three groups. We denote these as follows:

Group 1: Sunday.

Group 2: Monday and Saturday.

Group 3: Tuesday, Wednesday, Thursday, and Friday.

The smoothed histograms for each of these groups are then plotted on the right hand bottom plot. In general, such exploratory data analysis of time series can be conducted in several steps. Yet here, for the purpose of creating a single figure, all steps are aggregated into Listing 8.20. The key finding from this analysis is the existence of 3-day groups, and this is then used in the next code listing for creating a forecasting model.

Also central to the study of time series is the concept of *stationarity*, a concept that we briefly discuss in the final code example of this section.

Listing 8.20: Exploratory data analysis of a time series

```
using CSV, TimeSeries, Dates, Statistics, Measures, Plots, StatsPlots; pyplot()

df = CSV.read("../data/oneOnEpsilonBlogs.csv",copycols = true)

tsA = TimeArray(Date.(df.Day,Dates.DateFormat("m/d/y")),df.Users)
tsB = moving(mean,tsA,7,padding = true)
tsC = TimeArray(timestamp(tsA), values(tsA) - values(tsB))

dow = dayofweek.(timestamp(tsA));
dayDiv = [filter((x)->!isnan(x) && x >= -50 && x <= 50, values(tsC)[dow .== d])
        for d in 1:7]

start2020 = findfirst((d)->Year(d)==Year(20),timestamp(tsA))
indexLast = length(tsA)
indexes2020 = start2020:indexLast

dayGroup1 = [7] #Sun
dayGroup2 = [1,6] #Mon,Sat
dayGroup3 = [2,3,4,5] #Tue, Wed, Thu, Fri
dayGroups = [dayGroup1, dayGroup2, dayGroup3]
groupDivs = [vcat(dayDiv[g]...) for g in dayGroups]

default(legend = :topleft)
labels = ["Daily" "7 Day Average"]
p1 = plot(1:indexLast, [values(tsA) values(tsB)], c=[:blue :red],label = labels,
        xlabel = "Day", ylabel = "Daily Users", ylim = (0,200))

p2 = plot(indexes2020, [values(tsA)[indexes2020] values(tsB)[indexes2020]],
        c=[:blue :red], label=labels, xlabel="Day in 2020",
        ylabel="Daily Users",ylim=(0,200))

p3 = plot(1:indexLast, values(tsC),label = labels, c=:black,
        xlabel="Day",ylabel="Variation",ylim=(-50,50),legend=false)

p4 = plot(indexes2020, values(tsC)[indexes2020],label = labels, c=:black,
         xlabel="Day in 2020",ylabel="Variation",ylim=(-50,50),legend=false)

dayNames = dayname.(timestamp(tsA)[4:10])
p5 = density(dayDiv, label = hcat(dayNames...),legend = :topright,
            xlabel = "Variation", ylabel = "Frequency",)

dayGroupNames = ["Sun", "Mon+Sat", "Tue+Wed+Thu+Fri"]
p6 = density(groupDivs,label = hcat(dayGroupNames...), legend = :topright,
            xlabel = "Variation", ylabel = "Frequency",c=[:blue :red :green])

plot(p1, p2, p3, p4, p5, p6, layout=(3,2), size = (900,600), margin = 5mm)
```

In line 3, we read the input file into the data frame `df`. In line 5, we create a `TimeArray()`, `tsA`, based on the `Day` column for a time stamp, and the `Users` column as the data. We convert `df.Day` into a type `Date` by specifying the `DateFormat`. In line 6, we run a 7-day moving average on the `tsA`. Specification of `padding = true` implies that `tsB` has the same length as `tsA`. In line 7, we create, `tsC` with the same time stamps as the previous time arrays, using `timestamp()`, and a difference between the values of `tsA` and `tsB`. In line 9, we create the array `dow` by broadcasting `dayofweek()` onto the time stamps. In lines 10–11, we create an array of 7 arrays, `dayDiv`. Each of the elements of `dayDiv` contains the values from `tsC` for a specific day of the week. We filter the values that are not in the range [-50,50] and are not numbers. For that we use the `isnan()` function with a not (`!`). In line 13, we set `start2020` as the index of the first sample in the year 2020. This is via the `findfirst()` function and the `Year()` function. Line 14 finds the ending index and line 15 defines the range of values in 2020. Lines 17–19 defines indices of days split into 3 groups. We use these in line 21 to create an array of arrays of day measurements, one for each group. The remainder of the code uses the processed data to create Figure 8.15. Observe the use of `density()` from `StatsPlots.jl` and the use of the `dayname()` function.

Using Linear Regression for Time Series

We now continue with the `oneOnEpsilonBlogs.csv` dataset and create a very simple time-series forecasting model based on linear regression. This by no means the "state of the art" of time-series modeling for forecasting. Nevertheless, our model illustrates how some of the linear regression and least squares ideas from this chapter may be employed for forecasting.

Looking at the top left plot of Figure 8.15, it is obvious that there isn't one simple trend in the time series. That is, visits to the website are not simply increasing or decreasing over the whole time period. Hence with our predictive model, we don't aim to use all of the time-series data. Instead, we only train on small short-term chunks and then create short-term predictions moving forward. As we illustrate, such short-term predictions sometimes work well.

Key to our predictive model is the analysis from the previous code listing where we postulate that there are three groups of days. We then devise a regression model with time being the main independent variable but also using the categorical variable of day group. As described in Section 8.4, we then use dummy variables for two out of the three groups to arrive at the model,

$$Y_t = \beta_0 + \beta_1 t + \beta_2 \mathbf{1}_{\text{Group 1}}(t) + \beta_3 \mathbf{1}_{\text{Group 2}}(t) + \varepsilon_t. \tag{8.52}$$

Here $\mathbf{1}_{\text{Group 1}}(t)$ is 1 when day t is a Sunday and is 0 otherwise. Similarly, $\mathbf{1}_{\text{Group 2}}(t)$ takes the value of 1 when day t is a Monday or a Saturday. Note that there is no need for such a dummy variable for Group 3 because it is taken as the base level.

With a model such as (8.52), we can now observe data over some training range, say $\{T_0, \ldots, T_1\}$ and fit the model parameters. We can then use these parameters to make a prediction over a testing range such as $\{T_1 + 1, \ldots, T_2\}$. We do so in Listing 8.22 for four different training range - testing range pairs. The prediction results are presented in Figure 8.16 where the training data is in blue, the model fit over the training data is in red, the future testing data is in green and the future prediction (forecast) is in purple. It is apparent that for some train-test pairs this predictive model works well, e.g. bottom right plot, but for others not so well, e.g. bottom left plot. We also print the model output for the top left fit, so as to get a sense of the magnitude of the coefficient estimates.

Listing 8.21: Using linear regression for forecasting in a time series

```
using CSV,DataFrames,Dates,GLM,Statistics,LinearAlgebra,Measures,Plots; pyplot()

df = CSV.read("../data/oneOnEpsilonBlogs.csv",copycols = true)
len = size(df)[1]

dow = dayofweek.(Date.(df.Day,Dates.DateFormat("m/d/y")))
dayGroups = [[7], [1,6], [2,3,4,5]]
inds = [[in(d,grp) for d in dow] for grp in dayGroups]
df2 = DataFrame(Time=1:len,Users=df.Users,
                Group1=inds[1],Group2=inds[2])

trainRange1, futureRange1 = 100:130, 130:180
trainRange2, futureRange2 = 200:300, 300:320
trainRange3, futureRange3 = 400:500, 500:600
trainRange4, futureRange4 = 560:600, 600:630

function forecast(trainRange,futureRange)
    model = lm(@formula(Users ~ Time + Group1 + Group2),df2[trainRange,:])
    pred(time) =  dot(coef(model),[1,time,inds[1][time],inds[2][time]])
    model, pred.(trainRange), pred.(futureRange)
end

function forecastPlot(train,future)
    p = plot(train[1], df.Users[train[1]], label = "Observed Users",
            xlabel = "Day", ylabel = "Daily Users", c=:blue)
        plot!(train[1], train[2], label = "Train", c=:red)
        plot!(future[1], future[2], label = "Forecast",c=:magenta)
        plot!(future[1], df.Users[future[1]], label = "Actual Users",c=:green)
    return p
end

model1, train1, fcst1 = forecast(trainRange1,futureRange1)
println(model1)

default(legend = :topleft)
p1 = forecastPlot((trainRange1, train1), (futureRange1, fcst1))

_, train2, fcst2 = forecast(trainRange2, futureRange2)
default(legend = false)
p2 = forecastPlot((trainRange2, train2), (futureRange2, fcst2))

_, train3, fcst3 = forecast(trainRange3, futureRange3)
p3 = forecastPlot((trainRange3, train3), (futureRange3, fcst3))

_, train4, fcst4 = forecast(trainRange4, futureRange4)
p4 = forecastPlot((trainRange4, train4), (futureRange4, fcst4))
plot(p1, p2, p3, p4, size = (900,600), margin = 5mm)
```

```
Users ~ 1 + Time + Group1 + Group2
---------------------------------------------------------------------------------
                 Estimate  Std. Error    t value  Pr(>|t|)    Lower 95%  Upper 95%
---------------------------------------------------------------------------------
(Intercept)  -28.8327      55.541    -0.519124    0.6079  -142.793      85.1281
Time           0.996371     0.48016   2.07508     0.0476     0.0111641   1.98158
Group1       -25.55        12.2511   -2.08552     0.0466   -50.6873     -0.412742
Group2       -14.05         9.63914  -1.4576      0.1565   -33.8279      5.72789
---------------------------------------------------------------------------------
```

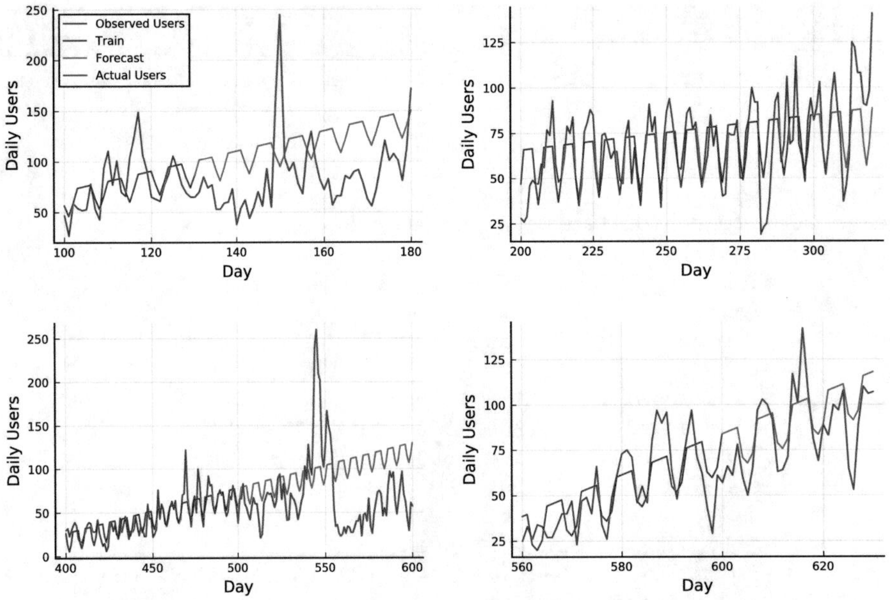

Figure 8.16: Using linear regression for time-series forecasting.

In line 6, we set `dow` to match the day of the week of each time point. In line 7, we set an array of arrays indicating groups of days. Then line 8 sets an array of arrays of indicator variables. We use the `in()` function to see if each day `d` from `dow` is in the desired group. We now create the data frame `df2` with indicators for `Group1` and `Group2` (there is no need for an indicator for the third group. In lines 12–15, we set four scenarios of training range and a future range over which to predict. The function `forecast()` defined in lines 17–21 creates a linear model, `model` based on `trainRange`. The prediction function `pred()` defined in line 19 is then used to make a prediction as part of the return value tuple in line 20. The function `forecastPlot()` in lines 23–30 assumes that the inputs `train` and `future` are each tuples and plots a plot for these tuples. In lines 32–33, we carry out forecasting for `futureRange1` based on `trainRange1`. We prin tout the model in line 33 and then create the first plot for this range in line 36. Similar plots for the other ranges are created in the remainder of the code.

Stationarity, Autocorrelation and Time-series Models

The regression-based prediction model (8.52) is a very crude way to analyze time series because it assumes that the time series is a deterministic model of time,

$$Y_t = \beta_0 + \beta_1 t + \beta_2 \mathbf{1}_{\text{Group 1}}(t) + \beta_3 \mathbf{1}_{\text{Group 2}}(t),$$

with an addition of an independent error term ε_t. More advanced models assume that the daily noise component of the time series has more sophisticated statistical properties and is not just an independent sequence.

We don't delve into such models but rather refer the reader to [SS17]. Nevertheless, we present a simple *autocorrelation* analysis of the time series, as this is often the gateway for fitting more advanced models. In general, such analysis can only take place on *stationary* sequences. A concept that we define now.

A sequence of random variables is said to be *(strictly) stationary* if the distribution of the sequence does not vary when shifted by time. More specifically, consider any subset $(X_{t_1}, X_{t_2}, \ldots, X_{t_k})$ and compare the joint distribution of this random vector with the joint distribution of the random vector $(X_{t_1+h}, X_{t_2+h}, \ldots, X_{t_k+h})$ where h is some *time lag*. If the distributions are identical for any subset and any time lag, then the sequence of random variables is stationary. Sequences of independent random variables are clearly stationary, however, many other sequences can also be stationary, even without independence.

When the sequence is stationary the distribution of each X_t is identical and the same as its mean. Further, we can consider the correlation between two random variables X_t and X_{t+h} and observe that it only depends on h. We then denote it $\gamma(h)$ and call it the *autocorrelation function*:

$$\text{Corr}(X_t, X_{t+h}) = \text{Corr}(X_0, X_h) := \gamma(h).$$

Many attributes of a stationary sequence can then be described via its autocorrelation function. Note that $\gamma(0) = 1$ and $\gamma(h) = \gamma(-h)$ and thus practically we are only interested in $\gamma(h)$ for $h = 1, 2, 3, \ldots$. Note also that if we wish to analyze $\gamma(h)$, the sequence does not have to be strictly stationary and a weaker version suffices. A sequence is called *weakly stationary* or *wide sense stationary* if the mean $\mathbb{E}[X_t]$ is independent of t, and $\text{Cov}(X_t, X_{t+h})$ only depends on h.

In practice, when dealing with time-series data we often take steps to convert the time series into what we believe to be a stationary (or weakly stationary) sequence. There are multiple methods for doing so and, in Listing 8.22, we follow one approach for the `oneOnEpsilonBlogs.csv` dataset. We first remove the 7 day moving average, similar to what was done for the middle row plots of Figure 8.15. We then further remove the weakly cyclic component by subtracting the daily mean (for each of the 7 days of the week). Finally we take an additional step called *differencing*. This creates a new times series, $\{\tilde{x}_t\}$, where,

$$\tilde{x}_t = x_t - x_{t-1}.$$

This time series, plotted on the left plot of Figure 8.17 appears to be stationary. We then plot a *correlogram* on the right plot of Figure 8.17, which is a plot of the estimated autocorrelation function $\gamma(h)$. Here, the autocorrelation is estimated via `autocor()` from `StatsBase.jl` using similar formulas to those used for estimating correlations (and covariances) in Section 4.2. The result of this estimation exercise is the autocorrelation function as appearing in the correlogram. Now it can be further used in more advanced time-series models. A topic that we don't cover further. Nevertheless, as it is a central topic in the analysis of time series, we chose to present it here as we close this chapter.

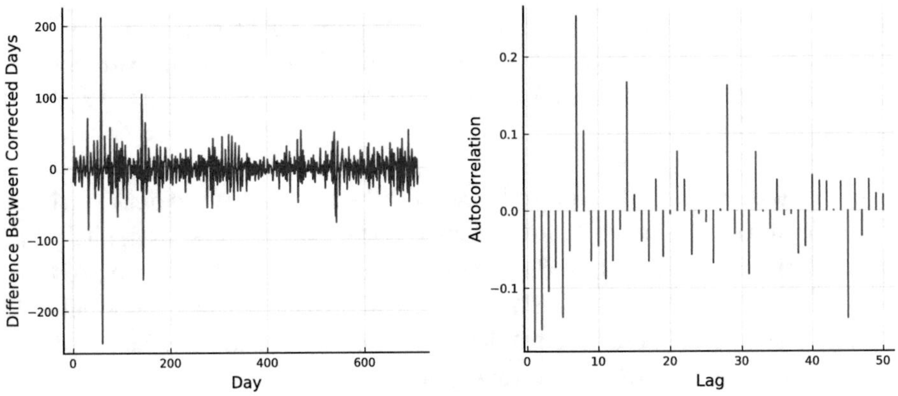

Figure 8.17: Left: Difference between corrected days. Right: A correlogram for this data.

Listing 8.22: Differencing, autocorrelation and a correlogram of a time series

```
1  using CSV, TimeSeries, Dates, Statistics, StatsBase, Measures, Plots; pyplot()
2
3  df = CSV.read("../data/oneOnEpsilonBlogs.csv",copycols = true)
4  tsA = TimeArray(Date.(df.Day,Dates.DateFormat("m/d/y")),df.Users)
5  tsB = moving(mean,tsA,7,padding = true)
6  tsC = TimeArray(timestamp(tsA), values(tsA) - values(tsB))
7
8  dow = dayofweek.(timestamp(tsA))
9  dayDiv = [filter((x)->!isnan(x), values(tsC)[dow .== d]) for d in 1:7]
10 dayMeans = mean.(dayDiv)
11 dailyCorrection = dayMeans[dow]
12 errs = filter((x)->!isnan(x),values(tsC) - dailyCorrection)
13 diffs = diff(errs)
14
15 lags = 1:50
16 acc = autocor(diffs,lags)
17
18 default(legend = false)
19 p1 = plot(diffs, c=:blue,
20         xlabel="Day",ylabel="Difference Between Corrected Days")
21 p2 = plot(lags,acc,line=:stem, c=:blue,
22         xlabel="Lag",ylabel="Autocorrelation")
23 plot(p1,p2,size=(900,400),margin = 5mm)
```

Lines 3–6 are similar to Listing 8.20. In line 8, we set `dow` which records a day of week for every entry in the time series. In line 9, we create `dayDiv`, an array of arrays with each subarray matching the values of a specific day of the week. We then broadcast `mean()` onto these arrays in line 10 to arrive at the array, `dayMeans` of length 7. We then create `dailyCorrection` which is an array that has the daily mean in a cyclic pattern. This allows us to create `errs` in line 12 which is the difference between `tsC` and the daily correction. Now we use the `diff()` function to create the differenced time series `diffs`. This is the data that we treat as stationary. In line 16, we compute the autocorrelation for `diffs` using the `autocor()` function from `StatsBase.jl`. The remainder of the code creates Figure 8.17.

Chapter 9

Machine Learning Basics

The previous chapters covered some of the key concepts of classical statistics. These include point estimation, confidence intervals, hypothesis testing, and regression analysis. With such tasks, the focus is often on the model, its properties, and interpretation of inference. However, paradigms arising in *computer science* introduce additional data focused tasks, some of which are related to classical statistical tasks and others involving automated analysis of *big data*. We now focus on such methods which are generally called *machine learning*.

There is not a clear boundary between statistics and machine learning and there are several competing terms in this area. Some of these include *statistical learning* often employed in the statistics community, as well as the related terms *data mining* and even *artificial intelligence*. These terms are used in computer science and business contexts. Further, one key paradigm that has emerged within machine learning in the past decade is *deep learning*, an area which deals with creating machine learning models that involve multi-layer neural networks. Such methods often yield models that integrate well into other automatic systems including *apps* for both personal and business use.

In this chapter we present an overview of several machine learning methods ranging from classical methods to deep learning. We clearly cannot cover everything that exists in the machine learning world because such content will require multiple chapters or even multiple books. Our purpose here is merely to present the reader with a taste of the associated problems and methods from the field. We hope that by considering the 20 code listings of this chapter together with the associated descriptions, the reader may begin to get a taste for the nature of machine learning. For further reading dealing with some theoretical aspects of machine learning, mixed with (Python) code we recommend [KBTV19] and references there-in. Further, a very popular reference focusing almost entirely on deep learning is [GBC16]. A recent Julia centric book discussing some aspects of machine learning is [V20]. Finally a book which deals with classical statistics, computational methods in statistics, and machine learning is [EH16].

Julia has dozens of packages related to machine learning and here we only use a few. The main deep learning library is `Flux.jl` and indeed this chapter makes quite a bit of use of the `Flux` package. Further there are frameworks such as `MLJ.jl` (Machine Learning Julia) and an adaptation of Python's scikit-learn via the `ScikitLearn.jl` Julia package. We don't use `MLJ` or `ScikitLearn` per-se, but recommend that the reader investigate them independently.

Y. Nazarathy and H. Klok, *Statistics with Julia*, Springer Series in the Data Sciences,
https://doi.org/10.1007/978-3-030-70901-3_9

Machine learning problems, tasks, and activities can be divided into several categories. Some of the major categories include:

Supervised learning - This suite of problems deals with a situation similar to the regression problems of Chapter 8. Data is available in the form of (x_i, y_i) and the goal is to learn how to predict Y based on X. In this sense, the regression analysis methods of Chapter 8 already serve a basis for machine learning. However in machine learning, each x_i is often very high dimensional. The elements of x_i are called *features* and the variable y_i is referred to as the *label*. When the labels are only 0 or 1, or come from a finite discrete set, the supervised learning problem is called a *classification problem*. As opposed to that, when the labels are continuous as in the models of Chapter 8, this is called a *regression problem*.

Unsupervised learning - In this case there are only features X but no labels Y. Think, for example, of a baby that learns about the world without receiving explicit feedback and direction. In basic unsupervised learning, one important task is creating clusters of points and recognizing the clusters. This falls in the realm of *clustering* algorithms. Another task is reducing the dimension of the data, i.e. *data reduction*.

Reinforcement learning - In this case an *agent* makes decisions dynamically over time, aiming to maximize some objective or achieve some desired behavior. In certain cases the agent has some knowledge about the way the world responds to the decisions, but this knowledge is often lacking or very partial. The methods of reinforcement learning allow us to control such systems in a near-optimal manner. Notable examples include playing games such as chess or Alpha-Go. Other examples include playing an unknown video game against a computer and eventually improving. Practically there are often applications in robotics related to reinforcement learning. See, for example, the (now classic) website `http://heli.stanford.edu/`.

Generative modeling - This is the task of observing data similar to the unsupervised case and creating a model that can then create additional similar data. One general application of these types of models is the *deep fake* technology where images and movies can be modified to look differently yet appear natural. For example, one face can be implanted on another. Such technology is not necessarily positive and has been put to some negative use in recent years. Nevertheless, understanding the basics of how it works is important.

This chapter provides an overview of specific methods for each of the above tasks with simple examples. We mainly use a very basic classic dataset, *MNIST digits dataset* (Modified National Institute of Standards and Technology database). We begin with Section 9.1 where we introduce some basic concepts of machine learning, mostly related to supervised classification but also dealing with stochastic optimization techniques that are common in machine learning. In Section 9.2 we present several concrete methods for supervised classification. These include basic least squares, logistic regression, support vector machines, random forests, and deep learning. We continue with Section 9.3 where we present the concept of regularization, focusing on ridge regression optimized via cross validation and dropout in deep learning. In Section 9.4 we explore some unsupervised learning techniques including clustering and Principle Component Analysis (PCA). Then in Section 9.5 we explore the basics of Markov decision processes and reinforcement learning. We close with Section 9.6 where we briefly demonstrate how to train and use Generative Adversarial Networks (GANs).

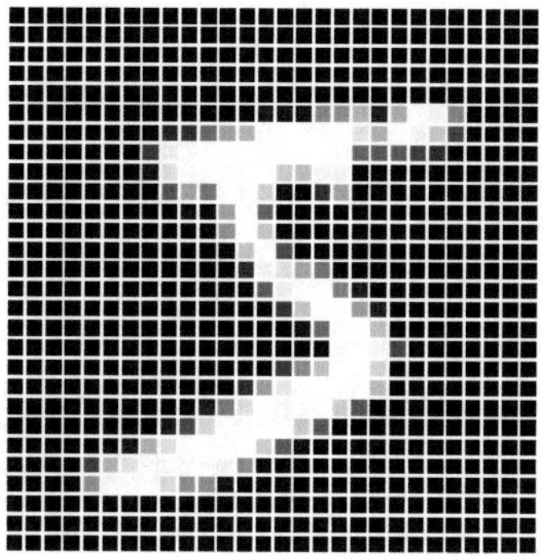

Figure 9.1: The first image of the MNIST dataset. Each image in MNIST is 28×28, i.e. containing 784 grayscale pixels.

9.1 Training, Testing, and Tricks of the Trade

In this section and the two sections that follow, we focus on *supervised learning* and mostly on *classification*. Aspects of the other major type of supervised learning, *regression*, were covered in the previous chapter, mostly in the context linear regression.

The general data setup of supervised learning is as follows. We have n observations where each observation $i \in \{1, \ldots, n\}$ consists of a pair (x_i, y_i) where $\{x_i\}_{i=1}^n$ is generally called the *data* and $\{y_i\}_{i=1}^n$ are the *labels*. Each data point x_i is often a high dimensional vector representing text, recorded voice, images, video, or heterogeneous data. Labels are often taken from a much smaller finite set of values.

When presented with data comprised of x and y, the *classification problem* is to create a model which we denote by $\hat{f}(\cdot)$ that maps data to labels. If the labels are assumed to take on a continuum of values then the task is sometimes called a *regression problem*, however we focus on classification problems where the labels are either $\{0, 1\}$ or falling in some other small set, such as $\{0, 1, 2, 3, 4, 5, 6, 7, 8, 9\}$. The case of $\{0, 1\}$ labels is called *binary classification*, and often the labels $\{-1, +1\}$ are used instead, referring to "negative" or "positive" labels.

Many of the examples in this chapter use the popular MNIST digits dataset where each x_i is an image of a digit as in Figure 9.1 and the corresponding label, y_i, records the actual digit that the image represents. This is a very common dataset for introductory machine learning as well as experimentation of more advanced machine learning concepts.

As with statistical inference, when carrying out machine learning, the general assumption is that the dataset presents a good representation of a broader (typically infinite) collection of $(\tilde{x}_i, \tilde{y}_i)$ pairs that still haven't been encountered but follow the same distributional law as the observed data. Once a model $\hat{f}(\cdot)$ is learned, it can be used in production by being applied to new data points that were not available during the learning phase. If we denote these new data points by $\{\tilde{x}_i\}_{i=1}^{\infty}$, then $\hat{y}_i = \hat{f}(\tilde{x}_i)$ is the predicted value (label) for each such new datapoint. The implementation of the learned model $\hat{f}(\cdot)$ is often employed as part of greater automated systems. For example, one may have a trained model $\hat{f}(\cdot)$ for classifying images of handwritten letters. Then this model is used again and again as part of bigger systems and apps, some of which have been branded *artificial intelligence* when they perform impressive tasks that do not just involve well defined procedural and logical operations.

Creating a model $\hat{f}(\cdot)$ for such data, or any other dataset, typically implies *training*, or *learning* a function, $\hat{f}(\cdot)$ that generally adheres to the following two broad objectives:

$$\text{TRAIN DATA FIT}: \quad \hat{f}(x_i) = y_i, \quad \text{for as many } i \text{ as possible}, \tag{9.1}$$

$$\text{UNKNOWN DATA FIT}: \quad \hat{y}_i = \hat{f}(\tilde{x}_i) = \tilde{y}_i, \quad \text{as often as possible}. \tag{9.2}$$

The first objective, TRAIN DATA FIT, is that the model will fit the training data accurately. The second objective UNKNOWN DATA FIT, is that for new unknown production data points, $\tilde{x}_i$, the model will also work well. That is, once the model is used as part of a greater system it should work well for newly encountered data points. Note that depending on the system at hand, the labels in production, $\tilde{y}_i$, are rarely known. An exception is when a setup called *active learning* is employed in which additional data that appears in production is used to update model parameters.

A problem with such a formulation is that the objective (9.1) is fully attainable by encoding the dataset exactly in $\hat{f}(\cdot)$ and forcing $\hat{f}(x_i) = y_i$ to hold for all i. Doing so, would generally be at the cost of the second objective (9.2). This is called *overfitting*. Hence we generally seek a model $\hat{f}(\cdot)$ that aims towards the first objective while remaining not too complicated so that it doesn't overfit. See also Figure 8.1 in Chapter 8 to get a feel for the idea of overfitting (in the context of regression).

Since evaluating the second objective is not possible as it deals with unknown data, we often split our data into a *training set* and a *test set*. For example, the standard MNIST example has $n = 60,000$ images treated as a training set and an additional $10,000$ images treated as a test set. This allows us to train the model on the training set aiming to achieve the TRAIN DATA FIT objective of (9.1), and then later to simulate the model as though it is in production by evaluating how well the UNKNOWN DATA FIT objective of (9.2) performs. However, modifying the model and then repeatedly carrying out tests on the same test set is generally not accepted because it can again lead to overfitting. Hence in practice we often even further split out a chunk of the training set and call it the *validation set* or *development set*. We can do so repeatedly with different models and compare. Other related methods that we discuss in the sequel include *cross validation*.

There are clearly an infinite number of options for models $\hat{f}(\cdot)$ and the field of machine learning presents algorithms for training models with desirable attributes. These types of models include support vector machines, neural networks, random forests, and many other types of classifiers. In each such case, there are parameters to the classifier which we denote via θ and the process of training is the process of finding the model parameters which fit the data the best way. However there are also secondary parameters which are either fixed in the model or specify the nature of the training algorithm. These are called *hyper-parameters*. As a starting point, we first demonstrate how to use a specific model in production after it was trained. In this case we use a deep convolutional neural network.

Without going into excessive detail, look at Figure 9.7 in Section 9.2 and the associated code Listing 9.9. The figure illustrates an architecture of the deep convolutional neural network that we use, and the listing trains this network (in parallel to training another neural network). The output of the training is then stored in the file `mnistConv.bson` (binary JSON). The data stored in that file represents the parameters θ. It spans about 23 kilobytes which is equivalent to more than $2,500$ (64 bit) real numbers. There are also hyper-parameters at play in the training of Listing 9.9. We

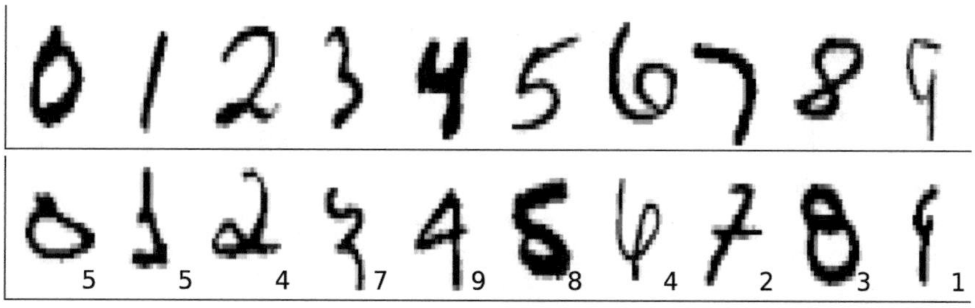

Figure 9.2: Top: Digits correctly classified. Bottom: Incorrectly classified digits with the correct classification indicated.

discuss these later when we discuss that code listing. However, let us first use this model as though we are in production. That is, let us carry out the *testing phase.*

For this consider Listing 9.1. In this listing we use the model parameters from `mnistConv.bson` to create the model $\hat{f}(\cdot)$. We then apply it to the $10,000$ testing images and asses how well the objective UNKNOWN DATA FIT of (9.2) is adhered to. The natural way for such an assessment is to compute the *classification accuracy*, or simply *accuracy*, by evaluating the proportion or percentage of correctly classified digits. As is evident from the output of Listing 9.1, the accuracy is 96.93%. This level is not yet the *human level accuracy* which for this dataset is effectively at 100%, nevertheless it is impressive enough to be integrated in automated applications. In fact, training (occurring in Listing 9.9) only used $5,000$ samples, and with more training samples, requiring more time, the accuracy would further improve.

In addition to just considering accuracy, there are other ways of quantifying performance. One such way is the *confusion matrix* which we present later in this chapter with the output of Listing 9.5. Also, in certain cases the classifier $\hat{f}(\cdot)$ can be modified to yield a probabilistic answer instead of a fixed hard classification. In certain production situations, probabilistic answers are more fruitful as they contain more information. For many types of models such probabilistic answers are already internally part of the model. This means that for each image in our example, instead of obtaining a predicted label we obtain a probability distribution over $\{0, 1, 2, \ldots, 9\}$.

In the case of this example, we create the illustrative Figure 9.2. The digit images at the top of the figure are examples of images that were correctly classified, ordered 0 to 9. As opposed to that, at the bottom we present images of example digits that are not correctly classified, where the correct label order is again 0 to 9. We also indicate the classified label that was obtained for each image. While a human may most probably classify each of these (tougher) images correctly, you may agree that the misclassification associated with 3 (classified as 7), 6 (classified as 4), and 9, classified as 1 is perhaps somewhat expected.

Listing 9.1 uses the `Flux.jl` package. The key lines defining the mathematical structure of $\hat{f}(\cdot)$ are lines 4-6. Then once the pre-trained parameters are loaded from `mnistConv.bson`, the `predictor()` function is the equivalent of $\hat{f}(\cdot)$. The remainder of the code is straightforward and simply applies `predictor()` to each image in the MNIST test set. Finding correctly classified examples, and incorrectly classified examples for the creation of Figure 9.2 requires a few lines of code and some basic logic.

Listing 9.1: Using a pre-trained neural network for classification

```
using Flux, Flux.Data.MNIST, Statistics, BSON, StatsBase, Plots; pyplot()
using Flux: onecold

model= Chain(Conv((5, 5), 1=>8, relu), MaxPool((2,2)),
             Conv((3, 3), 8=>16, relu), MaxPool((2,2)),
                flatten, Dense(400, 10), softmax)

BSON.@load "../data/mnistConv.bson" modelParams
Flux.loadparams!(model, modelParams)

function predictor(img)
    whcn = ones(Float32,28,28, 1, 1)
    whcn[:,:,1,1] = Float32.(img)
    onecold(model(whcn),0:9)[1]
end

testLabels = Flux.Data.MNIST.labels(:test)
testImages = Flux.Data.MNIST.images(:test)
nTest = length(testLabels)

iC, iR = 0, 0
nCorrect = 0
goodExamples = zeros(Int,10)
badExamples = zeros(Int,10)
predictedBad = zeros(Int,10)
for i in 1:nTest
    prediction = predictor(testImages[i])
    trueLabel = testLabels[i]
    predictionIsCorrect = (prediction == trueLabel)
    global nCorrect += predictionIsCorrect
    global iC; global iR
    if predictionIsCorrect && trueLabel == iC
        goodExamples[iC+1] = i
        iC += 1
    end
    if !predictionIsCorrect && trueLabel == iR
        badExamples[iR+1] = i
        predictedBad[iR+1] = prediction
        iR += 1
    end
end

println("Percentage correctly classified: ", 100*nCorrect/nTest)

default(yflip = true, size = (1000,300),
        legend=false,color = :Greys,ticks=false)
p1 = heatmap(hcat(float.(testImages[goodExamples])...))
p2 = heatmap(hcat(float.(testImages[badExamples])...))
for i in 1:10
    annotate!(28i-3,25,text("$(predictedBad[i])",18))
end
plot(p1,p2,layout=(2,1))
```

```
Percentage correctly classified: 96.93
```

In lines 4-6 we define the neural network model. The `Flux` constructor function `Chain()` returns a function represented via `model`. In line 8 we load the data using the BSON (Binary JSON format). We then use `loadparams!()` from the `Flux` package in line 9 to set the parameters into `model`. In lines 11-15 we define the function `predictor()` which operates on an image, `img`, converts it into WHCN format and then uses `model()` to make a prediction in line 14. The *WHCN format* (Width, Heights, #Channels, Batch Size) is standard in convolutional neural networks and is further discussed in Listing 9.9. The `ones()` function in line 12 creates a *tensor* of dimensions $28 \times 28 \times 1 \times 1$ as needed as input to `model`. Line 13 converts the image to `Float32` and places it in `whcn`. Line 14 applies `model()` to `whcn`. The result is a *one hot* vector of length 10 that is all 0's with the exception of a 1 at the index matching the classified digit. We then use the `onecold()` function to represent it as a digit in the range `0:9`. The result is an array of length 1 and we use `[1]` to extract the element in it. Lines 17-19 use `MNIST` supplied with `Flux.jl` to obtain the test labels and test images. The number of images, $10,000$, is set in `nTest`. The loop in lines 26-41 applies `predictor()` to each image and checks if the classification is correct or not. The logic in lines 32-40 is for collecting `goodExamples` and `badExamples` for the figure. The remainder of the code prints the output and creates the figure.

Performance Measures for Binary Classification

In the previous example, the classification accuracy was 96.93% and this sounds impressive. However, treating accuracy on its own as a performance measure can sometimes be misleading. For example, assume that we are trying to classify a simpler problem, determining if a digit is "1" or not. Call a case of "1" as "positive" $(+1)$ and otherwise "negative" (-1). It turns out that for MNIST there are roughly the same number of labels for each of the 10 possible digits. In such a case, we say the data is *balanced.* However once we create new labels which are either -1 or $+1$, then this binary classification problem is heavily *skewed* as there are roughly 90% negatives and only 10% positives. Now if, for example, we have a degenerate classifier such as $f(\text{image}) \equiv -1$ for any image, it will reach an accuracy of 90%! This might sound good, but such a classifier is clearly useless. Similarly, even if the classifier is not degenerate but is still very heavily biased towards -1, a similar problem will appear.

Hence there is sometimes a need for more appropriate performance measures, especially when there are skewed classes. This scenario is somewhat reminiscent of the tradeoff between the significance level α and power $1 - \beta$ as apparent in classic statistical tests. See Section 5.6. In machine learning the setup is similar to the tradeoff of Type I and Type II errors of statistics, but not exactly identical. It is common to use the pair of performance measures, *precision* and *recall* which we describe below.

Consider first Table 9.1 which is somewhat similar to Table 5.1 of Chapter 5. The counts in the table are for $10,000$ images of which $1,035$ are the digit "1" (positive) and the remainder are negative. The decision is based on the classifier presented in Listing 9.2 below and the numbers of true negative, false positive, false negative, and true positive cases are presented based on the output of that listing.

It is clear that ideally we wish the classifier to have as few false positives and false negatives as possible, but these are often competing objectives. Precision and recall help to quantify these

Table 9.1: Counts in binary classification from Listing 9.2.

		Decision	
		Decide −1 (8,916)	Decide +1 (1,084)
Reality	Label is −1 (8,865)	True negative (8,781)	False positive (84)
	Label is +1 (1,035)	False negative (135)	True Positive (1000)

objectives and are computed as follows where $|\cdot|$ stands for the number of elements in a collection.

$$\text{Precision} = \frac{|\text{true positive}|}{|\text{true positive}| + |\text{false positive}|}, \qquad \text{Recall} = \frac{|\text{true positive}|}{|\text{true positive}| + |\text{false negative}|}.$$

In the case of Table 9.1,

$$\text{Precision} = \frac{1000}{1000 + 84} = 92.25\%, \qquad \text{Recall} = \frac{1000}{1000 + 135} = 88.11\%.$$

The closer we are to 100%, the better. Comparing again with Section 5.6 observes that the recall is essentially equivalent to the power of a statistical test as it quantifies the chance of detecting a positive in case the label is actually positive. This is also called *sensitivity* in biostatistics. However the precision does not agree with the *specificity* from biostatistics (or $1 - \alpha$ as denoted in statistical tests). Nevertheless, a high precision implies a low rate of false positives similarly to the fact that in statistical testing, a high $1 - \alpha$ value (or specificity) also means a low number of false positives.

In evaluating and designing classifiers, precision and recall typically yield competing objectives where we can often tune parameters and improve one of the measures at the expense of the other. In the context of statistical testing (refer again to Section 5.6), things are more prescribed since we fix α, the probability of Type I error, at an acceptable level, and then search for a statistical test that minimizes β, the probability of Type II error (this is maximization of power, sensitivity, or recall). However in machine learning, there is more freedom. We don't necessarily discriminate between the cost of false positives and false negatives, and even if we do, it can be by considering the application at hand.

Precision and recall are two numbers and in searching for an optimal classifier, one often needs a way to consider both precision and recall jointly. A popular way to average precision and recall is by considering their harmonic mean, see Section 4.2. This is called the *F_1score* and is computed as follows:

$$F_1 = \frac{2}{\frac{1}{\text{Precision}} + \frac{1}{\text{Recall}}} = 2\frac{\text{Precision} \times \text{Recall}}{\text{Precision} + \text{Recall}}. \tag{9.3}$$

In the case of the example counts in Table 9.1, we have $F_1 = 90.13\%$.

While we don't cover it further, we note that sometimes you can use a generalization of F_1 called the *$F_\beta$score* where β determines how much more important is recall in comparison to precision (do not confuse with β of hypothesis testing). However in general, if there is not a clear reason to

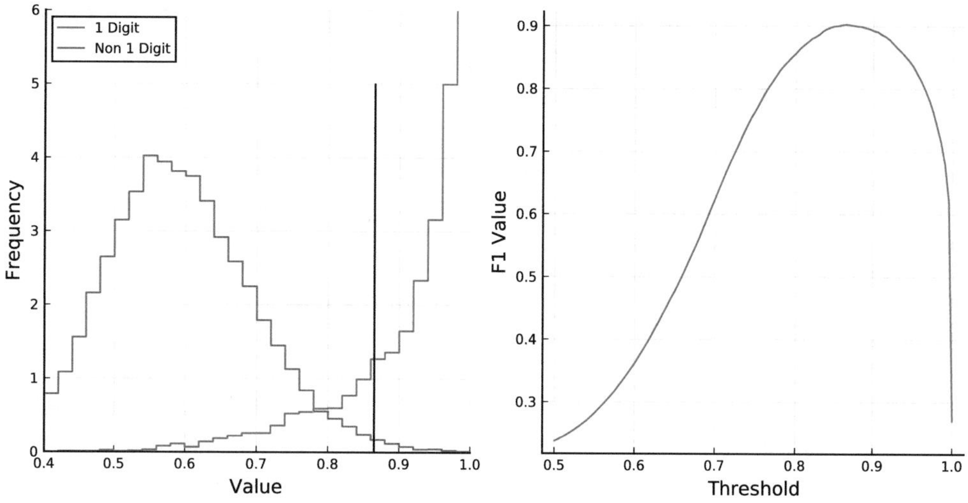

Figure 9.3: Left: The distribution of `peakProp()` with the optimal threshold marked by the black line. Right: The F_1 score as a function of the threshold.

allocate costs for false positives and false negatives differently, use of the F_1 score as a single measure of performance is sensible.

An Ad hoc Binary Classification Example

Many of the methods in this chapter are general and work well on a variety of datasets. Such generality is useful as it relieves one from having to custom fit the machine learning model to the exact nature of the problem at hand. Nevertheless as a vehicle for illustration of the aforementioned concepts of binary classification, we now present a model that is only suited for a specific problem (and even for that problem it is not the best model that one can find).

Assume that with the MNIST data assume we only wish to classify an image as either the digit "1", or not. How would we do that? The support vector machine, random forest, or neural network models presented in the sequel would work well for such binary classification, however here we go for a method that is based on the way in which digits are drawn. We rely on the fact that "1" digits roughly appear as a vertical line while other digits have more lit up pixels not just on a vertical line within the image. This is clearly only a rough description because the character "1" involves more than just a vertical line.

With this observation, our classifier works by considering each of the 28 rows of the image and locating the maximal (brightest) pixel out of the 28 pixels in that row. If there are multiple pixels with the same maximal intensity, ties are broken arbitrarily. We reason that if the image is a "1" digit, then on each row, there will generally only be a few neighboring bright pixels to the maximal pixel. However, if it is not a 1 digit, there may potentially be other bright pixels on the row. For this we include two pixels to the left of the brightest pixels and two pixels to the right as long as we don't overshoot the boundary. We then sum up these 5 pixels and do so for every row. Finally we divide this total by the total sum of the pixels and call this function of the image the "peak proportion" and represent it via `peakProp()`.

We may expect that for images representing a "1" digit, the peak proportion is high (close to unity), while for other images it isn't. This is indeed the case as presented on the left of Figure 9.3. In that plot, we plot the distribution of the peak proportion for these two classes of images. This indicates that `peakProp()` is a sensible statistic to consider because there is a distinct statistical separation between "1" digits and other digits. Note that such a statisticwould be useless when trying to compare other digits, as it is tailor made for the digit "1". We are now left with the task of choosing the threshold θ for the following classifier:

$$\hat{f}_\theta(\text{image}) = \begin{cases} -1 & \texttt{peakProp}(\text{image}) \leq \theta, \\ +1 & \texttt{peakProp}(\text{image}) > \theta. \end{cases} \tag{9.4}$$

We do this by plotting the F_1 score as a function of θ and choosing the optimal θ which turns out to be 0.865. This then defines the classifier. All these computations as well as testing the performance of the classifier on a test set are carried out in Listing 9.2. The listing also creates Figure 9.3.

Listing 9.2: Attempting hand crafted machine learning

```
using MLDatasets, StatsBase, Measures, Plots; pyplot()

xTrain, yTrain = MLDatasets.MNIST.traindata(Float32)
xTest, yTest = MLDatasets.MNIST.testdata(Float32)
nTrain, nTest = size(xTrain)[3], size(xTest)[3]
trainData = [xTrain[:,:,k]' for k in 1:nTrain]
testData = [xTest[:,:,k]' for k in 1:nTest]
positiveTrain = trainData[yTrain .== 1]
negativeTrain = trainData[yTrain .!= 1]
testLabels = yTest .== 1

function peakProp(img)
        peakSum = 0.0
        for j in 1:28
                m = argmax(img[j,:])
                (m <=2 || m >= 26) && continue
                peakSum += sum(img[j,m-2:m+2])
        end
        peakSum/sum(img)
end
predict(img,theta) = peakProp(img) <= theta ? false : true
function F1value(theta)
        predictionOnPositive = predict.(positiveTrain,theta)
        predictionOnNegative = predict.(negativeTrain,theta)
        TP = sum(predictionOnPositive)
        FN = sum(1 .- predictionOnPositive)
        FP = sum(predictionOnNegative)
        TN = sum(1 .- predictionOnNegative)
        recall, precision = TP/(TP + FN), TP/(TP + FP)
        return 2*(precision*recall)/(precision+recall)
end

psPositive, psNegative = peakProp.(positiveTrain), peakProp.(negativeTrain)
thetaRange = 0.5:0.005:1
f1Values = F1value.(thetaRange)

bestF1, bestIndex = findmax(f1Values)
bestTheta = thetaRange[bestIndex]
println("Best theta = ", bestTheta, " with F1 value of ", bestF1)

println("On test set:")
testPredictions = predict.(testData,bestTheta)
TP, FN = sum(testPredictions[testLabels]), sum(.!testPredictions[testLabels])
FP, TN = sum(testPredictions[.!testLabels]), sum(.!testPredictions[.!testLabels])
recall, precision = TP/(TP + FN), TP/(TP + FP)
F1test = harmmean([precision,recall])
@show TP, FN, FP, TN; @show recall, precision; @show F1test

p1 = stephist(psPositive, normed = true, label="1 Digit",bins=50)
stephist!(psNegative, normed = true, xlim=(0.4,1), ylim=(0,6),bins=50,
                        xlabel = "Value", ylabel = "Frequency",
                        label="Non 1 Digit")
plot!([bestTheta,bestTheta],[0,5], c =:black, label = :none)
p2 = plot(thetaRange,f1Values, legend = false,
        xlabel = "Threshold", ylabel = "F1 Value")
plot!([bestTheta],[bestF1], c=:black)
plot(p1,p2,size=(800,400))
```

```
Best theta = 0.865 with F1 value of 0.9028719433229222
On test set:
(TP, FN, FP, TN) = (1000, 135, 84, 8781)
(recall, precision) = (0.8810572687224669, 0.922509225092251)
F1test = 0.9013068949977467
```

We use the `MLDatasets` package to obtain the MNIST data. This differs from the previous example and most other examples in this chapter that get the data from `Flux.jl`. Lines 3-4 retrieve the data into `xTrain` and `xTest` in the form of a 3 dimensional tensor. We then put it in the form of an array of images, also transposing, in lines 6-7. Then lines 8-9 break up the training data into positive examples and negative examples. Line 12-20 implement `peakProp()` as described above. Note that in line 16 if the index of the maximal row pixel, `m` is not in the range $\{3, \ldots, 25\}$ we use `continue` to jump to the top of the loop for the next iteration. The accumulation of pixel intensities is in line 17 and in lines 19 after iterating over all rows `j` we return the proportion. In line 21 we implement the `predict()` function as in (9.4). Lines 22-31 implement the `F1value()` function for assessing the F_1 score for a given `theta` value. Line 33 broadcasts `peakProp()` on the positive training examples and negative examples for the purpose of creating the data for the left-hand plot of Figure 9.3. Lines 35 creates `f1Values` used for the right-hand plot. The optimal θ threshold is found in lines 37-38 and printed in line 39. The remainder of the code creates the output and figure.

Gradient Descent

Fitting procedures in machine learning almost always involve iterative algorithms. The nature of the iteration is typically that in each step, or at least in most steps, a better model fit is obtained. This is often carried out by getting closer to the minimum of a loss function which captures the difference between predicted labels, and the actual labels of the training data. Such a paradigm of using loss functions was first explained in the context of least squares fitting and linear regression in Section 8.1. We now elaborate.

In general, when presented with data $\{(x_i, y_i)\}_{i=1}^n$, a model $\hat{f}_\theta(x)$, and a function $\ell(u, v)$, we aim to find θ that minimizes the loss function,

$$L(\theta) = \sum_{i=1}^{n} \ell(\hat{f}_\theta(x_i), y_i). \tag{9.5}$$

In Section 8.1 we used $\ell(u, v) = (u - v)^2$ and $\ell(u, v) = |u - v|$, see, for example, Equation (8.3). In this chapter other options for $\ell(u, v)$ will also be useful. In any case, in trying to minimize $L(\theta)$ we often rely on the gradient $\nabla L(\theta)$. This a vector of the same dimension as θ where the j'th coordinate is the derivative of $L(\theta)$ with respect to θ_j. For any θ, the gradient vector, $\nabla L(\theta)$, points towards the direction of steepest ascent. This means that the opposite direction, $-\nabla L(\theta)$ points in the direction of steepest descent. Further, when we modify θ by taking a small step in the steepest descent direction of magnitude δ then the decrease in the loss function is approximately $\delta||\nabla L(\theta)||$.

A gradient descent approach for minimizing $L(\theta)$ is based on taking many such small steps. The common algorithm for this is repeats iterations of the form

$$\theta(t+1) = \theta(t) - \eta \nabla L\big(\theta(t)\big), \tag{9.6}$$

where $\eta > 0$. This was already presented in the context of linear regression in Equation (8.16) of Section 8.1. In fact, a simple gradient descent implementation was used in Listing 8.3.

Many machine learning algorithms are based on variants of (9.6). Such methods are in general called *gradient-based learning* and the hyper-parameter η is often called the *learning rate*. Before we outline the types of variants employed and discuss the *ADAM algorithm* (Adaptive Moment Estimation), let us first explore a very elementary example of gradient descent learning.

For this we momentarily return to the basic statistical inference setting of Chapter 5 and assume we are presented with univariate data of a random sample, $x_1, \ldots, x_n$. In the context of machine learning, this can be viewed as unsupervised learning because there are only features and no labels. Assume now that we wish to find a single number, x^* that summarizes $x_1, \ldots, x_n$ as best as possible. One way to specify this in terms of a loss function is to seek a value x^* that minimizes

$$L(u) = \sum_{i=1}^{n} (x_i - u)^2. \tag{9.7}$$

Analytically, it is very easy to show that $x^* = \overline{x}$, the sample mean. This can be done by taking the derivative (gradient in one dimension) of the loss function. The gradient is

$$\nabla L(u) = -2 \sum_{i=1}^{n} (x_i - u) = -2\Big(\sum_{i=1}^{n} x_i\Big) + 2\, n\, u. \tag{9.8}$$

If we equate it to 0 and solve for u, we obtain $u = \overline{x}$. Further, the second derivative is $2n > 0$ (positive) indicating that $\overline{x}$ is a minimum. Alternatively, without using calculus, we may represent $L(u)$ as an upward facing parabola in u via,

$$L(u) = \underbrace{n}_{a} u^2 + \underbrace{\Big(-2\sum_{i=1}^{n} x_i\Big)}_{b} u + \underbrace{\sum_{i=1}^{n} x_i^2}_{c}.$$

This then allows us to read off the minimum value of the parabola at $u = -\frac{b}{2a} = \overline{x}$. In any case we see that the sample mean has the interpretation of minimizing the sum of squared deviations in the data.

With such a simple solution for the minimization of (9.7) there is no practical reason to execute gradient descent for this problem. All one needs to do is compute the sample mean. Nevertheless, to get a better feel for gradient descent it is useful to consider how the algorithm performs on this problem. For this consider (9.6) and use the gradient expression (9.8). This means, that,

$$\begin{aligned}
\theta(t+1) &= \theta(t) - \eta\Big(-2\Big(\sum_{i=1}^{n} x_i\Big) + 2\, n\, \theta(t)\Big) \\
&= \underbrace{\big(1 - 2n\eta\big)}_{\alpha}\, \theta(t) + \underbrace{2\eta \sum_{i=1}^{n} x_i}_{\beta}.
\end{aligned}$$

Now a recursion of the form $\theta(t+1) = \alpha\theta(t) + \beta$, starting at some value $\theta(0)$ yields

$$\theta(t) = \alpha^t \theta(0) + \beta \frac{1-\alpha^t}{1-\alpha}.$$

Such a form can be obtained by iterating the recursion and summing up a geometric sum. This means that if $|\alpha| < 1$ then,

$$\lim_{t\to\infty} \theta(t) = \frac{\beta}{1-\alpha} = \frac{2\eta \sum_{i=1}^n x_i}{1-(1-2n\eta)} = \overline{x}.$$

The condition $|\alpha| < 1$ is equivalent to $\eta < n^{-1}$. That is we see that if η is not too large, the sequence converges to the minimum from any starting point $\theta(0)$. Otherwise it does not. Further it can be seen that the fastest convergence occurs when η is arbitrarily close to n^{-1} from below. This type of behavior of the learning rate is typical: On the one hand, too large of a learning rate implies that gradient descent does not converge. On the other hand, too low of a learning rate implies that convergence is very slow. In this simple toy example we can analytically analyze the learning rate. However in more realistic examples that follow, experimentation is needed. This falls under the umbrella of *hyper-parameter tuning.*

Improvements to Basic Gradient Descent

For practical, real sized, machine learning problems, the basic gradient descent update rule (9.6) is often enhanced to improve performance. In some cases, such modifications are critical to success and without such modifications, the algorithm practically fails. Some of the problems that can occur in gradient-based learning include:

1. The data (x, y) may be so large so that evaluation of the gradient is extremely costly in terms of computation time and resources.
2. During the optimization, certain coordinates of the gradient may be almost zero and the algorithm can get stuck. This is effect is termed *vanishing gradient.* It can happen due to saddle-points, near saddle-points, or local minima.
3. In the basic rule (9.6), the same learning rate η is used for all coordinates of θ. However in certain cases, some coordinates may require a high learning rate while others may require a low learning rate. Further this requirement may vary depending on the location of $\theta(t)$ during the learning process.
4. While certain problems are *convex* and thus have a single *local minimum* that is also the *global minimum*, many problems are *non-convex* and of these many do not have a single local minimum point. Hence there is no guarantee that the algorithm will reach the global minimum point, and it may potentially get stuck in local minima that are far from the global optimum.

Problem 1 is often alleviated by computing the gradient with only a subset of the data in each iteration. In deep learning this is done by breaking up the data into batches of smaller size, called *mini-batches.* The learning algorithm then cycles through all the mini-batches, where a complete cycle is often termed a learning *epoch.* This is similar to *stochastic gradient descent* (SGD) in which a random subset of the data is used in each iteration. We present a simple example of SGD in Listing 9.4.

Interestingly, using *noisy gradients* as with SGD or mini-batches can often help with problem 2 (flat points) and problem 4 (getting stuck in local minima). This is because noisy gradients add a random component to the optimization that sometimes helps to nudge $\theta(t)$ out of problem spots.

Problem 2 is also sometimes handled by modifying certain attributes of the model $\hat{f}(\cdot)$. For example in deep learning, discussed in the sequel, using the *relu* activation function as opposed to other activation functions sometimes helps to combat vanishing gradient problems. More details about activation functions are in the description near Listing 9.9. An additional general mechanism that is used is *momentum*. The basic idea of momentum is to use *exponential smoothing* based on past gradients via a recursion of the form,

$$m(t+1) = \beta\, m(t) + (1-\beta)\, \nabla L\big(\theta(t)\big). \tag{9.9}$$

Here β is often taken at a value of 0.9 or similar. Then the momentum, $m(t)$ is used in place of the gradient within the recursion (9.6). This then allows us to keep some history of the gradients similarly to how a rolling ball down a hill keeps some momentum and rolls down the smoothest path in the right direction. The idea of momentum can be further refined via *Nesterov acceleration*, and *momentum bias correction* which we do not discuss further.

Problem 3 is often alleviated by usage of *adaptive gradient techniques*. This is a broad class of algorithms of which a simple version is called *RMS prop* standing for (Root Mean Square propagation). With RMS prop, each coordinate of θ will have its own learning rate scaled proportional to the smoothed root sum of squares of the gradient of that coordinate.

Within these methods, one algorithm that has emerged to become the most popular in recent years is the *ADAM* (Adaptive Moment Estimation) algorithm, see [KB14]. Much of the success of ADAM is due to practical, empirical evidence with deep learning on real datasets. It combines the ideas of momentum with bias correction, and RMS prop, into a simple algorithm that in general works very well. In addition to the learning rate, η, the ADAM algorithm also has other hyper-parameters that can be tuned. These are generally called β_1 and β_2. Here β_1 is effectively the β parameter of the momentum update (9.9). It is taken as 0.9 by default. Further, β_2 is a similar parameter used for RMS prop and is taken as 0.999 by default. In practice, β_1 and β_2 often don't need to deviate from their defaults, while the learning rate η should often be fine tuned.

When carrying out gradient-based learning, we can often use explicit expressions for the gradient. This was, for example, the case in the simple toy example above where a formula for the gradient is in (9.8). The same holds for least squares presented in Chapter 8 where the formula for the gradient is in (8.14) of that chapter. However in other cases, there is not a simple closed form formula. This is especially true when the loss function $L(\theta)$ depends on a complicated *computation graph* where differentiation requires repeated application of the *chain rule* from calculus. In such a case we often make use of a technique called *automatic differentiation* where the parts of the computation graph that evaluate the loss function (and the model) are evaluated in a recursive automatic matter. In Julia, this feature is particularly developed using an automatic differentiation package called `Zygote.jl`. That package, implementing *reverse mode automatic differentiation* converts the Julia code into gradients using advanced techniques as described here [I18b]. You may also refer to the survey, [BPRS18], for a general overview of automatic differentiation.

As a basic example of using `Flux.jl` (implicitly using `Zygote.jl`), we present Listing 9.3. In this example we revisit the least squares fitting of Listing 8.3. For such a small example, advanced machinery of ADAM and automatic differentiation are clearly overkill. Still, we present this example here as a minimal illustration of how one of the ways to interface with automatic differentiation using the `gradient()` function call in line 19, and with the ADAM optimizer, constructed in line 15, which is used for making a gradient descent step in line 20. Compare the output of this listing with the output of Listing 8.3 to see that the resulting coefficients are the same.

Listing 9.3: Using `Flux.jl` and ADAM for optimization

```
1  using Flux, Random, LinearAlgebra, CSV
2  using Flux.Optimise: update!
3  Random.seed!(0)
4
5  data = CSV.read("../data/L1L2data.csv")
6  xVals, yVals = Array{Float64}(data.X), Array{Float64}(data.Y)
7
8  eta = 0.05
9  epsilon = 10^-7
10
11 b = rand(2)
12
13 predict(x)  = b[1] .+ b[2]*x
14 loss(x,y)  = sum((y .- predict(x)).^2)
15 opt = ADAM(eta)
16
17 iter, gradNorm = 0, 1.0
18 while gradNorm >= epsilon
19     gs = gradient(()->loss(xVals,yVals),params(b))
20     update!(opt,b,gs[b])
21     gradNorm = norm(gs[b])
22     global iter += 1
23 end
24
25 println("Number of iterations: ", iter)
26 println("Coefficients:", b)
```

```
Number of iterations: 274
Coefficients:[0.9449341659265621, 0.7164973148378927]
```

We read the data in line 5. In line 8 we define the learning rate η and in line 9 we define `epsilon` used for the stopping criterion. In line 11 we initialize the parameters `b` with random values. In line 13 we define the `predict()` function which makes a prediction of $\hat{f}(x)$ using the basic simple regression model. In line 14 we define the loss function. It assumes a global `b` used in `predict()` and can operate on vector `x` and vector `y`. In line 15 we define the optimizer `opt` using the `ADAM()` constructor with a specified learning rate `eta`. The gradient-based learning loop is in lines 18-23 with a termination condition when the norm of the gradient is less than `epsilon`. Line 19 is important as it generates a gradient object which is of type `Zygote.Grads`. Here `Zygote.jl` which is `Flux`'s automatic differentiation tool computes the gradient of the anonymous function `()->loss(xVals,yVals)`. The gradient is computed with respect to `b` as specified by wrapping it with `params()` in the second argument. Line 20 invokes an update of the parameters `b` using the optimizer, `opt` and the gradient. The norm is computed in line 21 to be used as the stopping criterion of the `while` loop. The remainder of the code prints the output.

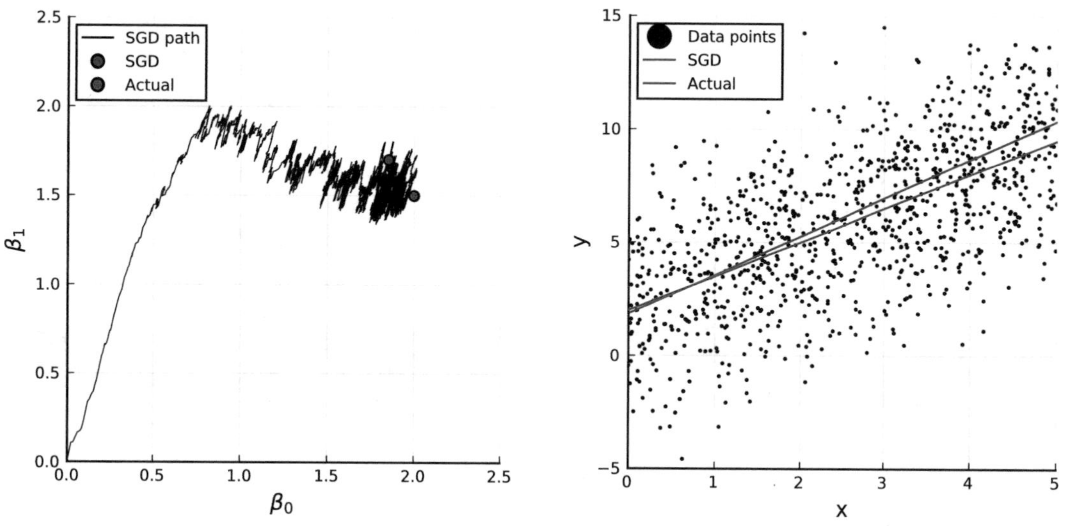

Figure 9.4: An application of stochastic gradient descent for solving least squares. Left: The path starting at (0, 0) ends at the blue point while the red point marks the actual parameter values. Right: The data with the fit line.

Stochastic Gradient Descent

As stated previously, there are multiple problems with standard gradient descent, some of which can be overcome by using *Stochastic Gradient Descent* (SGD). In its purest form, SGD operates as follows. Let the data points be $\{(x_i, y_i)\}_{i=1}^n$ where n a large number. The loss function (9.5) has gradient,

$$\nabla L(\theta) = \sum_{i=1}^{n} \nabla \ell(\hat{f}_\theta(x_i), y_i), \tag{9.10}$$

where the individual elements in the summation are the gradient of $\ell(\hat{f}_\theta(x_i), y_i)$ with respect to θ associated with each observation (x_i, y_i). The idea of SGD is at each step to choose some random subset of $\{1, \ldots, n\}$ of size k, where $k \ll n$ (k is much smaller than n). Denoting this subset by $\{i_1, \ldots, i_k\}$, the *stochastic gradient* is then evaluated as

$$\widetilde{\nabla L(\theta)} = \sum_{i \in \{i_1, \ldots, i_k\}} \nabla \ell(\hat{f}_\theta(x_i), y_i). \tag{9.11}$$

Clearly, computation of (9.11) can be carried out in a much quicker manner than (9.10) because $k \ll n$. Further, SGD sometimes allows the optimization to escape local minima. Once a stochastic gradient is available, gradient descent variants such as ADAM are implemented as described previously.

One extreme form of SGD is to have $k = 1$. Other forms keep k at some sensible size such as 100 or 1,000, often depending on the computational architecture at use (e.g. GPU, TPU, etc.). Further in practice, instead of picking independent random subsets $\{i_1, \ldots, i_k\}$, we cycle through the data via mini-batches.

We now present a simple SGD example, attempting to solve a least squares problem. In practice, you would not use SGD for such a simple problem, however our presentation is here for pedagogical purposes. In Listing 9.4 we generate a dataset with $1,000$ observations coming from the model $Y = \beta_0 + \beta_1 x_i + \varepsilon$ with $\beta_0 = 2.0$, $\beta_1 = 1.5$. The mean of ε is 0 and the variance is 2.5^2, and the points x_i uniformly generated from a set of discrete points in the range $[0, 5]$. The listing generates Figure 9.4 which plots the trajectory of SGD. The key step in the code is in lines 14-16 where for a random index `i`, we use the least squares gradient formula similarly to (8.14) from Chapter 8.

Listing 9.4: Using SGD for least squares

```
1  using Random, Distributions, Plots, Measures, LaTeXStrings; pyplot()
2  Random.seed!(1)
3
4  n = 10^3
5  beta0, beta1, sigma = 2.0, 1.5, 2.5
6  eta = 10^-3
7
8  xVals = rand(0:0.01:5,n)
9  yVals = beta0 .+ beta1*xVals + rand(Normal(0,sigma),n)
10
11 pts, b = [], [0, 0]
12 push!(pts,b)
13 for k in 1:10^4
14     i = rand(1:n)
15     g = [   2(b[1] + b[2]*xVals[i]-yVals[i]),
16             2*xVals[i]*(b[1] + b[2]*xVals[i]-yVals[i])  ]
17     global b -= eta*g
18     push!(pts,b)
19 end
20
21 p1 = plot(first.(pts),last.(pts), c=:black,lw=0.5,label="SGD path")
22      scatter!([b[1]],[b[2]],c=:blue,ms=5,label="SGD")
23      scatter!([beta0],[beta1],
24          c=:red,ms=5,label="Actual",
25          xlabel=L"\beta_0", ylabel=L"\beta_1",
26          ratio=:equal, xlims=(0,2.5), ylims=(0,2.5))
27
28 p2 = scatter(xVals,yVals, c=:black, ms=1, label="Data points")
29      plot!([0,5],[b[1],b[1]+5b[2]], c=:blue,label="SGD")
30      plot!([0,5],[beta0,beta0+5*beta1], c=:red, label="Actual",
31              xlims=(0,5), ylims=(-5,15), xlabel = "x", ylabel = "y")
32
33 plot(p1, p2, legend=:topleft, size=(800, 400), margin = 5mm)
```

In lines 8-9 we setup synthetic data for this problem based on the parameters set in lines 4-5. The x-values fall uniformly over the discrete grid `0:0.01:5` and for every x-value, the y-value follows $y = \beta_0 + \beta_1 x + \varepsilon$ where ε is normally distributed with a standard deviation of 2.5. Lines 13-19 implement stochastic gradient descent for 10^4 iterations. The starting value is $(\beta_0, \beta_1) = (0, 0)$. The learning rate is 10^{-3}. Lines 15-16 compute the stochastic gradient `g` for a random index `i`. For plotting purposes, every additional point is pushed into the array, `pts`. The index `i` for the random data observation of each iteration is obtained in line 14. Then lines 15-16 evaluate the gradient based on that observation. Finally, line 20 makes the step in the direction `g`. The remainder of the code creates Figure 9.4.

9.2 Supervised Learning Methods

We now focus on common supervised learning methods, a topic that sits at the heart of machine learning. For simplicity, all the examples that we present in this section deal with classification of MNIST digit images. For some methods we only present minimal descriptions and for other methods we dive deeper. These are the methods that we cover

Linear least squares classifiers - These simple methods use threshold functions together with least squares as covered in Chapter 8.

Logistic softmax regression - The logistic regression model is viewed in statistics as the first motivation for GLM (Generalized Linear Models). It was covered in Section 8.6. In machine learning it is the most basic neural network and is often used as a benchmark for comparing how more involved models perform. Basic logistic regression is used for binary outcomes, however when combined with a *softmax* layer it can be used for multi-class classification.

Support Vector Machines (SVM) - This class of classifiers naturally sits within the suite of machine learning algorithms as it has very nice theoretical properties and is practically very useful as well. The method uses *separating hyperplanes* to create classifiers and also allows us to use *kernel functions* to carry out transformations.

Decision trees and random forest - These methods create decision trees for classifying data. Random forest is a *bagging algorithm* applied to decision trees which yields a powerful and very generic method for classification (and regression).

Neural networks - These methods create non-linear compositions of functions that allow us to express a variety of relationships. Often called *deep neural networks*, these methods have gained massive popularity in recent years. We also look at the popular form of deep neural networks called *convolutional neural networks* that are very useful for image classification.

In addition to the above methods, one may also wish to investigate other classic machine learning classification algorithms that we don't cover here. These include *Naive Bayes* and *k-nearest neighbors*. See some of the references introduced at the onset of this chapter for further details.

Linear Least Squares Classification

One of the most simple classifiers that we can create is based on least squares. We consider each image as a vector and obtain different least squares estimators for each type of digit. For digit $\ell \in \{0, 1, 2, \ldots, 9\}$ we collect all the training data vectors, with $y_i = \ell$. Then for each such i, we set $y_i = +1$ and for all other i with $y_i \neq \ell$, we set $y_i = -1$. This labels our data as classifying "yes digit ℓ" vs. "not digit ℓ". Call this vector of -1 and $+1$ values $y^{(\ell)}$ for every digit ℓ. This type of strategy is called a *one vs. rest* (or *one vs. all*) because it breaks up the multi-class classification problem with 10 classes, into 10 separate binary classification problems. An alternative that we do not cover here is the *one vs. one* strategy where we would compare each pair of digits, in which case there would be $10 \times (10-1)/2 = 45$ different binary classification problems. Now sticking with the one vs. rest strategy we compute

$$\beta^{(\ell)} = A^{\dagger} y^{(\ell)} \qquad \text{for} \qquad \ell = 0, 1, 2, \ldots, 9, \tag{9.12}$$

where $A^{\dagger}$ is the $785 \times 60,000$ dimensional pseudo-inverse associated with the 60,000 (training) images. It is the pseudo-inverse of the $60,000 \times 785$ matrix A (allowing also a first column of 1's for a *bias term*). The matrix A is constructed with a first column of 1's, the second column having the first pixel of each image, the third column having the second pixel of each image, up to the last column having the 784'th pixel of each image. The pixel order is not really important and can be row major, column major, or any other order, as long as it is consistent. You may review Section 8.1 for details about the pseudo-inverse.

Now for every image i, the inner product of $\beta^{(\ell)}$ with the image (augmented with a 1 for the constant term) yields an estimate of how likely this image is for the digit ℓ. A very high value indicates a high likelihood and a low value is a low likelihood. We then classify an arbitrary image $\tilde{x}$ by selecting,

$$\hat{y}(\tilde{x}) = \text{argmax}_{\ell=0,\dots,9} \ \beta^{(\ell)} \cdot \begin{bmatrix} 1 \\ \tilde{x} \end{bmatrix}. \tag{9.13}$$

That is, the classifier chooses the digit ℓ that maximizes the inner product. Observe that during training, this classifier only requires calculating the pseudo-inverse of A once. It then only needs to remember 10 vectors of length 785, $\beta^{(0)}, \dots, \beta^{(9)}$. Then based on these 10 vectors, a decision rule is very simple to execute in (9.13).

We illustrate this in Listing 9.5 where we achieve 86% accuracy. We also output the *confusion matrix* in the output. This matrix shows for each real label (row), how many labels were classified (column). It is a count over the 10,000 test images.

Listing 9.5: Linear least squares classification

```
using Flux, Flux.Data.MNIST, LinearAlgebra
using Flux: onehotbatch

imgs   = Flux.Data.MNIST.images()
labels = Flux.Data.MNIST.labels()
nTrain = length(imgs)
trainData = vcat([hcat(float.(imgs[i])...) for i in 1:nTrain]...)
trainLabels = labels[1:nTrain]
testImgs = Flux.Data.MNIST.images(:test)
testLabels = Flux.Data.MNIST.labels(:test)
nTest = length(testImgs)
testData = vcat([hcat(float.(testImgs[i])...) for i in 1:nTest]...)

A = [ones(nTrain) trainData]
Adag = pinv(A)
tfPM(x) = x ? +1 : -1
yDat(k) = tfPM.(onehotbatch(trainLabels,0:9)'[:,k+1])
bets = [Adag*yDat(k) for k in 0:9]

classify(input) = findmax([([1 ; input])'*bets[k] for k in 1:10])[2]-1

predictions = [classify(testData[k,:]) for k in 1:nTest]
confusionMatrix = [sum((predictions .== i) .& (testLabels .== j))
                                for i in 0:9, j in 0:9]
accuracy = sum(diag(confusionMatrix))/nTest

println("Accuracy: ", accuracy, "\nConfusion Matrix:")
show(stdout, "text/plain", confusionMatrix)
```

```
Accuracy: 0.8603
Confusion Matrix:
 944     0    18     4     0    23    18     5    14    15
   0  1107    54    17    22    18    10    40    46    11
   1     2   813    23     6     3     9    16    11     2
   2     2    26   880     1    72     0     6    30    17
   2     3    15     5   881    24    22    26    27    80
   7     1     0    17     5   659    17     0    40     1
  14     5    42     9    10    23   875     1    15     1
   2     1    22    21     2    14     0   884    12    77
   7    14    37    22    11    39     7     0   759     4
   1     0     5    12    44    17     0    50    20   801
```

In lines 4-6 we load the training images, the labels and determine `nTrain` as the number of training images. Line 7 converts the training images into a big matrix `trainData`. Lines 9-12 deal with the test images in a similar manner. In line 14 we construct the matrix A. This is followed by line 15 where we compute $A^{\dagger}$. In line 16 we construct a simple function, `tfPM()`, that converts true or false values to $+1$, -1, respectively. Then in line 17 we use the `onehotbatch()` function from `Flux`. It converts each of the training labels into a array of length 10 comprised of true/false values where only a single entry of the array is true, matching the location of the digit. This then creates `yDat(k)` once `tfPM()` is applied. It is $y^{(\ell)}$ as described above. Line 18 executes the estimation implementing (9.12). Our classifier is then implemented in line 20 according to (9.13). In line 22 we use the classifier to create predictions for the test data. We then compute `confusionMatrix` in lines 23-24 and `accuracy` in line 25.

Logistic Softmax Regression

We can progress beyond basic least squares and try to use the same classification idea with logistic regression. See Section 8.6. Here we can again use a one vs. rest approach and utilize the `GLM` package in a similar way to how we carried out logistic regression fitting with a single independent variable in Listing 8.18. Now there are 784 independent variables and a single y-intercept also called a *bias term*. To carry out learning in such a case, we would run 10 such logistic regression fitting procedures to arrive at the parameters of logistic regression $w^{(\ell)}$ with $b^{(\ell)}$ for each $\ell \in \{0, \dots, 9\}$.

With these estimated parameters the predictor is very similar to (9.13) and can be written as

$$\hat{y}(\tilde{x}) = \text{argmax}_{\ell=0,\dots,9}\ \sigma(w^{(\ell)} \cdot \tilde{x} + b^{(\ell)}). \tag{9.14}$$

Here $\sigma(\cdot)$ is the sigmoid function. Note that in the language of neural networks, the $\sigma(\cdot)$ is often called the *activation function*. Remember that the sigmoid activation function, $\sigma(u)$, operating on a scalar u is

$$\sigma(u) = \frac{1}{1+e^{-u}} = \frac{e^u}{e^u+1}. \tag{9.15}$$

Interestingly, once the weights and bias terms are known, due to the monotonicity of $\sigma(\cdot)$, (9.14) works the same even if we remove $\sigma(\cdot)$. However, the training that yields the parameters for (9.14) relies on the fact that this is a logistic regression problem and not a linear regression problem. In fact, such a one vs. all logistic regression approach performs better than the linear classifier.

We can even go a step further and construct a multi-class "logistic regression like" classifier. This type of classifier is called *multinomial logistic regression* in the world of statistics. In machine learning it is sometimes simply called *logistic regression* or *logistic softmax regression*. The name comes from the fact that we replace the sigmoid function $\sigma(\cdot)$ with the *softmax* function. The softmax function $s(z)$ operates on an input vector z of length K and returns a probability vector of the same length that is a probability distribution. It is defined as follows:

$$s(z) = \frac{1}{\sum_{j=1}^{K} e^{z_j}} \begin{bmatrix} e^{z_1} & e^{z_2} & \dots & e^{z_K} \end{bmatrix}^{\top}. \tag{9.16}$$

Hence the i'th entry of $s(z)$ is e^{z_i} normalized by the sum $\sum_{j=1}^{K} e^{z_j}$. This function is useful because the input vector z can include both positive, zero, and negative entries and the sum of the entries does not need to be unity. After application of the softmax function, the result $s(z)$ has only positive numbers that sum to unity, and the index ℓ with the largest entry in z is the index in $s(z)$ that has the highest probability.

Now the predictor of logistic softmax regression classifier can be written in compact form via

$$\hat{y}(\tilde{x}) = \text{argmax}_{\ell=0,\dots,9}\ s_\ell\big(W\tilde{x} + b\big). \tag{9.17}$$

Here $s_\ell(\cdot)$ is the ℓ'th entry of the softmax function. Further W is a 10×784 matrix and b is a 10 dimensional vector. This is no longer a one vs. rest predictor. It is rather a predictor that couples all 10 possible outputs obtained via the affine transformation $W\tilde{x}+b$. Interestingly, just like (9.14) would essentially predict the same even if $\sigma(\cdot)$ is removed, the predictor (9.17) would predict the same if the softmax $s_\ell(\cdot)$ is removed. However, just like the one vs. all logistic regression predictor requires $\sigma(\cdot)$ for training, training of the current predictor also requires the softmax element. Further, with this predictor, if we were to consider $s(W\tilde{x} + b)$ it would give a probability distribution over $\{0, 1, \dots, 9\}$ for the belief about the digit represented by the image $\tilde{x}$. This, is in its own right, sometimes an interesting output measure.

The question now remains on how to find the weight matrix W and bias vector b of (9.17). For this we carry out gradient-based learning with a *cross entropy* loss function. In general, this function operates on two probability distributions denoted via $\mathcal{Y}$ and $\hat{\mathcal{Y}}$. The former is the "desired distribution" and the latter is the "estimated distribution". Representing these distribution as K dimensional vectors, where K is the number of labels, the cross entropy is defined as

$$\text{CE}(\mathcal{Y}, \hat{\mathcal{Y}}) = -\sum_{j=1}^{K} \mathcal{Y}_j \log(\hat{\mathcal{Y}}_j). \tag{9.18}$$

Now in classification, whenever we have a label $y \in \{0, \dots, 9\}$ we create a *one hot encoded* vector $\mathcal{Y}$ which is a vector of all zeros except for the $y + 1$'st coordinate which is 1. The addition of 1 to the index is to offset for the fact that digits range from 0 to 9 as opposed to 1 to K. This means that in classification with a label y, the cross entropy (9.18) reduces to $-\log(\hat{\mathcal{Y}}_{y+1})$. Hence when we consider all the labels, the cross entropy loss over all n data points yields

$$-\sum_{i=1}^{n} \log(\hat{\mathcal{Y}}_{y_i+1}).$$

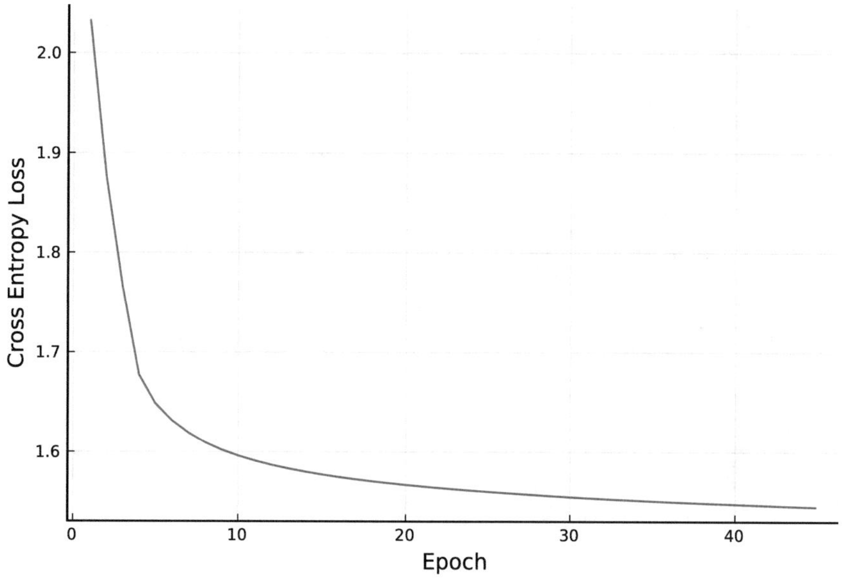

Figure 9.5: Progression of the loss function during training of a logistic softmax regression network.

Now from the design of the logistic softmax regression model, for any set of parameters W and b, and any input image $\tilde{x}$ we have $\hat{\mathcal{Y}} = s(W\tilde{x} + b)$. Hence putting these components together, the loss function is

$$L(W, b) = -\sum_{i=1}^{n} \log\big(s_{y_i+1}(W\tilde{x}_i + b)\big). \tag{9.19}$$

It can be shown that the loss function is convex with respect to its input (W, b). This then results in a monotone decrease of the loss during gradient-based learning. See Figure 9.5. In more involved neural networks, such convexity no longer holds and at certain times during the learning process, the loss may increase.

In practice as we execute gradient-based learning we break up the data into mini-batches. The indices of each mini-batch is a subset of $\{1, \ldots, n\}$ and the stochastic/partial gradient is computed similarly to (9.11). Each step of the gradient descent uses a different mini-batch. Then, a full pass on all mini-batches (on the whole dataset) is termed an *epoch*.

In Listing 9.6 we train such a logistic regression softmax classifier using `Flux.jl`. We use $20,000$ training images with a mini batch size of $1,000$. Hence each epoch is composed of 20 iterations where we use the ADAM optimizer with a learning rate $\eta = 0.01$. Training is pursued until the reduction in loss is less than 5×10^{-4}. This translates to 45 epochs for this dataset. The resulting classifier achieves an accuracy of about 90.08% and the trajectory of the loss is plotted in Figure 9.5. This code listing uses `Flux.jl` in a very similar way to Listing 9.3. Here we also use the `softmax()`, `crossentropy()` functions. Other listings in this chapter that use `Flux.jl` such as Listing 9.9, execute training using a different to the `gradient()` and `update!()` interface that we use here and in Listing 9.3.

Listing 9.6: Logistic softmax regression for classification

```
using Flux, Flux.Data.MNIST, Statistics, BSON, Random, StatsBase, Plots; pyplot()
using Flux: params, onehotbatch, crossentropy, update!
Random.seed!(0)

nTrain = 20000
miniBatchSize = 1000
imgs   = Flux.Data.MNIST.images()[1:nTrain]
labels = Flux.Data.MNIST.labels()[1:nTrain]

trainData = hcat([vcat(float.(imgs[i])...) for i in 1:nTrain]...)
trainLabels = labels

testImgs = Flux.Data.MNIST.images(:test)
testLabels = Flux.Data.MNIST.labels(:test)
nTest = length(testImgs)
testData = hcat([vcat(float.(testImgs[i])...) for i in 1:nTest]...)

W = randn(10,28*28)
b = randn(10)

logisticM(imgVec) = softmax(W*imgVec .+ b)
logisticMclassifier(imgVec) = argmax(logisticM(imgVec))-1
loss(x,y) = crossentropy(logisticM(x),onehotbatch(y,0:9))
opt = ADAM(0.01)

lossValue = 0.0
lossArray = []
epochNum = 0
while true
    global lossValue
    prevLossValue = lossValue
    for batch in Iterators.partition(1:nTrain,miniBatchSize)
      gs = gradient(()->loss(trainData[:,batch],trainLabels[batch]),params(W,b))
      for p in (W,b)
          update!(opt,p,gs[p])
      end
    end
    global epochNum += 1
    lossValue = loss(trainData,trainLabels)
    push!(lossArray,lossValue)
    print(".")
    abs(prevLossValue-lossValue) < 5e-4 && break
end

println("\nNumber of epochs: ", epochNum)
acccuracy = mean([logisticMclassifier(testData[:,k]) for k in 1:nTest]
                .== testLabels)
println("Accuracy: ", acccuracy)

plot(lossArray, xlabel = "Epoch", ylabel = "Cross Entropy Loss", legend = false)
```

```
.............................................
Number of epochs: 45
Accuracy: 0.9008
```

In line 5 we define the number of images to use for training via `nTrain` . In line 6 we specify the mini batch size `miniBatchSize`. Lines 7-16 setup the training and testing data. In lines 18-19 we set random initial values for the weights and biases. Line 21 defines the function `logisticM()` that implements $s(W\tilde{x} + b)$ where $\tilde{x}$ is `imgVec`. The line 22 defines the actual classifier as in (9.17). A loss function, `loss()` based on data `x` and labels `y` is defined in line 23. It uses `crossentropy()` function from the `Flux` package. Lines 29-43 are the main gradient-based learning loop. In lines 32-37 we loop over all of the batches where in each iteration `batch` is a range within `1:nTrain` of size `miniBatchSize`. Then in line 33 the gradient is computed via automatic differentiation yielding `gs` that can be indexed via `W` and `b`. The update is carried out in line 35, by iterating with `p` over each of the two parameters. Line 42 is the stopping criterion where the `break` statement is evaluated only once the first condition is met. The remainder of the code creates Figure 9.5 and outputs the accuracy estimate.

Support Vector Machines

Support Vector Machine (SVM) algorithms are directly suited for classification and are considered by many as the "most elegant" machine learning algorithms for that task. We don't get into the full details here, but only demonstrate a Julia implementation using `LIBSVM.j` which wraps the well known `LIBSVM` package [CC11]. We recommend Section 7.7 of [KBTV19] for a comprehensive description of SVM theory and practice. Computationally, training an SVM classifier can be done by solution of a *quadratic programming* optimization problem. However, in tuning parameters, dealing with multi-class classification, and adjusting basis functions, the process is more involved and the `LIBSVM` package handles this.

In their most basic form, SVMs deal with binary classification by finding an optimal *separating hyperplane* in the feature space. The main idea is to split the feature vectors $\{x_i\}_{i=1}^n$ into the $+1$ and -1 cases, and then to find a hyperplane that separates the positive and negative cases in the best way possible. The data is said to be *linearly separable* if one can find a hyperplane such that all positive points lie on one side and all negative points lie on the other side. In such a case a *hard margin* approach is used. The alternative is called the *soft margin* approach. Moving from binary classification to multi-class classification follows methods such as the one vs. rest approach using in linear least classification above.

SVMs are one class of machine learning algorithms that strongly benefit from the so-called *kernel method*, also known as the *kernel trick*. The general idea of such methods is to implicitly transform the features into a different space represented by a *basis*. In practice, this allows us to transform the seemingly straight-line (hyperplane) based classifier into a classifier that can fit arbitrary patterns to the data. The idea and mathematics of such kernel methods are beyond our scope, but we refer the reader to [KBTV19] for a comprehensive overview.

The kernel trick can also be applied to other machine learning algorithms, however in SVMs it is most popular. The most commonly used kernel is the *radial basis function kernel* and indeed it is the default kernel used in `LIBSVM.jl`.

In Listing 9.7 we use the `LIBSVM` package for a classification experiment on $10,000$ MNIST images. The listing also times the computation and redirects the verbose output (`stdout`) to a log file. As can be observed, the prediction accuracy (measured on the test set) is 91.73%. In this specific example we use a *linear kernel* as opposed to the default *radial basis* kernel.

Listing 9.7: Support vector machines

```
1  using Flux.Data.MNIST, LIBSVM, Plots
2
3  logFilePath = "../data/svmlog.txt"
4  nTrain = 10^4
5
6  trainImgs   = MNIST.images()[1:nTrain]
7  trainLabels = MNIST.labels()[1:nTrain]
8  trainData = hcat([vcat(float.(trainImgs[i])...) for i in 1:nTrain]...)
9
10 testImgs = MNIST.images(:test)
11 testLabels = MNIST.labels(:test)
12 nTest = length(testImgs)
13 testData = hcat([vcat(float.(testImgs[i])...) for i in 1:nTest]...)
14
15 @info "Training model with verbose output to $logFilePath."
16 @time begin
17     sOut = stdout
18     logF = open(logFilePath, "w")
19     redirect_stdout(logF)
20     model = svmtrain(trainData, trainLabels,
21                 kernel = Kernel.Linear, verbose=true)
22     close(logF)
23     redirect_stdout(sOut)
24     @info "Training complete."
25 end
26
27 predicted_labels, _ = svmpredict(model, testData)
28
29 accuracy = sum(predicted_labels .== testLabels)/nTest
30 println("Prediction accuracy (measured on test set of size $nTest): ", accuracy)
```

```
Info: Training model with verbose output to ../data/svmlog.txt.
@ Main In[6]:15
Info: Training complete.
@ Main In[6]:24
48.904273 seconds (662 allocations: 135.949 MiB, 0.12% gc time)
Prediction accuracy (measured on test set of size 10000): 0.9173
```

In line 3 we define a file path for creating a log of the output. In line 4 we set the training set data. The training and testing data and labels are set in lines 6-13. Line 15 uses the `@info` macro for output printing. This macro is common for usage as part of machine learning scripts. We then carry out the training wrapped within the `@time` macro. Lines 17-19 redirect the *standard output*, `stdout`, to file. This is after keeping references to the environment's standard output via `sOut` in line 17. The training using the `LIBSVM` package is invoked in lines 20-21 using `svmtrain()`. We use `Kernel.Linear` as the kernel and indicate that output of the training should be printed via `verbose = true`. However the output is redirected to file as it is lengthy. In line 22 we close the output file and then in line 23 we restore the standard output. The `svmpredict()` function is used in line 27 and the accuracy is then computed and printed.

Decision Trees and Random Forests

An alternative general purpose classifier is the *random forest* algorithm. It is a *bagging algorithm* applied to *decision trees*. We omit most of the details of how random forests are constructed and refer the reader to the literature such as Chapter 8 of [KBTV19].

In Listing 9.8 we use the `DecisionTree` package to create a random forest classifier. The `build_forest()` function accepts a few meters including the number of features in every tree, the number of trees, the portion of samples per tree, and the maximal tree depth. The classifier achieves 92.95% accuracy on the test set.

Listing 9.8: Random forest

```
1  using Flux.Data.MNIST, DecisionTree, Random
2  Random.seed!(0)
3
4  trainImgs   = MNIST.images()
5  trainLabels = MNIST.labels()
6  nTrain = length(trainImgs)
7  trainData = vcat([hcat(float.(trainImgs[i])...) for i in 1:nTrain]...)
8
9  testImgs = MNIST.images(:test)
10 testLabels = MNIST.labels(:test)
11 nTest = length(testImgs)
12 testData = vcat([hcat(float.(testImgs[i])...) for i in 1:nTest]...)
13
14 numFeaturesPerTree = 10
15 numTrees = 40
16 portionSamplesPerTree = 0.7
17 maxTreeDepth = 10
18
19 model = build_forest(trainLabels, trainData,
20                     numFeaturesPerTree, numTrees,
21                     portionSamplesPerTree, maxTreeDepth)
22 println("Trained model:")
23 println(model)
24
25 predicted_labels = [apply_forest(model, testData[k,:]) for k in 1:nTest]
26 accuracy = sum(predicted_labels .== testLabels)/nTest
27 println("\nPrediction accuracy (measured on test set of size $nTest): ",accuracy)
```

```
Trained model:
Ensemble of Decision Trees
Trees:      40
Avg Leaves: 852.325
Avg Depth:  10.0

Prediction accuracy (measured on test set of size 10000): 0.9295
```

Lines 4-12 setup the train and the test data in a similar manner to previous examples. This time all 60,000 training MNIST images are used. Lines 14-17 setup hyper-parameters for the random forest algorithm. In lines 19-21 we use the `build_forest()` function to run a random forest algorithm on the `trainData` with `trainLabels`. In line 25 we use the `apply_forest()` function with the `model` created on `testData`. This creates the `predicted_labels` array.

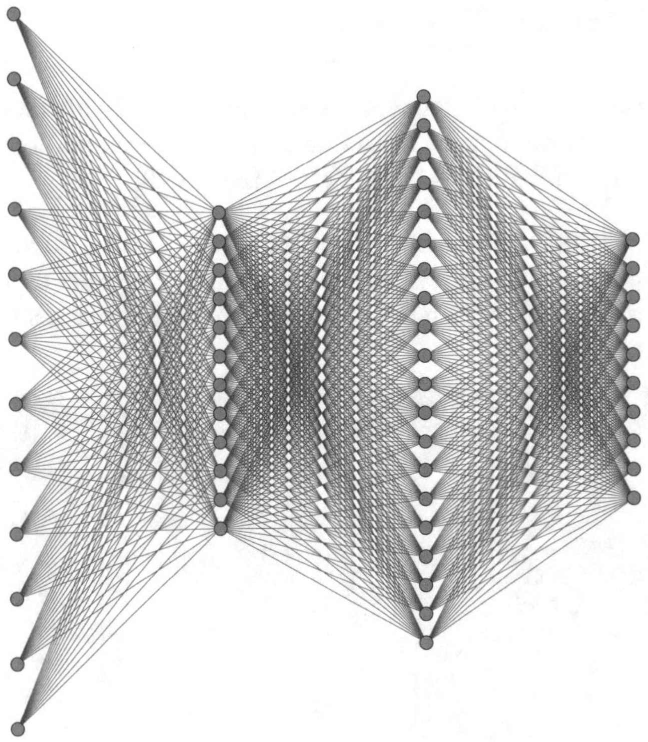

Figure 9.6: A dense (fully connected) neural network similar to `model1` of Listing 9.9. The left most nodes are the input layer. The right most nodes are the neurons of the output layer. The middle nodes are neurons of inner layers.

Deep and Convolutional Neural Networks

Deep learning is probably the most impactful machine learning method of recent years. *Neural networks*, with multiple layers called *deep neural networks* are now present in many software applications, often branded as artificial intelligence. The basic ideas of neural networks and deep learning have actually been around since the 1950s with the introduction of the *perceptron* algorithm. However, in the past decade, the application of these methods has become considerably more popular due to computational and algorithmic advances, as well as due to availability of huge training datasets. These days using deep learning is both very powerful and easily accessible.

Mathematically, deep learning models are actually not difficult to understand. Similarly to other classifiers, a model is a function from the features to the labels (or to the set of probability distributions over the labels). For example, for the case of MNIST, a trained neural network is a function,

$$\tilde{f} : \mathbb{R}^{784} \to \mathbb{R}^{10}.$$

This means that each (vectorized) input image x of dimension $28 \times 28 = 784$ can be transformed to a probability distribution over the 10 labels given via $\tilde{f}(x)$. Further, as with several of the other classifiers described previously, the actual classifier for this neural network, denoted $\hat{f}(\cdot)$, chooses the label that has maximal probability,

$$\hat{f}(x) = \text{argmax}_{\ell=0,\dots,9}\ \tilde{f}_\ell(x). \tag{9.20}$$

Compare (9.20) with the linear classifier (9.13), the logistic regression classifier (9.14) and the logistic softmax regression classifier (9.17). In all cases an affine (linear plus bias) transformation is applied to the input followed by some (potentially) non-linear transformation. All these previous classifiers are examples of neural networks. However with deep learning there is more. In general, a deep learning model allows us to compose such *layers* one after another.

As an example consider Figure 9.6[1]. The network in that figure is called a *dense neural network* (also known as a *fully connected neural network*). Each *neuron* is a dot in one of the inner layers or outer layer of the figure and involves a weight and a bias. It can be mathematically represented via

$$\tilde{f}(x) = \underbrace{s\Big(W_3\ \underbrace{\rho_2\big(W_2\ \underbrace{\rho_1(W_1 \overbrace{x}^{\text{Input layer}} + b_1)}_{\text{First hidden layer}} + b_2\big)}_{\text{Second hidden layer}} + b_3\Big)}_{\text{Output layer}}. \tag{9.21}$$

Here the feature vector x is called the *input layer* and then the various components of (9.21) are the first *hidden layer*, second hidden layer, and *output layer*. The parameters of the neural network are the weight matrices W_1, W_2, and W_3, together with the three corresponding bias vectors b_1, b_2, and b_3. The *architecture* of the neural network is a specification of the dimensions of the weight matrices, the bias vectors, the number of layers, any restrictions on these matrices, and also the non-linear functions, $\rho_1(\cdot)$, $\rho_2(\cdot)$, and $s(\cdot)$. The first two functions are called *activation functions* and are applied element wise on their inputs. The third function $s(\cdot)$ is the softmax function (9.16), often used in the output layer of classification networks.

Classically the common choice of an activation function $\rho_i(\cdot)$ is the sigmoid function (9.15). However, other alternatives include the *relu* activation function and *tanh* activation function,

$$\rho_{\text{relu}}(u) = \begin{cases} u & \text{if } u \geqslant 0, \\ 0 & \text{if } u < 0, \end{cases} \qquad \text{and} \qquad \rho_{\tanh}(u) = \frac{e^u - e^{-u}}{e^u + e^{-u}}.$$

Specifically, the very simple relu, while not differentiable at $u = 0$, has been found to perform well in the training of very deep networks. Using relu helps mitigate a computational training phenomena called the *vanishing gradient problem*. This is when the gradient of certain weights is so small that the weights effectively don't update with gradient-based learning. Such a problem is especially present in very deep neural networks. This is where relu is often most useful. As opposed to that, in shallow and simpler networks, using sigmoid or tanh often yields better result. The best choice of activation function is generally not an exact science but rather a matter of experimentation.

Neural networks can involve a large number of parameters. Consider, for example, the dense network illustrated in Figure 9.6 and represented in (9.21). The figure is only a schematic and does not represent the exact number of neurons. For one thing, in the MNIST images, the number of

[1]The Figure is generated via `http://alexlenail.me/NN-SVG/index.html`.

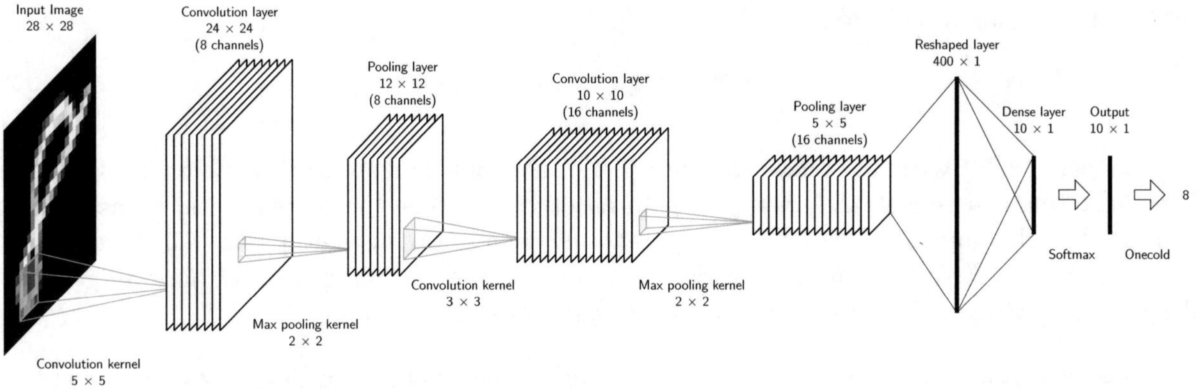

Figure 9.7: The architecture of the convolutional neural network (`model2`), used in Listing 9.9.

elements in the input layer should be 784 (in the figure there are only 12 input elements). It is a matter of architecture choice to determine how many neurons are in each hidden layer. As an example, consider lines 27-28 of Listing 9.9 where `model1` is defined. Here there are `200` neurons in the first hidden layer, and `100` neurons in the second hidden layer. Such an architecture implies the following for the dimensions of the matrices in (9.21). W_1 is 784×200, W_2 is 200×100, and W_3 is 100×10. Further, the bias vectors b_1, b_2, and b_3 are of dimensions 200, 100, and 10, respectively. This means that this neural network involves a total of 178,100 parameters! While this is already a large number, larger neural networks used in applications may be much deeper, deal with much more complicated feature vectors, and may thus end up having tens of millions of parameters.

Effectively training neural networks with a huge number of parameters is often practically impossible. This is where great advances have been made by introducing special structures, the most notable of which is the suite of *convolutional neural network* architectures which is motivated by more classic *image processing* techniques. In these networks, mathematical structures similar to (9.21) may still be used to describe the network, however the matrices at play involve a much smaller number of parameters as they define *convolution* operations. A convolution is mathematical operation involving signals, vectors, distributions, or images, and it may be interpreted in different ways depending on the context. In the context of convolutional neural networks, it is a linear operation dealing with neighboring pixels in the image. We have already implicitly presented a convolution example back in Listing 1.13 of Chapter 1. In that case, neighboring pixels were averaged. A convolution generalizes such averaging, and in convolutional neural networks, the exact nature of the convolutions is automatically learned during training.

Since images often involve color, the natural input structure for neural networks dealing with images involves *tensors* (an image is a 3-tensor because there are three color layers). Further dimensions are often added to the tensors to allow other features, multiple images (movies), and to collect mini-batched images together. The practice of convolutional neural networks has quickly evolved and standardized over the past few years and there are now clear descriptions on how to specify such networks. As our example, consider Figure 9.7 which exactly matches `model2` defined in lines 30-32 of Listing 9.9.

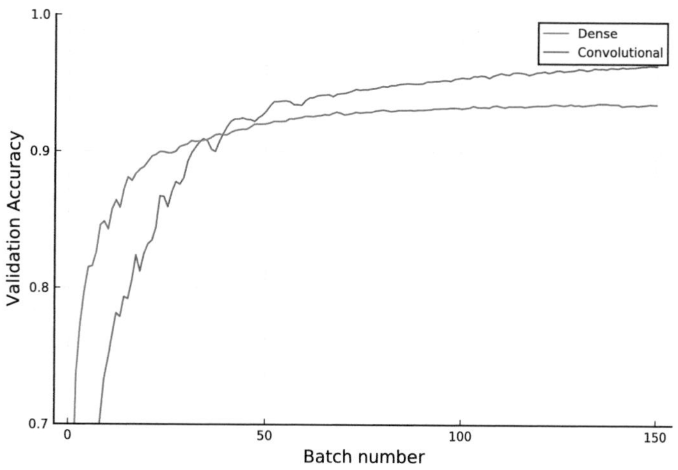

Figure 9.8: The progression of loss accuracy (on the validation set) while training two neural network architectures.

The key objects in Figure 9.7 include *convolutional layers*, *pooling layers*, and a *dense layer*. Each convolutional layer can be mathematically represented as $\rho(Wz + b)$ where z is the input, W is a matrix operation, b is a bias vector, and $\rho(\cdot)$ is the activation function. Importantly, the matrix represents a convolutional operation localized on the pixels of the image. This implies that the matrix W is very structured with a small number of parameters. The pooling layer, also known as *max pooling*, does not involve any parameters, but simply operates by taking local maxima of neighboring pixels. It is common to have convolutional deep neural networks with alternating convolution-pooling layers, with a dense layer at the end of the network for combining different parts of the image.

The first convolutional layer is specified as `Conv((5, 5), 1=>8, relu)` - see Line 30 of Listing 9.9. This implies that a convolution kernel with a width and height of 5 pixels is "passed" over the image. Initially since the image is monochrome it has a single *channel*, however the `1=>8` syntax specifies creating a tensor with 8 channels at the output of that layer. Creating multiple channels allows the neural network to capture both localized pixel relationships and relationships between different locations in the image. Due to the edge effects of the convolution, the resulting images are 24×24 because $28 - 5 + 1 = 24$. Alternatively, *padding* could have been specified, but we did not do so here. The number of parameters in the matrix that (mathematically) specifies such a convolution is only $5 \times 5 \times 8 = 200$. Compare this with $3,612,672$ which is the number of entries in a $784 \times 4,608$ matrix (where $4,608 = 24 \times 24 \times 8$). Note that neither the user nor the implementation never needs to "write out" the entries of this (*block Toeplitz*) matrix. It is rather a description for understanding the mathematical description of the network. You may find an elementary introduction to such matrix-based convolution operations in Section 7.4 of [BL18]. Our key point here is simply that a very complicated operation can be summarized by a (relatively) small number of parameters.

Following the first convolutional layer, the pooling layer, specified via `MaxPool((2,2))` in line 30 converts the pixels of every 2×2 block of each channel to a single pixel by taking the maximum. Hence as appearing in Figure 9.7, there are now 8 channels of 12×12 images. This $8 \times 12 \times 12$ tensor is then the input to the next layer, `Conv((3, 3), 8=>16, relu)` which carries out 3×3 convolutions. Since these convolutions are again carried out without padding, the output of this layer has 10×10 pixels per channel with $10 = 12 - 3 + 1$. Note that the architecture specifies 16 layers here via the `8=>16` pair. Max pooling is then carried out again yielding 16 channels each of 5×5 pixels. This is a total of 400 neurons which are then flattened onto a single layer of $400 = 16 \times 5 \times 5$ neurons.

The architecture of this model may seem somewhat arbitrary. Such is often the case for convolutional deep neural networks. The key point is that one may experiment with several architectures and then choose the one with the best performance. You may observe a variety of examples using `Flux.jl` in `https://github.com/FluxML/model-zoo`.

Consider now Figure 9.8 created by Listing 9.9. We train both `model1` and `model2` on 5,000 images for 30 epochs. We use a batch size of 1,000 hence there are 5 batches per epoch and a total of 150 iterations. In each iteration, a validation set composed of 5,000 images is used to compute the accuracy of the classifiers. As can be seen from the output of Listing 9.9, `model1` achieves 93.5% accuracy and `model2` achieves 96.3% accuracy. The strength of `model2` is also apparent in Figure 9.8. In training both of these models we use cross entropy loss, and the ADAM optimizer with a learning rate of 0.005. We also save the parameters of `model2` and use them in Listing 9.1.

Listing 9.9: Training dense and convolutional neural networks

```
using Flux, Flux.Data.MNIST, Statistics, BSON, Random, Plots; pyplot()
using Flux: onehotbatch, onecold, crossentropy
Random.seed!(0)

epochs = 30
eta = 5e-3
batchSize = 1000
trainRange, validateRange = 1:5000, 5001:10000

function minibatch(x, y, indexRange)
    xBatch = Array{Float32}(undef, size(x[1])..., 1, length(indexRange))
    for i in 1:length(indexRange)
        xBatch[:, :, :, i] = Float32.(x[indexRange[i]])
    end
    return (xBatch, onehotbatch(y[indexRange], 0:9))
end

trainLabels = MNIST.labels()[trainRange]
trainImgs = MNIST.images()[trainRange]
mbIdxs = Iterators.partition(1:length(trainImgs), batchSize)
trainSet = [minibatch(trainImgs, trainLabels, bi) for bi in mbIdxs]

validateLabels = MNIST.labels()[validateRange]
validateImgs = MNIST.images()[validateRange]
validateSet = minibatch(validateImgs, validateLabels, 1:length(validateImgs))

model1= Chain(flatten, Dense(784, 200,relu),Dense(200, 100,tanh),
                                Dense(100, 10,sigmoid), softmax)

model2= Chain(Conv((5, 5), 1=>8, relu), MaxPool((2,2)),
                Conv((3, 3), 8=>16, relu), MaxPool((2,2)),
                flatten, Dense(400, 10), softmax)

opt1 = ADAM(eta); opt2 = ADAM(eta)
accuracyPaths = [[],[]]
accuracy(x, y, model) = mean(onecold(model(x)) .== onecold(y))
loss(x, y, model) = crossentropy(model(x), y)
cbF1() = push!(accuracyPaths[1],accuracy(validateSet..., model1))
cbF2() = push!(accuracyPaths[2],accuracy(validateSet..., model2))

model1(trainSet[1][1]); model2(trainSet[1][1])
for _ in 1:epochs
    Flux.train!((x,y)->loss(x,y,model1), params(model1), trainSet, opt1, cb=cbF1)
    Flux.train!((x,y)->loss(x,y,model2), params(model2), trainSet, opt2, cb=cbF2)
        print(".")
end

println("\nModel1 (Dense) accuracy = ", accuracy(validateSet..., model1))
println("Model2 (Convolutional) accuracy = ", accuracy(validateSet..., model2))
cd(@__DIR__)
BSON.@save "../data/mnistConv.bson" modelParams=cpu.(params(model2))
plot(accuracyPaths,label = ["Dense" "Convolutional"],
        ylim=(0.7,1.0), xlabel="Batch number", ylabel = "Validation Accuracy")
```

```
..............................
Model1 (Dense) accuracy = 0.935
Model2 (Convolutional) accuracy = 0.963
```

We set the number of epochs in line 5, the learning rate in line 6, and the batch size in line 7. In line 8 we define a range within the data for training, `trainRange`, and similarly a range for validation, `validateRange`. Lines 10-16 define the `minibatch()` function which returns a tuple of batched data and labels. It works on the data `x`, the labels `y`, and a range specification of indices, `indexRange`. In lines 18-19 we obtain the training data and labels. In line 20 we create a collection of indices, one for each mini batch. The `trainSet` array is created in line 21 by invoking the `minibatch()` function on each element of `mbIdxs`. The validation set is treated similarly in lines 23-25, however it is composed of a single mini batch. Lines 27-28 defines `model1` which works on flattening the input image and passing it through 3 dense layers followed by a `softmax`. Lines 30-32 define model 2 via a convolutional network as appears in Figure 9.7. Line 34 defines the optimizers, one for each network. Line 35 defines an array of arrays that will record the accuracy. The `accuracy()` function is defined in line 36. In addition to the data and labels, It accepts `model` as input and computes the proportion of labels that are predicted by `model`. For this it uses `onecold()` to convert from a one hot encoded vector to a label, and the `mean()` function for computing the sample proportion after `.==` is used. The loss function for a given `model` is defined in line 37 using the `crossentropy()` function. Lines 38-39 define *call back functions* which are to be invoked during training. Each of these functions appends the current accuracy to `accuracyPaths`. The call to `model1()` and `model2()` in line 41 allows the Julia compiler to precompile the models prior to beginning the training loop. In certain cases this can speed up performance. The main training loop is in lines 42-46. The `train!()` function of the `Flux` package is called. Since the data is already in the mini batch ready form of `trainSet`, the function does not require many arguments when called in lines 43 and 44 for `model1` and `model2`, respectively. The final accuracy is computed and printed in lines 48-49 and plotted in lines 52-53. Line 50 sets the current directory to `@__DIR__` which is the directory of the current code listing and then uses the `@save` macro from package `BSON` for saving the parameters of `model2`. These parameters can be used in production as illustrated in Listing 9.1.

9.3 Bias, Variance, and Regularization

When we introduced supervised learning in Section 9.1, we presented two competing objectives: TRAIN DATA FIT (9.1) and UNKNOWN DATA FIT (9.2). The first deals with the goal of fitting the model to the training data. The second deals with the goal of the model working well on *unseen data*. These objectives often compete because a very tight fit to the data can be achieved by *overfitting* a model and then for unseen data the model does not perform well.

We now discuss how these objectives are quantified via *model bias*, *model variance*, and the *bias-variance tradeoff*. We also present practical *regularization* techniques for optimizing the bias-variance tradeoff. Back in Section 5.4, when discussing point estimation in the context of classic statistical inference, we saw in (5.6) that the mean squared error (loss) of an estimator can be represented as the sum of the variance of the estimator and the square of the bias. Now in a supervised machine learning context, somewhat similar analysis can be carried out. Another similar analysis appears in the decomposition of the sum of squares in ANOVA (7.28) from Section 7.3.

The key is to consider the data $\{(x_i, y_i)\}_{i=1}^n$ as a random sample, where each pair (x_i, y_i) is independent of all other data pairs. Similar to previous chapters we use capital letters for the random variables representing these data points. Hence X is a random feature vector and Y is the associated random label. Note that clearly X and Y themselves should not be independent because then predicting Y based on X would not be possible.

The underlying assumption is that there exists some unknown relationship between x_i and y_i of the form $y_i = f(x_i) + \text{noise}$. Then the *inherent noise level* can be represented via $\mathbb{E}[(Y - f(X))^2]$. Now when presented with a classification or regression algorithm, $\hat{f}(\cdot)$, we consider the expected loss

$$\mathbb{E}[L] = \mathbb{E}\Big[\big(\tilde{Y} - \hat{f}(\tilde{X}\,;\,D)\big)^2\Big]. \tag{9.22}$$

Here we use D to represent the random dataset used to train $\hat{f}(\cdot)$ and the individual pair $(\tilde{X}, \tilde{Y})$ is some fixed point from the unseen (future) data. This makes $\hat{f}(\cdot\,;\,D)$ a random model with randomness generated from D (as well as possible randomness from the training algorithm). The expectation in (9.22) is also with respect to the random unseen pair $(\tilde{X}, \tilde{Y})$.

With these assumptions, a sum of squares decomposition can be computed and yields the *bias-variance-noise decomposition equation*,

$$\mathbb{E}[L] = \underbrace{\Big(\mathbb{E}[\hat{f}(\tilde{X}\,;\,D)] - \mathbb{E}[f(\tilde{X})]\Big)^2}_{\text{Bias squared of } \hat{f}(\cdot)} + \underbrace{\text{Var}\big(\hat{f}(\tilde{X}\,;\,D)\big)}_{\text{Variance of } \hat{f}(\cdot)} + \underbrace{\mathbb{E}[(Y - f(X))^2]}_{\text{Noise of } (X,Y)}. \tag{9.23}$$

The main takeaway from (9.23) is that if we ignore the inherent noise, the loss of the model has two key components, *bias*, and *variance*. The bias is a measure of how the model $\hat{f}(\cdot)$ misclassifies the correct relationship $f(\cdot)$. That is in models with high bias, $\hat{f}(\cdot)$ does not accurately describe $f(\cdot)$. That is high bias generally implies *under-fitting*. Similarly, models with low bias are detailed descriptions of reality. The variance is a measure of the variability of the model $\hat{f}(\cdot\,;\,D)$ with respect to the random sample D. Models with high variance are often *overfit* (to the training data) and do not *generalize* (to unseen data) well. Similarly, models with low variance are much more robust to the training data and generalize to the unseen data much better.

Understanding these concepts, even if qualitatively, allows the machine learning practitioner to tune the models to the data. The perils of high bias (under-fitting) and high variance (overfitting) can sometimes be quantified and even tuned via *tuning parameters* as we present in the examples below. Similar analysis to the derivation that leads to (9.23) can also be attempted for other loss functions. However, the results are generally less elegant. Nevertheless, the concepts of model bias, model variance, and the bias-variance tradeoff still persist. For example, in a classification setting you may compare the accuracy obtained on the training set to that obtained on a validation set. If there is a high discrepancy where the train accuracy is much higher than the validation accuracy, then there is probably a variance problem indicating that the model is overfitting.

There are several approaches for controlling model bias, model variance, and optimizing the bias-variance tradeoff. One key approach is called *regularization.* In regression models, it is common to add *regularization terms* to the loss function. In deep learning, the simple randomized technique of *dropout* has become popular. Both of these approaches are illustrated below.

The *hyper-parameter tuning* process which often (implicitly) deals with the bias-variance tradeoff can be carried out in multiple ways. One notable way which we explore here is *k-fold cross validation.* Here the idea is to break up the training set into k groups. Then use $k-1$ of the groups for training and 1 group for validation. However this is repeated k times so that each of the k groups gets to be a validation set once. Then the results over all k training sessions are averaged. This process can then be carried out over multiple values of the hyper-parameter in question and the loss can be plotted or analyzed. In fact, one may even estimate the bias and the variance terms in this way but we do not do so here. Still, we use k-fold cross validation in the ridge regression example that follows.

Addition of Regularization Terms

The addition of a regularization term to the loss function is the most classic and common regularization technique. Denote now the parameters of the model as β. The main idea is to take the data fitting objective

$$\min_{\beta} L(\text{data}, \beta), \tag{9.24}$$

and augment the loss function $L(\cdot,\cdot)$ with an additional regularization term $R(\lambda, \beta)$ that depends on a *regularization parameter* λ and the estimated parameters β. The optimization problem then becomes

$$\min_{\beta} \; L(\text{data}, \beta) + R(\lambda, \beta). \tag{9.25}$$

Now λ, often a scalar in the range $[0, \infty)$ but also sometimes a vector, is a *hyper-parameter* that allows us to optimize the bias-variance tradeoff. A common general regularization technique is *elastic net* where $\lambda = [\lambda_1 \; \lambda_2]^\top$ and,

$$R(\lambda, \beta) = \lambda_1 ||\beta||_1 + \lambda_2 ||\beta||^2.$$

Here $||\beta||^2 = \sum_{i=1}^{p+1} \beta_i^2$ and $||\beta||_1 = \sum_{i=1}^{p+1} |\beta_i|$ when $p+1$ is the dimension of the parameter space. Hence the values of λ_1 and λ_2 determine what kind of penalty the objective function will pay for high values of β_i. Clearly with $\lambda_1, \lambda_2 = 0$ the original objective (9.24) isn't changed. Further as $\lambda_1, \lambda_2 \to \infty$ the estimates $\beta_i \to 0$ and the data is fully ignored. As λ_1 or λ_2 grows the bias in the model grows, however the variance is decreased as overfitting is mitigated. The virtue of regularization is that there is often a magical "sweet spot" for λ where the objective (9.25) does a much better job than the non-regularized (9.24).

Particular cases of elastic net are the classic *ridge regression* (also called *Tikhonov regularization*) and *LASSO* standing for *least absolute shrinkage and selection operator.* In the former $\lambda_1 = 0$ and only λ_2 is used, and in the latter $\lambda_2 = 0$ and only λ_1 is used. One of the virtues of LASSO (also present in the more general elastic net case) is that the $||\beta||_1$ cost allows the algorithm to knock out variables by "zeroing out" their β_i values. Hence LASSO is very useful as an advanced model selection technique. We have seen LASSO in Listing 8.17 of Chapter 8. For arbitrary elastic net regularization we refer the reader to explore the Julia package `GLMNet`. We now focus on the more elementary ridge regression case and illustrate how to carry out *k-fold cross validation* for tuning the regularization parameter.

The case of ridge regression is slightly simpler to analyze than LASSO or the general elastic net. In this case the data fitting problem can be represented as

$$\min_{\beta \in \mathbb{R}^{p+1}} ||A\beta - y||^2 + \lambda ||\beta||^2,$$

where we now consider λ as a scalar (previously λ_2) in the range $[0, \infty)$. Note also that here y denotes a vector of all the labels $\{y_i\}_{i=1}^n$. The squared norm, $||\cdot||^2$ of a vector is just its inner product with

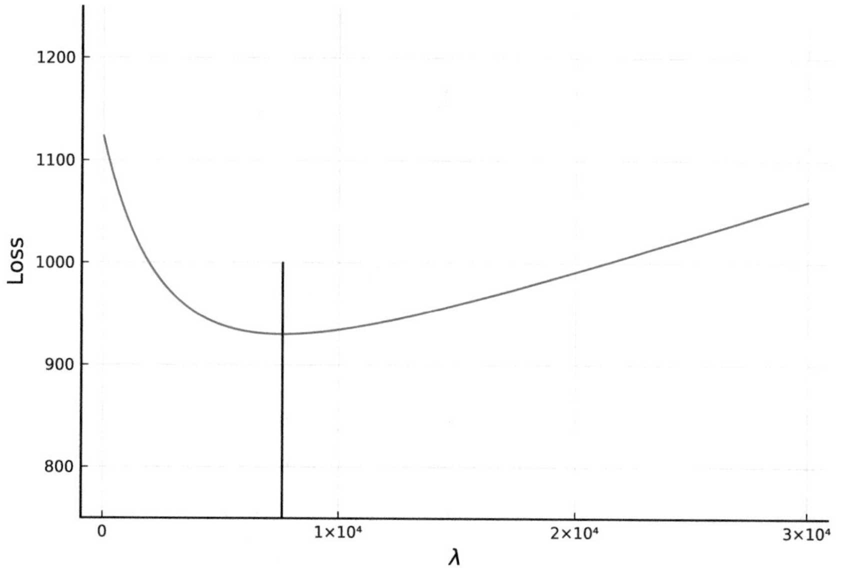

Figure 9.9: Tuning λ in ridge regression via k-fold cross validation. The optimal λ is marked via the black line.

itself and A is the $n \times (p+1)$ design matrix 8.40 comprised of x values and the first column is a column of ones. The problem can just be recast as

$$\min_{\beta \in \mathbb{R}^{p+1}} \left\| \begin{bmatrix} A \\ \sqrt{\lambda} I \end{bmatrix} \beta - \begin{bmatrix} y \\ 0 \end{bmatrix} \right\|^2.$$

The pseudo-inverse (see Section 8.1) associated with the augmented matrix,

$$\tilde{A} = \begin{bmatrix} A \\ \sqrt{\lambda} I \end{bmatrix},$$

is $\tilde{A}^\dagger = (A^\top A + \lambda I)^{-1} [A^\top \quad \sqrt{\lambda} I]$. Hence the parameter estimate is $\hat{\beta} = (A^\top A + \lambda I)^{-1} A^\top y$.

Incidentally, for any $\lambda > 0$ it can be shown that $\tilde{A}$ has linearly independent columns. This even holds if A has some dependent columns (and isn't full rank). Thus perfect collinearity is alleviated by adding a regularization term. See Listing 8.11 for an illustration of collinearity. It can even be shown that $A^\dagger$ of (8.13) satisfies

$$A^\dagger = \lim_{\lambda \to 0} (A^\top A + \lambda I)^{-1} A^\top$$

Similar (and much more practically), collinearity that isn't perfect can also be alleviated by considering non-zero λ values. Hence this type of *shrinkage estimator* is very popular for a variety of reasons.

Listing 9.10 presents an example of ridge regression where we carry out *k-fold cross validation* to find a good λ value. Here we use the `cpus`, for `MASS` from `RDatasets.jl`. A search over a range of λ values is presented in Figure 9.9 where we see that the optimal λ is at around $\lambda = 7,600$.

Listing 9.10: Ridge regression with k-fold cross validation

```
using RDatasets, DataFrames, Random, Statistics, LinearAlgebra
using MultivariateStats, LaTeXStrings, Plots; pyplot()
Random.seed!(0)

df = dataset("MASS", "cpus")
n = size(df)[1]
df = df[shuffle(1:n),:]

K = 10
nG = Int(floor(n/K))
n = K*nG
println("Losing $(size(df)[1] - n) observations.")

lamGrid = 0:100:30000

devSet(k) = collect(1+nG*(k-1):nG*k)
trainSet(k) = setdiff(1:n,devSet(k))

xTrain(k) = convert(Array{Float64,2},df[trainSet(k),[:Cach, :ChMin]])
xDev(k) = convert(Array{Float64,2},df[devSet(k),[ :Cach, :ChMin]])

yTrain(k) = convert(Array{Float64,1},df[trainSet(k),:Perf])
yDev(k) = convert(Array{Float64,1},df[devSet(k),:Perf])

errVals = zeros(length(lamGrid))
for (i,lam) in enumerate(lamGrid)
    errSamples = zeros(K)
    for k in 1:K
        beta = ridge(xTrain(k),yTrain(k),lam)
        errSamples[k] = norm([ones(nG) xDev(k)]*beta - yDev(k) )^2
    end
    errVals[i] = sqrt(mean(errSamples))
end

i = argmin(errVals)
bestLambda = lamGrid[i]

betaFinal = ridge(convert(Array{Float64,2},df[:,[:Cach, :ChMin]]),
                  convert(Array{Float64,1},df[:,:Perf]),bestLambda)

macro RR(x) return:(round.($x,digits = 3)) end
println("Found best lambda for regularization: ", bestLambda)
println("Beta estimate: ", @RR betaFinal)

plot(lamGrid, errVals,legend = false,
     xlabel = L"\lambda", ylabel = "Loss")
plot!([bestLambda,bestLambda],[0,10^3], c = :black, ylim = (750, 1250))
```

```
Losing 9 observations.
Found best lambda for regularization: 7600
Beta estimate: [2.207, 3.754, 32.339]
```

In lines 5-7 we read the data frame, get the number of observations `n` and then shuffle the observations. In lines 9-12 we determine parameters associated with the cross validation. The variable `K` determines the number of groups and then `nG` determines the number of observations per group. We then reassign a value to `n` as the number of effective observations. A few observations are lost as a remainder and we print this in line 12. In line 14 we set the range of λ values for ridge regression over which we optimize. In lines 16-17 we define the functions `devSet()` and `trainSet()` for determining indices of these sets. Then in lines 19-20 and 22-23 we set up functions that actually extract the features (x) and labels (y) from the dataframe using these sets. The features are the dataframe variables `:Cach` and `:ChMin` and the label (in this case a continuous explanatory variable) is `:Perf`. In lines 26-33 we carry out ridge regression over each λ value in `lamGrid`. Each time we estimate `beta` using the function `ridge()` from `MultivariateStats.jl` in line 29 and then evaluate the error in line 30. This is done for each group in the k-fold cross validation. The final error for that λ is estimated in line 32. In lines 35-36 we find the best λ. We then carry out ridge regression one last time in lines 38-39, using all of the data. Line 41 is simply a short macro for rounding, `RR()`. We use it in line 43 for printing the β estimate. The remainder of the code creates Figure 9.4.

Dropout for Neural Networks

For neural networks, one may use an additive regularization term similar to ridge regression, LASSO, or elastic nets. However *dropout* is a much more elementary technique that has become popular in recent years. The basic mechanism of dropout is very simple and may even initially appear counter intuitive. However empirical evidence of the performance of dropout has shown that it is a very effective method and today it has become the standard approach for neural network regularization.

The basic operation of dropout is that during the training of a neural network, individuals nodes are "zeroed out", randomly and independently. This means that on every iteration (mini-batch), some nodes will not be used and the gradients of their weights and biases will be zero. Such randomization enforces learning that is similar to an *ensemble* of models. That is, even though there is only a single neural network, there is the effect of averaging a combination of many neural networks together which each uses a different random subset of the nodes.

When applying dropout, the hyper-parameter of choice is the *dropout probability.* A typical value is 0.5, however like any other parameter in a machine learning model, it may be tuned. In certain deep networks, one may even apply different dropout probabilities for different layers.

In Listing 9.11 we illustrate the effect of dropout by considering a network that involves both convolutional components and dense components. As you may observe from the model defined in lines 30-37, there are two convolutional layers followed by four dense layers, each involving a considerable number of neurons. For each such layer we define dropout by using `Dropout()` with a specified dropout probability, `dropP`.

As an example of *hyper-parameter tuning* we try dropout probabilities 0.0 (no dropout), 0.25, 0.5, and 0.75. For each such probability we carry out 10 training sessions. The resulting validation accuracy is the plotted in the box-plot of Figure 9.10. As the results of this experiment show, having dropout at 0.25 performs best.

Listing 9.11: Tuning the dropout probability for deep learning

```
using Flux, Flux.Data.MNIST, Statistics, BSON, Random, StatsPlots; pyplot()
using Flux: onehotbatch, onecold, crossentropy, @epochs

epochs = 30
eta = 1e-3
batchSize = 200
trainRange, validateRange = 1:1000, 1001:5000

function minibatch(x, y, idxs)
    xBatch = Array{Float32}(undef, size(x[1])..., 1, length(idxs))
    for i in 1:length(idxs)
        xBatch[:, :, :, i] = Float32.(x[idxs[i]])
    end
    return (xBatch, onehotbatch(y[idxs], 0:9))
end

trainLabels = MNIST.labels()[trainRange]
trainImgs = MNIST.images()[trainRange]
mbIdxs = Iterators.partition(1:length(trainImgs), batchSize)
trainSet = [minibatch(trainImgs, trainLabels, i) for i in mbIdxs]

validateLabels = MNIST.labels()[validateRange]
validateImgs = MNIST.images()[validateRange]
validateSet = minibatch(validateImgs, validateLabels, 1:length(validateImgs))

accuracy(x, y, model) = mean(onecold(model(x)) .== onecold(y))
loss(x, y, model) = crossentropy(model(x), y)

function evalAccuracy(dropP)
    model= Chain(Conv((5, 5), 1=>8, relu), MaxPool((2,2)),
                Conv((3, 3), 8=>16, relu), MaxPool((2,2)),
                flatten,
                Dense(400, 200,relu), Dropout(dropP),
                Dense(200, 200,relu), Dropout(dropP),
                Dense(200, 200,relu), Dropout(dropP),
                Dense(200, 10,relu), Dropout(dropP),
                softmax)
    opt = ADAM(eta);
    @epochs epochs Flux.train!((x,y)->loss(x,y,model),params(model),trainSet,opt)
    accuracy(validateSet..., model)
end

pToTest = [0.0, 0.25, 0.5, 0.75]
n = 10
results = [[evalAccuracy(p) for _ in 1:n] for p in pToTest]
bestAcc, bestI = findmax(median.(results))
println("The best dropout probability is $(pToTest[bestI]).")
println("It achieves $(bestAcc) accuracy on average.")

boxplot(results,label="",
        xticks=([1:1:4;],string.(pToTest)),
        xlabel="Dropout Probability", ylabel = "Accuracy", legend = false)
```

```
The best dropout probability is 0.25.
It achieves 0.941125 accuracy on average.
```

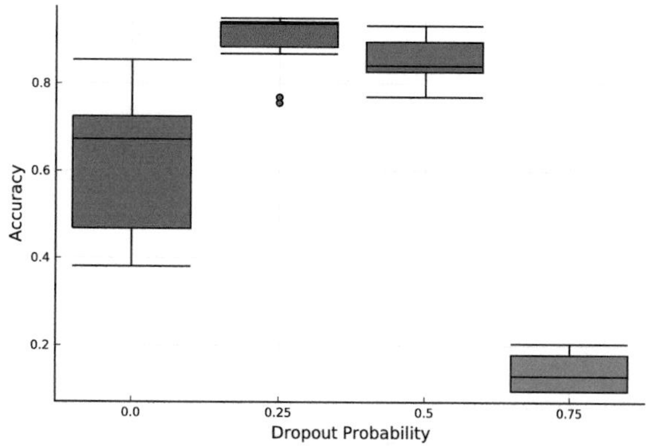

Figure 9.10: Hyper-parameter tuning for the dropout probability of a neural network. Ten training runs were conducted for each dropout probability and the best one appears to be at 0.25.

Lines 1-24 are similar to Listing 9.9. We set basic parameters, define the `minibatch()` function, load the data, and define the `accuracy()` and `loss()` functions. In lines 29-41 we define the `evalAccuracy()` function which trains a network with a given drop out probability, `dropP` and evaluates the obtained accuracy. In this case, the network model is defined in lines 30-37 and is comprised of several dense layers after initial convolutional layers. The actual training is in line 39 using the `@epochs` macro from `Flux.jl`. This macro executes the second argument, `Flux.train!()` for `epochs` times. In line 43 we set drop out probabilities that we try in `pToTest`. In line 44, `n`, is the number of training runs to carry out for each drop out probability. Line 45 is the main experiment evaluation, keeping the recorded accuracies in `results`. In line 46 we find the best accuracy and display the results in lines 47-48. Lines 50-52 create Figure 9.10.

9.4 Unsupervised Learning Methods

Data is not always labeled. That is, the features of a dataset are not always classified, or the labels themselves may be unknown. In these cases of *unlabelled data*, a learning goal is to identify patterns in the underlying features of observations, such that these or, new observations with similar features can be grouped accordingly, and some overall conclusions drawn. This is known as *unsupervised learning*, and common types of tasks in this space include various forms of *clustering* as well as *dimension reduction.*

For a dataset $x_1, \ldots, x_n$, clustering is the act of associating a cluster ℓ with each observation, where ℓ comes from a small finite set. That is, clustering considers the data and outputs a function $c(x)$ which maps data points to $\{1, \ldots, k\}$. The ℓ'th cluster is then,

$$C_\ell = \{x_i \ \text{ with } \ i \in \{1, \ldots, n\} \mid c(x_i) = \ell\}.$$

A clustering algorithm attempts to choose the clusters such that the elements of each C_ℓ are as homogenous as possible.

A dimension reduction algorithm attempts to create a transformed dataset $\tilde{x}_1, \ldots, \tilde{x}_n$ where each $\tilde{x}_i$ is of lower dimension than x_i, yet the information embodied by the new dataset is similar to the information in the original dataset. Good dimension reduction is able to significantly reduce the size of each x_i while at the same time maintaining the main attributes of the dataset.

For clustering we consider two types of algorithms: *k-means* and *hierarchical clustering*. For dimension reduction we consider *Principal Component Analysis (PCA)*.

Clustering with k-Means

When using k-means clustering, we assume that the data points, $x_1, \ldots, x_n$ are vectors in p-dimensional Euclidean space. We then specify a number k, determining the number of clusters that we wish to find. We then seek the function $c(x)$ (or a partition $C_1, \ldots, C_k$) together with means of clusters, $J_1, \ldots, J_k$. The ideal aim is to minimize

$$\sum_{\ell=1}^{k} \sum_{x \in C_\ell} ||x - J_\ell||^2. \tag{9.26}$$

Such a minimization is generally computationally difficult, however it can be approximately achieved via the k-means algorithm. It separates the problem into two sub-problems or sub-tasks: mean computation, and labeling.

Mean computation: Given $c(x)$, finding the means $J_1, \ldots, J_k$ is simply done by setting,

$$J_\ell = \frac{1}{|C_\ell|} \sum_{x \in C_\ell} x, \qquad \text{for} \qquad \ell = 1, \ldots, k. \tag{9.27}$$

This is the element-wise average over all the vectors in C_ℓ where each of the p coordinates is averaged separately. The mean J_ℓ is called the *centroid* for cluster ℓ.

Labeling: Given, $J_1, \ldots, J_k$ finding $c(x)$ that minimizes (9.26) is done by setting,

$$c(x) = \text{argmin}_\ell \; ||x - J_\ell||, \tag{9.28}$$

That is, the label of each element is determined by the closest centroid in Euclidean space.

The k-means algorithm operates by iterating over the mean computation step, (9.27), followed by the labeling step (9.28). This is done until no more changes are made to the labels and the means. Such an iteration generally doesn't find the absolute minimum of the objective (9.26), however the approximation found often satisfactory. As the algorithm is initialized, random means, $J_1, \ldots, J_k$ are selected. These initial random means are then used to determine initial labels according to (9.28). After such initialization, the process repeats until convergence is attained.

The process of selecting the hyper-parameter k is external to the k-means algorithm. One way to do so is to run k-means for increasing values of k and seek a *knee point* or *elbow* in the plot of (9.26). As k increases the objective (9.26) will decrease, however beyond a certain k the value of adding further clusters quickly diminishes. In other cases, such as the illustrative example we present now, there is a natural value of k, sometimes known based on external information.

We now consider the `xclara` dataset from the `clusters` `RDatasets` package. This dataset comprises observations consisting of two variables `V1` and `V2`. We set $k = 3$ and carry out k-means

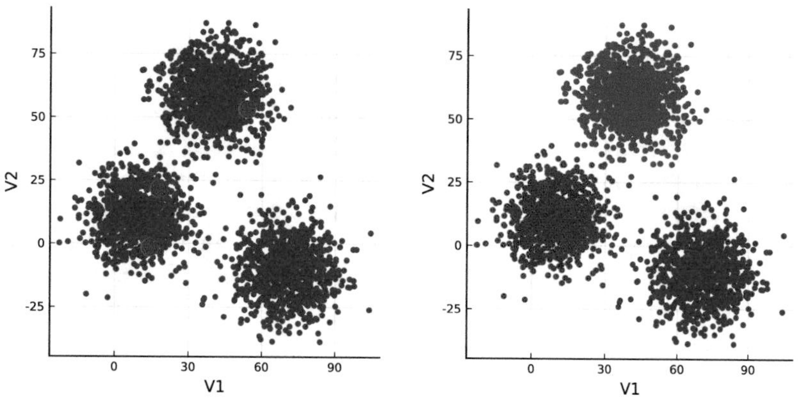

Figure 9.11: k-means clustering, for $k = 3$. The red dots on the left indicate random initial centroids.

in two separate code examples. In Listing 9.12 we use the `Clustering` package. This listing also generates Figure 9.11 where the left plot indicates the initial state where prior to clustering and the right plot indicates the result. We then implement k-means from scratch in Listing 9.13. Even though we start with different initial conditions in both code examples, both examples yield the same clustering (up to ordering of the labels). This is evident via the number of observations found in each cluster.

Listing 9.12: Carrying out k-means via the `Clustering` package

```
using Clustering, RDatasets, Random, Measures, Plots; pyplot()
Random.seed!(0)

K = 3
df = dataset("cluster", "xclara")
data = copy(convert(Array{Float64}, df)')

seeds = initseeds(:rand, data, K)
xclaraKmeans = kmeans(data, K, init = seeds)

println("Number of clusters: ", nclusters(xclaraKmeans))
println("Counts of clusters: ", counts(xclaraKmeans))

df.Group  = assignments(xclaraKmeans)

p1 = scatter(df[:, :V1], df[:, :V2], c=:blue, msw=0)
     scatter!(df[seeds, :V1], df[seeds, :V2], markersize=12, c=:red, msw=0)

p2 = scatter( df[df.Group .== 1, :V1], df[df.Group .== 1, :V2], c=:blue, msw=0)
     scatter!( df[df.Group .== 2, :V1], df[df.Group .== 2, :V2], c=:red, msw=0)
     scatter!( df[df.Group .== 3, :V1], df[df.Group .== 3, :V2], c=:green, msw=0)

plot(p1,p2,legend=:none,ratio=:equal,
    size=(800,400), xlabel="V1", ylabel="V2", margin = 5mm)
```

```
Number of clusters: 3
Counts of clusters: [952, 899, 1149]
```

In line 5 the dataset `xclara` from the `clusters` package from `RDatasets` is stored as `df`. In line 6 the data frame `df` is converted to an array of `Float64` type in order to remove the missing type of the data frame for compatibility. The array is then transposed, as it needs to be in this format for the `kmeans()` function from `Clustering.j` which is used in line 8-9. The `initseeds()` function call in line 8 sets `seeds` as 3 randomly selected data points. In line 11 we use `ncluster()` to query the clustering result on the number of clusters. This was just input to the algorithm. On line 12 we use `counts()` to query how many data points are in each cluster. In line 14 we use `assignments()` to find out the cluster id of each observation. These are stored as a new column in the dataframe, `Group`. The reminder of the code creates Figure 9.11.

The manual implementation of Listing 9.13 is instructive because the mean computation (9.26) and labeling step (9.28) are both immediately evident. Labeling is carried out in line 19 and mean computation is carried out in line 20. This implementation identifies three clusters of the same size as the previous one. However, in k-means there is never control over the exact labeling of the clusters, hence in this case, the ordering of the clusters is different.

Listing 9.13: Manual implementation of k-means

```
1   using RDatasets, Distributions, Random, LinearAlgebra
2   Random.seed!(0)
3
4   K = 3
5   df = dataset("cluster", "xclara")
6   n,_ = size(df)
7   dataPoints = [convert(Array{Float64,1},df[i,:]) for i in 1:n]
8   shuffle!(dataPoints)
9
10  xMin,xMax = minimum(first.(dataPoints)),maximum(first.(dataPoints))
11  yMin,yMax = minimum(last.(dataPoints)),maximum(last.(dataPoints))
12
13  means = [[rand(Uniform(xMin,xMax)),rand(Uniform(yMin,yMax))]  for _ in 1:K]
14  labels = rand(1:K,n)
15  prevMeans = -means
16
17  while norm(prevMeans - means) > 0.001
18      prevMeans = means
19      labels = [argmin([norm(means[i]-x) for i in 1:K]) for x in dataPoints]
20      means = [sum(dataPoints[labels .== i])/sum(labels .==i) for i in 1:K]
21  end
22
23  countResult = [sum(labels .== i) for i in 1:K]
24  println("Counts of clusters (manual implementation): ", countResult)
```

```
Counts of clusters (manual implementation): [899, 1149, 952]
```

In line 7 we create an array of data points. We then shuffle this array in place in line 8. Lines 10-11 find a bounding box for the data points for selecting random starting means. In line 13 we randomly initialize the array `means` and in line 14 we randomly allocate labels. The main algorithm loop is in lines 17-21. In line 19 we recompute the labels based on (9.28). In line 20 we recompute the means based on (9.27). We count the number of elements per cluster in line 23 and then print the output in line 24.

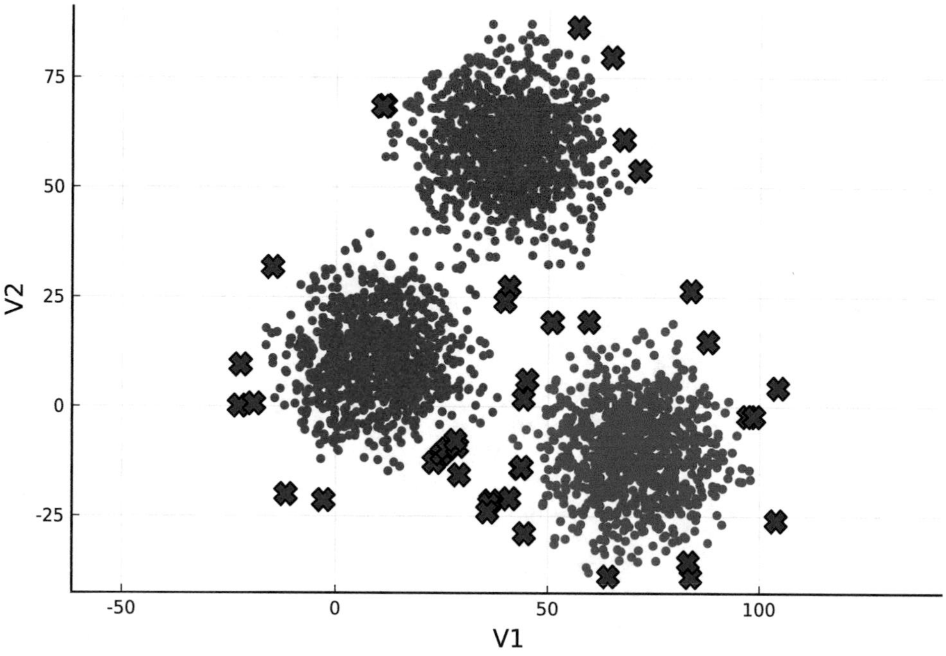

Figure 9.12: Three main clusters arising after 30 iterations of hierarchical clustering with several other small clusters (marked by "X").

Hierarchical Clustering

An alternative form of clustering is *hierarchical clustering*. This type of algorithm comes in several variants, including *agglomerative*, *divisive* and several variants for measuring the distance between observations. We focus on the divisive case where in each iteration an additional cluster is created. This is done iteratively by considering all of the current clusters and splitting the cluster that is most heterogeneous into two "child" clusters.

Consider the textual output of Listing 9.14. This listing uses the `hclust()` function from the `Clustering` package. The output presents the number of observations in each cluster. At `K=1` all `3000` observations are in a single cluster. Then in the next iteration a split is carried out, after which a single observation is in a second cluster and `2999` observations are still in the first cluster. These splits continue as the algorithm proceeds where only by the `K=18` iteration a significant second cluster is formed. Then finally by `K=28` a third significant cluster is formed and from there afterwards, with the exception of a few observations, the bulk of the observations are in the three big clusters. The results are presented in Figure 9.12 where the observations that are not in the three big clusters are specifically marked.

While we do not present one here, a common diagram associated with hierarchical clustering is a *dendrogram*. Such a diagram uses the fact, that clusters have "child clusters", and allows us to present the data in a natural hierarchical manner. Further, as is evident from Figure 9.12, hierarchical clustering allows us to identify outliers in the data. Also note that hierarchical clustering is better suited for dealing with non-numerical data than k-means. This can be done by choosing appropriate distance metrics.

Listing 9.14: Carrying out hierarchical clustering

```
using RDatasets, Clustering, Random, LinearAlgebra, Plots; pyplot()
Random.seed!(0)

df = dataset("cluster", "xclara")
n,_ = size(df)
dataPoints = [convert(Array{Float64,1},df[i,:]) for i in 1:n]
shuffle!(dataPoints)
D = [norm(pt1 - pt2) for pt1 in dataPoints, pt2 in dataPoints]

result = hclust(D)
for K in 1:30
    clusters = cutree(result,k=K)
    println("K=$(K): ",[sum(clusters .== i) for i in 1:K])
end

cluster(ell,K) = (1:n)[cutree(result,k=K) .== ell]

C1, C2, C3 = cluster(1,30),cluster(2,30),cluster(3,30)

plt = scatter( first.(dataPoints[C1]),last.(dataPoints[C1]),c=:blue, msw=0)
      scatter!( first.(dataPoints[C2]),last.(dataPoints[C2]), c=:red, msw=0)
      scatter!( first.(dataPoints[C3]),last.(dataPoints[C3]), c=:green, msw=0)
for ell in 4:30
    clst = cluster(ell,30)
    scatter!(first.(dataPoints[clst]),last.(dataPoints[clst]),
        ms=10, c=:purple, shape=:xcross, ratio=:equal, legend=:none,
        xlabel="V1", ylabel="V2")
end
plot(plt)
```

```
K=1: [3000]
K=2: [2999, 1]
K=3: [2997, 2, 1]
K=4: [2997, 1, 1, 1]
K=5: [2996, 1, 1, 1, 1]
K=6: [2995, 1, 1, 1, 1, 1]
K=7: [2994, 1, 1, 1, 1, 1, 1]
K=8: [2993, 1, 1, 1, 1, 1, 1, 1]
K=9: [2989, 1, 1, 1, 4, 1, 1, 1, 1]
K=10: [2988, 1, 1, 1, 1, 4, 1, 1, 1, 1]
K=11: [2986, 1, 1, 1, 2, 1, 4, 1, 1, 1, 1]
K=12: [2985, 1, 1, 1, 1, 2, 1, 4, 1, 1, 1, 1]
K=13: [2984, 1, 1, 1, 1, 2, 1, 4, 1, 1, 1, 1, 1]
K=14: [2983, 1, 1, 1, 1, 2, 1, 4, 1, 1, 1, 1, 1, 1]
K=15: [2982, 1, 1, 1, 1, 2, 1, 4, 1, 1, 1, 1, 1, 1, 1]
K=16: [2981, 1, 1, 1, 1, 2, 1, 4, 1, 1, 1, 1, 1, 1, 1, 1]
K=17: [2980, 1, 1, 1, 1, 2, 1, 4, 1, 1, 1, 1, 1, 1, 1, 1, 1]
K=18: [2039, 941, 1, 1, 1, 1, 2, 1, 4, 1, 1, 1, 1, 1, 1, 1, 1, 1]
K=19: [2039, 939, 1, 1, 1, 1, 2, 2, 1, 4, 1, 1, 1, 1, 1, 1, 1, 1, 1]
K=20: [2039, 937, 1, 1, 1, 2, 1, 2, 2, 1, 4, 1, 1, 1, 1, 1, 1, 1, 1, 1]
K=21: [2038, 937, 1, 1, 1, 2, 1, 2, 2, 1, 4, 1, 1, 1, 1, 1, 1, 1, 1, 1, 1]
K=22: [2037, 937, 1, 1, 1, 2, 1, 2, 2, 1, 4, 1, 1, 1, 1, 1, 1, 1, 1, 1, 1, 1]
K=23: [2032, 937, 1, 1, 1, 5, 2, 1, 2, 2, 1, 4, 1, 1, 1, 1, 1, 1, 1, 1, 1, 1, 1]
K=24: [2032, 935, 1, 1, 1, 5, 2, 2, 1, 2, 2, 1, 4, 1, 1, 1, 1, 1, 1, 1, 1, 1, 1, 1]
K=25: [2030, 935, 1, 1, 1, 5, 2, 2, 1, 2, 2, 2, 1, 4, 1, 1, 1, 1, 1, 1, 1, 1, 1, 1, 1]
K=26: [2030, 933, 1, 1, 1, 5, 2, 2, 1, 2, 2, 2, 2, 1, 4, 1, 1, 1, 1, 1, 1, 1, 1, 1, 1, 1]
K=27: [2028, 933, 1, 1, 1, 5, 2, 2, 1, 2, 2, 2, 2, 1, 4, 2, 1, 1, 1, 1, 1, 1, 1, 1, 1, 1, 1]
K=28: [882, 1146, 933, 1, 1, 1, 5, 2, 2, 1, 2, 2, 2, 2, 1, 4, 2, 1, 1, 1, 1, 1, 1, 1, 1, 1, 1, 1]
K=29: [882, 1146, 932, 1, 1, 1, 5, 2, 2, 1, 2, 2, 2, 2, 1, 4, 2, 1, 1, 1, 1, 1, 1, 1, 1, 1, 1, 1, 1]
K=30: [882, 1146, 932, 1, 1, 1, 5, 2, 2, 1, 2, 2, 2, 2, 1, 1, 2, 3, 1, 1, 1, 1, 1, 1, 1, 1, 1, 1, 1, 1]
```

Line 4-7 are similar to the previous clustering listings. In line 8 we create the matrix `D` which measures the pairwise distances between each of the data points. It is then fed to the `hclust()` function from `Clustering.jl` in line 10. The clustering output is then stored in `result`. In lines 11-14 we print the representation of the clusters for clusters 1 to 30. Each time, the clustering is obtained via `cutree()` with a specified number of clusters `k=K`. The number of elements in each cluster is printed in line 13. In line 16 we define our function `cluster()` which returns the points that are set in cluster `ell` when the number of clusters is `K`. We then use this function in line 18 to create the sets of points (cluster), `C1`, `C2`, and `C3`. The remainder of the code creates Figure 9.14.

Principal Component Analysis

Say we are presented with a dataset of vector observations $x_1, \ldots, x_n$, each of dimension p, where $n > p$. The archetypical algorithm for dimensionality reduction is *Principal Component Analysis (PCA)*. The idea is to choose a suitable dimension $k < p$ and create k-dimensional vectors $\tilde{x}_1, \ldots, \tilde{x}_n$ where for every observation i,

$$\tilde{x}_i = V x_i, \tag{9.29}$$

with V a $k \times p$ matrix. Each row of V is called a principal component as it takes a linear combination of x_i and creates a coordinate in $\tilde{x}_i$. The act of carrying out PCA is the act of determining k, finding V and analyzing the reduced dataset $\tilde{x}_1, \ldots, \tilde{x}_n$.

One way to consider PCA is as an advanced *data visualization* technique. For example, consider the MNIST dataset where $p = 784$. We can view one image, x_i by plotting the image as in Figure 9.1, however how can we view thousands of images together? For this, if we reduce the dimension of an image from 784 to $k = 2$ we are able to create plots like Figure 9.14. In such a plot, the full information of every image is clearly not present, still we may see how the image compares to others. Here the matrix V of (9.29) is 2×784 dimensional and it consists of two principal components. It is computed from the full data consisting of $60,000$ images, each with 784 pixels.

There are different ways to compute V and determine k. One way is to consider the sample covariance matrix, $\hat{\Sigma}$, associated with the data as presented in Listing 4.13 of Chapter 4. Since it is a symmetric matrix, all eigenvalues are real and there are corresponding orthonormal eigenvectors. Hence we may *diagonalize* it via $\hat{\Sigma} = M \text{diag}(\lambda_1, \ldots, \lambda_p) M^\top$, where M is a vector of column eigenvectors, $v_1, \ldots, v_p$, with corresponding eigenvalues $\lambda_1, \ldots, \lambda_p$. Assume also that $\lambda_1 \geq \lambda_2 \geq \ldots \geq \lambda_p$. We may then represent $\hat{\Sigma}$ via the sum of *rank one matrices* and for $k < p$

$$\hat{\Sigma} = \sum_{i=1}^{p} \lambda_i v_i v_i^\top \approx \sum_{i=1}^{k} \lambda_i v_i v_i^\top.$$

We then set the matrix V to consist of the rows $v_1^\top, \ldots, v_k^\top$. By ordering the eigenvalues, we are able to choose the significant ones first and make a judgement call on k. This is sometimes done with an aid of a *scree plot* that plots the diminishing contribution of each λ_i.

Before considering PCA on the MNIST data, Listing 9.15 carries out PCA on the simple `iris` datset. It applies `MultivariateStats` package to carry out PCA as well as a covariance-based computation. It creates Figure 9.13 which presents both a scree plot and a scatter of the transformed

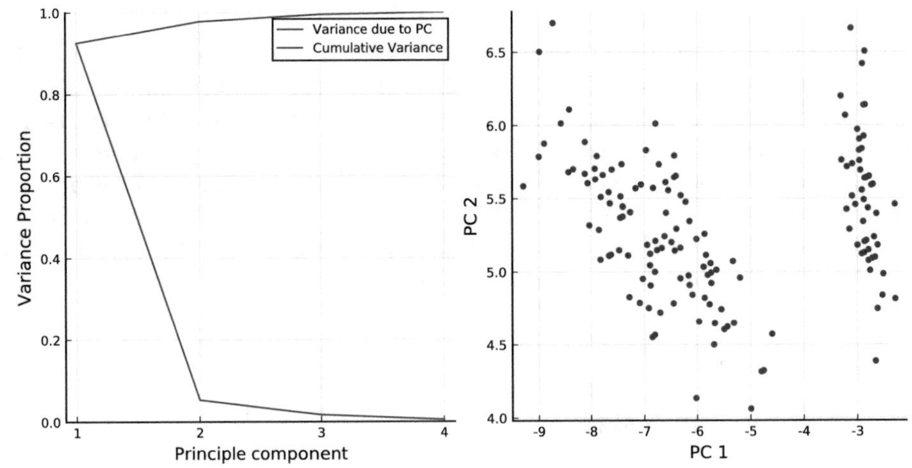

Figure 9.13: A scree plot, along with the result of principal component analysis.

coordinates based on their principle components. The output illustrates that our manual computation using a covariance matrix agrees with the computation carried out by the `fit(PCA,...)` function.

Listing 9.15: Principal component analysis on a simple dataset

```
using Statistics, MultivariateStats, RDatasets, LinearAlgebra, Plots; pyplot()

data = dataset("datasets", "iris")
data = data[:,[:SepalLength,:SepalWidth,:PetalLength,:PetalWidth]]
x = convert(Array{Float64,2},data)'

model = fit(PCA, x, maxoutdim=4, pratio = 0.999)
M = projection(model)

function manualProjection(x)
    covMat = cov(x')
    ev = eigvals(covMat)
    eigOrder = sortperm(eigvals(covMat),rev=true)
    eigvecs(covMat)[:,eigOrder]
end

println("Manual vs. package: ",maximum(abs.(M-manualProjection(x))))

pcVar = principalvars(model) ./ tvar(model)
cumVar = cumsum(pcVar)
pcDat = M[:,1:2]'*x

p1 = plot(pcVar, c=:blue, label="Variance due to PC")
        plot!(1:length(cumVar), cumVar, label="Cumulative Variance", c=:red,
        xlabel="Principle component",ylabel="Variance Proportion",ylims=(0,1))
p2 = scatter(pcDat[1,:],pcDat[2,:], c=:blue, xlabel="PC 1", ylabel="PC 2",
            msw=0, legend=:none)
plot(p1, p2, size=(800,400))
```

```
Manual vs. package: 4.551914400963142e-15
```

In line 3 we read the `iris` dataset. In line 4 we limit the dataframe, `data` to the four numerical variables in the dataset. We then create a 4×150 dimensional matrix of the data, `x` where each column is a 4 dimensional data point. Line 7 uses the `fit()` function on `PCA`, as defined in the `MultivariateStats` package. The `maxoutdim` setting is redundant in our case, however in other cases can be used to limit the number of principal components obtained. The `pratio` setting indicates to stop when the cumulative variance is greater than 0.999. The default is 0.99. The resulting `model` object can then be queried. Line 8 uses the `projection()` function from the `MultivariateStats` package. It returns a matrix where each column is a principal component. We define our own function `manualProjection()` in lines 10-15. This function creates a matrix with columns as principal components, analogously to the matrix `M` created in line 8. We compute the sample covariance matrix of the data `x` in line 11 and then compute its eigenvalues in line 12. Then our use of `sortperm()` with `rev=true` returns the permutation of eigenvalues from highest to smallest (the covariance matrix is guaranteed to be symmetric and hence the eigenvalues are real). We then compute corresponding eigenvectors with `eigvecs()` in line 14. These are in a matrix with each column an eigenvector. We then reshuffle the columns according to the `eigOrder` previously computed. This is the matrix of principal components. The matrix of principal components resulting from `manualProjection()` is the same as the matrix that was generated in line 8. We illustrate this in line 17 where we print the maximum absolute difference of entries. In line 19 we use the `principalvars()` function on `model`. It returns the variances associated with each principal component. This can also be obtained within `manualProjection()` if we wished by evaluating `ev[eigOrder]`. These values are then normalized by dividing by the scalar `tvar(model)`. We accumulate these values in line 20 when we use the builtin `cumsum()` function. In line 21 we decide to use 2 principal components and hence select the first two columns of `M` via `1:2`. Applying the transpose of this matrix to the data `x` yields a 2×150 matrix, `pcDat`, where each column represents a data point via two principal components. The remainder of the code creates Figure 9.13.

PCA is often used in conjunction with other algorithms, including supervised learning algorithms. We may often pre-process the data and train an algorithm to classify based on principal components instead of the original data. We don't illustrate interaction of PCA and supervised learning here, however we hint at the power of PCA by applying it to the MNIST dataset.

Listing 9.16 extracts the first two principal components for the 784 long vectors describing images of digits. That is, each image is described only by two coordinates. In doing so, we create Figure 9.14 which hints at some interesting patterns. We plot the principal component clouds for certain combinations of digits together (there are about $6,000$ points for each digit). It is evident that even with two principal components, separation between certain digits is possible. For example, the digits 8 and 9 are well separated and so are the digits 0 and 1. However for other pairs of digits such as 2 and 3, we see that two principle components are not enough to obtain good separation.

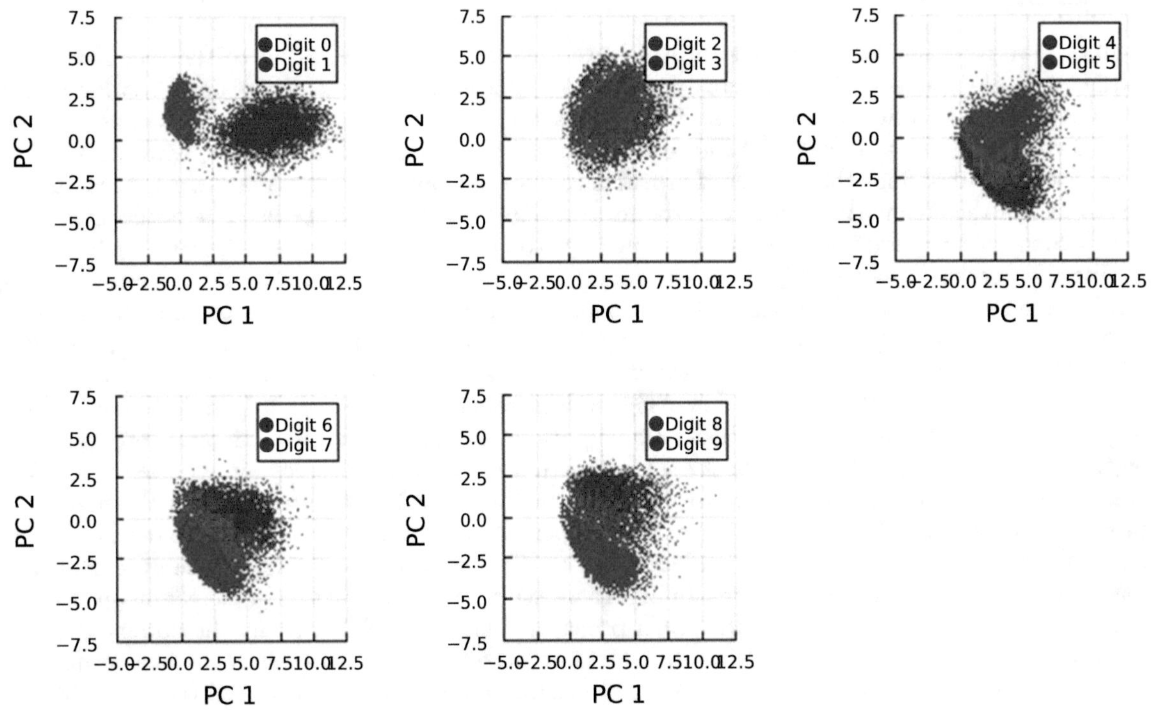

Figure 9.14: PCA on MNIST. As is evident, it is easy to separate the digits 8 and 9, however digits 2 and 3 are much harder to separate using only two principal components.

Listing 9.16: Principal component analysis on MNIST

```
using MultivariateStats,RDatasets,LinearAlgebra,Flux.Data.MNIST,Measures,Plots
pyplot()

imgs, labels   = MNIST.images(), MNIST.labels()
x = hcat([vcat(float.(im)...) for im in imgs]...)
pca = fit(PCA, x; maxoutdim=2)
M = projection(pca)

function compareDigits(dA,dB)
    imA, imB = imgs[labels .== dA], imgs[labels .== dB]
    xA = hcat([vcat(float.(im)...) for im in imA]...)
    xB = hcat([vcat(float.(im)...) for im in imB]...)
    zA, zB = M'*xA, M'*xB
    default(ms=0.8, msw=0, xlims=(-5,12.5), ylims=(-7.5,7.5),
            legend = :topright, xlabel="PC 1", ylabel="PC 2")
    scatter(zA[1,:],zA[2,:], c=:red,  label="Digit $(dA)")
    scatter!(zB[1,:],zB[2,:], c=:blue, label="Digit $(dB)")
end

plots = []
for k in 1:5
    push!(plots,compareDigits(2k-2,2k-1))
end
plot(plots...,size = (800, 500), margin = 5mm)
```

In lines 4-7 we compute the first 2 principal components for the MNIST dataset yielding the projection matrix `M`. In lines 9-18 we create the function `compareDigits()` designed to create a plot of the first two principal components associated with the images containing the digits `dA` and `dB`. First the images matching those digits are filtered in line 10. Then datasets `xA` and `xB` are created. Then the transpose of the projection matrix `M` is applied to the data points, yielding the collections of 2-dimensional points `zA`, `zB`. That is, each 784 dimensional vector is collapsed to 2 dimensions. The remainder of the lines in the function create a plot of the points. The return value of `scatter!()` is a plot and it is returned from the function in line 17. Lines 20-24 plot the 5 sub-plots, comparing neighboring digits.

9.5 Markov Decision Processes and Reinforcement Learning

Almost all of the content of this book focuses on methods for gaining information. In some cases we explore statistical inference, in others machine learning, and in other cases system performance via simulation or numerical computation. In each case the objective is to gain additional information about the system at hand. However, why do we need such information? The most common answer is that information is needed for decision-making that will affect the future of the system. In certain cases the decision-making process can be detached from precise details of information retrieval and inference, and this is the mode of operation implied with most methods in this book: one obtains information, and perhaps later makes decisions based on that information. Hence in general, for most methods explored in the book, decision-making is implicit and not part of the demonstrated methodology.

However there are cases where we observe the system, gain information and make decisions simultaneously. This is where techniques such as *Markov Decision Processes* (MDP), *Partially Observable Markov Decision Processes* (POMDP) and *Reinforcement Learning* (RL) play a role. We now briefly explore such methods which classically fall under the area of *stochastic dynamic programming* or *stochastic optimal control*, and more recently under *artificial intelligence*. These areas are sometimes closely related to *robotics*.

As an example, consider a scenario focusing on the engagement level of a student. Assume that there are L levels of engagement, $1, 2, \ldots, L$ where at level 1 the student is not engaged at all and at the other extreme, at level L she is maximally engaged. Our goal is to maintain engagement as high as possible over time. We do this by choosing one of two actions at any time instant. (0): "do nothing" and let the student operate independently. (1): "stimulate" the student. In general, stimulating the student has a higher tendency to increase her engagement level, however this isn't without cost as it requires resources.

We may denote the engagement level at (discrete) time t by $X(t)$, and for simplicity we assume here that at any time $X(t)$ either increases or decreases by 1. An exception exists at the extremes of $X(t) = 1$ and $X(t) = L$. In these cases the student engagement either stays the same or increases/decreases, respectively, by 1. The actual transition of engagement level is random, however we assume that if our action is "stimulate" then there is more likely to be an increase of engagement than if we "do nothing".

The control problem is the problem of deciding when to "do nothing" and when to "stimulate".

For this we formulate a *reward function* and assume that at any time t our reward is

$$R(t) = X(t) - \kappa\, A(t). \tag{9.30}$$

Here κ is some positive constant and $A(t) = 0$ if the action at time t is to "do nothing", while $A(t) = 1$ if the action is to "stimulate". Hence the constant κ captures our relative cost of stimulation effort in comparison to the benefit of a unit of student engagement. We see that $R(t)$ depends both on our action and the state.

The reward is accumulated over time into an *optimization objective* via,

$$\mathbb{E}\Big[\sum_{t=0}^{\infty} \beta^{\top} R(t)\Big], \tag{9.31}$$

where $\beta \in (0,1)$ and is called the *discount factor*. Through the presence of β, future rewards are discounted with a factor of $\beta^{\top}$, indicating that in general the present is more important than the future. There can also be other types of objectives, for example, *finite horizon* or *infinite horizon average reward*, however we focus on this *infinite horizon expected discounted reward* case here.

A *control policy* is embodied by the sequence of actions $\{A(t)\}_{t=0}^{\infty}$. If these actions are chosen independently of observations then it is an *open loop* control policy. However, for our purposes, things are more interesting in the *closed loop* or *feedback control* case in which at each time t we observe the state $X(t)$, or some noisy version of it. This state feedback helps us decide on $A(t)$.

We encode the effect of an action on the state via a family of Markovian transition probability matrices. For each action a, we set a transition probability matrix $P^{(a)}$. In our case, as there are two possible actions, we have two matrices such as, for example,

$$P^{(0)} = \begin{bmatrix} 1/2 & 1/2 & & & & & \\ 1/2 & 0 & 1/2 & & & & \\ & 1/2 & 0 & 1/2 & & & \\ & & \ddots & \ddots & \ddots & & \\ & & & \ddots & \ddots & \ddots & \\ & & & & 1/2 & 0 & 1/2 \\ & & & & & 1/2 & 1/2 \end{bmatrix}, \quad P^{(1)} = \begin{bmatrix} 1/4 & 3/4 & & & & & \\ 1/4 & 0 & 3/4 & & & & \\ & 1/4 & 0 & 3/4 & & & \\ & & \ddots & \ddots & \ddots & & \\ & & & \ddots & \ddots & \ddots & \\ & & & & 1/4 & 0 & 3/4 \\ & & & & & 1/4 & 3/4 \end{bmatrix}. \tag{9.32}$$

Hence, for example, in state 1 (first row), if we choose action 0 then the transitions follow $[1/2 \;\; 1/2 \;\; 0 \;\ldots]$, whereas if we choose action 1 then the transitions follow $[1/4 \;\; 3/4 \;\; 0 \;\ldots]$. That is for a given state i, choosing action a implies that the next state is distributed according to $[P_{i1}^{(a)} \;\; P_{i2}^{(a)} \;\; \cdots]$ (where most of the entries of this probability vector are 0 in this example).

In the case of MDP and POMDP these matrices are assumed known, however in the case of RL these matrices are unknown. The difference between MDP and POMDP is that in MDP we know exactly in which state we are in, whereas in POMDPs our observations are noisy or partial and only hint at the current state. Hence in general we can treat MDP as the basic case and POMDP and RL can be viewed as two variants of MDP. In the case of POMDP the state isn't fully observed, while in the case of RL the transition probabilities are not known.

We don't focus on POMDPs further in this book (one can follow [L09] for a simple tutorial), but rather we continue with an MDP example and then explore a basic RL example. For this, we first need to understand more technical aspects of MDP.

Optimal Policies, Value Functions and Bellman Equations

The theory of MDPs (see, for example, [P14] or [B11]) shows that under general conditions, for such time-homogenous infinite horizon discounted cost problems, it is enough to consider *stationary deterministic Markov policies.* In this case, a policy is a function mapping every state to an action. If we denote the set of all such policies by Π, then an optimal policy is an element, $\pi \in \Pi$ that maximizes the objective, (9.31) for any initial state. As such, the *value function* is a function defined as

$$V(i) = \max_{\pi \in \Pi} \mathbb{E}\Big[\sum_{t=0}^{\infty} \beta^{\top} R(t) \mid X(0) = i\Big],$$

for any initial state i. It defines the best possible total discounted reward (value) for any particular initial situation (state i).

We often don't know the value function for a given problem, nevertheless it is a tool that helps find optimal policies. This is because the value function appears in the *Bellman equation*:

$$V(i) = \max_{a \in \mathcal{A}}\{Q(i,a)\}, \qquad \text{with} \qquad Q(i,a) = r(i,a) + \beta \sum_{j} P_{i,j}^{(a)} V(j), \tag{9.33}$$

where $r(i,a)$ is the expected reward (with cost deducted) for state i and action a, and the set $\mathcal{A}$ is the set of possible actions. In our example, $\mathcal{A} = \{0,1\}$ and $r(i,a)$ is based on (9.30) yielding,

$$r(i,0) = i, \qquad \text{and} \qquad r(i,1) = i - \kappa.$$

Notice that the value function $V(\cdot)$ appears in both sides of the Bellman equation. To understand the basic idea, consider first the *Q-function* in (9.33). It measures the "quality" of being in state i and applying action a. If such a state-action pair is exhibited then the immediate expected reward $r(i,a)$ is obtained followed by a transition to some random state j. This happens with probability $P_{i,j}^{(a)}$, at which point the problem continues and has value $V(j)$. However, the transition is at the next time step and hence multiplication by the discount factor β presents the value in terms of the current time step.

The Bellman equation uses the *dynamic programming principle* to determine the optimal cost in terms of maximization of the Q-function by maximizing over all actions $a \in \mathcal{A}$. It yields an equation where the "unknown" is the value function, $V(\cdot)$.

Some MDP theory deals with the validity and properties of the Bellman equation. Then, much of the study of MDP deals with methods of solving the Bellman equation (9.33), or analyzing properties of the solution. Observe that if we knew the value function $V(\cdot)$ or the Q-function $Q(\cdot,\cdot)$, then we would also know an optimal policy, as for every state and action, i and a, we would seek to set $\pi(i) = a^*$ where a^* is the action that maximizes the right-hand side of the Bellman equation.

Basic Value Iteration for MDP

In basic MDP, when confronted with a Bellman equation, there are typically several types of methods that can be used to solve it. Solving it implies finding the value function and with it

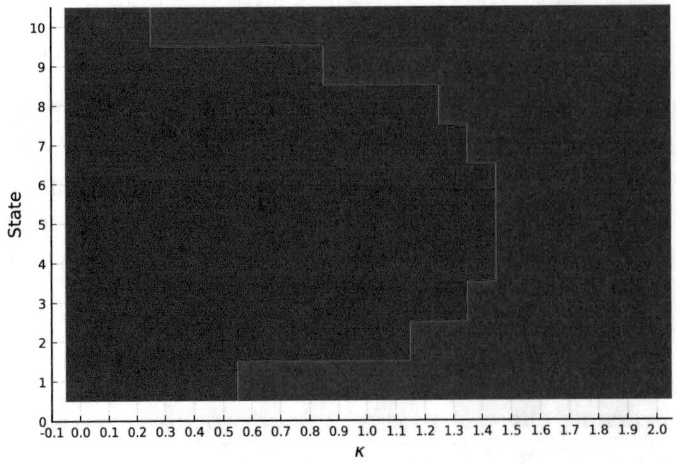

Figure 9.15: The optimal policy as a function of κ (horizontal axis) and the current state (vertical axis). Red is "stimulate" and blue is "do nothing".

an optimal policy. The main known methods include *value iteration*, *policy iteration* and *linear programming*. Here we only focus on the most basic of these methods: value iteration.

Observing that the Bellman equation (9.33) contains $V(\cdot)$ both in the left-hand side and the right-hand side, value iteration iterates over successive value functions, $V_0(\cdot), V_1(\cdot), V_2(\cdot), \ldots$, until convergence. That is we begin with some arbitrary value function, $V_0(\cdot)$ and repeatedly apply

$$V_{t+1}(i) = \max_{a \in \mathcal{A}} \big\{ r(i,a) + \beta \sum_j P_{i,j}^{(a)} V_t(j) \big\}, \qquad \text{for all } \ i. \tag{9.34}$$

Convergence is mathematically guaranteed (at least with finite state spaces) because the *Bellman operator*,

$$\mathcal{O}\big(V(\cdot)\big) = \max_{a \in \mathcal{A}} \big\{ r(i,a) + \beta \sum_j P_{i,j}^{(a)} V_t(j) \, \big\} \tag{9.35}$$

is a *contraction* in the mathematical sense. See [P14] for more details.

Programmatically, implementing the value iteration in (9.34) is straightforward for small state space examples. We iterate and stop when the difference between iterates under some sensible *norm* is smaller than a prescribed level, ε. This is implemented in Listing 9.17 where we actually consider a collection of problems characterized by the cost parameter κ. For each problem, once the value function is found, we use it to determine the optimal policy. The policies are then plotted in Figure 9.15 for $L = 10$ and $\beta = 0.75$.

Listing 9.17: Value iteration for an MDP

```
using LinearAlgebra, LaTeXStrings, Plots; pyplot()

L = 10
p0, p1 = 1/2, 3/4
beta = 0.75
epsilon = 0.001

function valueIteration(kappa)
    P0 = diagm(1=>fill(p0,L-1)) + diagm(-1=>fill(1-p0,L-1))
    P0[1,1], P0[L,L] = 1 - p0, p0

    P1 = diagm(1=>fill(p1,L-1)) + diagm(-1=>fill(1-p1,L-1))
    P1[1,1], P1[L,L] = 1 - p1, p1

    R0 = collect(1:L)
    R1 = R0 .- kappa

    bellmanOperator(Vprev)=
        max.(R0 + beta*P0*Vprev, R1 + beta*P1*Vprev)
    optimalPolicy(V,state)=
        (R0+beta*P0*V)[state] >= (R1+beta*P1*V)[state] ? 0 : 1

    V, Vprev = fill(0,L), fill(1,L)
    while norm(V-Vprev) > epsilon
        Vprev = V
        V = bellmanOperator(Vprev)
    end

    return [optimalPolicy(V,s) for s in 1:L]
end

kappaGrid = 0:0.1:2.0
policyMap = zeros(L,length(kappaGrid))

for (i,kappa) in enumerate(kappaGrid)
    policyMap[:,i] = valueIteration(kappa)
end
heatmap(policyMap, fill=cgrad([:blue, :red]),
        xticks=(0:1:21, -0.1:0.1:2), yticks=(0:L, 0:L),
        xlabel=L"\kappa", ylabel="State", colorbar_entry=false)
```

Lines 3-6 define the model and algorithm parameters, including the discount factor `beta`, and a stopping threshold for value iteration `epsilon`. In lines 8-30 we implement the function `valueIteration()` which depends on a specified cost, `kappa`. The value iteration method is performed in lines 24-27, where we iterate until the normed difference between two value functions is less than or equal to `epsilon`. In line 26 we apply the Bellman operator, (9.35) via the function `bellmanOperator()` which we define in lines 18-19. Note the use of `max()` function with the broadcast dot operator (`.`), which allows us to find the element-wise maximum. The function `optimalPolicy()` defined in lines 20-21 returns the optimal action to be taken given a current state. Through the use of this function along with a comprehension in line 29, `valueIteration()` returns the optimal policy for a given `state`. Note that an alternative method would be to continue discovering the optimal policy during the value iteration process by considering the actions that maximize the Bellman operator. However we didn't use such an implementation here. The remainder of the code applies value iteration over a gird of κ values, `kappaGrid`. Note the use of `enumerate()` in the `for` loop of lines 35-37. Lines 38-40 create Figure 9.15.

Reinforcement Learning via Q-Learning

In many practical situations there isn't a clear model for the transition probability matrices $P^{(a)}$, $a \in \mathcal{A}$. For example, in our engagement level example, the matrices (9.32) are a postulated model of reality. In some situations the parameters of such matrices may be estimated from previous experience, however often this isn't feasible due to changing conditions or lack of data.

Such situations are handled by reinforcement learning. The class of RL methods is a broad class of models dealing with control of systems for which we lack parameter knowledge. In classic control theory this situation falls under the umbrella of *adaptive control.* However in contemporary robotics, self-driving cars, and artificial neural networks, RL has become the key term.

Here we explore one class of RL algorithms called *Q-learning.* The main idea of this method is to learn the Q-function as in (9.33) without explicitly decomposing $Q(i,a)$ into P, V, and r. Observe from the Bellman equation, that if we were to know $Q(i,a)$ for every state i and action a, then we can also compute the optimal policy by selecting the a that maximizes $Q(i,a)$ for every i.

The key of Q-learning is to continuously learn $Q(\cdot,\cdot)$ while using the learned estimates to select actions as we go. For this, denote by $\hat{Q}_t(\cdot,\cdot)$ the estimate of $Q(\cdot,\cdot)$ we have at time t. At any given time we attempt to balance *exploration and exploitation.* With a high probability, we decide on action a that maximizes $\hat{Q}_t(i,a)$ - this is exploitation. However, we leave some possibility to explore other actions, and occasionally decide on an arbitrary (random) action a - this is exploration. In our example, as time progresses we reduce the probability of exploration. For example, we use $t^{-0.2}$ for this probability, which implies that as time evolves we slowly explore less and less.

As we operate our system with Q-learning, after an action is chosen, reward r is obtained and the system transitions from state i to state j. At that point we update the (i,a) entry of the Q-function estimate as follows:

$$\hat{Q}_{t+1}(i,a) = (1-\alpha_t)\,\hat{Q}_t(i,a) + \alpha_t\Big(r + \beta \max_{a\in\mathcal{A}_s} \hat{Q}_t(j,a)\Big). \tag{9.36}$$

Here α_t is a decaying (or constant) sequence of probabilities (in the example below we use $\alpha_t = t^{-0.2}$). The key of the *Q-learning update equation* (9.36) is a weighted average of the previous estimate

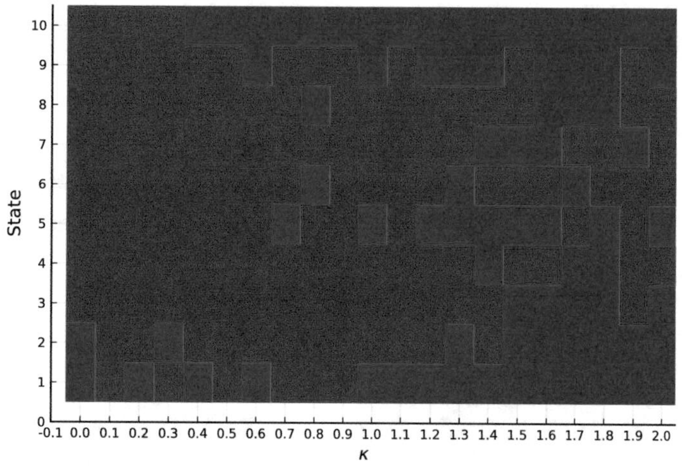

Figure 9.16: The policy learned via Q-Learning as a function of κ over long time horizons. Compare to Figure 9.15.

$\hat{Q}_t(i,a)$ and a single sample of the right-hand side of the Bellman equation (9.33). Miraculously as the system progresses under such a control, this scheme is able to estimate the Q-function and hence control the system well.

Note that ideally we would set $\{\alpha_t\}_{t=1}^{\infty}$ to satisfy,

$$\sum_{t=1}^{\infty} \alpha_t = \infty, \qquad \text{and} \qquad \sum_{t=1}^{\infty} \alpha_t^2 < \infty.$$

With such a condition, based on the theory of *stochastic approximation*, it is guaranteed that as $t \to \infty$, $\hat{Q}_t(i,a) \to Q(i,a)$. This property shows that (at least in principle), systems controlled via Q-learning may still be controlled in an asymptotically optimal manner, even without explicit knowledge of the underlying transition matrices $P^{(a)}$.

While we don't demonstrate an example, we also mention *deep reinforcement learning*. The main idea of such methods is to approximate functions with neural networks and train the neural networks in parallel to controlling the network. For example, with *deep Q-learning*, the Q-function estimate in (9.36) is represented via a neural network. Then with each iteration of (9.36), the neural network is further trained.

In Listing 9.18 below we simulate the engagement level model under Q-learning using the same parameters as before. Just as in the previous value iteration example of Listing 9.17, we do so for a range of cost parameters κ. The resulting policy is presented in Figure 9.16 which can be compared to Figure 9.15, with 10^6 time steps used for each value of κ. The control policies obtained (one for every κ) are similar to the optimal policies in Figure 9.15, but not identical. This is because (for this example) the difference between $Q(i,0)$ and $Q(i,1)$ is negligible for many values of i.

Listing 9.18: A Q-Learning example

```
using LinearAlgebra, StatsBase, Random, LaTeXStrings, Plots; pyplot()
Random.seed!(0)

L = 10
p0, p1 = 1/2, 3/4
beta = 0.75
pExplore(t) = t^-0.2
alpha(t) = t^-0.2
T = 10^6

function QlearnSim(kappa)
    P0 = diagm(1=>fill(p0,L-1)) + diagm(-1=>fill(1-p0,L-1))
    P0[1,1], P0[L,L] = 1 - p0, p0

    P1 = diagm(1=>fill(p1,L-1)) + diagm(-1=>fill(1-p1,L-1))
    P1[1,1], P1[L,L] = 1 - p1, p1

    R0 = collect(1:L)
    R1 = R0 .- kappa

    nextState(s,a) =
        a == 0 ? sample(1:L,weights(P0[s,:])) : sample(1:L,weights(P1[s,:]))

    Q = zeros(L,2)
    s = 1
    optimalAction(s) = Q[s,1] >= Q[s,2] ? 0 : 1
    for t in 1:T
        if rand() < pExplore(t)
            a = rand([0,1])
        else
            a = optimalAction(s)
        end
        sNew = nextState(s,a)
        r = a == 0 ? R0[sNew] : R1[sNew]
        Q[s,a+1]=(1-alpha(t))*Q[s,a+1]+alpha(t)*(r+beta*max(Q[sNew,1],Q[sNew,2]))
        s = sNew
    end
    [optimalAction(s) for s in 1:L]
end

kappaGrid = 0.0:0.1:2.0
policyMap = zeros(L,length(kappaGrid))

for (i,kappa) in enumerate(kappaGrid)
    policyMap[:,i] = QlearnSim(kappa)
end

heatmap(policyMap, fill=cgrad([:blue, :red]),
        xticks=(0:1:21, -0.1:0.1:2), yticks=(0:L, 0:L),
        xlabel=L"\kappa", ylabel="State", colorbar_entry=false)
```

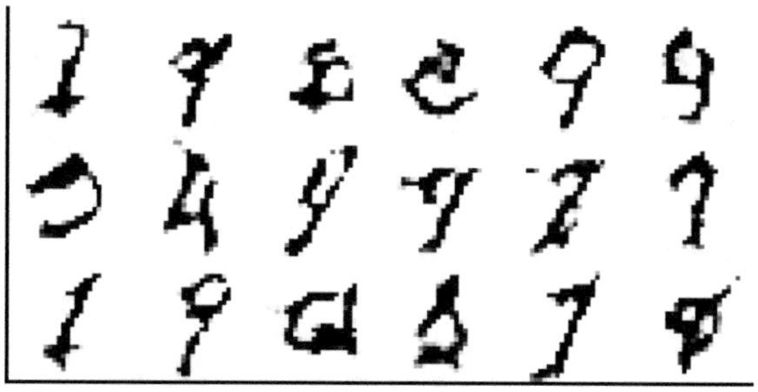

Figure 9.17: Digit images generated from a generative adversarial network (GAN).

In lines 4-8 we set the basic parameters as well as the functions `pExplore()` and `alpha()`, which are used for the probability of exploration and α_t, respectively. In lines 11-39 we implement the function `QlearnSim()`, which simulates the system controlled via Q-learning. The main simulation loop is in lines 27-37. Here we choose a random action (either 0 or 1) with probability `pExplore()`, or otherwise we use the *Q-table*, `Q[]` to select an `optimalAction()`. Then line 36 updates the Q-table as per the Q-learning update equation (9.36). Note that indexation into actions in the Q-table is via 1 and 2 as Julia arrays begin with index 1), and since our action space is $\{0, 1\}$, `a+1` is used in line 35. The remainder of the code is similar to the previous Listing 9.17.

9.6 Generative Adversarial Networks

Generative models allow us to create synthetic data that is similar to a given dataset. One such framework that has emerged in recent years is the framework of *Generative Adversarial Networks* (GANs). See [G14]. The main components of a GAN are two competing neural network models, a *generator*, and a *discriminator*. Both models are trained simultaneously with the generator creating fake data and the discriminator attempting to classify if the data is real or fake. As this process continues, the generator is eventually able to create fake data that "fools" the discriminator. The output of this process is then the generator model which can be used for creating fake data and integrated in part of other applications.

We present a minimal example, based on the MNIST dataset. We first consider a pre-trained generator and show how it can be used to generate fake images. This is in Listing 9.19 where the file `mnistGAN40.bson` is read into the agreed upon model, similarly to the first example of this chapter, Listing 9.1. In this model, *latent variables* of dimension 100 are randomly generated and fed into the pre-trained generator. Each such vector of latent variables then creates an arbitrary artificial digit. The output is in Figure 9.17 which presents 18 such digits.

The model appearing in lines 7-10 is a convolutional neural network similar in nature to `model2` of Listing 9.9. We mention some attributes of this model that have not appeared earlier in this chapter. These include *batch normalization* using `BatchNorm()`, and padding in the convolutional layers specified via `pad`.

Listing 9.19: Generating images from a pre-trained generative adversarial network

```
1  using Flux, BSON, Random, Plots; pyplot()
2  Random.seed!(0)
3
4  latentDim = 100
5  outputX, outputY = 6, 3
6
7  gen =  Chain(Dense(latentDim,7*7*256),BatchNorm(7*7*256,relu),
8          x->reshape(x,7,7,256,:),ConvTranspose((5,5),256=>128;stride=1,pad=2),
9          BatchNorm(128,relu),ConvTranspose((4,4),128=>64;stride=2,pad=1),
10         BatchNorm(64,relu),ConvTranspose((4,4),64=>1,tanh;stride=2,pad=1))
11
12 cd(@__DIR__)
13 BSON.@load "../data/mnistGAN40.bson" genParams
14 Flux.loadparams!(gen, genParams)
15
16 fixedNoise = [randn(latentDim, 1) for _ in 1:outputX*outputY]
17 fakeImages = @. gen(fixedNoise)
18 imageArray = permutedims(dropdims(reduce(vcat,
19             reduce.(hcat, Iterators.partition(fakeImages, outputY)));
20              dims=(3, 4)), (2, 1))
21
22 heatmap(imageArray, yflip = true, color = :Greys,
23         size = (300,150), legend=false, ticks=false)
```

In line 4 we define `latentDim` which is the agreed upon latent dimension of the generator network. The variables `outputX` and `outputY` in line 5 specify the dimensions of the grid of images that is to be created. The generator, `gen`, is defined in lines 7-10 and is in agreement with the training in Listing 9.20. After setting the working directory as the directory of the current file in line 12, the network's BSON file is loaded in line 13 and set onto `gen` in line 14. This is similar to the first listing of this chapter, Listing 9.1. Noise components for each of the images are created in line 16. Then in line 17 the generator, `gen()` is invoked on `fixedNoise`. The use of the macro `@.` implies that all operations in the expression are carried out element wise. This allows us to apply `gen()` on a whole array of noise elements. Lines 18-20, create a single rectangular image from `fakeImages` and it is presented via `heatmap()` in lines 22-23 to appear in Figure 9.17.

The training of the GAN is executed in Listing 9.20. In contrast to most code listings in the book, this listing is time consuming and may take hours, or days depending on the hardware. In our case it was executed in about 18 hours on a laptop without GPUs. Using GPUs or TPUs would greatly accelerate the process. The generator defined in lines 16-19 is identical to the generator in Listing 9.19. The discriminator defined in lines 13-15 is designed to detect if an image is fake or real. It is a convolutional neural network, similar to the other convolutional networks in this chapter with a few notable new features not seen yet. These include the *leakyrelu* activation function as well as the `stride` flag in the convolutional layers.

Critical to the GAN training is the loss function of the discriminator and the generator. The discriminator loss defined in lines 21-22 and used in line 29 is a sum of two penalties. A cross entropy penalty for not classifying real images as 1 together with a cross entropy penalty for not classifying fake images as 0. The generator loss is defined in line 23 and used in line 38.

Listing 9.20: Training a generative adversarial network

```
using Flux, MLDatasets, Statistics, Random, BSON
using Flux.Optimise: update!
using Flux: logitbinarycrossentropy

batchSize, latentDim = 500, 100
epochs = 40
etaD, etaG = 0.0002, 0.0002

images, _ = MLDatasets.MNIST.traindata(Float32)
imageTensor = reshape(@.(2f0 * images - 1f0), 28, 28, 1, :)
data = [imageTensor[:, :, :, r] for r in Iterators.partition(1:60000, batchSize)]

dscr = Chain(Conv((4,4),1=>64;stride=2,pad=1),x->leakyrelu.(x,0.2f0),
        Dropout(0.25),Conv((4,4),64=>128;stride=2,pad=1),x->leakyrelu.(x,0.2f0),
        Dropout(0.25), x->reshape(x, 7 * 7 * 128, :), Dense(7 * 7 * 128, 1))
gen =  Chain(Dense(latentDim,7*7*256),BatchNorm(7*7*256,relu),
        x->reshape(x,7,7,256,:),ConvTranspose((5,5),256=>128;stride=1,pad=2),
        BatchNorm(128,relu),ConvTranspose((4,4),128=>64;stride=2,pad=1),
        BatchNorm(64,relu),ConvTranspose((4,4),64=>1,tanh;stride=2,pad=1))

dLoss(realOut,fakeOut) =    mean(logitbinarycrossentropy.(realOut,1f0)) +
                            mean(logitbinarycrossentropy.(fakeOut,0f0))
gLoss(u) = mean(logitbinarycrossentropy.(u, 1f0))

function updateD!(gen, dscr, x, opt_dscr)
    noise = randn!(similar(x, (latentDim, batchSize)))
    fakeInput = gen(noise)
    ps = Flux.params(dscr)
    loss, back = Flux.pullback(()->dLoss(dscr(x), dscr(fakeInput)), ps)
    grad = back(1f0)
    update!(opt_dscr, ps, grad)
    return loss
end

function updateG!(gen, dscr, x, optGen)
    noise = randn!(similar(x, (latentDim, batchSize)))
    ps = Flux.params(gen)
    loss, back = Flux.pullback(()->gLoss(dscr(gen(noise))),ps)
    grad = back(1f0)
    update!(optGen, ps, grad)
    return loss
end

optDscr, optGen = ADAM(etaD), ADAM(etaG)
cd(@__DIR__)
@time begin
    for ep in 1:epochs
        for (bi,x) in enumerate(data)
            lossD = updateD!(gen, dscr, x, optDscr)
            lossG = updateG!(gen, dscr, x, optGen)
            @info "Epoch $ep, batch $bi, D loss = $(lossD), G loss = $(lossG)"
        end
        @info "Saving generator for epcoh $ep"
        BSON.@save "../data/mnistGAN$(ep).bson" genParams=cpu.(params(gen))
    end
end
```

```
[ Info: Epoch 1, batch 1, D loss = 1.3508931, G loss = 0.67289555
[ Info: Epoch 1, batch 2, D loss = 1.2776518, G loss = 0.608905
[ Info: Epoch 1, batch 3, D loss = 1.2069604, G loss = 0.5492967
[ Info: Epoch 1, batch 4, D loss = 1.147316, G loss = 0.49714798
[ Info: Epoch 1, batch 5, D loss = 1.0984138, G loss = 0.44755268
.
.
.
[ Info: Epoch 40, batch 117, D loss = 1.0338308, G loss = 1.4138021
[ Info: Epoch 40, batch 118, D loss = 0.867478, G loss = 1.4979299
[ Info: Epoch 40, batch 119, D loss = 1.6245712, G loss = 1.3874383
[ Info: Epoch 40, batch 120, D loss = 1.1869614, G loss = 1.2456459
[ Info: Saving generator for epcoh 40
64570.222768 seconds (678.26 G allocations: 29.082 TiB, 11.30% gc time)
```

Both the batch size and latent dimension are set in line 5. Training is set to run for up to 40 epochs in line 6. The learning rates for the discriminator and generator are in set in line 7. In line 9 we obtain the images via the `MLDatasets` package. In line 10 we set a tensor for the images and also rescale the level by multiplying by 2 and subtracting 1. Note that the suffix `f0` is for indicating that number literals are `Float32` (as opposed to `Int64` or `Float64` by default). Also note the use of the `@.` macro for making sure all operations are element wise. In line 11 we set the training data in the array `data` where each element of `data` is a mini batch. The discriminator, `dscr`, is defined in lines 13-15. It involves convolutional layers, the leaky relu activation function, `leakyrelu()`, dropout, and a dense layer. The generator, `gen`, is defined in lines 16-19. It involves a dense layer, batch normalization with the relu activation function, and convolutional transpose layers, with the tanh activation function. The loss function for the discriminator is defined in lines 21-22. The loss function for the generator is in line 23. The update function for the discriminator is in lines 25-33. In line 26 noise is created in a matrix based on the latent dimension, `latentDim` and the batch size, `batchSize`. The function `similar()` simply creates an array similar to x in type of the specified size. The function `randn!()` fills that array in place. Input is then created by the generator in line 27. Line 28 gets the parameters of the discriminator and then `Flux`'s `pullback()` function is called in line 29. The gradient is computed in line 30 and a gradient descent `update!()` is executed in line 31. Lines 35-42 implement `updateG!()` similarly. The ADAM optimizers are created in line 44 and in lines 47-55 the actual training loop takes place. The loop in lines 48-52 loops over all of the batches each one calling an update on the discriminator in line 49 and the generator in line 50. In each epoch, the parameters of the generator are saved to a specific file in line 54.

Chapter 10

Simulation of Dynamic Models

Most of the statistical methods presented in the previous chapters deal with inherently static data. With the exception of a few time-series examples, there is rarely a time component involved and typically observed random variables or vectors are assumed independent. We now move on to a different setting that involves a time component and/or dependent random variables. In general, such models are called "dynamic" as they describe changes over time or space. A consequence of dynamic behavior is dependence between random variables at different points in time or space.

Our focus in this chapter is not on statistical inference for such models, but rather on model construction, simulation, and analysis. Understanding the basics that we present here can help readers understand more complex systems and examples from *applied probability*, *stochastic operations research*, and methods of *stochastic control* such as *reinforcement learning*, already covered in Section 9.5 of Chapter 9. Dynamic stochastic models are a vast and exciting area and here we only touch the tip of the iceberg.

A basic paradigm is as follows: in discrete time $t = 0, 1, 2, \ldots$, one way to describe a random dynamical system is via the recursion,

$$X(t+1) = f\big(X(t), \xi(t)\big), \tag{10.1}$$

where $X(t)$ is the *state* of the system at time t, $\xi(t)$ is some random perturbation noise and $f(\cdot, \cdot)$ is a function that yields the next state as a function of the current state and the noise component. Continuous time and other variations also exist. Simulation of such a dynamic model then refers to the act of using Monte Carlo to generate trajectories,

$$X(0), X(1), X(2), \ldots,$$

for the purpose of evaluating performance and deciding on good control methods.

In this chapter we focus on a few elementary cases. In Section 10.1 we consider deterministic dynamical systems. We also present the very topical SEIR epidemic model as it received much attention in the era of COVID-19. In Section 10.2 we discuss simulation of Markov Chains both in discrete and continuous time. In Section 10.3 we discuss discrete event simulation, which is a general method for simulating processes that are subject to changes over discrete time points. In Section 10.4 we discuss models with additive noise and present a simple case of the Kalman

Y. Nazarathy and H. Klok, *Statistics with Julia*, Springer Series in the Data Sciences,
https://doi.org/10.1007/978-3-030-70901-3_10

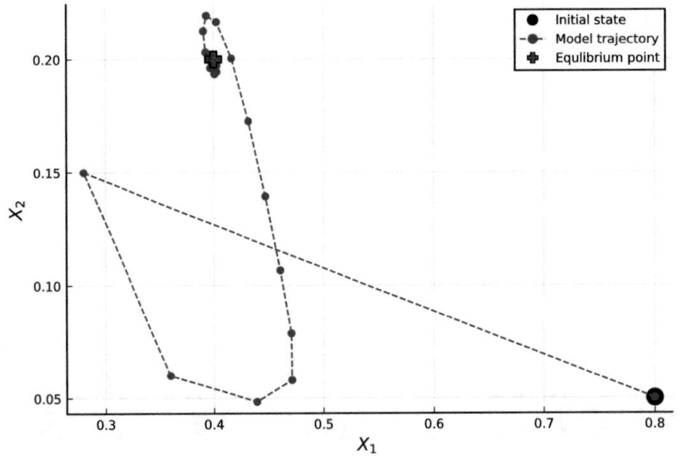

Figure 10.1: Trajectory of the predator-prey model of (10.4) and (10.5) for $X_1 = 0.8$ (prey), $X_2 = 0.05$ (predator), and parameters $a, c, d = 2, 1, 5$. Given these values the system converges to the equilibrium point in red.

filter. Then in Section 10.5 we briefly discuss network reliability and touch on elementary examples from *reliability theory*. We close with a discussion an example of common random numbers in Section 10.6. This Monte Carlo implementation strategy has been used in quite a few examples throughout our book, and our purpose here is to understand it a bit better.

10.1 Deterministic Dynamical Systems

Before we consider systems such as (10.1), we first consider systems without a noise component. In discrete time these can be described via the *difference equation*,

$$X(t+1) = f\big(X(t)\big), \tag{10.2}$$

and in continuous time via the *Ordinary Differential Equation* (ODE),

$$\frac{d}{dt}X(t) = f\big(X(t)\big). \tag{10.3}$$

These are generally called *dynamical systems* as they describe the evolution of the "dynamic" *state* $X(t)$ over time. Many physical, biological, and social systems may be modeled in this way, and a common objective is to obtain the *trajectory* of the system over time, given an *initial state* $X(0)$. In the case of a difference equation this is straightforward via recursion of equation (10.2). In continuous time we use ODE solution techniques to find the solutions of (10.3).

Discrete Time

The state $X(t)$ can take on different forms. In some cases it is a scalar, in other cases a vector, and yet in other cases it is an element from an arbitrary set. As a first example, assume that it is a two-dimensional vector representing normalized quantities of animals living in a competitive environment. Here $X_1(t)$ is the number of "prey" animals and $X_2(t)$ is the number of "predators". The species then affect each other via natural growth, natural mortality, and the hunting of the prey by the predators.

One very common model for such a population is the *predator-prey model*, described by the *Lotka-Volterra equations*:

$$X_1(t+1) = aX_1(t)\big(1 - X_1(t)\big) - X_1(t)X_2(t), \tag{10.4}$$

$$X_2(t+1) = -cX_2(t) + dX_1(t)X_2(t). \tag{10.5}$$

Here a, c, and d are positive constants that parameterize the evolution of this system. For parameter values in a certain range, there exists an *equilibrium point*. For example, if $a = 2$, $c = 1$ and $d = 5$ an equilibrium point is obtained via,

$$X^* = \big(X_1^*, X_2^*\big) = \Big(\frac{1+c}{d}, \frac{d(a-1) - a(c+1)}{d}\Big) = \big(0.4, 0.2\big). \tag{10.6}$$

To see that this is an equilibrium point, observe that using X^* for both $X(t)$ and $X(t+1)$ in (10.4) and (10.5) satisfies the equations. Hence, according to the model, once the predator and prey populations reach this point they will never move away from it. This is the definition of an equilibrium point.

Listing 10.1 simulates the trajectory of the predator-prey model by carrying out straightforward iteration over (10.4) and (10.5) given an initial state, and specific values of a, c, and d. The trajectory can be seen in Figure 10.1, along with the equilibrium point.

Listing 10.1: Trajectory of a predator-prey model

```
using Plots, LaTeXStrings; pyplot()

a, c, d = 2, 1, 5
next(x,y) = [a*x*(1-x) - x*y, -c*y + d*x*y]
equibPoint = [(1+c)/d ,(d*(a-1)-a*(1+c))/d]

initX = [0.8,0.05]
tEnd = 100

traj = [[] for _ in 1:tEnd]
traj[1] = initX

for t in 2:tEnd
    traj[t] = next(traj[t-1]...)
end

scatter([traj[1][1]], [traj[1][2]],
        c=:black, ms=10,
        label="Initial state")
plot!(first.(traj),last.(traj),
        c=:blue, ls=:dash, m=(:dot, 5, Plots.stroke(0)),
        label="Model trajectory")
scatter!([equibPoint[1]], [equibPoint[2]],
        c=:red, shape=:cross, ms=10, label="Equlibrium point",
        xlabel=L"X_1", ylabel=L"X_2")
```

In line 4 we define the function `next()` that implements the recursion of (10.4) and (10.5). In line 5 the equilibrium point is calculated via the closed form formula in (10.6). The initial state of the system is set in line 7, and the total number of discrete time points to iterate over is set in line 8. In line 10 we pre-allocate an array of arrays of length `tEnd`, where each sub-array is an array of two elements representing values of X_1 and X_2, respectively. The first element of the array is then initialized in line 11. Lines 13–15 loop over the time horizon and the `next()` function is applied at each time to obtain the state evolution. Note the use of the splat operator `...` in line 14 for transforming the two elements of `traj[t-1]` into distinct input arguments to `next()`. The remainder of the code plots Figure 10.1.

Continuous Time

We now look at the continuous time case through a physical example. Consider a block of mass M which rests on a flat surface. A spring horizontally connects the block to a near-by wall. The block is then horizontally displaced a distance z from its equilibrium position and then released. Figure 10.2 illustrates this scenario. The question is then how to describe the state of this system over time.

For this example we first make several assumptions. We assume that the spring operates elastically, and therefore the force generated by the spring on the block is given by

$$F_s = -kz,$$

where k is the spring constant of the particular spring, and z is the displacement of the spring from its equilibrium position. Note that the force acts in the opposite direction of the displacement. In addition, we assume that dry friction exists between the block and the surface it rests on, therefore the frictional force is given by

$$F_f = -bV,$$

where b is the coefficient of friction between the block and the surface, and V is the velocity of the block. Again note that the frictional force acts in the opposite direction of the force applied, as it resists motion.

With these established we can now describe the system. Let $X_1(t)$ denote the location of the mass and $X_2(t)$ the velocity of the mass. Using basic dynamics, these can then be described via,

$$\begin{bmatrix} \dot{X}_1(t) \\ \dot{X}_2(t) \end{bmatrix} = A \begin{bmatrix} X_1(t) \\ X_2(t) \end{bmatrix} \qquad \text{where} \qquad A = \begin{bmatrix} 0 & 1 \\ -\frac{k}{M} & -\frac{b}{M} \end{bmatrix}. \tag{10.7}$$

The first equation of (10.7) simply indicates that $X_2(t)$ is the derivative of $X_1(t)$ (the notation of a "dot" over a variable denotes the derivative). The second equation can be read as

$$M\dot{X}_2(t) = F_s + F_f. \tag{10.8}$$

Here the right-hand side is the sum of the forces described above and the left-hand side is "mass multiplied by acceleration". Equation (10.8) arises from basic laws of *Newtonian physics* or *classical*

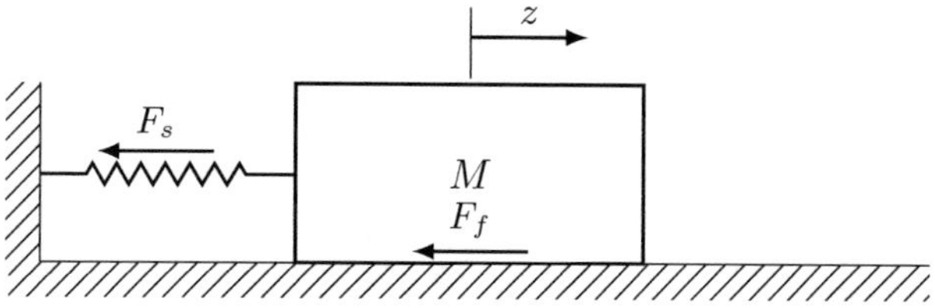

Figure 10.2: Spring and mass system, with spring force F_s, friction force F_f and applied displacement z.

mechanics. With such an ODE (sometimes called a linear system of ODEs), it turns out that given initial conditions $X(0)$, a solution to this ODE is,

$$X(t) = e^{At} X(0), \tag{10.9}$$

where e^{At} is a *matrix exponential*. Hence using the matrix exponential is one way of obtaining solutions to the trajectory of $X(t)$.

Many other alternative methods are implemented in Julia's `DifferentialEquations` package. We use both approaches in Listing 10.2 where we compute the evolution of this system given a starting velocity of zero, and a displacement of 8 units to the right of the equilibrium point. The changing state of the system is shown in the resulting Figure 10.3.

Listing 10.2: Trajectory of a spring and mass system

```
using DifferentialEquations, LinearAlgebra, Plots; pyplot()

k, b, M = 1.2, 0.3, 2.0
A = [0 1;
    -k/M -b/M]

initX = [8., 0.0]
tEnd = 50.0
tRange = 0:0.1:tEnd

manualSol = [exp(A*t)*initX for t in tRange]

linearRHS(x,Amat,t) = Amat*x
prob = ODEProblem(linearRHS, initX, (0,tEnd), A)
sol = solve(prob)

p1 = plot(first.(manualSol), last.(manualSol),
        c=:blue, label="Manual trajectory")
p1 = scatter!(first.(sol.u), last.(sol.u),
        c=:red, ms = 5, msw=0, label="DiffEq package")
p1 = scatter!([initX[1]], [initX[2]],
        c=:black, ms=10, label="Initial state", xlims=(-7,9), ylims=(-9,7),
        ratio=:equal, xlabel="Displacement", ylabel="Velocity")
p2 = plot(tRange, first.(manualSol),
        c=:blue, label="Manual trajectory")
p2 = scatter!(sol.t, first.(sol.u),
        c=:red, ms = 5, msw=0, label="DiffEq package")
p2 = scatter!([0], [initX[1]],
        c=:black, ms=10, label="Initial state", xlabel="Time",
        ylabel="Displacement")
plot(p1, p2, size=(800,400), legend=:topright)
```

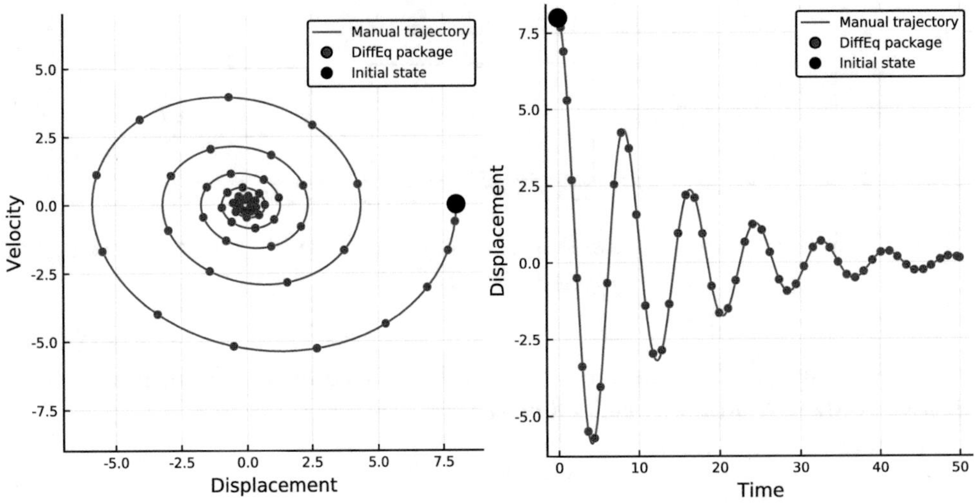

Figure 10.3: Trajectory of a spring and mass system.

In line 3 we set the values for the spring constant `k`, the friction constant `b` and the mass `M`. In lines 4–5 the matrix `A` is defined as in (10.7). In line 7 the initial conditions of the system are set, with the mass displaced 8 units to the right of the equilibrium point and the velocity set to zero. In line 11 we compute the trajectory of the system via the brute-force approach of (10.9). Here we use `exp()` from the `LinearAlgebra` package to evaluate the matrix exponential in (10.9). The resulting array `manualSol` is an array of two-dimensional arrays (state vectors), one for each point in time in `tRange`. In lines 13–15 the `DifferentialEquations` package is used to solve the ODE. In line 13 a function which is the right-hand side of the ODE of (10.7) is defined. Line 14 defines an `ODEProblem` object as `prob`. This object is defined by the right-hand side function `linearRHS`, the initial condition `initX`, a tuple of a time horizon `(0,tEnd)`, and a parameter to pass to the right-hand side function, `A`. Finally line 15 uses `solve()` from the `DifferentialEquations` package to obtain a numerical solution of the ODE. The remaining code generates Figure 10.2, which shows the manual solution of the trajectory in blue, and discrete points along the trajectory obtained by `solve()` of `DifferentialEquations` in red. Observe that in line 19, `sol.u` is used to get an array of the trajectory of state from the ODE solution. Similarly, in line 26 `sol.t` is used to get the time points matching `sol.u`.

The SEIR Epidemic Model

As far as things appear in mid 2021, the COVID-19 pandemic is a major historical event affecting human life, societies, and economies. With such an event, dynamic mathematical models are playing a major role in aiding policy makers for prediction and analysis. Often the models employed are quite complex, yet a basic deterministic dynamical system that is often used as a first step is the *SIR (Susceptible-Infected-Removed) model* as well as the slightly more detailed *SEIR (Susceptible-Exposed-Infected-Removed) model*. These types of models have existed since the 1920s, [KM1927] when they were developed after the major Spanish Influenza epidemic of 1918-1920. While there are more advanced *epidemiological* models in *mathematical biology*, understanding SIR and SEIR is often a first step for quantification of epidemics as well as understanding phenomena such as *flattening the curve* and *heard immunity*.

The deterministic dynamical system versions of SIR and SEIR involve a state $X(t)$ that is composed of three elements in the case of SIR and four elements in the case of SEIR. Each element is sometimes called a compartment and hence these models are called *compartmental models*. A large finite population is assumed to be distributed among the compartments *susceptible*, *exposed*, *infected*, and *removed*. That is each individual is assumed to be in one of these compartments. In SIR the exposed compartment is not present.

Susceptible individuals are those that are not yet ill and can become potentially ill if in contact with infected individuals. Exposed individuals are those that have already been in contact with infected individuals however their infection is currently incubated and they cannot still infect others. Infected individuals are those that are ill and can also infect others. The removed compartment is sometimes called *recovered* (although unfortunately in the case of COVID-19, the former term is more suitable because some removed individuals die). In any case, these models assume that recovered individuals have full immunity, as those that are removed/recovered do not affect the epidemic further.

At any time, the counts of individuals in each of the four compartments (three in the case of SIR) is given via $S(t), E(t), I(t)$, and $R(t)$. However as this is a differential equation model, these counts are generally not integer. A population of size M is assumed and hence

$$S(t) + E(t) + I(t) + R(t) = M.$$

Hence practically, it is sufficient to describe the system state via only three coordinates (two in the case of SIR because $E(t) \equiv 0$).

The model can be parameterized in several ways and here we choose a representation based on the non-negative rates, β, γ, and δ. Practically if the time unit is taken as days, then β^{-1} can be considered as the mean number of days between contacts of individuals and hence β is the *contact rate*. The value of γ^{-1} can be considered as the mean disease duration and hence γ is the *recovery rate*. The value of δ^{-1} can be considered as the mean incubation period of the disease during which an individual is exposed to the virus but is still not infecting others. Hence we call δ the *de-incubation rate*.

As is apparent at the time of writing this book, a disease such as COVID-19 incubates for about 5 days and hence $\delta = 1/5$. The mean disease duration is about 10 days and hence $\gamma = 0.1$. Finally, in a society without special social distancing, we take $\beta = 0.25$. There is not strong justification for the magnitude of this value, however one way is to consider the *basic reproduction number*, R_0. This elusive quantity is central to the study and discussion of epidemics and constitutes the mean number of individuals that an infected individual infects at onset of the epidemic. For COVID-19 without special *social distancing* measures, a commonly assumed value for R_0 is 2.5. Now for SEIR (and SIR) models it can be shown that

$$R_0 = \frac{\beta}{\gamma} = \frac{\gamma^{-1}}{\beta^{-1}} = \frac{\text{Mean disease duration}}{\text{Mean time between contacts}}.$$

Hence $\beta = 0.25$ yields the desired $R_0 = 2.5$ when $\gamma = 0.1$. We should mention that if one was to try and fit an SEIR model to real data, then tuning the β parameter is generally a difficult task.

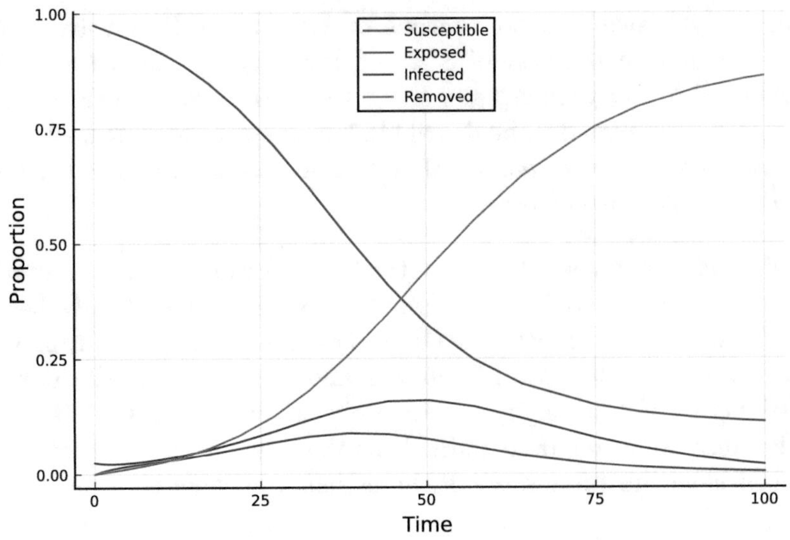

Figure 10.4: A trajectory of a deterministic SEIR model.

Now with the parameters and state of the model in place we can now present the model as a system of differential equations:

$$\begin{aligned}\dot{S}(t) &= -\beta\tfrac{1}{M}S(t)I(t)\\ \dot{E}(t) &= \beta\tfrac{1}{M}S(t)I(t)-\delta E(t)\\ \dot{I}(t) &= \delta E(t)-\gamma I(t)\\ \dot{R}(t) &= \gamma I(t)\end{aligned} \tag{10.10}$$

As can be observed from these equations there are three types of transitions between compartments: $S \to E$, $E \to I$, and $I \to R$. The latter two occur at rates proportional to the number of individuals in the source compartment, $\delta E(t)$ and $\gamma I(t)$. However the $S \to E$ transition is slightly more involved. The main driving force of infection is interaction between individuals and this is assumed to follow the general *law of mass action*. The idea is that if at time t there are $S(t)$ susceptible individuals and $I(t)$ infected individuals then new infections will occur at a rate proportional to the product $S(t)I(t)$. This is because each individual in $S(t)$ comes into contact with infected individuals at a rate proportional to $I(t)$.

Listing 10.3 generates a trajectory of the SEIR model with the aforementioned parameters which appears in Figure 10.4. In this case we assume that at onset 2.5% of the population is infected. In this case the final number of infected ends up with about 86%. Also of interest is the height of the red infection curve. Many of the social distancing measures imposed in 2020/2021 to combat COVID-19 were imposed with a view of reducing β and hence reducing the height of the infection curve as well as the final proportion of infected individuals. You may try to modify β in the code and see the effect on R_0, the final proportion of infected, and the shape of the infected curve. Note that in the code we take $M = 1$ to obtain proportions.

Listing 10.3: Trajectory of a deterministic SEIR epidemic

```
1  using DifferentialEquations, Plots; pyplot()
2
3  beta, delta, gamma = 0.25, 0.2, 0.1
4  initialInfect = 0.025
5  println("R0 = ", beta/gamma)
6
7  initX = [1-initialInfect, 0.0, initialInfect, 0.0]
8  tEnd = 100.0
9
10 RHS(x,parms,t) = [  -beta*x[1]*x[3],
11                     beta*x[1]*x[3] - delta*x[2],
12                     delta*x[2] - gamma*x[3],
13                     gamma*x[3] ]
14
15 prob = ODEProblem(RHS, initX, (0,tEnd), 0)
16 sol = solve(prob)
17 println("Final infected proportion= ", sol.u[end][4])
18
19 plot(sol.t,((x)->x[1]).(sol.u),label = "Susceptible", c=:green)
20 plot!(sol.t,((x)->x[2]).(sol.u),label = "Exposed", c=:blue)
21 plot!(sol.t,((x)->x[3]).(sol.u),label = "Infected", c=:red)
22 plot!(sol.t,((x)->x[4]).(sol.u),label = "Removed", c=:yellow,
23     xlabel = "Time", ylabel = "Proportion",legend = :top)
```

```
R0 = 2.5
Final infected proportion= 0.862203941883436
```

The parameters of SEIR are set in line 3 and the initial number of infected $I(0)$ is set in line 3. These parameters agree with an R_0 value similar to what is believed for COVID-19. The initial state of the system is set in line 7 and a maximal duration in line 8. Lines 10–13 implement the right-hand side of the SEIR system of ODEs from (10.10). The ODE is setup in line 15 and is solved in line 16. The final number of infected is printed line 17 and the remainder of the code creates Figure 10.4.

10.2 Markov Chains

In the previous section we considered systems that evolve deterministically. However sometimes it is more natural and applicable to model systems and assume that they have a built-in stochastic component. We now introduce and explore one such broad class of models called *Markov chains*. We first consider discrete time models and then move on to continuous time.

With a rich enough state space, many natural phenomena can be described via Markov chain models. Furthermore, in certain cases such models are artificially constructed as an aid for computation. We saw such a use of Markov chains Monte Carlo (MCMC) in Section 5.7, and also briefly considered simulation of a simple discrete time Markov chain in Listing 1.8 of Section 1.3. We now dive into further details.

The basic model evolution introduced in the previous section followed $X(t+1) = f\big(X(t)\big)$ where $X(t)$ is the state. That is, the next state is a direct deterministic function of the current state. Markov chains behave similarly, however in the case of a Markov chain $X(t+1)$ depends on $X(t)$ probabilistically. That is, the next state $X(t+1)$ is drawn randomly, based on a probability distribution that depends on the value of $X(t)$. For this, the model specification is typically based on a *probability transition law*,

$$p_{i,j} := \mathbb{P}\big(X(t+1) = j \mid X(t) = i\big) \qquad \text{for all states } \ i, j. \tag{10.11}$$

Here $p_{i,j}$ specifies the probability of transitioning from a current state i to a next state j. For every i,

$$\sum_j p_{i,j} = 1,$$

and hence the sequence $(p_{i,1}, p_{i,2}, \ldots)$ specifies a probability distribution. The actual *state space* where i and j take values can vary depending on context. If the state space is countable, then the transition probabilities for all i and j describe the Markov chain. Furthermore, if the state space is finite, then the probabilities may be organized in a *transition probability matrix*, $P = [p_{i,j}]$, where each row specifies a probability distribution (or probability vector). In other cases where the state space is uncountable, it isn't possible to only consider events such as $X(t+1) = j$ and therefore the definition of (10.11) is varied slightly to allow $X(t+1) \in A$ for a rich collection of sets A. We don't discuss such situations further here, as we assume that the state space is at most countable.

At the onset of this chapter in (10.1), we specified the equation $X(t+1) = f\big(X(t), \xi(t)\big)$, where $\xi(t)$ is some random perturbation noise. One may ask: How does the evolution of a Markov chain fit this description? For this, assume that you are given the probabilities in (10.11). Now by setting the random perturbation noise ξ as a uniform $[0,1]$ random variable, we are able to specify $f(i, \xi)$ as a function that evaluates the inverse CDF associated with the distribution $(p_{i,1}, p_{i,2}, \ldots)$ at the point ξ. This ensures that the probabilities in (10.11) are adhered to based on the inverse probability transform (see Section 3.4). For illustration, we implement such a function $f(\cdot, \cdot)$ in Listing 10.4, where we specify a transition probability matrix (see the function `f1()` in the listing).

Alternatively, in certain cases it is more natural to first consider the *stochastic recursive sequence* $X(t+1) = f\big(X(t), \xi(t)\big)$ and to construct the associated transition probability matrix from it as needed. For example, assume that $f(\cdot, \cdot)$ is specified as follows

$$f(x, u) = x + u \mod 5, \tag{10.12}$$

for $x \in \{0, 1, 2, 3, 4\}$ and $u \in \{-1, 0, +1\}$. This describes a situation where the state is decremented, stays the same or incremented, all modulo 5, meaning that decrementing from 0 yields 4 and incrementing from 4 yields 0. By using this $f(\cdot, \cdot)$ in (10.1), and assuming some probability law for $\xi(t)$, we arrive at a stochastic model specifying random movement (with "wrap around") on $\{0, 1, 2, 3, 4\}$. It turns out that if we assume the noise component $\xi(t)$ is i.i.d, then such a stochastic sequence may be encoded via a transition probability matrix, and that the model is a Markov chain even though it wasn't initially specified via P.

For example, say that $\xi(t)$ takes values $\{-1, 0, +1\}$ uniformly. Then using (10.11), you may see

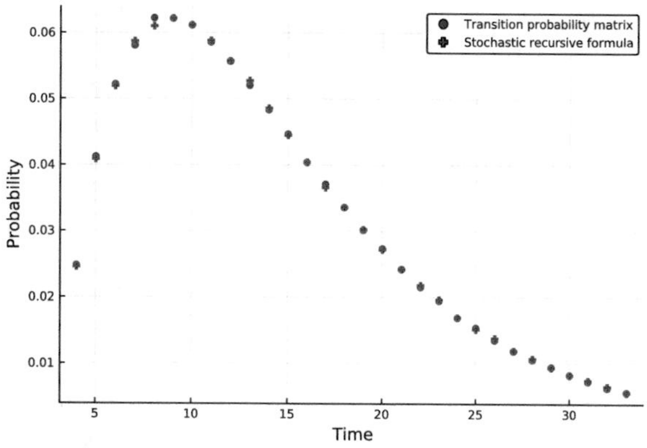

Figure 10.5: Estimates of the distribution of the time until all states in the Markov chain are visited. The blue dots are generated using the transition probability matrix, while the red dots are generated using a stochastic recursive formula.

that the corresponding transition probability matrix is

$$P = \begin{bmatrix} 1/3 & 1/3 & 0 & 0 & 1/3 \\ 1/3 & 1/3 & 1/3 & 0 & 0 \\ 0 & 1/3 & 1/3 & 1/3 & 0 \\ 0 & 0 & 1/3 & 1/3 & 1/3 \\ 1/3 & 0 & 0 & 1/3 & 1/3 \end{bmatrix}.$$

Thus we see that the dynamics of a Markov chain can be described by either a transition probability matrix, or by a stochastic recursive sequence as in (10.1). In both cases, if we specify the initial distribution $\mathbb{P}(X(0) = i)$, the evolution of the sequence of random variables, $X(0), X(1), X(2), \ldots$ is well defined.

Given the Markov chain sequence $\{X(t)\}_{t=0}^{\infty}$, we are sometimes interested in its limiting statistical behavior, and at other times we use this sequence to construct another random variable and are interested in the distribution of this variable, or just in its mean. As an example, for the Markov chain described above, let τ be the minimal time such that all states have been visited:

$$\tau = \inf\{t \ : \ \exists \ t_0, t_1, t_2, t_3, t_4 \leq t \text{ with } X(t_i) = i\}. \tag{10.13}$$

It is clear that τ is a random quantity because depending on the realization of $\{X(t)\}_{t=0}^{\infty}$, τ may obtain different values. For example, if we start with $X(0) = 0$ and then for the first 4 transitions $X(t)$ increases, then $\tau = 4$. However, it may also be that τ is a bigger number, for example, if the sequence of states happens to be $0, 1, 2, 1, 2, 1, 0, 4, 0, 1, 2, 1, 0, 4, 3, \ldots$, then $\tau = 14$ because this is the first time where all states have been covered.

In Listing 10.4 we illustrate both alternatives to generating a Markov chain. The function `f1()` uses the transition probability matrix, and the function `f2()` implements (10.12) directly. For both cases we assume that $\mathbb{P}(X(0) = 0) = 1$, i.e. we start in state 0 with certainty. We then estimate $\mathbb{E}[\tau]$ and plot estimates of the distribution of τ in Figure 10.5. It can be observed from the output that both methods are statistically identical. Note that it is possible to use *first step analysis*, a concept that we don't cover further here, to analytically show that $\mathbb{E}[\tau] = 15$.

Listing 10.4: Two different ways of describing Markov chains

```
using LinearAlgebra, Statistics, StatsBase, Plots; pyplot()

n, N = 5, 10^6
P = diagm(-1 => fill(1/3,n-1),
           0 => fill(1/3,n),
           1 => fill(1/3,n-1))
P[1,n], P[n,1] = 1/3, 1/3

A = UpperTriangular(ones(n,n))
C = P*A

function f1(x,u)
    for xNew in 1:n
        if u <= C[x+1,xNew]
            return xNew-1
        end
    end
end

f2(x,xi) = mod(x + xi , n)

function countTau(f,rnd)
    t = 0
    visits = fill(false,n)
    state = 0
    while sum(visits) < n
        state = f(state,rnd())
        visits[state+1] |= true
        t += 1
    end
    return t-1
end

data1 = [countTau(f1,rand) for _ in 1:N]
data2 = [countTau(f2,()->rand([-1,0,1]) ) for _ in 1:N]
est1, est2 = mean(data1), mean(data2)
c1, c2 = counts(data1)/N,counts(data2)/N
println("Estimated mean value of tau using f1: ",est1)
println("Estimated mean value of tau using f2: ",est2)
println("\nThe matrix P:", P)
scatter(4:33,c1[1:30],
        c=:blue, ms=5, msw=0,
        label="Transition probability matrix")
scatter!(4:33,c2[1:30],
        c=:red, ms=5, msw=0, shape=:cross,
        label="Stochastic recursive formula", xlabel="Time", ylabel="Probability")
```

```
Estimated mean value of tau using f1: 15.0134
Estimated mean value of tau using f2: 15.00187

The matrix P:
5x5 Array{Float64,2}:
 0.333333  0.333333  0.0       0.0       0.333333
 0.333333  0.333333  0.333333  0.0       0.0
 0.0       0.333333  0.333333  0.333333  0.0
 0.0       0.0       0.333333  0.333333  0.333333
 0.333333  0.0       0.0       0.333333  0.333333
```

In line 3 we set `n` as the number of states and `N` as the number of simulation runs to carry out. Lines 4–7 construct the transition probability matrix `P` by using `diagm()` to fill the diagonals of the matrix, and by assigning values to the north-east and south-west entries as well. In line 9 we construct an upper triangular matrix, `A`, and when it is right multiplied by `P` in line 10, we obtain a matrix of cumulative distribution vectors `C`. Lines 12–18 implement the function `f1()`. It assumes a uniform random variable `u` and returns a state using the inverse probability transform using the matrix `C`. Note that `x+1` in line 14 is because we treat the states as being `0...n` while the matrix indices are shifted by 1. For the same reason, we subtract 1 in line 15. Line 20 implements the function `f2()` as per (10.12). The function `countTau()` in lines 22–32 operates on two input arguments `f` and `rnd`, each of which is assumed to be a function. It then iterates (10.1) using the input arguments, and as it does so checks for the condition defining τ in (10.13). Note that we can use it with both types of $f(\cdot)$ functions, each with their respective type of random variable. The actual simulation time step is in line 27 and then we use the "(self) logical or operator", `|=` in line 28 to record a visit to the current state. Here again, `state+1` is due to the discrepancy between the state space and array indexing. Lines 34 and 35 exhibit calls to `countTau()` where in line 34, the input argument `f1` is augmented with the systems `rand` function, and in line 35 we create an anonymous function, `()->rand([-1,0,1])` as a second input argument. Lines 40–44 produce Figure 10.5, along with textual output showing that both methods estimate $\mathbb{E}[\tau]$ similarly.

A few more comments about discrete time Markov chains are in order. First, note that any process, $\{X(t)\}_{t=0}^{\infty}$ that satisfies this property,

$$\mathbb{P}\big(X(t+1) = j \mid X(t) = i, X(t-1) = i_{-1}, X(t-2) = i_{-2}, \dots\big) = \mathbb{P}\big(X(t+1) = j \mid X(t) = i\big) \quad (10.14)$$

is called a Markov chain. This *Markov property* indicates that given the current state ($X(t) = i$), any previous states, $i_{-1}, i_{-2}, \dots$ do not affect the evolution of the system. This is sometimes called the *memoryless property* or *Markov property.* Furthermore, all of the Markov chains that we consider in this chapter are *time homogenous.* This property states that for any times t_1 and t_2,

$$\mathbb{P}\big(X(t_1 + 1) = j \mid X(t_1) = i\big) = \mathbb{P}\big(X(t_2 + 1) = j \mid X(t_2) = i\big).$$

If this were not the case, then the transition probability matrix, P would not be sufficient for describing the evolution of the Markov chain. Instead we would need a time-dependent family of matrices, $P(t)$. Also note that Markov chains possess a variety of elegant mathematical properties that extended well beyond our examples here. See [N97] for an extensive introduction.

Further Discrete Time Modeling, Analysis, and Simulation

Modeling using Markov chains sometimes involves constructing the state space and the associated transition probability matrix for a given scenario. In some cases this is straightforward, while in others some modeling insight is required. We now explore another example to illustrate this.

Consider the following fictional scenario. A series of boxes are connected in a row, with each adjacent box accessed via a sliding door, as in Figure 10.6. In the left most box there is a cat, and in the right most box a mouse. Then, at discrete points in time, $t = 0, 1, 2, \dots,$ the doors connecting the boxes open, and both the cat and mouse migrate from their current positions, to directly adjacent boxes. They always move from their current box, randomly, with equal probability of going either left or right one box at a time.

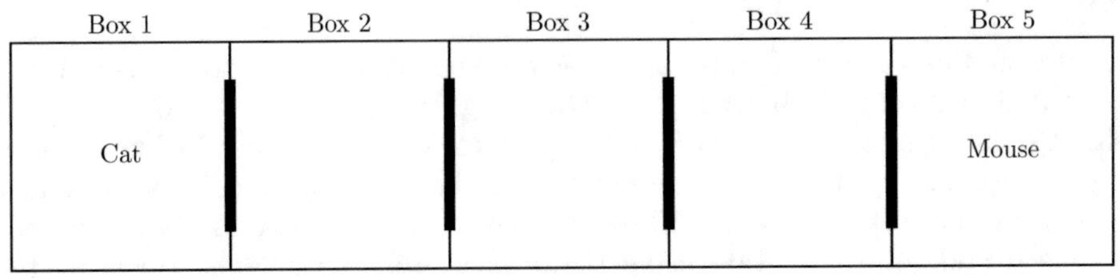

Figure 10.6: Illustration of the setup, consisting of adjacent boxes, and the starting positions of the cat and mouse, for $n = 5$ boxes.

At $t = 1$, both the cat and mouse must move to box 2 and 4, respectively. However at $t = 2$, the cat may move to either 1 or 3, and the mouse to either 3 or 5. This process of opening and closing the sliding doors repeats until eventually the cat and mouse are in the same box, at which point the mouse is eaten by the cat and the game ends.

This situation is different from the type of Markov chain described in the previous section and from the weather chain described in Listing 1.8 of Section 1.3. In these earlier cases, the processes are *recurrent* and go on forever. In the current case the process appears to be *transient* since at a given (random) point of time, the mouse is eaten. For recurrent Markov chains typical questions often deal with the steady state stationary distribution. However in a situation such as the one we describe here, a typical question may be: how long until the mouse is eaten? As this is a random variable, we may be interested in its distribution, or at least its expected value.

When modeling such a scenario using a Markov chain there are many options because we have freedom as to how to describe the states. For example, one way is to describe the states as tuples (x, y) where x is the location of the cat and y is the location of the mouse. However, we don't have to consider all possible combinations of x and y because it always holds that $x \leq y$. We may also observe that at any given time, both the mouse and the cat are either both in odd locations or both in even locations. This is because they are forced to move at each step, and the process alternates between odd and even. Such *periodic* phenomena can be studied further in Markov chains, however for our purposes we use this knowledge to set a small state space as follows:

State 1: (1,5). The game starts in this state. The game continues.

State 2: (2,4). The game continues.

State 3: (1,3). The game continues.

State 4: (3,5). The game continues.

State 5: (2,2), (3,3), and (4,4). The game ends.

With the states defined, we set the state space to consist of states $\{1, 2, 3, 4, 5\}$ where each state describes a situation as depicted above. From this, the stochastic matrix P is then constructed as follows:

$$P = \begin{bmatrix} 0 & 1 & 0 & 0 & 0 \\ 1/4 & 0 & 1/4 & 1/4 & 1/4 \\ 0 & 1/2 & 0 & 0 & 1/2 \\ 0 & 1/2 & 0 & 0 & 1/2 \\ 0 & 0 & 0 & 0 & 1 \end{bmatrix}. \tag{10.15}$$

With such a representation of this Markov chain, we are now interested in the *hitting time* of state 5. That is, the time until state 5 is reached, denoted via $\tau = \inf\{t \,:\, X(t) = 5\}$. It turns out that the theory of Markov chains goes a long way in computing expressions such as $\mathbb{E}[\tau]$. One way this can be done is by considering

$$p_0 = [1 \;\; 0 \;\; 0 \;\; 0], \qquad \text{and} \qquad T = \begin{bmatrix} 0 & 1 & 0 & 0 \\ 1/4 & 0 & 1/4 & 1/4 \\ 0 & 1/2 & 0 & 0 \\ 0 & 1/2 & 0 & 0 \end{bmatrix}.$$

Here p_0 is an initial distribution vector over the states $\{1, 2, 3, 4\}$ and T is part of the transition probability matrix P that relates to states $\{1, 2, 3, 4\}$. It can be shown using probabilistic arguments that

$$\mathbb{E}[\tau] = p_0 \left(I + T + T^2 + \ldots\right) \mathbf{1},$$

where $\mathbf{1}$ is a vector of 1's. This is done by considering all possible paths that can lead to the absorbing state 5. Here, for each $k = 0, 1, 2, \ldots$, each term $p_0 T^k \mathbf{1}$ describes the probability of reaching state 5 for the first time in k steps. Now by the *theory of non-negative matrices* it holds that

$$I + T + T^2 + \ldots = (I - T)^{-1}, \tag{10.16}$$

and the inverse exists (T is a sub-stochastic matrix with maximal eigenvalue strictly inside the unit circle). This can now be computed to find the analytic solution,

$$\mathbb{E}[\tau] = p_0 \left(I + T + T^2 + \cdots\right) \mathbf{1} = p_0 \, (I - T)^{-1} \mathbf{1} = 4.5. \tag{10.17}$$

Hence the mean time until the cat catches the mouse is 4.5. Listing 10.5 illustrates this computation, as well as the validity of the *infinite matrix geometric series*, (10.16), sometimes called a *Leontief series*. It also shows that the maximal eigenvalue of T is in the unit circle.

Listing 10.5: Calculation of a matrix infinite geometric series

```
using LinearAlgebra
P = [ 0   1   0   0   0;
      1/4 0   1/4 1/4 1/4;
      0   1/2 0   0   1/2;
      0   1/2 0   0   1/2;
      0   0   0   0   1]
T = P[1:4,1:4]
p0 = [1 0 0 0]
for n in 1:10
    println(first(p0*sum([T^k for k in 0:n])*ones(4)))
end
println("Using inverse: ", first(p0*inv(I-T)*ones(4)))
println("Eigenvalues of T: ", sort(eigvals(T)))
```

```
2.0
2.75
3.25
3.625
3.875
4.0625
4.1875
4.28125
4.34375
4.390625
Using inverse: 4.5
Eigenvalues of T: [-0.7071067811865, 0.0, 2.862293735361e-17, 0.7071067811865]
```

In line 7 we construct the matrix `T` as the sub-matrix of the matrix `P`. In lines 9–11 we consider the LHS series in (10.16) for increasing values of `n`. In line 12 the RHS of (10.17) is calculated. Note the use of the `inv()` function to calculate the inverse of `I-T`. Line 13 prints the sorted eigenvalues and shows that the largest eigenvalue has magnitude less than 1 and hence all eigenvalues lie in the unit circle.

Continuing with this cat and mouse example, in Listing 10.6 we arrive at the same result via alternative methods. One method is via a first principles implementation of the scenario, which is done in function `cmHitTime()`. The two other alternative methods make use of the `mcTraj()` function which we implement. This is a much more generic function, which creates a trajectory of a Markov chain with an arbitrary transition probability matrix `P`, given a starting state `initState`. It runs either for a duration of `T`, or stops when hitting state `stopState`. Note that by default `stopState = 0`, indicating the simulation only stops after `T` steps.

For illustration we use `mcTraj()` in two alternative ways. One way is by invoking it many times over (`N`) as follows: `mcTraj(P,1,10^6,5)`, where `P` is the transition probability matrix in (10.15), the second and fourth arguments are the initial and stopping states, respectively, and the third argument, `10^6`, is intended to be a high enough `T` such that the simulation only stops due to hitting state 5. Then averaging the lengths of all `N` trajectories yields an estimate of $\mathbb{E}[\tau]$.

The second way in which we use `mcTraj()` is related to the concept of *regenerative simulation*. We modify the final row of the transition probability matrix (10.15) by setting $P_{5,1} = 1$ and $P_{5,5} = 0$. This implies that once state 5 is reached, instead of the processes being absorbed in that state, it regenerates and starts afresh in state 1. In the language of Markov chains, this makes the transition probability matrix *irreducible* and hence (as it is a finite state space) *positive recurrent*. This then means that it possess a *stationary distribution* (or *limiting distribution*). It then holds that the inverse of the limiting probability of state 5 is the number of steps that are required to revisit the state. This allows us to generate one long trajectory of this Markov chain, estimate the limiting probability in state 5, and then obtain an estimate for $\mathbb{E}[\tau]$.

Listing 10.6: Markovian cat and mouse survival

```
using Statistics, StatsBase, Random, LinearAlgebra
Random.seed!(1)

function cmHitTime()
    catIndex, mouseIndex, t = 1, 5, 0
        while catIndex != mouseIndex
            catIndex += catIndex == 1 ? 1 : rand([-1,1])
            mouseIndex += mouseIndex == 5 ? -1 : rand([-1,1])
            t += 1
        end
    return t
end
function mcTraj(P,initState,T,stopState=0)
    n = size(P)[1]
    state = initState
    traj = [state]
    for t in 1:T-1
        state = sample(1:n,weights(P[state,:]))
        push!(traj,state)
        if state == stopState
            break
        end
    end
    return traj
end
N = 10^6
P = [  0   1   0   0   0;
      1/4  0  1/4 1/4 1/4;
       0  1/2  0   0  1/2;
       0  1/2  0   0  1/2;
       0   0   0   0   1]

theor = [1 0 0 0] * (inv(I - P[1:4,1:4])*ones(4))
est1 = mean([cmHitTime() for _ in 1:N])
est2 = mean([length(mcTraj(P,1,10^6,5))-1 for _ in 1:N])

P[5,:] = [1 0 0 0 0]
pi5 = sum(mcTraj(P,1,N) .== 5)/N
est3 = 1/pi5 - 1

println("Theoretical: ", theor)
println("Estimate 1: ",est1)
println("Estimate 2: ",est2)
println("Estimate 3: ",est3)
```

```
Theoretical: 4.5
Estimate 1: 4.497357
Estimate 2: 4.501016
Estimate 3: 4.507305440667045
```

In lines 4–12 we define the function `cmHitTime()` which returns a random time until the cat catches the mouse. The initial positions of the cat and mouse (`catIndex` and `mouseIndex`, respectively) are set in line 5. The while loop in lines 6–10 then updates these position indexes until the `catIndex` and `mouseIndex` are the same. Note that in line 7, if the cat is in position/box 1, then it moves to box 2 with certainty (+1), else its position index is uniformly and randomly incremented either up or down by 1. A similar approach is used for the index/position of the mouse in line 8. In lines 13–25 we define the function `mcTraj()`. As opposed to `cmHitTime()`, this function generates a trajectory of a general finite state discrete time Markov chain. The argument matrix `P` is the transition probability matrix; the argument `initState` is an initial starting state; the argument `T` is a maximal duration of a simulation; and the argument `stopState` is an index of a state to stop on if reached before `T`. The default value of `0` specified indicates that there is no stop state because the state space is taken to be `1,...n` (the dimension of `P`). The logic of the simulation is similar to the simulation in Listing 1.8. The key is line 18 where the `sample` function samples the next state from `1:n` based on probabilities determined by the respective row of the matrix `P`. Note that the iteration over the time horizon `1:T` can stop if the `stopState` is reached and the `break` statement of line 21 is executed. In lines 27–31 we define the transition probability matrix `P` as in (10.15). In line 33 we calculate the analytic solution to the average life expectancy of the mouse according to (10.17). In line 34 we use the `cmHitTime()` function to generate `N` i.i.d. random variables and compute their mean as `est1`. In line 35 we use the `mcTraj()` function setting a time horizon of 10^6 (effectively unbounded for this example) and a `stopState` of `5`. We then generate trajectories and subtract 1 from their length to get a hitting time. Averaging this over `N` trajectories creates `est2`. Lines 37–39 create the third estimate, `est3` via regenerative simulation as described above. Here we estimate the long term proportion of being in state 5 in line 38.

Continuous Time Markov Chains

A *continuous time Markov chain* also known as a *Markov jump process* is a stochastic process $X(t)$ with a discrete state space operating in continuous time t, satisfying the property,

$$\mathbb{P}\big(X(t+s)=j \mid X(t)=i \ \text{ and information about } X(u) \text{ for } u<t\big) = \mathbb{P}\big(X(t+s)=j \mid X(t)=i\big). \tag{10.18}$$

That is, only the most recent information (at time t) affects the distribution of the process at a future time $(t+s)$. Other definitions can also be stated, however (10.18) captures the essence of the Markov property, similar to (10.14) for discrete time Markov chains. An extensive account of continuous time Markov chains can be found in [N97].

While there are different ways to parameterize continuous time Markov chain models, a very common way is by using a so-called *generator matrix.* Such a square matrix, with dimension matching the number of states, has non-negative elements on the off-diagonal and non-positive diagonal values where each entry in the diagonal is the negative of the sum of the other entries on the same row. This ensures that the sum of each row is 0. For example, for a chain with three states, a generator matrix may be

$$Q = \begin{bmatrix} -3 & 1 & 2 \\ 1 & -2 & 1 \\ 0 & 1.5 & -1.5 \end{bmatrix}. \tag{10.19}$$

The values Q_{ij} for $i \neq j$ indicate the *intensity* of transitioning from state i to state j. In this example, since $Q_{12} = 1$ and $Q_{13} = 2$, there is an intensity of 1 for transitions from state 1 to state 2, and an intensity of 2 for transitions from state 1 to state 3. This implies that when $X(t) = 1$, during the time interval $t + \Delta$, for small Δ, there is a chance of approximately $1 \times \Delta$ for transitioning to state 2 and a chance of approximately $2 \times \Delta$ for transitioning to state 3. Furthermore there is a (significant) chance of approximately $1 - 3 \times \Delta$ for not making a transition at all.

An attribute of continuous time Markov chains is that when $X(t) = i$, the distribution of time until a state transition occurs is exponentially distributed with parameter $-Q_{ii}$. In the case of the example above, when $X(t) = 1$ the mean duration until a state change is $1/3$. Furthermore, upon a state transition, the transition is to state j with probability $-Q_{ij}/Q_{ii}$. In addition, the target state j is independent of the duration spent in state i. These properties are central to continuous time Markov chains. See [N97] for more details.

We can also associate some discrete time Markov chains with the continuous time models. One way to do this is to fix some time step Δ (not necessarily small), and define for $t = 0, 1, 2, 3, \ldots$,

$$\widetilde{X}(t) = X(t\Delta).$$

The discrete time process, $\widetilde{X}(\cdot)$ is sometimes called the *skeleton* at time steps of Δ of the continuous time process $X(\cdot)$. It turns out that for continuous time Markov chains,

$$\mathbb{P}\big(X(t) = j \mid X(0) = i\big) = [e^{Qt}]_{ij},$$

i.e. the above is given by the i, j'th entry of the matrix exponential. Hence the transition probability matrix of the discrete time Markov chain $\widetilde{X}(t)$ is the matrix exponential $e^{Q\Delta}$. This hints at one way of approximately simulating a continuous time Markov chain: set Δ small and simulate a discrete time Markov chain with transition probability matrix $e^{Q\Delta}$. Note also that if Δ is small then,

$$e^{Q\Delta} \approx I + \Delta Q. \tag{10.20}$$

However, a much better algorithm exists. For this, consider another discrete time Markov chain associated with a continuous time Markov chain: the *embedded Markov chain* or *jump chain*. This is a process that samples the continuous time Markov chain only at jump times. It has a transition probability matrix P, with $P_{ii} = 0$ (as there isn't a transition from a state to itself), and for $i \neq j$, $P_{ij} = -Q_{ij}/Q_{ii}$. The well-known *Gillespie algorithm*, which we call here the *Doob-Gillespie algorithm*, simulates a discrete time jump chain and stretches the intervals between the jumps by exponential random variables to yield a trajectory of the continuous time Markov chain. At each iteration of the algorithm, if we are in state i, we increment time by an exponential random variable with rate $-Q_{ii}$ and choose the next state based on P_{ij}.

In Listing 10.7 we consider a continuous time Markov chain with three states, starting with initial probability distribution $[0.4 \quad 0.5 \quad 0.1]$ and with generator matrix (10.19). The code determines the probability distribution of the state at time $T = 0.25$ showing that it is approximately $[0.27 \quad 0.43 \quad 0.3]$. This is achieved in three different ways. The first method is via the `crudeSimulation()` function, which is an inefficient simulation of a discrete time Markov chain *skeleton* with transition probability matrix $P = I + \Delta Q$, where Δ is taken as a small scalar value. The second method is via the `doobGillespie()` function, which is an implementation of the Doob-Gillespie algorithm presented above. Finally, the matrix exponential `exp()` is used as a non-Monte Carlo evaluation.

Listing 10.7: Simulation and analysis using a generator matrix

```
using StatsBase, Distributions, Random, LinearAlgebra
Random.seed!(1)

function crudeSimulation(deltaT,T,Q,initProb)
    n = size(Q)[1]
    Pdelta = I + Q*deltaT
    state  = sample(1:n,weights(initProb))
    t = 0.0
    while t < T
        t += deltaT
        state = sample(1:n,weights(Pdelta[state,:]))
    end
    return state
end

function doobGillespie(T,Q,initProb)
    n = size(Q)[1]
    Pjump  = (Q-diagm(0 => diag(Q)))./-diag(Q)
    lamVec = -diag(Q)
    state  = sample(1:n,weights(initProb))
    sojournTime = rand(Exponential(1/lamVec[state]))
    t = 0.0
    while t + sojournTime < T
        t += sojournTime
        state = sample(1:n,weights(Pjump[state,:]))
        sojournTime = rand(Exponential(1/lamVec[state]))
    end
    return state
end

T, N = 0.25, 10^5

Q = [-3 1 2
    1 -2 1
    0 1.5 -1.5]

p0 = [0.4 0.5 0.1]

crudeSimEst = counts([crudeSimulation(10^-3., T, Q, p0) for _ in 1:N])/N
doobGillespieEst = counts([doobGillespie(T, Q, p0) for _ in 1:N])/N
explicitEst = p0*exp(Q*T)

println("CrudeSim: \t\t", crudeSimEst)
println("Doob Gillespie Sim: \t", doobGillespieEst)
println("Explicit: \t\t", explicitEst)
```

```
CrudeSim:               [0.26845, 0.43054, 0.30101]
Doob Gillespie Sim:     [0.26709, 0.43268, 0.30023]
Explicit:               [0.269073 0.431815 0.299112]
```

In lines 4–14 we define the `crudeSimulation()` function, which approximately simulates a continuous time Markov chain through the implementation of (10.20). Observe that in line 10, time is increment by the discrete (small) interval `deltaT`. In lines 16–29 we define the `doobGillespie()` function which approximates the long term distribution of the state by simulating exponentially spaced discrete jumps according to the logic described above. Key here is that in every iteration there are two random number generations. In line 25, the next state is generated according to the embedded Markov chain, `Pjump`. In line 26 an exponential random variable is generated. In line 31 we set the time horizon `T` and the number of repetitions `N`. In lines 33–35 we set the generator matrix, `Q`. In line 37 we set the initial probability vector, `p0`. In lines 39–41 we evaluate the probability distribution of the state at time `T` via three alternative ways yielding the result in `crudeSimEst`, `doobGillespieEst` and `explicitEst`.

A Simple Markovian Queue

We now briefly explore *queueing theory*, which is the mathematical study of queues and congestion. See, for example, [HB13] for an elegant introduction to the field. This field of *stochastic operations research* and *applied probability* is full of mathematical models for modeling queues, waiting times, and congestion. One of the most basic models in the field is called the M/M/1 queue. In this model a single server (this is the "1" in the model name) serves customers from a queue, where each customer arrives according to a Poisson process and each one has independent exponential service times. The "M"'s in the model name indicate Poisson arrivals and exponential service times where "M" stands for "Markovian", or "memoryless".

The number of customers in the system can be represented by $X(t)$, a continuous time Markov chain taking on values in the state space $\{0, 1, 2, \dots\}$. In this case the (infinite) tridiagonal generator matrix is given by

$$Q = \begin{bmatrix} -\lambda & \lambda & & & \\ \mu & -(\lambda+\mu) & \lambda & & \\ & \mu & -(\lambda+\mu) & \lambda & \\ & & \mu & -(\lambda+\mu) & \ddots \\ & & & \ddots & \ddots \end{bmatrix}. \qquad (10.21)$$

Here λ indicates the rate of arrival, changing $X(t)$ from state i to state $i+1$ and μ indicates the rate of service, changing $X(t)$ from state i to state $i-1$. A common important parameter is called the *offered load*,

$$\rho = \frac{\lambda}{\mu}.$$

When $\rho < 1$ the process $X(t)$ is stochastically stable, in which case there is a stationary distribution for the continuous time Markov chain with

$$\lim_{t\to\infty} \mathbb{P}(X(t) = k) = (1-\rho)\rho^k, \qquad k = 0, 1, 2, \dots. \qquad (10.22)$$

As this is simply the geometric distribution (see Section 3.5), it isn't hard to see that the steady state mean (which we denote by L) is,

$$L_{\text{M/M/1}} = \frac{\rho}{1-\rho}. \qquad (10.23)$$

In Listing 10.8 we implement a Doob-Gillespie simulation of the M/M/1 queue. First we plot a trajectory of the queue length process $X(t)$ over $t \in [0, 200]$ in Figure 10.7. Then we simulate the queue for a long time horizon and check that the empirically observed mean queue length agrees with the analytic solution from (10.23).

Listing 10.8: M/M/1 queue simulation

```
using Distributions, Random, Plots; pyplot()
Random.seed!(4)

function simulateMM1DoobGillespie(lambda,mu,Q0,T)
    t, Q = 0.0 , Q0
    tValues, qValues = [0.0], [Q0]
    while t<T
        if Q == 0
            t += rand(Exponential(1/lambda))
            Q = 1
        else
            t += rand(Exponential(1/(lambda+mu)))
            Q += 2(rand() < lambda/(lambda+mu)) -1
        end
        push!(tValues,t)
        push!(qValues,Q)
    end
    return [tValues, qValues]
end

function stichSteps(epochs,q)
    n = length(epochs)
    newEpochs  = [ epochs[1] ]
    newQ = [ q[1] ]
    for i in 2:n
        push!(newEpochs,epochs[i])
        push!(newQ,q[i-1])
        push!(newEpochs,epochs[i])
        push!(newQ,q[i])
    end
    return [newEpochs, newQ]
end

lambda, mu = 0.7, 1.0
Tplot, Testimation = 200, 10^7
Q0 = 20

eL,qL = simulateMM1DoobGillespie(lambda, mu ,Q0, Testimation)
meanQueueLength = (eL[2:end]-eL[1:end-1])'*qL[1:end-1]/last(eL)
rho = lambda/mu
println("Estimated mean queue length: ", meanQueueLength )
println("Theoretical mean queue length: ", rho/(1-rho) )

epochs, qValues = simulateMM1DoobGillespie(lambda, mu, Q0,Tplot)
epochsForPlot, qForPlot = stichSteps(epochs,qValues)
plot(epochsForPlot,qForPlot,
        c=:blue, xlims=(0,Tplot), ylims=(0,25), xlabel="Time",
        ylabel="Customers in queue", legend=:none)
```

```
Estimated mean queue length: 2.33569071839852
Theoretical mean queue length: 2.333333333333333
```

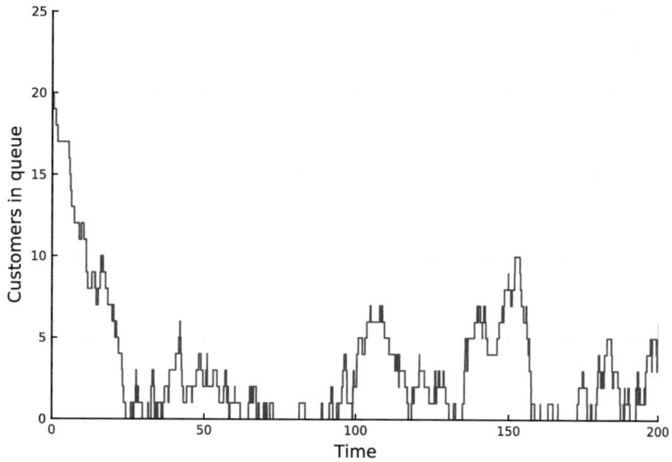

Figure 10.7: A queue length process of the M/M/1 queue starting with 20 customers in the system and with $\rho = 0.7$.

In lines 4–19 we implement the `simulateMM1DoobGillespie()` function. This function uses the Doob-Gillespie algorithm to create a trajectory of the M/M/1 queue. In contrast to the `doobGillespie()` function defined in Listing 10.7, our current function records the whole trajectory of the continuous time Markov chain. That is, the return value consists of `tValues` indicating times and `qValues` indicating state values (the state is held constant between times). Observe that in line 9 of the function implementation, the state sojourn time of rate λ is used at it matches state 0. Then in line 12, the state sojourn time has rate $\lambda + \mu$ and in line 13 there is a state transition either up or down, independently of the state sojourn time. In lines 21–32 we define the `stichSteps()` function, which creates a trajectory that can be plotted based on an array of time epochs, called `epochs`, and an array of queue lengths at each epoch `q`. The parameters of the queue and of the simulation are set in lines 34–36. Note that two separate times are set. The first, `Tplot = 200`, is used to plot a trajectory starting with `Q0 = 20` customers in the system. The second much longer duration, `Testimation`, is used for a simulation run that estimates the mean queue length. In lines 38–42 we handle the long time horizon simulation to print the estimate of the mean queue length compared to the theoretical value from (10.23). Importantly, in line 39 the difference sequence of time jumps is calculated via `eL[2:end]-eL[1:end-1]`. By taking the inner product of this vector with the queue lengths we are able to integrate over the queue length from time 0 until the last time, `eL`, and obtain the average queue length. In lines 44–48 we run a simulation for the short time horizon, apply `stichSteps()` to it, and plot the trajectory in Figure 10.7.

A Stochastic SEIR Model

We now return to the epidemic scenario modeled in Listing 10.3. There is a stochastic continuous time Markov chain version of this epidemic model which in many ways is more natural than its deterministic counterpart and can provide more information than a deterministic model. The basic idea is to define a continuous time Markov chain where the state $X(t) = \big(S(t), E(t), I(t), R(t)\big)$ takes values in the discrete set $\{0, 1, 2, \dots, M\}^4$ where M is the number of individuals in the epidemic and the components of $X(t)$ have the same interpretation as in the deterministic model of Listing 10.3.

Now transitions between states follow intensities parameterized by β, γ, and δ, similarly to the

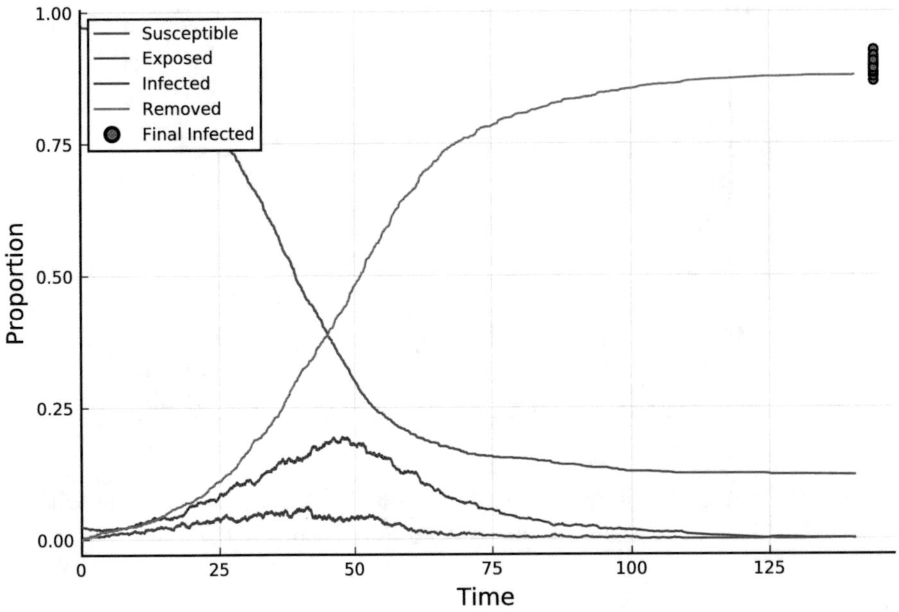

Figure 10.8: A single trajectory of a stochastic SEIR model. The "Final Infected" points result from 30 other independent trajectories.

deterministic model of Listing 10.3. That is, given that the state is $(s, e, i, r) \in \{0, 1, 2, \ldots, M\}^4$ at a given time, the following transitions can occur

$$
\begin{aligned}
(s, e, i, r) &\to (s-1, e+1, i, r) && \text{at rate} && \frac{\beta}{M} \times s \times i, \\
(s, e, i, r) &\to (s, e-1, i+1, r) && \text{at rate} && \delta \times e, \\
(s, e, i, r) &\to (s, e, i-1, r+1) && \text{at rate} && \gamma \times i.
\end{aligned}
$$

This type of model as well as similar stochastic epidemic models is analyzed in [DG01]. See also the documentation for `DifferentialEquations.jl` where methods and specialized code for defining and simulating *chemical reactions* models using are introduced.

In Listing 10.9 we execute a Doob-Gillespie simulation of the stochastic SEIR model. The nature of the simulation code is similar to the M/M/1 simulation in Listing 10.8 even though the underlying system and model is very different. The listing generates a single trajectory of $S(t), E(t), I(t), R(t)$ plotted in Figure 10.4. It also generates 30 additional trajectories for which we only plot the end value of the number of removed. This allows us to obtain an assessment of the variability of the results predicted by the model. Compare the stochastic Figure 10.8 with Figure 10.4.

Listing 10.9: Stochastic SEIR epidemic simulation

```julia
using Distributions, Random, Plots; pyplot()
Random.seed!(0)

beta, delta, gamma = 0.25, 0.4, 0.1
initialInfect = 0.025
M = 1000
I0 = Int(floor(initialInfect*M))
N = 30

function simulateSIRDoobGillespie(beta,delta,gamma,I0,M,T)
    t, S, E, I, R = 0.0, M-I0, 0, I0, 0
    tValues, sValues, eValues, iValues, rValues = [0.0], [S], [E], [I], [R]
    while t<T
        infectionRate = beta*I*S
        symptomRate = delta*E
        removalRate = gamma*I
        totalRate = infectionRate + symptomRate + removalRate
        probs = [infectionRate, symptomRate, removalRate]/totalRate
        t += rand(Exponential(1/(totalRate)))
        u = rand()
        if u < probs[1]
            S -= 1; E += 1
        elseif u < probs[1] + probs[2]
            E -=1; I+=1
        else
            I -= 1; R += 1
        end
        push!(tValues,t)
        push!(sValues,S);push!(eValues,E);push!(iValues,I);push!(rValues,R)
        I == 0 && break
    end
    return [tValues, sValues, eValues, iValues, rValues]
end

tV,sV,eV,iV,rV = simulateSIRDoobGillespie(beta/M,delta,gamma,I0,M,Inf)
lastT = tV[end]

finals = [simulateSIRDoobGillespie(beta/M,delta,gamma,I0,M,Inf)[5][end]
                for _ in 1:N]/M

p1 = plot(tV,sV/M,label = "Susceptible", c=:green)
plot!(tV,eV/M,label = "Exposed", c=:blue)
plot!(tV,iV/M,label = "Infected",c=:red)
plot!(tV,rV/M,label = "Removed", c=:yellow,
    xlabel = "Time", ylabel = "Proportion",
    legend = :topleft, xlim = (0,lastT*1.05))
scatter!(lastT*1.025*ones(N),finals, c = :yellow,label= "Final Infected")
```

The model parameters are set in line 4 and the initial infected proportion, population size, and initial number of infected are set in lines 5–6. The number of replicates for observing the end behavior is set in line 8. The function `simulateSIRDoobGillespie()` in lines 9–32 simulates the epidemic where the three driving rates that may occur are set in lines 13–15 and then the probabilities of transition are `probs` in line 17. The code in line 29 executes the `break` statement only if `I==0`. This use of *short circuit evaluation* is a common Julia idiom. In line 34 we run a single trajectory which is later plotted. Then in lines 35–36 we run `N` trajectories only for the purpose of evaluation the end size of the epidemic. Both the single run and a scatter plot of the `finals` array are plotted.

10.3 Discrete Event Simulation

We now introduce the concept of *discrete event simulation.* This is a way of simulating dynamic systems that are subject to changes occurring over discrete points of time. The basic idea is to consider discrete time instances, $T_1 < T_2 < T_3 < \ldots$, and assume that in between T_i and T_{i+1} the system state model $X(t)$ remains unchanged, or follows a deterministic path. At each discrete time point T_i the system state is modified due to an *event* that causes such a state change. This type of simulation is often suitable for models occurring in logistics, social service, and communication.

As an illustrative hypothetical example, consider a health clinic with a waiting room for patients. Assume that two doctors are operating in their own rooms and there is a secretary administrating patients. The state of the system can be represented by the combination of the number of patients in the waiting room; the number of patients (say 0 or 1) speaking with the secretary; the number of patients engaged with the doctors; the activity of the doctors (say administrating aid to patients, on a break, or not engaged); and the activity of the secretary (say engaged with a patient, speaking on the phone, on a break, or not engaged).

Some of the events that may take place in such a clinic may include: a new patient arrives to the clinic; a patient enters a doctors' room; a patient leaves the doctors' room and goes to speak with the secretary; the secretary answers a phone call; the secretary completes a phone call, etc. The occurrence of each event causes a state change, and these events appear over discrete time points $T_1 < T_2 < T_3 < \ldots$. Hence to simulate such a health clinic, we advance simulation time, t, over discrete time points.

The question is then, at which time points do events occur? The answer depends on the simulation scenario since the time of future events depends on previous events that have occurred. For example, consider the event "a patient leaves the doctors' room and goes to speak with the secretary". This type of event will occur after the patient entered the doctors' room and is implemented by *scheduling the event* just as the patient entered the doctors' room. That is, in a discrete event simulation, there is typically an *event schedule* that keeps track of all future events. Then, the simulation algorithm advances time from T_i to T_{i+1}, where T_{i+1} is the time corresponding to the next event in the schedule. Hence a discrete event simulation maintains some data structure for the event schedule that is dynamically updated as simulation time progresses. General commercial simulation software such as AnyLogic, Arena, and GoldSim do this in a generic manner, however in the examples that we present below, the event schedule is implemented in a way that is suited for our example simulation problem. For other applications, one can also look at the `SimJulia` package, which is for *process oriented* simulation in Julia, and is briefly mentioned in Appendix C.

We now return to the single server queue, similar to the M/M/1 queue that was simulated as a continuous time Markov chain in Section 10.2 with the Doob-Gillespie algorithm. In cases where inter-arrival or processing times in the queue are no longer exponentially distributed, modeling the system as a continuous time Markov chain is not easily possible (it is possible by means of extension of the state space, however this is not always the easiest implementation). Instead, simulating the system using discrete event simulation is straightforward.

In the case of a single server queue there are two types of events: (i) Customer arrives to the system, and (ii) Service completion of a customer. In this case, a discrete event simulation only needs to maintain a schedule of when each of these events is to occur in the future. We now elaborate on this via two simple variants of the M/M/1 queue.

M/M/1 vs. M/D/1 and M/M/1/K

We now consider two variants of the M/M/1 queue model covered in (10.2), namely, the M/D/1 and M/M/1/K models. In the M/D/1 model, the "D" stands for deterministic service times. This is a model where there is no variability of service durations, i.e. all customers require a service of duration exactly μ^{-1}. In a sense, such a model appears simpler than M/M/1, however mathematically it is slightly more challenging for analysis. Nevertheless, in queueing theory it is a special case of the M/G/1 queue, where "G" stands for a general distribution of service time. For this, the Khinchine-Pollatzek formula (see, for example, [HB13]) may be used to obtain the steady state mean number of customers in a system, which exists when $\rho = \lambda/\mu < 1$.

The second M/M/1 variant that we consider, M/M/1/K, is actually mathematically simpler. This model assumes that the system has finite capacity of size K. That is, at times when there are $K-1$ customers in the queue and one is being served (a total of K in the system), then any arriving customers are lost and never return. From a mathematical perspective, this actually implies that M/M/1/K systems are finite state continuous time Markov chains with generator matrix,

$$Q = \begin{bmatrix} -\lambda & \lambda & & & & \\ \mu & -(\lambda+\mu) & \lambda & & & \\ & \mu & -(\lambda+\mu) & \lambda & & \\ & & \ddots & \ddots & \ddots & \\ & & & \mu & -(\lambda+\mu) & \lambda \\ & & & & \mu & -\mu \end{bmatrix}. \tag{10.24}$$

Compare (10.24) with the (infinite size) matrix (10.21) of the standard M/M/1 queue. For any $\rho = \lambda/\mu \neq 1$ this generator matrix possess a truncated geometric steady state distribution (and for $\rho = 1$ a uniform distribution). In this case, it is easy to compute the steady state mean queue length.

Based on the above and after some analytic calculations, we have that mean queue lengths for all three systems are as follows:

$$L_{\text{M/M/1}} = \frac{\rho}{1-\rho}, \tag{10.25}$$

$$L_{\text{M/D/1}} = \frac{\rho}{1-\rho}\Big(\frac{2-\rho}{2}\Big), \tag{10.26}$$

$$L_{\text{M/M/1/K}} = \frac{\rho}{1-\rho}\Big(\frac{1-(K+1)\rho^K + K\rho^{K+1}}{1-\rho^{K+1}}\Big). \tag{10.27}$$

The first is the mean queue length of an M/M/1 queue in steady state. The second refers to an M/D/1 queue where the service times are deterministic. The third is for an M/M/1/K queue (finite capacity). It may be interesting to compare (10.25) to both (10.26) and (10.27). From the formulas, keeping in mind that $\rho < 1$, it isn't hard to see that each of $L_{\text{M/D/1}}$ and $L_{\text{M/M/1/K}}$ are lower than $L_{\text{M/M/1}}$. Interestingly when $\rho \approx 1$ (but smaller than 1), the M/D/1 case has queue lengths that are approximately half as long on average than M/M/1.

We now compare these theoretical formulas and observations to averages obtained via discrete event simulation. In Listing 10.10 we implement a function `queueDES()`, which performs discrete event simulation for a finite or infinite capacity queue. The simulation considers these three queue variants with $\rho \approx 0.63$, and the queue length estimates obtained for a long time horizon are shown to closely match the analytic formulas of (10.25), (10.26), and (10.27).

Listing 10.10: Discrete event simulation of queues

```
using Distributions, Random
Random.seed!(1)

function queueDES(T, arrF, serF, capacity = Inf, initQ = 0)
    t, q, qL = 0.0, initQ, 0.0

    nextArr, nextSer = arrF(), q == 0 ? Inf : serF()
    while t < T
        tPrev, qPrev = t, q
        if nextSer < nextArr
            t = nextSer
            q -= 1
            if q > 0
                nextSer = t + serF()
            else
                nextSer = Inf
            end
        else
            t = nextArr
            if q == 0
                nextSer = t + serF()
            end
            if q < capacity
                q += 1
            end
            nextArr = t + arrF()
        end
        qL += (t - tPrev)*qPrev
    end
    return qL/t
end

lam, mu, K = 0.82, 1.3, 5
rho = lam/mu
T = 10^6

mm1Theor = rho/(1-rho)
md1Theor = rho/(1-rho)*(2-rho)/2
mm1kTheor = rho/(1-rho)*(1-(K+1)*rho^K+K*rho^(K+1))/(1-rho^(K+1))

mm1Est = queueDES(T,()->rand(Exponential(1/lam)),
                            ()->rand(Exponential(1/mu)))
md1Est = queueDES(T,()->rand(Exponential(1/lam)),
                            ()->1/mu)
mm1kEst = queueDES(T,()->rand(Exponential(1/lam)),
                            ()->rand(Exponential(1/mu)), K)

println("The load on the system: ",rho)
println("Queueing theory: ", (mm1Theor,md1Theor,mm1kTheor) )
println("Via simulation: ", (mm1Est,md1Est,mm1kEst) )
```

```
The load on the system: 0.6307692307692307
Queueing theory: (1.7083333333333333, 1.169551282051282, 1.3050346932359453)
Via simulation: (1.7134526994574817, 1.1630297930829645, 1.302018728470463)
```

In lines 4–31 we implement the function `queueDES()` which carries out a discrete event simulation queue for up to `T` time units. The arguments `arrF` and `serF` are functions that present `queueDES()` with the next inter-arrival time and next service time, respectively. The argument `capacity` (with default value ∞) sets a queue limit to the queue (as needed for the M/M/1/K model). The argument `initQ` (with default value 0) is the initial queue length. In line 5 the initial time and queue length are set, along with the variable `qL` which is used later to calculate the average queue length. It is essentially a running *Riemann sum*, i.e. the sum of products of the time between each event by the length of the queue in between each event, as calculated in line 28. The main simulation loop is in lines 8–29. If the next service time occurs before the next arrival, the queue is decremented by one, and the service time is updated. If the next arrival occurs before the next service time, the queue is increased by one (as long as the queue is not at capacity) and the next arrival time is updated. Regardless of which occurs, `qL` is updated in line 28. This process continues until the time exceeds `T`. In line 30 the average queue length for the simulation is calculated and returned. In lines 33–35 the parameters of our three different queues are set, along with the maximum time units to be simulated `T`, and in lines 37–39 the analytic solutions of the three queues are calculated as per (10.25), (10.26), and (10.27). In lines 41–46 the three queues are simulated via `queueDES()`, and the numerically estimated mean queue lengths printed alongside their analytic counterparts in lines 48–50.

Waiting Times in Queues

The previous example of discrete event simulation maintained the state of the queueing system only via the number of items in the queue and the scheduled events. However, in some situations we need to maintain a more detailed state representation. For example, instead of just keeping track of "how many customers" are in the system we may want to keep "individual information about each customer". We now consider such a case with an example of waiting times in an M/M/1 queue operating under a first come first served policy. We have already touched such a case in Listing 3.6 of Chapter 3, and in that example we implicitly used the formula,

$$\mathbb{P}\big(\mathrm{W} \le x\big) = 1 - \rho e^{-(\mu-\lambda)x}, \qquad \text{for} \qquad x \ge 0, \tag{10.28}$$

where W is a random variable representing the *waiting time* of a customer arriving to a system in steady state. Observe that for $x = 0$, $\mathbb{P}\big(\mathrm{W} \le 0\big) = \mathbb{P}\big(\mathrm{W} = 0\big) = 1 - \rho$. That is, the probability of not waiting at all is $1 - \rho$. This is in agreement with the steady state distribution (10.22) since by setting $k = 0$ in that equation we obtain we see that in steady state, the system is empty a fraction $1 - \rho$ of the time. Observe that W is a random variable that is neither purely discrete nor purely continuous. There is a "mass" at $x = 0$ and then for $x \ge 0$ it is continuous.

To get a feeling for the mathematical nature of queueing theory, we now present a derivation for (10.28). It is obtained by considering the random variable X, representing the number of customers in the queue in steady state. As in (10.22), X it has a geometric distribution. Now, by conditioning on the values of X we are able to use the law of total probability to derive the complement of (10.28) for strictly positive values of x.

In the second step of the derivation we assume that for $k = 1, 2, \ldots$ customers, the waiting time of the arriving customer is distributed as the sum of k independent exponential random variables, each with mean μ^{-1}. This is the density $f_k(u)$ which is a gamma (called *Erlang*) distribution. The

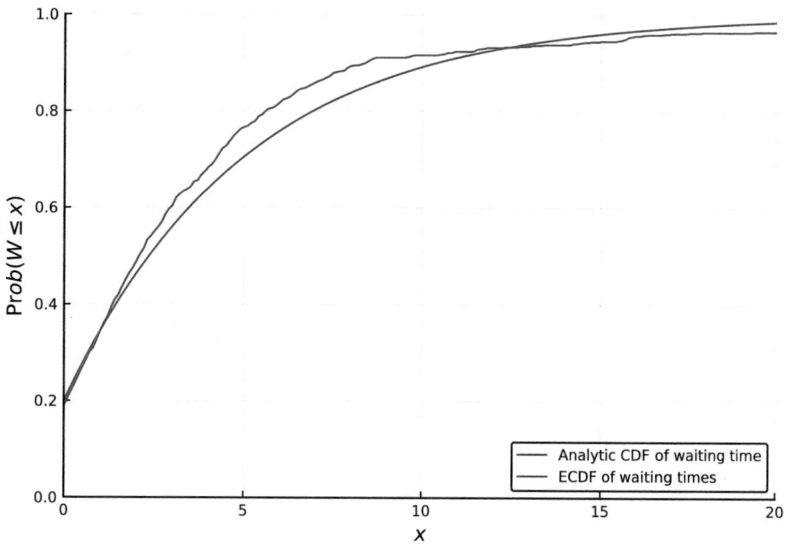

Figure 10.9: The CDF of the waiting time distribution in an M/M/1 queue with $\rho = 0.8$.

remainder of the calculation is slightly detailed, but straightforward. Here are the details:

$$\begin{aligned}
\mathbb{P}(\text{W} > x) &= \sum_{k=1}^{\infty} \mathbb{P}(\text{W} > x \mid X = k)\mathbb{P}(X = k) \\
&= \sum_{k=1}^{\infty} \int_x^{\infty} f_k(u)\, du\, (1-\rho)\rho^k \\
&= \sum_{k=1}^{\infty} \int_x^{\infty} \frac{\mu^k}{(k-1)!} u^{k-1} e^{-\mu u}\, du\, (1-\rho)\rho^k \\
&= (1-\rho)\lambda \int_x^{\infty} e^{-\mu u} \sum_{k=0}^{\infty} \frac{(\lambda u)^k}{k!}\, du \\
&= (1-\rho)\lambda \int_x^{\infty} e^{-(\mu-\lambda)u}\, du \\
&= (1-\rho)\frac{\lambda}{\mu-\lambda} e^{-(\mu-\lambda)x} \\
&= \rho e^{-(\mu-\lambda)x}.
\end{aligned}$$

The M/M/1 queue is one of a few special cases as we are able to use such probabilistic analysis to obtain an explicit formula for the distribution of the waiting time. However in *stochastic modeling*, if we modify the system even slightly, it is often the case that such an explicit performance measure is hard to come by, and hence discrete event simulation is often used. Sticking with M/M/1 so we can compare analytic and simulated solutions, in Listing 10.11 we carry out a simulation for the M/M/1 queue. A comparison between the ECDF obtained from the simulation and the analytic CDF (10.28) is shown in Figure 10.9. It can be observed that there isn't a perfect match because we use a short time horizon in the simulation. You may modify the code by increasing `T` in line 41, to observe a tighter fit.

Listing 10.11: Discrete event simulation for M/M/1 waiting times

```
using DataStructures,Distributions,StatsBase,Random,Plots, LaTeXStrings; pyplot()

function simMM1Wait(lambda,mu,T)
    tNextArr = rand(Exponential(1/(lambda)))
    tNextDep = Inf
    t = tNextArr

    waitingRoom = Queue{Float64}()
    serverBusy = false
    waitTimes = Array{Float64,1}()

    while t<T
        if t == tNextArr
            if !serverBusy
                tNextDep = t + rand(Exponential(1/mu))
                serverBusy = true
                push!(waitTimes,0.0)
            else
                enqueue!(waitingRoom,t)
            end
            tNextArr = t + rand(Exponential(1/(lambda)))
        else
            if length(waitingRoom) == 0
               tNextDep = Inf
               serverBusy = false
            else
               tArr = dequeue!(waitingRoom)
               waitTime = t - tArr
               push!(waitTimes, waitTime)
               tNextDep = t + rand(Exponential(1/mu))
            end
        end
        t = min(tNextArr,tNextDep)
    end

    return waitTimes
end

Random.seed!(1)
lambda, mu = 0.8, 1.0
T = 10^3

data = simMM1Wait(lambda,mu,T)
empiricalCDF = ecdf(data)

F(x) = 1-(lambda/mu)*MathConstants.e^(-(mu-lambda)x)
xGrid = 0:0.1:20

plot(xGrid, F.(xGrid),
        c=:blue,label="Analytic CDF of waiting time")
plot!(xGrid, empiricalCDF(xGrid),
        c=:red,label="ECDF of waiting times",
        xlabel=L"x", ylabel=L"\Prob(W \leq x)", xlims=(0,20),ylims=(0,1),
        legend=:bottomright)
```

In lines 3–37 we define the main function used in this simulation, `simMM1Wait()`. This function returns a sequence of `waitTimes` for consecutive customers departing from the queue simulated for a time horizon `T`. The `simMM1Wait()` function uses a `Queue` data structure from the `DataStructures` package. This `waitingRoom` variable, defined in line 8, represents the waiting room of customers, and its elements represent the arrival times of customers. The main simulation loop is in lines 12–34. Lines 14–21 handle an "arrival" event. while lines 23–31 handle a "departure" event. In line 19, when new arrivals to the busy server occur, new elements are added to `waitingRoom` via the `enqueue!()` function. In line 23, `length()` is applied to `waitingRoom` to see if the queue is empty. If it is empty, then lines 24–25 set the state of the system as "idle" by setting the next departure time, `tNextDep` to `Inf` and `serverBusy` to `false`. On the other hand, in lines 27–30 a new customer is pulled from the waiting room via `dequeue!()` while line 28 calculates the `waitTime` that customer has experienced. In line 29 that waiting time is pushed to `waitTimes`. In line 30 the service duration of that customer is randomly generated and `tNextDep` is set. Lines 39–41 set the parameters. Lines 43–44 execute the simulation and compute the ECDF via `ecdf()`. Line 46 implements (10.28) as `F()`. The remainder of the code generates Figure 10.9.

10.4 Models with Additive Noise

In Section 10.1 we considered deterministic models. We then followed with inherently random models, including Markov chains and discrete event simulation. We now look at a third class of models. These are based on deterministic models that have been modified to incorporate randomness. A basic mechanism for creating such models is to take a system equation such as (10.2), and augment it with a noise component in an additive form. Denoting the noise by $\xi(t)$ we obtain

$$X(t+1) = f\big(X(t)\big) + \xi(t). \tag{10.29}$$

A similar type of modification can be done to continuous time systems, yielding *stochastic differential equations*. However our focus here will be on the discrete case.

As an illustrative example, we revisit the predator-prey model explored in Listing 10.1. For this example we add i.i.d. random variables with zero mean and a standard deviation of 0.02 to the prey population. This is done in Listing 10.12, and the resulting stochastic trajectory is plotted alongside the previously calculated deterministic trajectory as shown in Figure 10.10. Note that this listing is very similar to Listing 10.1, with the main difference being in line 21 where the addition of the noise vector `rand(Normal(0,sig)),0.0]` applies normally distributed disturbances to the prey and no explicit disturbances to the predator population.

By adding such a noise component one may generate multiple trajectories of $X(t)$, each for a different point ω in the probability sample space. Then a mean trajectory may be estimated by considering an *ensemble average* over all the generated trajectories. Similarly, variability estimates and confidence bands can be obtained. Note that contrary to what some practitioners wrongly believe, even if the noise is zero mean, it does not generally hold that the expected value of $X(t)$ of (10.29) equals $X(t)$ of (10.2). That is, the nature of the noise modifies the expected trajectory even though at every step, the noise is on average zero.

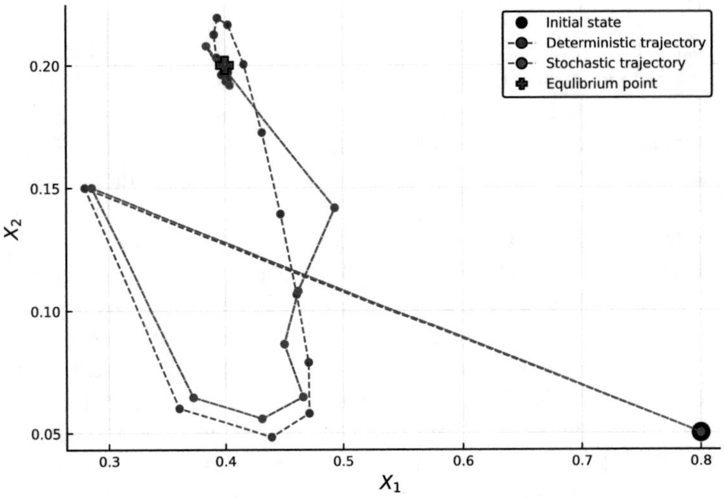

Figure 10.10: Trajectory of a stochastic predator-prey model together with a deterministic model.

Listing 10.12: Trajectory of a predator-prey model with noise

```
using Distributions, Random, Plots, LaTeXStrings; pyplot()
Random.seed!(1)

a, c, d = 2, 1, 5
sig = 0.02
next(x,y) = [a*x*(1-x) - x*y, -c*y + d*x*y]
equibPoint = [(1+c)/d ,(d*(a-1)-a*(1+c))/d]

initX = [0.8,0.05]
tEnd,tEndStoch = 100, 10

traj = [[] for _ in 1:tEnd]
trajStoch = [[] for _ in 1:tEndStoch]
traj[1], trajStoch[1] = initX, initX

for t in 2:tEnd
    traj[t] =  next(traj[t-1]...)
end

for t in 2:tEndStoch
    trajStoch[t] = next(trajStoch[t-1]...) + [rand(Normal(0,sig)),0.0]
end

scatter([traj[1][1]], [traj[1][2]],
        c=:black, ms=10, label="Initial state")
plot!(first.(traj), last.(traj),
        c=:blue, ls=:dash, m=(:dot, 5, Plots.stroke(0)),
        label="Deterministic trajectory")
plot!(first.(trajStoch),last.(trajStoch),
        c=:green, ls=:dash, m=(:dot, 5, Plots.stroke(0)),
        label="Stochastic trajectory")
scatter!([equibPoint[1]], [equibPoint[2]],
        c=:red, shape=:cross, ms=10, label="Equlibrium point",
        xlabel=L"X_1", ylabel=L"X_2")
```

State Tracking in Linear Systems

Many physical systems can be modeled by the evolution,

$$\begin{aligned} X(t+1) &= AX(t) + \xi(t), \\ Y(t) &= CX(t) + \zeta(t). \end{aligned} \tag{10.30}$$

Here, $X(t)$ and $Y(t)$ are the state and observation vectors, respectively, while $\xi(t)$ and $\zeta(t)$ are state and observation disturbances, respectively, and are often described by independent sequences of i.i.d. random variables. Such models are often used in *control theory* and *linear system theory*. See [AM10] for an overview description of control theory and [AM07] for a comprehensive introduction.

The matrix A describes the state evolution in a similar manner to the spring-mass example in (10.7), while the matrix C maps the current state to the measurement vector (prior to the addition of noise). That is, for such a system, the *sensors*' measurements are represented via $Y(t)$. In general, such systems are called *linear systems with additive noise*. One desire in such systems is to use the sensor measurements, $Y(0), Y(1), Y(2), \ldots$, to estimate (or track) the state as time progresses and the system is running.

Even if the number of sensors (dimension of $Y(t)$) is much smaller than the number of state variables (dimension of $X(t)$) we can often track the state $X(t)$ effectively. Furthermore, as we show below using *Kalman filtering*, we may even do so in the presence of the disturbance $\xi(t)$ and measurement noise $\zeta(t)$. To this end first assume that $\xi(t)$ and $\zeta(t)$ are both 0 vectors, i.e. there isn't any noise. In this case, the *Luenberger observer* is a state estimate $\hat{X}(t)$ which is parameterized by the *gain matrix* K, and operates as follows:

$$\hat{X}(t+1) = A\hat{X}(t) - K\big(\hat{Y}(t) - Y(t)\big), \qquad \text{with} \qquad \hat{Y}(t) = C\hat{X}(t). \tag{10.31}$$

Here, at time $t = 0$, $\hat{X}(0)$ is arbitrarily initialized, and then based on the observations $Y(0), Y(1), \ldots$ the state estimate is iterated as follows:

$$\begin{aligned} \hat{X}(t+1) &= A\hat{X}(t) - K\big(C\hat{X}(t) - Y(t)\big) \\ &= (A - KC)\hat{X}(t) + KY(t). \end{aligned}$$

In this case if we consider the *estimation error*, $e(t) = X(t) - \hat{X}(t)$, then we can show that

$$\begin{aligned} e(t+1) &= X(t+1) - \hat{X}(t+1) \\ &= AX(t) - \Big(A\hat{X}(t) - K(C\hat{X}(t) - Y(t))\Big) \\ &= A\big(X(t) - \hat{X}(t)\big) - K\big(Y(t) - C\hat{X}(t)\big) \\ &= A\big(X(t) - \hat{X}(t)\big) - K\big(CX(t) - C\hat{X}(t)\big) \\ &= Ae(t) - KCe(t) \\ &= (A - KC)e(t). \end{aligned} \tag{10.32}$$

Hence if we can design (or choose) a gain a matrix K such that $A - KC$ is a *stable matrix* (all eigenvalues are within the unit circle), then the Luenberger observer (10.31) will have $e(t) \to 0$ as $t \to \infty$. Remarkably, it turns out that if the pair A and C satisfy a rank condition called *observability* then we can always find such a matrix K, and hence always design a Luenberger observer to have asymptotically perfect tracking. If A is an $n \times n$ matrix and C is a $p \times n$ matrix, then the *observability matrix* is the $np \times n$ matrix,

$$\mathcal{O} = \begin{bmatrix} C \\ CA \\ CA^2 \\ \vdots \\ CA^{n-1} \end{bmatrix}.$$

The system is said to be *observable* if the matrix is *full rank*, i.e. in this case it requires $\mathcal{O}$ to have linearly independent columns. See [L79] for more details.

To appreciate the potential strength of the Luenberger observer, imagine a complex system where the state $X(t)$ is high dimensional, say 100, but the observations vector is much smaller, say only 3 dimensional (the matrix C would be 3×100). In such a system, subject to the technical observability condition on A and C, we can design Luenberger observer with matrix K such that after the system runs for a while, we have a near perfect representation of $X(t)$ via our $\hat{X}(t)$. This is only based on 3 dimensional measurements $Y(t)$ at each time point! See [AM07] for further details.

Now we allow noise $\xi(t)$ and $\zeta(t)$ and present Kalman filtering. Here we wish to find an optimal gain matrix K that will also take the statistical characteristics of the disturbance vectors $\xi(t)$ and $\zeta(t)$ into consideration. We do this based on the *Linear Minimum Mean Square Error* (LMMSE). Using the notation $||\cdot||$ for the L_2 norm, we try to set $\hat{X}(t)$ to be a linear function of the observed values which minimizes

$$\sum_{t=1}^{T} \mathbb{E}\Big[||X(t) - \hat{X}(t)||^2\Big], \tag{10.33}$$

for some time horizon T, or (often more practically), for the infinite horizon time average,

$$\lim_{T\to\infty} \frac{1}{T} \sum_{t=1}^{T} \mathbb{E}\Big[||X(t) - \hat{X}(t)||^2\Big].$$

The latter also generally equals the steady state expected mean squared error,

$$\mathcal{E}_\infty := \lim_{t\to\infty} \mathbb{E}\Big[||X(t) - \hat{X}(t)||^2\Big]. \tag{10.34}$$

For these cases, the Kalman filter is an algorithm that computes the gain matrix K (or sequence of matrices in the case of a finite horizon) which, if used in a Luenberger observer (10.31), yields a LMMSE solution. That is, a Kalman filter is a way to find a good gain matrix K.

Note that if the disturbances are assumed to be Gaussian then the LMMSE solution is also a *Minimum Mean Square Error (MMSE)* solution. That is, with Gaussian noise the Kalman filter is optimal in the MSE sense, while with non-gaussian noise it is optimal only within the class of linear estimators.

We skip the full details of implementing a Kalman filter. Nevertheless, we mention that a sequence of gain matrices for minimizing (10.33) can be computed recursively via the *Kalman filtering algorithm* or alternatively, the steady state Kalman gain K for minimizing (10.34), can be

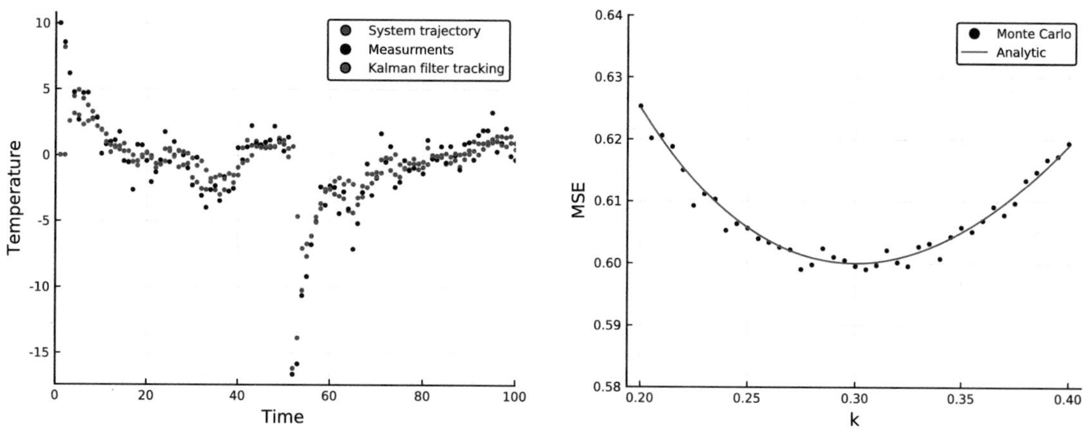

Figure 10.11: Left: Trajectory of a linear system with noise tracked by a Kalman filter. At time $t = 40$ the system is disturbed and it takes a few time epochs for the Kalman filter to catch up. Right: a plot of the steady state MSE as a function of gain. The optimal gain is at $k = 0.3$.

computed by solving a *Riccati equation* which considers the system matrices A and C, as well as the covariance matrices of $\xi(t)$ and $\zeta(t)$. We skip the details. If you are interested in the full details of Kalman filtering and ways to compute gain matrices, refer to [AM07]. We now present a simple scalar example similar to example 10.26 from [LG08].

A Scalar Example of Kalman Filtering

For this example, we construct a model based on (10.30), where all the variables are scalar,

$$\begin{aligned} X(t+1) &= aX(t) + \xi(t), \\ Y(t) &= X(t) + \zeta(t). \end{aligned} \tag{10.35}$$

We assume $a \in (0, 1)$ and hence this model describes a system that tends to revert towards 0 by a factor of a at each time unit. Also assume that $\xi(t)$ and $\zeta(t)$ are independent zero mean normal random variables with variances σ_ξ^2 and σ_ζ^2, respectively.

The process $X(t)$ is sometimes called an *autoregressive process* of order 1, denoted $AR(1)$ and among other phenomena, it can be used to describe the temperature of a system where "0" is taken as the reference point temperature. If undisturbed, the temperature $X(t)$ quickly converges to 0. However, since it is subject to temperature disturbances $\xi(t)$, there are fluctuations in the temperature.

The measurement is imprecise as there are measurement disturbances present, $\zeta(t)$, and the measured temperature $Y(t)$ deviates from the actual temperature $X(t)$. Our goal is then to estimate the current temperature at time t based on the measurement history, $Y(0), Y(1), \dots, Y(t-1)$. Following (10.31), the state estimate evolution follows:

$$\hat{X}(t+1) = a\hat{X}(t) - k(\hat{X}(t) - Y(t)), \tag{10.36}$$

where $\hat{X}(0)$ is some initial value.

Momentarily ignoring the noise components, consider the error dynamics (10.32) with $A = a$, $K = k$ (scalar gain parameter), and $C = 1$. We have that the matrix $A - KC$ in (10.32) is simply the scalar $a - k$ and hence if there isn't any noise, we expect $e(t) \to 0$ as long as $|a - k| < 1$. For example, we can set $k = a$ and just get from (10.36) $\hat{X}(t+1) = aY(t)$ as one would naively do without thinking about filtering measurements. However, in the presence of noise the error $e(t)$ will continue to fluctuate and hence there may be better choices of k that also take the variance parameters σ_ξ^2 and σ_ζ^2 into consideration.

Observe the left plot of Figure 10.11 for a system with $a = 0.8$, $\sigma_\xi^2 = 0.36$, and $\sigma_\zeta^2 = 1.0$. Here $X(t)$ is plotted in blue after starting at time $t = 0$ at a value of 10 and suffering an exogenous disturbance at time $t = 50$ to a level of -20 (this is a disturbance that isn't part of the model). The measurements $Y(t)$ are in black, and the trajectory of $\hat{X}(t)$ using a gain $k = 0.3$ is in red. The Kalman filter specifies this value of k, as we discuss below. You can visually observe that filtering generally does a better job in tracking the signal than just using $Y(t)$ as an estimate.

What plays a role in finding the optimal gain k? To get some sense into how this can be done, recall the computation in (10.32) and repeat it for (10.35) (again using the Luenberger observer (10.31)),

$$\begin{aligned} e(t+1) &= X(t+1) - \hat{X}(t+1) \\ &= aX(t) + \xi(t) - \Big(a\hat{X}(t) - k(\hat{X}(t) - X(t) - \zeta(t))\Big) \\ &= a\big(X(t) - \hat{X}(t)\big) - k\big(X(t) - \hat{X}(t)\big) + \xi(t) - k\zeta(t) \\ &= ae(t) - ke(t) + \xi(t) - k\zeta(t) \\ &= (a-k)e(t) + \xi(t) - k\zeta(t). \end{aligned} \tag{10.37}$$

Taking variance of both sides of the equation we obtain

$$\mathrm{Var}\big(e(t+1)\big) = (a-k)^2 \mathrm{Var}\big(e(t)\big) + \sigma_\xi^2 + k^2\sigma_\zeta^2. \tag{10.38}$$

Now observe that the steady state MSE, $\mathcal{E}_\infty$ of (10.34) equals $\lim_{t\to\infty} \mathrm{Var}\big(e(t)\big)$. Hence by taking $t \to \infty$ on (10.38), we have

$$\mathcal{E}_\infty = (a-k)^2\mathcal{E}_\infty + \sigma_\xi^2 + k^2\sigma_\zeta^2, \tag{10.39}$$

which after rearranging becomes

$$\mathcal{E}_\infty = \frac{\sigma_\xi^2 + k^2\sigma_\zeta^2}{1-(a-k)^2}. \tag{10.40}$$

This illustrates how the steady state MSE depends on k. One can then minimize $\mathcal{E}_\infty$, which in the case of our parameter settings is minimized at $k = 0.3$. Kalman filtering yields a way of carrying out this minimization in an efficient and systematic manner, even for much more complex systems. A plot of (10.40) is in red curve in the right hand plot of Figure 10.11. We also attempt using different values of k via Monte Carlo to see the validity of (10.40).These are the scattered points around the curve, each obtained from a simulation run of $T = 10^6$ time units where the MSE is estimated after throwing away the first 10^4 observations for "warm up".

Listing 10.13 creates Figure 10.11 with the main function, `luenbergerTrack()` used both for creating the short horizon trajectory on the left plot, and the long term Monte Carlo estimates on the right plot.

Listing 10.13: Kalman filtering

```
using Distributions, LinearAlgebra, Random, Measures, Plots; pyplot()
Random.seed!(1)

a, varXi, varZeta, = 0.8, 0.36, 1.0
X0, spikeTime, spikeSize = 10.0, 50, -20.0
Tsmall, Tlarge, warmTime = 100, 10^6, 10^4
kKalman = 0.3

function luenbergerTrack(k, T, spikeTime = Inf)
    X, Xhat = X0, 0.0
    xTraj, xHatTraj, yTraj = [X], [Xhat,Xhat], [X0]

    for t in 1:T-1
        X = a*X + rand(Normal(0,sqrt(varXi)))
        Y = X + rand(Normal(0,sqrt(varZeta)))
        Xhat =a*Xhat - k*(Xhat - Y)

        push!(xTraj,X)
        push!(xHatTraj,Xhat)
        push!(yTraj,Y)

        if t == spikeTime
            X += spikeSize
        end
    end
    deleteat!(xHatTraj,length(xHatTraj))
    xTraj, xHatTraj, yTraj
end

smallTraj, smallHat, smallY = luenbergerTrack(kKalman, Tsmall, spikeTime)

p1 = scatter(smallTraj, c=:blue,
    ms=3, msw=0, label="System trajectory")
p1 = scatter!(smallY, c = :black,
    ms=3, msw=0, label="Measurments")
p1 = scatter!(smallHat,c=:red,
    ms=3, msw=0, label="Kalman filter tracking",
    xlabel = "Time", ylabel = "Temperature",
    xlims=(0, Tsmall))

kRange = 0.2:0.005:0.4
errs = []
for k in kRange
    xTraj, xHatTraj, _ = luenbergerTrack(k, Tlarge)
    mse = norm(xTraj[warmTime:end] - xHatTraj[warmTime:end])^2/(Tlarge-warmTime)
    push!(errs, mse)
end

analyticErr(k) =(varXi + k^2*varZeta) / (1-(a-k)^2)

p2 = scatter(kRange,errs, c=:black, ms=3, msw=0,
    xlabel="k", ylabel="MSE", label = "Monte Carlo")
p2 = plot!(kRange,analyticErr.(kRange), c = :red,
    xlabel="k", ylabel="MSE", label = "Analytic", ylim =(0.58,0.64))

plot(p1, p2, size=(1000,400), margin = 5mm)
```

In lines 4–7 we define system, filtering, and simulation parameters. In lines 9–28 we define the `luenbergerTrack()` function for simulating a trajectory of time horizon `T` with gain `k`. The parameter `spikeTime` is used for introducing an exogenous spike, however with the default value `Inf`, such a spike doesn't occur. Lines 14–15 directly implement (10.35) and line 16 implements the filter (10.36). Line 26 uses `deleteat!()` to remove the last observation from `xHatTraj`. This agrees with the fact that it is initialized with two values in line 11. This is because it is a prediction of the next state X at every time. In line 30 we create the trajectory used for the left plot. It is uses the optimal gain, `kKalman`, specified earlier. Then lines 32–39 are used to create the left plot. Lines 41–47 empirically try a sequence of gain values over the range `kRange`. For each, a long term simulation is executed. Observe the use of `norm()` from `LinearAlgebra` for estimating `mse`. Line 49 defines `analyticErr()` that directly implements (10.40). The remainder of the code creates the right hand plot and combines the plot into Figure 10.11.

10.5 Network Reliability

We now briefly touch on the field of *network reliability* via simple examples. This discipline deals with the analysis of the reliability of systems composed of interconnected components. See, for example, [BCP95] for an introduction. Examples of systems that can be analyzed via network reliability models include road networks, electric power grids, computer networks, and other systems which can be described with the aid of (combinatorial) *graphs*. A graph is a collection of *vertices* and *edges*, where the edges describe connections between vertices. See, for example, Figure 10.12. As a simple application assume the graph represents a road network, where the edges represent roads and the vertices represent towns.

In the context of network reliability, after a graph is used to model relationships between components of the network, a probabilisitic model is imposed on the graph. With such a model, certain edges or nodes are subject to failure/repair, sometimes in a dynamic manner. The *reliability of the network* is then some statistical summary of the probability model quantifying performance measures.

As an example, consider the road network of Figure 10.12, and say we wish to have an active path between towns A and D. For this example there are three possible paths. However, what if the roads were subject to failure? In this case, a standard network reliability question may be: what is the probability of connectivity between towns A and D. Say we use a simplistic probability model which assumes that at a snapshot of time, each road is in a failed state with probability p, independently of all other roads. Hence the *reliability of the network* as a function of p is

$$r(p) = \mathbb{P}\big(\text{There is a path from } A \text{ to } D\big).$$

In the case of our simplistic network depicted in Figure 10.12 we can actually compute $r(p)$ analytically as follows:

$$\begin{aligned} r(p) &= 1 - \mathbb{P}\big(\text{There does not exist a path from } A \text{ to } D\big) \\ &= 1 - \mathbb{P}(A \to B \to D \text{ is broken})\ \mathbb{P}(A \to D \text{ is broken})\ \mathbb{P}(A \to C \to D \text{ is broken}) \\ &= 1 - (1-(1-p)^2)\ p\ (1-(1-p)^2) \\ &= 1 - p^3(p-2)^2. \end{aligned}$$

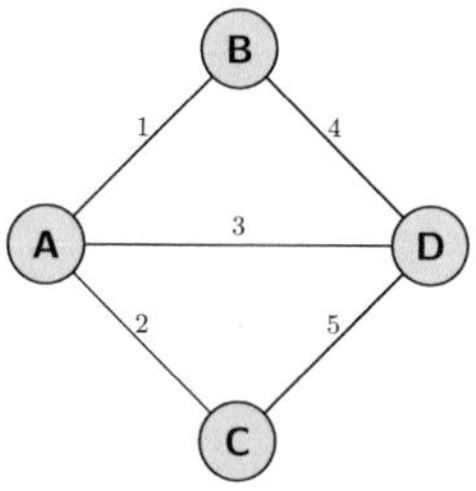

Figure 10.12: A graph with vertices $\{A, B, C, D\}$ and edges $1 = (A, B), 2 = (A, C), 3 = (A, D), 4 = (B, D), 5 = (C, D)$.

The key in the above computation is the fact that each path does not share edges with any other path. For example, the path $A \to B \to D$ and the path $A \to C \to D$ don't intersect on any edges. This allows us to move from the first line of the derivation to the second line. Afterwards, individual components can be calculated via

$$\mathbb{P}(A \to B \to D \text{ is broken}) = 1 - \mathbb{P}(A \to B \text{ is not broken})\mathbb{P}(B \to D \text{ is not broken}) = 1 - (1-p)^2,$$

and similarly for $\mathbb{P}(A \to C \to D$ is broken). Hence in such a simple example we can derive an analytic expression for the reliability of this network. However, for more complicated and interesting networks, this is not typically possible. This is because as redundancy emerges, strong dependencies exist between paths that share edges. A straightforward alternative approach to evaluate the reliability of the network is to use brute-force for generating many replications of random instances of the network, verifying if a path exists for each, and estimating the proportion from the Monte Carlo simulations.

We carry out an example of this brute-force method via Monte Carlo simulation in Listing 10.14. The estimates obtained are then compared with the solutions given by $r(p) = 1 - p^3(p-2)^2$, and the results plotted in Figure 10.13. Note that the functions defined in this code listing are not limited to the simple network of Figure 10.12, but are applicable to other networks through straightforward modifications of lines 19 and 20 by specifying a different adjacency list, source, and destination.

Graphs can be *directed* or *undirected*. Here we deal here with undirected graphs meaning that edges between vertices don't have a specified direction. In both cases, a common way to represent a graph is via an *adjacency matrix*. For a graph with L vertices, the $L \times L$ adjacency matrix R is defined to have entries,

$$R_{ij} = \begin{cases} 1 & \text{if edge } i \to j \text{ is in the graph,} \\ 0 & \text{if edge } i \to j \text{ is not in the graph.} \end{cases} \tag{10.41}$$

Since we are dealing with undirected graphs, the matrix R is always symmetric. With a graph represented via R, it turns out that for any integer $\ell \geq 1$, the i, j entry of the matrix power R^ℓ is the number of paths of length ℓ from vertex i to vertex j. You can verify this for $\ell = 2$ via,

$$[R^2]_{ij} = \sum_{k=1}^{L} R_{ik} R_{kj}.$$

We can then use this elegant property of adjacency matrices and their powers in Listing 10.14 to compute if a path exists between source and destination in a graph by checking for paths of length $\ell = 1, \ldots, L$. An alternative is to modify R by setting diagonal elements $R_{ii} = 1$. Then all that is needed to see if there is a path is to check from i to j is to consider the L'th power, R^L.

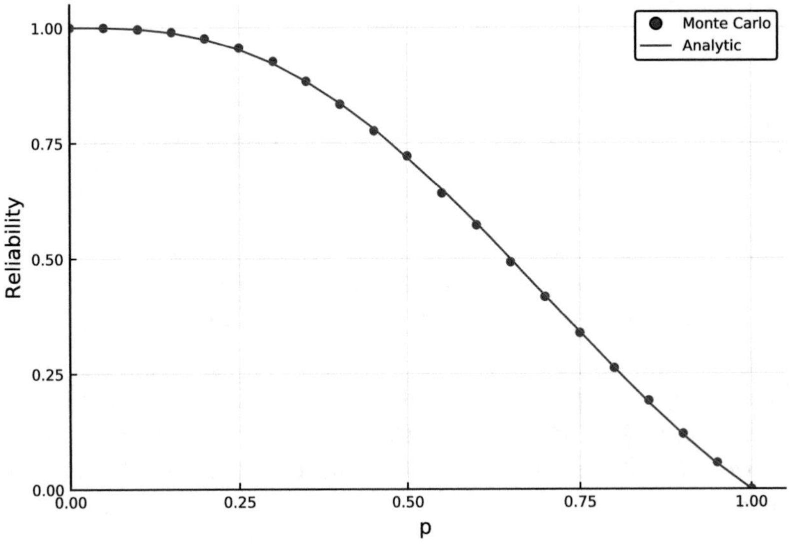

Figure 10.13: The reliability function of a simple network.

Listing 10.14: Simple network reliability

```
using LinearAlgebra, Random, Plots; pyplot()
Random.seed!(1)

N = 10^4
edges = [(1,2), (1,3), (2,4), (1,4), (3,4)]
L = maximum(maximum.(edges))
source, dest = 1, L

function adjMatrix(edges,L)
    R = zeros(Int, L, L)
    for e in edges
        R[ e[1], e[2] ], R[ e[2], e[1] ] = 1, 1
    end
    R
end

pathExists(R, source, destination) = sign.((I+R)^L)[source,destination]
randNet(p) = randsubseq(edges,1-p)

relEst(p) = sum([pathExists(adjMatrix(randNet(p),L),source,dest) for _ in 1:N])/N
relAnalytic(p) = 1-p^3*(p-2)^2

pGrid = 0:0.05:1
scatter(pGrid, relEst.(pGrid),
        c=:blue, ms=5, msw=0,label="Monte Carlo")
plot!(pGrid, relAnalytic.(pGrid),
        c=:red, label="Analytic", xlims=(0,1.05), ylims=(0,1.05),
        xlabel="p", ylabel="Reliability")
```

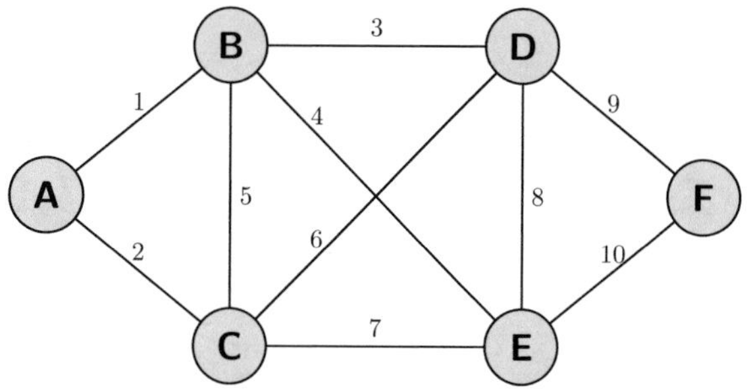

Figure 10.14: The undirected graph used in example Listing 10.15.

In lines 4–7 we define parameters including the list of edges via the `edges` array in accordance with Figure 10.12. In this array, we represent vertex A from the figure by `1`, vertex B by `2`, and so forth. The total number of vertices of this graph `L`, is set by first broadcasting `maximum()` onto each tuple and then taking the `maximum()`. In lines 9–15 we implement the `adjMatrix()` function, which takes an array of pairs (edges) as input, and from those edges creates an adjacency matrix. In line 17 we implement the `pathExists()` function, which checks if there is a path between the `source` and `destination` vertices in the graph represented by the adjacency matrix `R`. Prior to taking the L'th matrix power, we augment the adjacency matrix by setting 1 for entries on the diagonal via the addition of the identity matrix `I`. We then broadcast `sign()` to see if entries are 0 or greater than 0. In line 18 we use `randsubseq()` from `Random` to implement the `randNet()` function. It retains an edge from `edges` with probability `1-p` and otherwise removes it in accordance with our reliability model. This is carried out to each edge independently. In line 20 we implement the `relEst()` function which composes `randNet()`, `adjMatrix()`, and `pathExists()` to check if there is a path on a random instance of the network. This is repeated for `N` separate, independently simulated networks via a comprehension and the Monte Carlo proportion estimate is returned. In line 21 the analytic equation $r(p)$ is defined as `relAnalytic()`. The remainder of the code executes these `relEst()` and `relAnalytic()` over a grid of probabilities to create Figure 10.12 which compares the analytic solution with Monte Carlo based estimates.

A Dynamic Reliability Example

We now look at a dynamic reliability model. Instead of assuming a static setting where each edge fails with probability p, we introduce a time component to the model making it dynamic. We assume that at time 0 all edges of the network are operating (not broken) and that the lifetime of individual edges is i.i.d. exponentially distributed random variables, with parameter λ (assumed the same for all edges for simplicity). Then as time progresses, edges fail one after the other based on their lifetimes. At any given time, the network state $X(t)$ is the collection of edges that are still operating.

For a specific source and destination specification and given the network state $X(t)$, we can check at any time t if there is still a path between source and destination. We then define the failure time, or *network life time* via,

$$\tau = \inf\{t \geq 0 \; : \; \text{There isn't a path between source and destination in } X(t)\}.$$

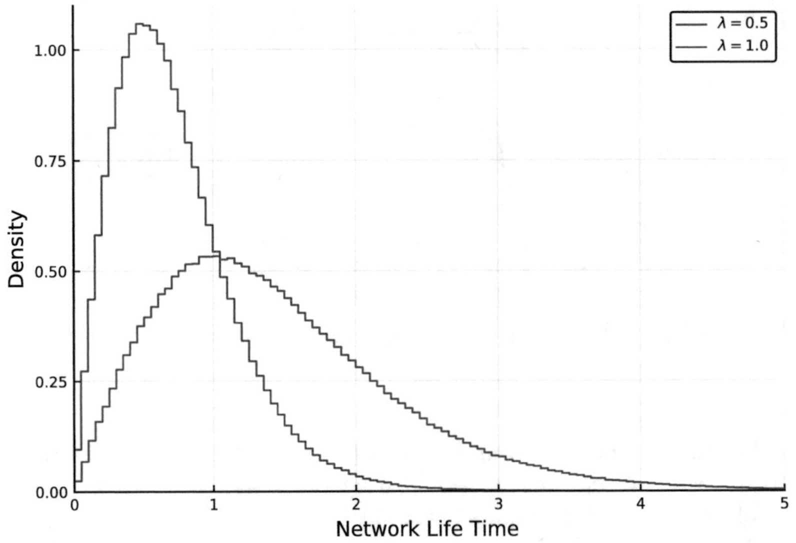

Figure 10.15: Comparison of the distribution of time until failure for $\lambda = 0.5$ and $\lambda = 1$.

One important reliability function is then the *expected network life time*,

$$r(\lambda) = \mathbb{E}[\tau].$$

Furthermore, we may be interested in the distribution of the network lifetime as influenced by λ. In general, we can expect that as λ increases, this distribution be more concentrated near 0 and that $r(\lambda)$ decreases. This is because with higher λ, individual edges tend to fail more quickly. An example of two such distributions are in Figure 10.15 generated by Listing 10.15 for the example network in Figure 10.14 with source being A and destination F.

Evaluating $r(\lambda)$ or the distribution of τ analytically is typically not possible. Instead, we resort to Monte Carlo simulation. With the i.i.d. assumption on edge life times, the network state, $X(t)$ is a well understood stochastic process as it can be described by a continuous time Markov chain (CTMC) where at any given time t, $X(t)$ denotes the set of operating edges. We can then use the Doob-Gillespie algorithm first introduced in Listing 10.7 for such a network. In doing so we observe that times between state changes are distributed exponentially, with a rate $\lambda \cdot E\big(X(t)\big)$, where $E(\cdot)$ counts the number of edges in the network. For example, to begin with $X(0) = \{1, 2, \ldots, 10\}$ which is the full edge set of Figure 10.14, and the time until the first failure event is distributed exponentially with parameter 10λ. After the first random edge fails, the time until the next failure event is distributed exponentially with parameter 9λ, and so forth. The failure time is then the first point in time t for which the set of edges $X(t)$ does not support a path from A to F. Also, in each iteration, in accordance with the CTMC theory that supports the Doob-Gillespie algorithm, we can uniformly select an edge from $X(t)$ to delete.

The implementation of Doob-Gillespie for this network in Listing 10.15 uses the `LightGraphs` package for handling the graph. This package allows us to encapsulate the representation, modification, and analysis of graphs. The implementation uses an *adjacency list* to represent a graph which is an alternative data structure to the adjacency matrix from (10.41). Here for each node, we keep a list of nodes that are adjacent to it. Our code uses the adjacency list representation as apparent via the use of the `fadjlist` field of `Graph` objects.

Listing 10.15: Dynamic network reliability

```
using LightGraphs, Distributions, StatsBase, Random, Plots, LaTeXStrings;pyplot()
Random.seed!(0)

function createNetwork(edges)
    network = Graph(maximum(maximum.(edges)))
    for e in edges
        add_edge!(network, e[1], e[2])
    end
    network
end

function uniformRandomEdge(network)
    outDegrees = length.(network.fadjlist)
    randI = sample(1:length(outDegrees),Weights(outDegrees))
    randJ = rand(network.fadjlist[randI])
    randI, randJ
end

function networkLife(network,source,dest,lambda)
    failureNetwork = copy(network)
    t = 0
    while has_path(failureNetwork, source, dest)
        t += rand(Exponential(1/(failureNetwork.ne*lambda)))
        i, j = uniformRandomEdge(failureNetwork)
        rem_edge!(failureNetwork, i, j)
    end
    t
end

lambda1, lambda2 = 0.5, 1.0
roads = [(1,2), (1,3), (2,4), (2,5), (2,3), (3,4), (3,5), (4,5), (4,6), (5,6)]
source, dest = 1, 6
network = createNetwork(roads)
N = 10^6

failTimes1 = [ networkLife(network,source,dest,lambda1) for _ in 1:N ]
failTimes2 = [ networkLife(network,source,dest,lambda2) for _ in 1:N ]

println("Edge Failure Rate = $(lambda1): Mean failure time = ",
        mean(failTimes1), " days.")
println("Edge Failure Rate = $(lambda2): Mean failure time = ",
        mean(failTimes2), " days.")

stephist(failTimes1, bins=200, c=:blue, normed=true, label=L"\lambda=0.5")
stephist!(failTimes2, bins=200, c=:red, normed=true, label=L"\lambda=1.0",
    xlims=(0,5), ylims=(0,1.1), xlabel="Network Life Time", ylabel = "Density")
```

```
Edge Failure Rate = 0.5: Mean failure time = 1.4471182849093784 days.
Edge Failure Rate = 1.5: Mean failure time = 0.48129663793885885 days.
```

In lines 4–10 we implement the `createNetwork()` function, which creates a `Graph` object from the `LighGraphs` package based on a list of edges. In line 5, the `Graph()` constructor is called, for which `maximum(maximum.(edges))` defines the number of vertices. Then in lines 6–8 a `for` loop is used to loop over each element of `edges` and add it to the graph via the `add_edge!()` function from `LightGraphs`. The return value in line 9, `network`, is a graph object. In lines 12–17 we implement `uniformRandomEdge()`, which takes a graph object from `LightGraphs` and returns a random uniformly selected edge (in the form of a tuple). In line 13, `outDegrees` is set by broadcasting `length()` to each element of `network.fadjlist`, i.e. to each element of the adjacency list. This sets `outDegrees` as an array counting how many edges point out from each of the vertices. In line 14 we set `randI` to be an index of a vertex by sampling with weights based on `outDegrees`. Then line 15 sets `randJ`. In line 16, the tuple, `(randI,randJ)` is returned which is guaranteed to be uniformly selected from the edges due to this sampling strategy. In lines 19–28 we implement the `networkLife()` function, which takes a `network` as input, and then degrades it according to a Poisson process at rate `lambda`. At each state it checks if a connection exists between `source` and `destination`, and returns the time when a path no longer exists. First, in line 20 the `copy()` function is used to create a copy of `network`. This is because `network` is passed by reference and we wish to degrade a copy of it, `failureNetwork`, and not the original network. Then in lines 22–26, the `LightGraphs` function `has_path()` is used to see if the network has a path from `source` to `dest`. Between each iteration, we wait for a duration that is exponentially distributed with a rate proportional to the number of edges (`failureNetwork.ne`). Then in line 24 `uniformRandomEdge()` is used to choose an edge, and in line 25 this is then removed via `rem_edge!()`. Two example λ values are set in line 30 and in line 31 the network shown in Figure 10.14 is defined. It is created into a `Graph` object in line 33. This simulations are executed in lines 36–37 by using the `networkLife()` function. The remaining lines summarize the simulation data in the form of text output and histograms.

10.6 Common Random Numbers and Multiple RNGs

More than half of the examples in this book involved some sort of (pseudo-) random number generation, often for the purpose of estimating some parameter, or performance measure. In such cases, one wishes to make the process as efficient as possible, i.e. one wishes to reduce the number of computations performed. However, there is an inherent tradeoff at play, since by reducing the number of computations one also reduces the confidence in the value of the parameter. Hence the concept of *variance reduction* is often employed to reduce the number of simulation runs, while maintaining the same precision of the parameter of interest. In this section we focus on one such technique called *common random numbers.*

We have actually already used this technique in several examples; see, for example, Listing 7.3. In these cases, the seed was fixed via `Random.seed!()` and a parameter was varied over some desired range. This approach often resulted in much smoother curves for the phenomena at interest.

In order to gain more insight into this common random numbers approach, consider the random variable with a distribution parameterized by λ,

$$X \sim \text{Uniform}\big(0,\ 2\lambda(1-\lambda)\big). \tag{10.42}$$

Clearly,

$$\mathbb{E}_\lambda[X] = \lambda(1-\lambda).$$

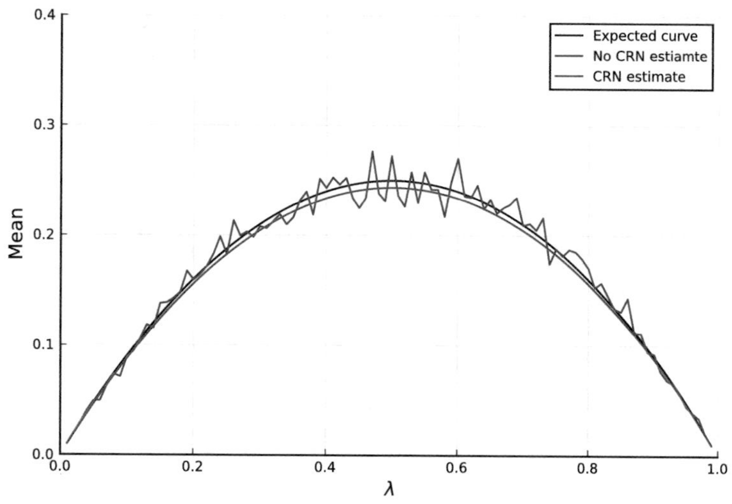

Figure 10.16: Using common random numbers.

Hence for this example, it is immediate that the expectation is maximized when $\lambda^* = 1/2$, which yields $\mathbb{E}_{\lambda^*}[X] = 1/4$. Now, for illustrative purposes, say we are not able to find λ^* analytically and wish to find this optimal λ using simulation.

To do so, we simulate n copies of X for each λ in some grid over $(0, 1)$, and for each λ obtain an estimate via,

$$\widehat{m}(\lambda) := \widehat{\mathbb{E}_\lambda[X]} = \frac{1}{n}\sum_{i=1}^{n} X_i^{(\lambda)}, \tag{10.43}$$

where $X_i^{(\lambda)}$ is a copy of the random variable with parameter λ. We then we choose $\widehat{\lambda^*}$ as the λ with maximal $\widehat{m}(\lambda)$.

Such a straightforward approach to simulation repeats the evaluation of $\widehat{m}(\lambda)$, and uses different independent random values each time. This is the behavior if `rand()` is simply used repetitively, and the seed is not set between each evaluation. Such an approach effectively implies (assuming ideal random numbers) that for each λ, each evaluation of $\widehat{m}(\lambda)$ is independent of the other evaluations.

The method of *common random numbers* is to use the same random numbers, i.e. a stream of random numbers for every λ over the grid. Mathematically this can be viewed as fixing an ω_0 in the probability sample space Ω (see Section 2.1) and re-evaluating the estimate $\widehat{m}(\lambda, \omega_0)$ for all values of λ. The idea is motivated by the assumption that for near-by parameter values, say λ_0 and λ_1, the estimate of $\widehat{m}(\lambda_0, \omega_0)$ and $\widehat{m}(\lambda_1, \omega_0)$ don't significantly differ. Hence by using the same ω_0 for both of these near values, a form of continuity on the estimated curve appears.

In Listing 10.16 we consider the example of estimating the maximizer λ^* from (10.43), and compare estimates obtained naively using different random numbers each time with estimates obtained via the use of common random numbers. The results shown in Figure 10.16 illustrate that for estimates obtained using common random numbers, the neighboring estimates do not differ greatly, (much less variance is observed), and the estimates are much closer to the true parameter values.

Listing 10.16: Variance reduction via common random numbers

```
1   using Distributions, Random, Plots, LaTeXStrings; pyplot()
2
3   seed = 1
4   N = 100
5   lamGrid = 0.01:0.01:0.99
6
7   theorM(lam) = mean(Uniform(0,2*lam*(1-lam)))
8   estM(lam) = mean(rand(Uniform(0,2*lam*(1-lam)),N))
9
10  function estM(lam,seed)
11      Random.seed!(seed)
12      estM(lam)
13  end
14
15  trueM = theorM.(lamGrid)
16  estM0 = estM.(lamGrid)
17  estMCRN = estM.(lamGrid,seed)
18
19  plot(lamGrid,trueM,
20          c=:black, label="Expected curve")
21  plot!(lamGrid,estM0,
22          c=:blue, label="No CRN estiamte")
23  plot!(lamGrid,estMCRN,
24          c=:red, label="CRN estimate",
25          xlims=(0,1), ylims=(0,0.4), xlabel=L"\lambda", ylabel = "Mean")
```

In line 7 we define the function `theorM()` which returns the theoretical mean, $\lambda(1-\lambda)$ by using the `mean()` method for a uniform random variable from `Distributions.jl`. In line 8 the function `estM()` is defined, which creates a sample of `n` random variables and computes their sample mean. In lines 10–13 we define an additional method for `estM()`. This method takes two arguments, the second one being `seed`. It sets the random seed in line 11 and then estimates the sample mean via the function of line 8. In line 15 the theoretical means are evaluated over the grid `lamGrid`, and the vector is set as `trueM`. In line 16 `estM()` is used to estimate the means over `lamGrid` without the use of common random numbers. In line 17 the second method of `estM()` is used to estimate the means over `lamGrid` through the use of common random numbers. This way the same stream of random numbers are used in each estimate. The remainder of the code is used to create Figure 10.16.

The Case for Using Multiple RNGs

We now consider another example, with the purpose of showing that in addition to the benefit of using common random numbers, there may sometimes be benefited from using multiple random number generators (RNGs) instead of a single RNG. Such practice is often employed in complex simulations, however here we illustrate a simple example.

Extend the previous example by considering a *random sum*. For this consider the random variable,

$$X = \sum_{i=1}^{N} Z_i, \tag{10.44}$$

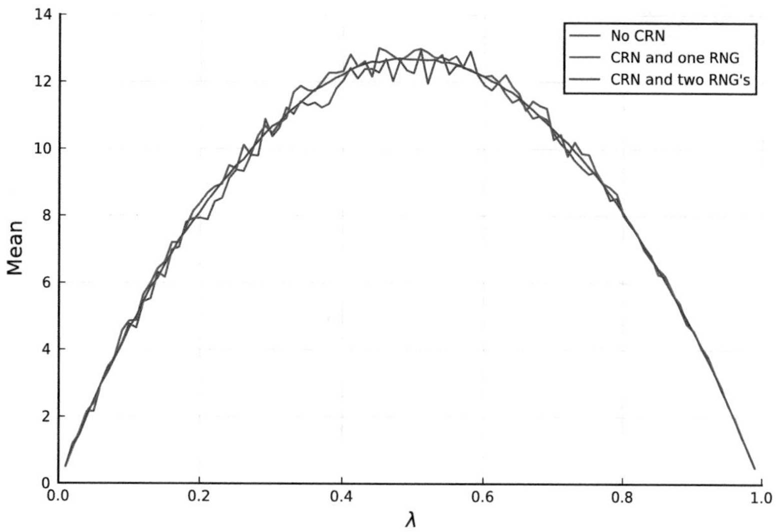

Figure 10.17: The effect of using two RNGs together with common random numbers: The blue curve is obtained with two RNGs and shows better performance than the red curve (no common random numbers) and the green curve (single RNG with common random numbers).

where $N \sim \text{Poisson}(K\lambda)$ and $Z_i \sim \text{Uniform}\big(0, 2(1-\lambda)\big)$ with $\lambda \in (0,1)$ and $K > 0$. In this case, it is possible to show that

$$\mathbb{E}_\lambda[X] = K\lambda(1-\lambda).$$

Like the previous example, here it is easy to see that the expectation is maximized when $\lambda^* = 1/2$. In which case, $\mathbb{E}_{\lambda^*}[X] = K/4$. However again, say that for illustration purposes, we wish to find this optimal λ using simulation. In this case we may again simulate n copies of X for each λ in some grid on $(0,1)$ and for each λ, obtain an estimate just like before via (10.43). We then choose $\widehat{\lambda^*}$ as the λ with maximal $\widehat{m}(\lambda)$.

Like the previous example, it may be of interest to employ common random numbers. However in this case, as we demonstrate, unless we maintain different random number streams for N and $\{Z_i\}$, the effect of common random numbers turns out to be almost insignificant. In this specific example this is due to the fact that the number of random numbers used to generate each X from (10.44) varies for each sample. Further if λ is modified and as a consequence samples of N are modified, the original random numbers that were used for a given Z_i are shifted. This can effectively break the desired effect of common random numbers. The phenomena is illustrated in Figure 10.17.

In Listing 10.17 we create Figure 10.17 where we also create our own Poisson random number generation function, `prn()` in line 6. It turns out that visually observing the benefit of multiple RNGs in this example works with our naive (quantile based) random number generator, but does not work with any Poisson random number generator. The listing also repeats the generation `M` times, to obtain estimates of the standard deviation of λ^*. This output serves as further proof that there is some benefit for using multiple random number generators.

Listing 10.17: A case for two RNGs

```
using Distributions, Random, Plots, LaTeXStrings; pyplot()

N, K, M = 10^2, 50, 10^3
lamRange = 0.01:0.01:0.99

prn(lambda,rng) = quantile(Poisson(lambda),rand(rng))
zDist(lam) = Uniform(0,2*(1-lam))

rv(lam,rng) = sum([rand(rng,zDist(lam)) for _ in 1:prn(K*lam,rng)])
rv2(lam,rng1,rng2) = sum([rand(rng1,zDist(lam)) for _ in 1:prn(K*lam,rng2)])

mEst(lam,rng) = mean([rv(lam,rng) for _ in 1:N])
mEst2(lam,rng1,rng2) = mean([rv2(lam,rng1,rng2) for _ in 1:N])

function mGraph0(seed)
    singleRng = MersenneTwister(seed)
    [mEst(lam,singleRng) for lam in lamRange]
end
mGraph1(seed) = [mEst(lam,MersenneTwister(seed)) for lam in lamRange]
mGraph2(seed1,seed2) = [mEst2(lam,MersenneTwister(seed1),
                MersenneTwister(seed2)) for lam in lamRange]

argMaxLam(graph) = lamRange[findmax(graph)[2]]

std0 = std([argMaxLam(mGraph0(seed)) for seed in 1:M])
std1 = std([argMaxLam(mGraph1(seed)) for seed in 1:M])
std2 = std([argMaxLam(mGraph2(seed,seed+M)) for seed in 1:M])

println("Standard deviation with no CRN: ", std0)
println("Standard deviation with CRN and single RNG: ", std1)
println("Standard deviation with CRN and two RNGs: ", std2)

plot(lamRange,mGraph0(1987),
        c=:red, label="No CRN")
plot!(lamRange,mGraph1(1987),
        c=:green, label="CRN and one RNG")
plot!(lamRange,mGraph2(1987,1988),
        c=:blue, label="CRN and two RNG's", xlims=(0,1),ylims=(0,14),
    xlabel=L"\lambda", ylabel = "Mean")
```

```
Standard deviation with no CRN: 0.037080520020152975
Standard deviation with CRN and single RNG: 0.03411444555309958
Standard deviation with CRN and two RNGs: 0.014645353747396726
```

In line 3 we define `N`, the number of repetitions to carry out for each value of λ; the constant `K`; and the number of repetitions to carry out in total for estimating the argmax, `M`. In line 6 we define our function `prn()`. It uses the inverse probability transform to generate a Poisson random variable with parameter `lambda` and with a random number generator `rng`. In line 7 we define a function for creating a $\text{Uniform}\big(0, 2(1-\lambda)\big)$ distribution. In lines 9–10 we create the two central functions for this example. The function `rv()` uses a single random number generator to generate the random variable (10.44). Then the function `rv2()` achieves this with two random variables. One for the uniform random variables and one for the Poisson random variable. Lines 12–13 create the functions `mEst()` and `mEst2()`. The first uses a single random number generator and the second uses two random number generators. Lines 15–21 define the functions `mGraph0()`, `mGraph1()` and `mGraph2()` for obtaining trajectories of the estimate for each λ in `lamRange`. The `mGraph0()` function uses a single RNG and no common random numbers, the function `mGraph1()` uses common random numbers reset each time on the same seed, and the `mGraph2()` function uses two RNGs. In line 23 `argMaxLam()` picks the maximum. The standard deviations are estimated in lines 25–27. The remainder of the code prints the output and creates Figure 10.17.

Appendix A

How-to in Julia

The code examples in this book are primarily designed to illustrate statistical concepts. However, they also have a secondary purpose. They serve a way of **learning how to use Julia by example**. Towards this end, the appendix links language features with specific code listings in the book. This appendix can be used on an ad-hoc basis to find code examples where you can see "how to" do specific things in Julia. Once you find the specific "how to" that you are looking for, you can refer to its associated code example, referenced via "⇒". This appendix is also available at

`https://statisticswithjulia.org/howto.html`.

The appendix is broken up into several subsections as follows. Basics (Section A.1) deal with basic language features. Text and I/O (Section A.2) deal with textual operations as well as input and output. Data Structures (Section A.3) deal with data structures and their use. This includes basic arrays as well as other structures. Data Frames, Time-Series, and Dates (Section A.4) deal with Data Frames and related objects for organizing heterogeneous data. Mathematics (Section A.5) covers various mathematical aspects of the language. Randomness, Statistics, and Machine Learning (Section A.6) deal with random number generation, elementary statistics, distributions, statistical inference, and machine learning. Graphics (Section A.7) deals with plotting, manipulation of figures, and animation.

A.1 Basics

Types

- ☐ Check the type of an object
  ⇒ **Listing 1.2**.
- ☐ Specify the type of an argument to a function
  ⇒ **Listing 1.11**.
- ☐ Specify the type of an array when initialized using `zeros()`
  ⇒ **Listing 1.8**.

Y. Nazarathy and H. Klok, *Statistics with Julia*, Springer Series in the Data Sciences,
https://doi.org/10.1007/978-3-030-70901-3

- ☐ Convert the type of a variable with `convert()`
  ⇒ **Listing 3.22**.
- ☐ Convert the type of a variable with a constructor like `Int()`
  ⇒ **Listing 1.11**.
- ☐ Use a 32 bit float instead of default 64 bit float with the `f0` float literal
  ⇒ **Listing 9.20**.
- ☐ Use big representation of numbers using `big()`
  ⇒ **Listing 2.3**.
- ☐ Check if a variable is immutable with `isimmutable()`
  ⇒ **Listing 4.1**.

Variables

- ☐ Modify a global variable inside a different scope by declaring `global`
  ⇒ **Listing 1.5**.
- ☐ Assign two values in a single statement (using an implicit tuple)
  ⇒ **Listing 1.6**.
- ☐ Copy a variable, array, or struct with `copy()`
  ⇒ **Listing 4.2**.
- ☐ Copy a variable, array, or struct with `deepcopy()`
  ⇒ **Listing 4.2**.

Conditionals and Logical Operations

- ☐ Use the conditional `if` statement
  ⇒ **Listing 1.6**.
- ☐ Use the conditional `else` statement
  ⇒ **Listing 1.11**.
- ☐ Use the conditional `elseif` statement
  ⇒ **Listing 1.17**.
- ☐ Use the shorthand conditional formatting operator `? :`
  ⇒ **Listing 2.5**.
- ☐ Carry out element-wise and using `.&`
  ⇒ **Listing 4.10**.
- ☐ Carry out element-wise negation using `.!`
  ⇒ **Listing 4.10**.
- ☐ Use logical or `||`
  ⇒ **Listing 1.13**.

- ☐ Use short circuit evaluation with logical and `&&`
  ⇒ **Listing 8.16**.

Loops

- ☐ Create a while loop
  ⇒ **Listing 1.11**.
- ☐ Loop over values in an array
  ⇒ **Listing 1.1**.
- ☐ Create nested for loops
  ⇒ **Listing 1.6**.
- ☐ Break out of a loop with `break`
  ⇒ **Listing 2.5**.
- ☐ Execute the next loop iteration from the top with `continue`
  ⇒ **Listing 2.5**.
- ☐ Loop over an enumeration of (Index, value) pairs created by `enumerate()`
  ⇒ **Listing 4.30**.

Functions

- ☐ Create a function
  ⇒ **Listing 1.6**.
- ☐ Create a one line function
  ⇒ **Listing 1.10**.
- ☐ Create function using `begin` and `end`
  ⇒ **Listing 3.32**.
- ☐ Create a function that returns a function
  ⇒ **Listing 1.7**.
- ☐ Pass functions as arguments to functions
  ⇒ **Listing 10.10**.
- ☐ Create a function with a multiple number of arguments
  ⇒ **Listing 1.7**.
- ☐ Use an anonymous function
  ⇒ **Listing 1.15**.
- ☐ Define a function inside another function
  ⇒ **Listing 2.4**.
- ☐ Create a function that returns a tuple
  ⇒ **Listing 7.10**.

- ☐ Setup default values to function arguments
  ⇒ **Listing 10.10**.

Other Basic Operations

- ☐ Check the running time of a block of code
  ⇒ **Listing 1.3**.
- ☐ Increment values using `+=`
  ⇒ **Listing 1.8**.
- ☐ Do element-wise comparisons such as, for example, using `.>`
  ⇒ **Listing 2.9**.
- ☐ Apply an element-wise computation to a tuple
  ⇒ **Listing 2.10**.
- ☐ Use the logical `xor()` function
  ⇒ **Listing 2.12**.
- ☐ Set a numerical value to be infinity with `Inf`
  ⇒ **Listing 3.6**.
- ☐ Include another block of Julia code using `include()`
  ⇒ **Listing 3.34**.
- ☐ Find the maximal value amongst several arguments using `max()`
  ⇒ **Listing 7.1**.
- ☐ Find the minimal value amongst several arguments using `min()`
  ⇒ **Listing 5.20**.

Metaprogramming

- ☐ Define a macro
  ⇒ **Listing 9.10**.

Interacting with Other Languages

- ☐ Copy data to the R environment with `@rput` from package `RCall`
  ⇒ **Listing 1.18**.
- ☐ Get data from the R environment with `@rget` from package `RCall`
  ⇒ **Listing 1.18**.
- ☐ Execute an R-language block with the command `R` from package `RCall`
  ⇒ **Listing 1.18**.
- ☐ Setup a Python object in Julia using `@pyimport` from package `PyCall`
  ⇒ **Listing 1.19**.

A.2 Text and I/O

Strings

- ☐ Split a string based on whitespace with `split()`
  ⇒ **Listing 1.9**.
- ☐ Use LaTeX formatting for strings
  ⇒ **Listing 2.4**.
- ☐ See if a string is a substring of another string with `occursin()`
  ⇒ **Listing 4.30**.
- ☐ Concatenate two strings using `*`
  ⇒ **Listing 1.9**.

Text Output

- ☐ Print text output including new lines, and tabs
  ⇒ **Listing 1.1**.
- ☐ Format variables within strings when printing
  ⇒ **Listing 2.1**.
- ☐ Display an expression to output using `display()`
  ⇒ **Listing 1.8**.
- ☐ Display an expression to output using `show(stdout,...)`
  ⇒ **Listing 9.5**.
- ☐ Present the value of an expression with `@show`
  ⇒ **Listing 4.1**.
- ☐ Display an information line with `@info`
  ⇒ **Listing 9.7**.
- ☐ Redirect the standard output to a file
  ⇒ **Listing 9.7**.

Reading and Writing From Files

- ☐ Open a file for writing with `open()`
  ⇒ **Listing 4.12**.
- ☐ Open a file for reading with `open()`
  ⇒ **Listing 4.30**.
- ☐ Write a string to a file with `write()`
  ⇒ **Listing 4.12**.

- ☐ Close a file after it was opened
  ⇒ **Listing 4.12**.
- ☐ Read from a file with `read()`
  ⇒ **Listing 4.12**.
- ☐ Find out the current working directory with `pwd()`
  ⇒ **Listing 4.31**.
- ☐ See the list of files in a directory with `readdir()`
  ⇒ **Listing 4.31**.
- ☐ See the directory of the current file with `@__DIR__`
  ⇒ **Listing 9.9**.
- ☐ Change the current directory with `cd()`
  ⇒ **Listing 9.9**.

CSV Files

- ☐ Read a CSV file to create a dataframe with a header
  ⇒ **Listing 4.3**.
- ☐ Read a CSV file to create a dataframe with without a header
  ⇒ **Listing 6.1**.
- ☐ Write to a CSV file with `CSV.write()`
  ⇒ **Listing 4.32**.

JSON

- ☐ Parse a JSON file with `JSON.parse()`
  ⇒ **Listing 1.9**.

BSON

- ☐ Write to a BSON file
  ⇒ **Listing 9.20**.
- ☐ Read from a BSON file
  ⇒ **Listing 9.19**.

HTTP Input

- ☐ Create an HTTP request
  ⇒ **Listing 1.9**.
- ☐ Convert binary data to a string
  ⇒ **Listing 1.9**.

A.3 Data Structures

Creating Arrays

- ☐ Create a range of numbers
  ⇒ **Listing 1.2**.
- ☐ Create an array of zero values with `zeros()`
  ⇒ **Listing 1.8**.
- ☐ Create an array of one values with `ones()`
  ⇒ **Listing 2.4**.
- ☐ Create an array with a repeated value using `fill()`
  ⇒ **Listing 7.10**.
- ☐ Create an array of strings
  ⇒ **Listing 1.1**.
- ☐ Create an array of numerical values based on a formula
  ⇒ **Listing 1.1**.
- ☐ Create an empty array of a given type
  ⇒ **Listing 1.3**.
- ☐ Create an array of character ranges
  ⇒ **Listing 2.2**.
- ☐ Create an array of tuples
  ⇒ **Listing 6.6**.
- ☐ Create an array of arrays
  ⇒ **Listing 1.15**.

Basic Array Operations

- ☐ Discover the `length()` of an array
  ⇒ **Listing 1.6**.
- ☐ Access elements of an array
  ⇒ **Listing 1.6**.
- ☐ Obtain the first and last elements of an array using `first()` and `last()`
  ⇒ **Listing 3.32**.
- ☐ Apply a function like `sqrt()` onto an array of numbers
  ⇒ **Listing 1.1**.
- ☐ Map a function onto an array with `map()`
  ⇒ **Listing 8.10**.

- ☐ Append with `push!()` to an array
  ⇒ **Listing 1.3**.
- ☐ Convert an object into an array with the `collect()` function
  ⇒ **Listing 1.9**.
- ☐ Pre-allocate an array of a given size
  ⇒ **Listing 1.16**.
- ☐ Delete an element from an array or collection with `deleteat!()`
  ⇒ **Listing 2.4**.
- ☐ Find the first element of an array matching a pattern with `findfirst()`
  ⇒ **Listing 2.4**.
- ☐ Append an array to an existing array with `append!()`
  ⇒ **Listing 2.5**.
- ☐ Sum up two equally size arrays element by element
  ⇒ **Listing 3.7**.
- ☐ Stick together several arrays into one array using `vcat()` and `...`
  ⇒ **Listing 7.9**.

Further Array Accessories

- ☐ Sum up values of an array with `sum()`
  ⇒ **Listing 1.7**.
- ☐ Search for a maximal index in an array using `findmax()`
  ⇒ **Listing 1.8**.
- ☐ Count the number of occurrence repetitions with the `count()` function
  ⇒ **Listing 1.9**.
- ☐ Sort an array using the `sort()` function
  ⇒ **Listing 1.9**.
- ☐ Filter an array based on a criterion using the `filter()` function
  ⇒ **Listing 1.15**.
- ☐ Find the maximal value in an array using `maximum()`
  ⇒ **Listing 2.3**.
- ☐ Count the number of occurrence repetitions with the `counts()` function from `StatsBase`
  ⇒ **Listing 2.3**.
- ☐ Reduce a collection to unique elements with `unique()`
  ⇒ **Listing 2.5**.
- ☐ Check if a an array is empty with `isempty()`
  ⇒ **Listing 3.6**.

- ☐ Find the minimal value in an array using `minimum()`
  ⇒ **Listing 3.6**.
- ☐ Accumulate values of an array with `accumulate()`
  ⇒ **Listing 3.30**.
- ☐ Sort an array in place using the `sort!()` function
  ⇒ **Listing 6.6**.

Sets

- ☐ Check if an element is an element of a set with `in()`
  ⇒ **Listing 2.6**.
- ☐ Check if a set is a subset of a set with `issubset()`
  ⇒ **Listing 2.6**.
- ☐ Obtain the set difference of two sets with `setdiff()`
  ⇒ **Listing 2.5**.
- ☐ Create a set from a range of numbers
  ⇒ **Listing 2.6**.
- ☐ Obtain the union of two sets with `union()`
  ⇒ **Listing 2.6**.
- ☐ Obtain the intersection of two sets with `intersect()`
  ⇒ **Listing 2.6**.

Matrices

- ☐ Obtain the dimensions of a matrix using `size()`
  ⇒ **Listing 10.6**.
- ☐ Define a matrix based on a set of values
  ⇒ **Listing 1.8**.
- ☐ Define a matrix based on side by side columns
  ⇒ **Listing 8.3**.
- ☐ Raise a matrix to a power
  ⇒ **Listing 1.8**.
- ☐ Access a given row of a matrix
  ⇒ **Listing 1.8**.
- ☐ Stick together two matrices using `vcat()`
  ⇒ **Listing 1.8**.
- ☐ Take a matrix and/or vector transpose
  ⇒ **Listing 1.8**.

- ☐ Modify the dimensions of a matrix with `reshape()`
  ⇒ **Listing 3.13**.
- ☐ Use an identity matrix with `I`
  ⇒ **Listing 1.8**.
- ☐ Setup a diagonal matrix with `diagm()` and a dictionary
  ⇒ **Listing 10.7**.
- ☐ Obtain the diagonal of a matrix with `diag()`
  ⇒ **Listing 10.7**.
- ☐ Create a matrix by sticking together column vectors
  ⇒ **Listing 1.8**.

Tensors

- ☐ Work with a tensor
  ⇒ **Listing 9.1**.

Dictionaries

- ☐ Access elements of a dictionary
  ⇒ **Listing 1.9**.
- ☐ Create a dictionary
  ⇒ **Listing 1.9**.

Graphs

- ☐ Create `Graph` objects from the package `LightGraphs`
  ⇒ **Listing 10.11**.
- ☐ Add edges to `Graph` objects using `add_edge!()`
  ⇒ **Listing 10.11**.
- ☐ Remove edges from `Graph` objects using `rem_edge!()`
  ⇒ **Listing 10.11**.

Other Data Structures

- ☐ Setup a `Queue` data structure from package `DataStructures`
  ⇒ **Listing 10.11**.
- ☐ Insert an element to a `Queue` data structure using `enqueue!()`
  ⇒ **Listing 10.11**.
- ☐ Remove an element from a `Queue` data structure using `dequeue!()`
  ⇒ **Listing 10.11**.

A.4 Data Frames, Time-Series, and Dates

Dataframe Basics

- ☐ Select certain rows of a `DataFrame`
  ⇒ **Listing 4.4**.

- ☐ Select certain columns of a `DataFrame`
  ⇒ **Listing 4.4**.

- ☐ Filter all rows of a `DataFrame` that using a boolean array
  ⇒ **Listing 4.4**.

- ☐ See if data all rows of a `DataFrame` that using a boolean array
  ⇒ **Listing 4.4**.

- ☐ Check for missing values using `dropmissing()`
  ⇒ **Listing 4.4**.

- ☐ Remove missing values using `dropmissing()`, removing any rows with missing values
  ⇒ **Listing 4.7**.

- ☐ Remove missing values using `skipmissing()` removing specific missing values
  ⇒ **Listing 4.7**.

- ☐ Sort a data frame based on a given column
  ⇒ **Listing 8.6**.

R Data Sets

- ☐ Obtain a data frame from `RDataSets` with `dataset()`
  ⇒ **Listing 8.8**.

Time-Series

- ☐ Create a time-series (`TimeArray`) object
  ⇒ **Listing 8.20**.

- ☐ Perform a moving average on a time-series
  ⇒ **Listing 8.20**.

- ☐ Compute the autocorrelation function estimate of a time-series
  ⇒ **Listing 8.20**.

Dates

- ☐ Parse dates using `Dates.DateFormat`
  ⇒ **Listing 8.20**.

- ☐ Obtain the day of week from a date object
  ⇒ **Listing 8.20**.
- ☐ Obtain the month from a date object
  ⇒ **Listing 4.19**.
- ☐ Obtain the year from a date object
  ⇒ **Listing 8.20**.
- ☐ Create a `Date` object based on day, month, and year
  ⇒ **Listing 4.19**.

A.5 Mathematics

Basic Math

- ☐ Compute the modulo (remainder) of integer division
  ⇒ **Listing 1.16**.
- ☐ Check if a number is even with `iseven()`
  ⇒ **Listing 2.1**.
- ☐ Take the product of elements of an array using `prod()`
  ⇒ **Listing 2.3**.
- ☐ Round numbers to a desired accuracy with `round()`
  ⇒ **Listing 2.8**.
- ☐ Compute the floor of value using `floor()`
  ⇒ **Listing 2.10**.
- ☐ Take the product of elements of an array using `*` with `...` as "product"
  ⇒ **Listing 5.18**.
- ☐ Represent π using the constant `pi`
  ⇒ **Listing 7.15**.
- ☐ Represent Euler's e using the constant `MathConstants.e`
  ⇒ **Listing 7.15**.

Math Functions

- ☐ Compute permutations using the `factorial()` function
  ⇒ **Listing 2.3**.
- ☐ Compute the absolute value with `abs()`
  ⇒ **Listing 2.3**.
- ☐ Compute the sign function with `sign()`
  ⇒ **Listing 8.7**.

- ☐ Create all the permutations of set with `permutations()` from `Combinatorics`
  ⇒ **Listing 2.5**.
- ☐ Calculate binomial coefficients with `binomial()`
  ⇒ **Listing 2.9**.
- ☐ Use mathematical special functions such as `zeta()`
  ⇒ **Listing 2.11**.
- ☐ Calculate the exponential function with `exp()`
  ⇒ **Listing 3.6**.
- ☐ Calculate the logarithm function with `log()`
  ⇒ **Listing 3.28**.
- ☐ Calculate trigonometric functions like `cos()`
  ⇒ **Listing 3.29**.
- ☐ Create all the combinations of set with `combinations()` from `Combinatorics`
  ⇒ **Listing 5.17**.

Linear Algebra

- ☐ Solve a system of equations using the backslash operator
  ⇒ **Listing 1.8**.
- ☐ Use `LinearAlgebra` functions such as `eigvecs()`
  ⇒ **Listing 1.8**.
- ☐ Carry out a Cholesky decomposition of a matrix
  ⇒ **Listing 3.32**.
- ☐ Calculate the inner product of a vector by multiplying the transpose by the vector
  ⇒ **Listing 3.33**.
- ☐ Calculate the inner product by using `dot()`
  ⇒ **Listing 8.3**.
- ☐ Compute a matrix exponential with `exp()`
  ⇒ **Listing 10.2**.
- ☐ Compute the inverse of a matrix with `inv()`
  ⇒ **Listing 10.6**.
- ☐ Compute the Moore-Penrose pseudo-inverse of a matrix with `pinv()`
  ⇒ **Listing 8.3**.
- ☐ Compute the L_p norm of a function with `norm()`
  ⇒ **Listing 8.2**.
- ☐ Compute the QR-factorization of a matrix with `qr()`
  ⇒ **Listing 8.3**.
- ☐ Compute the SVD-factorization of a matrix with `svd()`
  ⇒ **Listing 8.3**.

Numerical Math

- ☐ Find all roots of mathematical function using `find_zeros()`
  ⇒ **Listing 1.7**.
- ☐ Find a root of mathematical function using `find_zero()`
  ⇒ **Listing 5.10**.
- ☐ Carry out numerical integration using package `QuadGK`
  ⇒ **Listing 3.3**.
- ☐ Carry out numerical differentiation using package `Calculus`
  ⇒ **Listing 3.27**.
- ☐ Carry out numerical integration using package `HCubature`
  ⇒ **Listing 3.33**.
- ☐ Solve a system of equations numerically with `nlsolve()` from package `NLSolve`
  ⇒ **Listing 5.8**.
- ☐ Numerically solve a differential equations using the `DifferentialEquations` package
  ⇒ **Listing 10.2**.

A.6 Randomness, Statistics, and Machine Learning

Randomness

- ☐ Sample a random number using a prescribed weighting with `sample()`
  ⇒ **Listing 1.8**.
- ☐ Get a uniform random number in the range $[0, 1]$
  ⇒ **Listing 1.14**.
- ☐ Set the seed of the random number generator
  ⇒ **Listing 1.14**.
- ☐ Create a random permutation using `shuffle!()`
  ⇒ **Listing 2.8**.
- ☐ Generate a random number from a given range with `rand()`
  ⇒ **Listing 2.9**.
- ☐ Generate an array of random uniforms with `rand()`
  ⇒ **Listing 2.12**.
- ☐ Generate a random element from a set of values `rand()`
  ⇒ **Listing 2.13**.
- ☐ Generate an array of standard normal random variables with `randn()`
  ⇒ **Listing 9.6**.

☐ Generate an array of pseudorandom values from a given distribution
⇒ **Listing 3.4**.

☐ Generate multivariate normal random values via `MvNormal()`
⇒ **Listing 3.34**.

Distributions

☐ Creating a distribution object from the `Distributions` package
⇒ **Listing 3.4**.

☐ Evaluate the PDF (density) of a given distribution
⇒ **Listing 3.9**.

☐ Evaluate the CDF (cumulative probability) of a given distribution
⇒ **Listing 3.9**.

☐ Evaluate the CCDF (one minus cumulative probability) of a given distribution
⇒ **Listing 5.16**.

☐ Evaluate quantiles of a given distribution
⇒ **Listing 3.9**.

☐ Obtain the parameters of a given distribution
⇒ **Listing 3.10**.

☐ Evaluate the mean of a given distribution
⇒ **Listing 3.10**.

☐ Evaluate the median of a given distribution
⇒ **Listing 3.10**.

☐ Evaluate the variance of a given distribution
⇒ **Listing 3.10**.

☐ Evaluate the standard deviation of a given distribution
⇒ **Listing 3.10**.

☐ Evaluate the skewness of a given distribution
⇒ **Listing 3.10**.

☐ Evaluate the kurtosis of a given distribution
⇒ **Listing 3.10**.

☐ Evaluate the range of support of a given distribution
⇒ **Listing 3.10**.

☐ Evaluate the modes (or modes) of a given distribution
⇒ **Listing 3.10**.

Basic Statistics

- ☐ Calculate the arithmetic mean of an array
  ⇒ **Listing 1.3**.
- ☐ Calculate the geometric mean of an array
  ⇒ **Listing 4.11**.
- ☐ Calculate the harmonic mean of an array
  ⇒ **Listing 4.11**.
- ☐ Calculate a quantile
  ⇒ **Listing 1.3**.
- ☐ Calculate the sample variance of an array
  ⇒ **Listing 3.4**.
- ☐ Calculate the sample standard deviation of an array
  ⇒ **Listing 4.11**.
- ☐ Calculate the median of an array
  ⇒ **Listing 4.11**.
- ☐ Calculate the sample covariance from two arrays
  ⇒ **Listing 3.32**.
- ☐ Calculate the sample correlation from two arrays
  ⇒ **Listing 8.3**.
- ☐ Calculate the sample covariance matrix from a collection of arrays in a matrix
  ⇒ **Listing 4.13**.

Statistical Inference

- ☐ Use the `confint()` function on an hypothesis test
  ⇒ **Listing 6.1**.
- ☐ Carry out a one sample Z test using the `HypothesisTests` package
  ⇒ **Listing 7.1**.
- ☐ Carry out a one sample T-test using the `HypothesisTests` package
  ⇒ **Listing 7.1**.
- ☐ Carry out a two sample, equal variance, T-test using the `HypothesisTests` package
  ⇒ **Listing 7.6**.
- ☐ Carry out a two sample, non-equal variance, T-test using the `HypothesisTests` package
  ⇒ **Listing 7.7**.
- ☐ Carry out kernel density estimation using `kde()` from package `KernelDensity()`
  ⇒ **Listing 4.16**.
- ☐ Create and Empirical Cumulative Distribution Function using `ecdf()`
  ⇒ **Listing 4.17**.

Linear Models and Generalizations

☐ Create a formula for a (generalized) linear model with `@formula`
⇒ **Listing 8.4.**

☐ Fit a linear model with `fit()`, `lm()`, or `glm()`
⇒ **Listing 8.4.**

☐ Calculate the deviance of a linear model with `deviance()`
⇒ **Listing 8.4.**

☐ Get the standard error of a linear model with `stderror()`
⇒ **Listing 8.4.**

☐ Get the R^2 value of a linear model with `r2()`
⇒ **Listing 8.4.**

☐ Get the fit coefficients of a (generalized) linear model with `coef()`
⇒ **Listing 8.5.**

☐ Fit a logistic regression model using package `GLM`
⇒ **Listing 8.18.**

☐ Fit a GLMs with different link functions using package `GLM`
⇒ **Listing 8.19.**

☐ Fit a ridge regression model
⇒ **Listing 9.10.**

☐ Fit a LASSO model
⇒ **Listing 8.17.**

Supervised Classification

☐ Fit a Support Vector Machine (SVM) model
⇒ **Listing 9.7.**

☐ Fit a random forest model
⇒ **Listing 9.8.**

Unsupervised Learning

☐ Carry out k-means clustering
⇒ **Listing 9.12.**

☐ Carry out hierarchical clustering
⇒ **Listing 9.14.**

☐ Carry out principal component analysis
⇒ **Listing 9.15.**

Deep Learning using `Flux.jl`

- ☐ Use `Chain()` to construct a deep learning model
  ⇒ **Listing 9.1**.
- ☐ Use `loadparams!()` to fill the parameters of a model
  ⇒ **Listing 9.1**.
- ☐ Use `onecold()` to retrieve a label
  ⇒ **Listing 9.1**.
- ☐ Use `onehotbatch()` to create a one hot vector from a label
  ⇒ **Listing 9.9**.
- ☐ Train a model with `train!()`
  ⇒ **Listing 9.9**.
- ☐ Set callback functions to be called during training
  ⇒ **Listing 9.9**.
- ☐ Calculate a gradient using `gradient()`
  ⇒ **Listing 9.3**.
- ☐ Optimize using ADAM
  ⇒ **Listing 9.3**.
- ☐ Use the `softmax()` function
  ⇒ **Listing 9.6**.
- ☐ Use the `crossentropy()` function
  ⇒ **Listing 9.6**.
- ☐ Use dropout
  ⇒ **Listing 9.11**.
- ☐ Use the `@epochs` macro in training
  ⇒ **Listing 9.11**.
- ☐ Use batch normalization
  ⇒ **Listing 9.20**.

A.7 Graphics

The book contains 126 figures that are generated by the source code and presented in the book "as-is". These can also viewed on one page in this online gallery:

`https://statisticswithjulia.org/gallery.html`.

You can use this online gallery to then find an image that contains the type of features that you want to include in your figures and refer to the online source code.

Appendix B

Additional Julia Features

The code examples in the book use a variety of Julia language features. However these examples are purposefully short and do not exploit the full power of Julia as a general programming language. To fully explore Julia, you should be aware of other features not exploited in our code examples. Below is a list of key additional language features. For full documentation of each of these features, consult the official Julia documentation: `https://docs.julialang.org`.

Constant values: Global variables can be defined as constants with the `const` keyword.

Creation of packages: The nature of our code examples is illustrative, allowing them to run on a standard environment without requiring any special installation. However, once you create code that you wish to reuse, you may want to encapsulate it in a Julia package. This is done via the `generate` command in the package manager.

Documentation: It is easy to document a function by placing a markdown *docstring* above the function definition.

Environments: You may create different working environments where each environment has its own set of packages and versions installed. The key files that define an Environments are `Project.toml` and `Manifest.toml`.

Exception handling: Julia has built-in exception handling support. A key mechanism is the `try`, `throw` and `catch` construct, allowing functions to `throw()` an exceptions.

GPU support: There are plenty of mechanisms for integrating GPU support.

Interfaces: Much of Julia's power and extensibility comes from a collection of informal interfaces. By extending a few specific methods to work for a custom type, objects of that type not only receive those functionalities, but they are also able to be used in other methods that are written to generically build upon those behaviors. `Iterable` objects are particularly useful, and we have used them in several of our examples. In addition, there are methods for indexing, interfacing with *abstract arrays* and *strided arrays*, as well as ways of customizing broadcasting.

Low-level TCP/IP sockets: Julia supports TCP and UDP sockets via the `Sockets.jl` package, which is installed as part of Julia `Base`. The methods will be familiar to those who have used

Y. Nazarathy and H. Klok, *Statistics with Julia*, Springer Series in the Data Sciences,
https://doi.org/10.1007/978-3-030-70901-3

the Unix socket API. For example, `server = listen(ip"127.0.0.1", 2000)` will create a localhost socket listening on port 2000, `connect(ip"127.0.0.1", 2000)` will connect to the socket, and `close(server)` will disconnect the socket.

***Metaprogramming*:** Julia supports "Lisp like" metaprogramming, which makes it possible to create a program that generates some of its own code, and to create true Lisp-style macros which operate at the level of abstract syntax trees. As a brief example, `x = Meta.parse("1 + 2")` parses the argument string into an expression type object and stores it as `x`. This object can be inspected via `drop(x)` (note the `+` symbol, represented by `:+`). The expression can also be evaluated via `eval(x)`, which returns the numerical result of 3.

***Multiple dispatch*:** While some of our examples use multiple dispatch, we have not unleashed its full power. The ability to create different methods for the same function sits at the heart of the Julia programming paradigm and allows users to extend packages in a very productive manner.

***Modules*:** Modules in Julia are different workspaces that introduce a new global scope. They are delimited within `module Name ... end`, and they allow for the creation of top-level definitions (i.e. global variables) without worrying about naming conflicts when used together with other code. Within a module, you can control which names from other modules are visible via the `import` keyword, and which names are intended to be public via the `export` keyword.

***Parallel processing*:** Julia supports a variety of parallel computing constructs including green threads, tasks (known as coroutines in Julia), and communication channels between them. A basic macro is `@async` which when used via, for example, `@async myFunction()`, would execute `myFunction()` on its own thread.

***Profiling*:** There are a variety of profiling tools that allow to improve performance. One key aspect of performance that we have mostly ignored is *type stability*. See, for example, the `@code_warntype` macro which helps to check for type stability.

***Rational numbers*:** Julia supports rational numbers, along with arbitrary precision arithmetic. A rational number such as, for example, 2/3 is defined in Julia via `2//3`. Arithmetic with rational numbers is supported.

***Regular expressions*:** Julia supports regular expressions, allowing to match strings. For example, `occursin(r"^\s*(#)", "# a comment")` checks if # appears in the string and returns `true`.

***Running external programs*:** Julia borrows backtick notation for commands from the shell, Perl, and Ruby. However, the behavior of `'Hello world'` varies slightly from typical shell, Perl or Ruby behavior. In particular, the backticks create a `Cmd` object, which can be connected to other commands via pipes. In addition, Julia does not capture the output unless specifically arranged for it. And finally, the command is never run with a shell, but rather Julia parses the syntax directly, appropriately interpolating variables and splitting on words as the shell would, respecting shell quoting syntax. The command is run as Julia's immediate child process, using fork and exec calls. As a simple example, consider: `run(pipeline('echo world' & 'echo hello', 'sort'));`. This always outputs "Hello world" (here both echos are parsed to a singe UNIX pipe, and the other end of the pipe is read by the `sort` command).

Strings,: While some of our examples included string manipulation, we haven't delved into the subject deeply. Julia supports a variety of string operations for example, `occursin("world", "Hello, world")` returns `true`.

Unicode and character encoding: Most of the examples in the book were restricted to ASCII characters, however Julia fully supports Unicode. For example, `s = "\u2200 x \u2203 y"` yields the string $\forall\ x\ \exists\ y$.

User defined types: In addition to the basic types in the system (e.g. `Float64`), users and developers can create their own types via the `struct` keyword. In our examples, we have not created our own types, however many of the packages define new structs and in some examples of the book, we have referred directly to the fields of these structs. An example is in Listing 8.3 we use `F.Q` to refer to the field "`Q`" in the structure `F`.

Unit testing: As reusable code is developed it may also be helpful to create unit tests for verifying the validity of the code. This allows the code to be retested automatically every time it is modified or the system is upgraded. For this Julia supports unit testing via the `@test` macro, the `runtests()` function and other objects.

Appendix C

Additional Packages

We have used a variety of packages in this book. These were listed in Section 1.2. However there are many more. Currently, as of the time of writing, there are just over 4,000 registered packages in the Julia ecosystem. Many of these packages deal with numerical mathematics, scientific computing, or deal with some specific engineering or technical application. There are hundreds of packages associated with statistics and/or data-science, and we now provide an outline of some of the popular packages in this space that we didn't use in the code examples.

ARCH.jl is a package that allows for ARCH (Autoregressive Conditional Heteroskedasticity) modeling. ARCH models are a class of models designed to capture a features of financial returns data known as volatility clustering, i.e., the fact that large (in absolute value) returns tend to cluster together, such as during periods of financial turmoil, which then alternate with relatively calmer periods. This package provides efficient routines for simulating, estimating, and testing a variety of ARCH and GARCH models (with GARCH being Generalized ARCH).

AutoGrad.jl is an automatic differentiation package for Julia. It started as a port of the popular Python AutoGrad package and forms the foundation of the Knet Julia deep learning framework. AutoGrad can differentiate regular Julia code that includes loops, conditionals, helper functions, closures, etc. by keeping track of the primitive operations and using this execution trace to compute gradients. It uses reverse mode differentiation (a.k.a. back propagation) so it can efficiently handle functions with large array inputs and scalar outputs. It can compute gradients of gradients to handle higher order derivatives.

BayesNets.jl is a package implements Bayesian Networks for Julia through the introduction of the `BayesNet` type, which contains information on the directed acyclic graph, and a list of conditional probability distributions (CDP's). Several different CDP's are available. It allows us to use random sampling, weighted sampling, and Gibbs sampling for assignments. It supports inference methods for discrete Bayesian networks, parameter learning for an entire graph, structure learning, and the calculation of the Bayesian score for a discrete valued `BayesNet`, based purely on the structure and data. Visualization of network structures is also possible via integration with the `TikzGraphs.jl` package.

Bootstrap.jl is a package for statistical bootstrapping. It has several different resampling methods and also has functionality for confidence intervals.

Y. Nazarathy and H. Klok, *Statistics with Julia*, Springer Series in the Data Sciences,
https://doi.org/10.1007/978-3-030-70901-3

Convex.jl is a Julia package for Disciplined Convex Programming optimization problems. It can solve linear programs, mixed-integer linear programs, and DCP-compliant convex programs using a variety of solvers, including Mosek, Gurobi, ECOS, SCS, and GLPK, through the MathOptInterface interface. It also supports optimization with complex variables and coefficients.

CPLEX.jl is an unofficial interface to the IBM® ILOG® CPLEX® Optimization Studio. It provides an interface to the low-level C API, as well as an implementation of the solver-independent MathOptInterface.jl. You cannot use `CPLEX.jl` without having purchased and installed a copy of CPLEX Optimization Studio from IBM. This package is available free of charge and in no way replaces or alters any functionality of IBM's CPLEX Optimization Studio product.

CUDAnative.jl is a part of the JuliaGPU collection of packages, and provides support for compiling and executing native Julia kernels on CUDA hardware.

DataFramesMeta.jl is a package that provides a series of metaprogramming tools for `DataFrames.jl`, which improve performance and provide a more convenient syntax.

Distances.jl is a package for evaluating distances (metrics) between vectors. It also provides optimized functions to compute column-wise and pairwise distances. This is often substantially faster than a straightforward loop implementation.

FastGaussQuadrature.jl is a Julia package to compute n-point Gauss quadrature nodes and weights to 16 digit accuracy in $O(n)$ time. It includes several different algorithms, including `gausschebyshev()`, `gausslegendre()`, `gaussjacobi()`, `gaussradau()`, `gausslobatto()`, `gausslaguerre()`, and `gausshermite()`.

ForwardDiff.jl is a part of the JuliaDiff family and is a package that implements methods to take derivatives, gradients, Jacobians, Hessians, and higher order derivatives of native Julia functions (or objects) using forward mode automatic differentiation (AD).

GadFly.jl is a plotting and visualization system written in Julia and largely based on ggplot2 for R. it supports a large number of common plot types and composition techniques, along with interactive features, such as panning and zooming, which are powered by Snap.svg. It renders publication quality graphics in a variety of formats including SVG, PNG, Postscript, and PDF, and has tight integration with `DataFrames.jl`.

GLMNet.jl is a package that acts as a wrapper for Fortran code from glmnet. Also see `Lasso.jl` which is a pure Julia implementation of the glmnet coordinate descent algorithm that often achieves better performance.

Gurobi.jl is a wrapper for the Gurobi solver (through its C interface). Gurobi is a commercial optimization solver for a variety of mathematical programming problems, including linear programming (LP), quadratic programming (QP), quadratically constrained programming (QCP), mixed-integer linear programming (MILP), mixed-integer quadratic programming (MIQP), and mixed-integer quadratically constrained programming (MIQCP). It is highly recommend that the Gurobi.jl package is used with higher level packages such as `JuMP.jl` or `MathOptInterface.jl`.

Interpolations.jl is a package for fast, continuous interpolations of discrete datasets in Julia.

JuliaDB.jl is a package designed for working with large multidimensional datasets of any size. Using an efficient binary format, it allows data to be loaded and saved and efficiently, and quickly recalled later. It is versatile, and allows for fast indexing, filtering, and sorting operations, along with performing regressions. It comes with built-in distributed parallelism and aims to tie together the most useful data manipulation libraries for a comfortable experience.

JuliaDBMeta.jl is a set of macros that aim to simplify data manipulation with `JuliaDB.jl`.

JuMP.jl is a domain-specific modeling language for mathematical optimization embedded in Julia. It supports a number of open-source and commercial solvers (Artelys Knitro, BARON, Bonmin, Cbc, Clp, Couenne, CPLEX, ECOS, FICO Xpress, GLPK, Gurobi, Ipopt, MOSEK, NLopt, SCS) for a variety of problem classes, including linear programming, (mixed) integer programming, second-order conic programming, semi-definite programming, and non-linear programming (convex and non-convex). JuMP makes it easy to specify and solve optimization problems without expert knowledge, yet at the same time allows experts to implement advanced algorithmic techniques such as exploiting efficient hot-starts in linear programming or using callbacks to interact with branch-and-bound solvers. It is part of the JuliaOpt collection of packages.

Loess.jl is a pure Julia implementation of local polynomial regression (i.e. locally estimated scatterplot smoothing, known as LOESS).

LsqFit.jl is a package providing a small library of basic least-squares fitting in pure Julia. The basic functionality was originally in `Optim.jl`, before being separated. At this time, `LsqFit.jl` only utilizes the Levenberg-Marquardt algorithm for non-linear fitting.

Mamba.jl provides a pure Julia interface to implement and apply Markov chain Monte Carlo (MCMC) methods for Bayesian analysis. It provides a framework for the specification of hierarchical models, allows for block-updating of parameters, with samplers either defined by the user, or available from other packages, and allows for the execution of sampling schemes, and for posterior inference. It is intended to give users access to all levels of the design and implementation of MCMC simulators to particularly aid in the development of new methods. Several software options are available for MCMC sampling of Bayesian models. Individuals who are primarily interested in data analysis, unconcerned with the details of MCMC, and have models that can be fit in JAGS, Stan, or OpenBUGS are encouraged to use those programs. Mamba is intended for individuals who wish to have access to lower level MCMC tools, are knowledgeable of MCMC methodologies, and have experience, or wish to gain experience, with their application. The package also provides stand-alone convergence diagnostics and posterior inference tools, which are essential for the analysis of MCMC output regardless of the software used to generate it.

MLBase.jl aims to provide a collection of useful tools to support machine learning programs, including: Data manipulation and preprocessing, Score-based classification, Performance evaluation (e.g. evaluating ROC), Cross-validation, and Model tuning (i.e. searching for the best settings of parameters).

MLJ.jl stands for "Machine Learning Julia" and is a complete machine learning framework that encapsulates many other packages.

MXNet.jl is now part of the Apache MXNet project. It brings flexible and efficient GPU computing and state-of-art deep learning to Julia. Some of its features include efficient tensor/matrix computation across multiple devices, including multiple CPUs, GPUs and distributed server nodes, and flexible symbolic manipulation to composite and construction of state-of-the-art deep learning models.

NLopt.jl provides a Julia interface to the open-source NLopt library for non-linear optimization. NLopt provides a common interface for many different optimization algorithms, including, local and global optimization, algorithms that use function values only (no derivative) and those that exploit user-supplied gradients, as well as algorithms for unconstrained optimization, bound-constrained optimization, and general non-linear inequality/equality constraints. It can be used interchangeably with outer optimization packages such as those from JuMP.

OnlineStats.jl is a package which provides on-line algorithms for statistics, models, and data visualization. On-line algorithms are well suited for streaming data or when data is too large to hold in memory. Observations are processed one at a time and all algorithms use $O(1)$ memory.

Optim.jl is a package that is part of the JuliaNLSolvers family and provides support for univariate and multivariate optimization through various kinds of optimization functions. Since `Optim.jl` is written in Julia, it has several advantages: it removes the need for dependencies that other non-Julia solvers may need, reduces the assumptions the user must make, and allows for user controlled choices through Julia's multiple dispatch rather than relying on predefined choices made by the package developers. As it is written in Julia, it also has access to the automatic differentiation features via packages in the JuliaDiff family.

Plotly.jl is a Julia interface to the plot.ly plotting library and cloud services and can be used as one of the plotting backends of the `Plots.jl` package.

POMDPs.jl is part of the JuliaPOMDP collection of packages and aims to provide an interface for defining, solving, and simulating discrete and continuous, fully and partially observable Markov decision processes. Note that `POMDP.jl` only contains the interface for communicating MDP and POMDP problem definitions. For a full list of supporting packages and tools to be used along with POMDPs.jl, see JuliaPOMDP. These additional packages include simulators, policies, several different MDP and POMDP solvers, along with other tools.

ProgressMeter.jl is a package that enables the use of a progress meter for long-running Julia operations.

Reinforce.jl is an interface for *reinforcement learning*. It is intended to connect modular environments, policies, and solvers with a simple interface. Two packages build on `Reinforce.jl`: `AtariAlgos.jl`, which is an Arcade Learning Environment (ALE) wrapped as `Reinforce.jl` environment, and the `OpenAIGym.jl`, which wraps the open-source Python library gym, released by OpenAI.

ReinforcementLearning.jl is a reinforcement learning package. It features many different learning methods and has support for many different learning environments, including a wrapper for the Atari ArcadeLearningEnvironment, and the OpenAI Gym environment, along with others.

ScikitLearn.jl implements the popular scikit-learn interface and algorithms in Julia. It supports both models from the Julia ecosystem and those of the scikit-learn library via PyCall.jl.

Its main features include approximately 150 Julia and Python models accessed through a uniform interface, Pipelines and FeatureUnions, Cross-validation, hyper-parameter tuning, and DataFrames support.

SimJulia.jl is a discrete event process oriented simulation framework written in Julia. It is inspired by the Python SimPy library.

StatsFuns.jl is a package that provides a collection of mathematical constants and numerical functions for statistical computing, including various distribution related functions.

StatsKit.jl is a convenience meta-package which allows loading of essential packages for statistics in one command. It currently loads the following statistics packages: `Bootstrap`, `CategoricalArrays`, `Clustering`, `CSV`, `DataFrames`, `Distances`, `Distributions`, `GLM`, `HypothesisTests`,
`KernelDensity`, `Loess`, `MultivariateStatsStatsBase`, and `TimeSeries`.

Tables.jl combines the best of the `DataStreams.jl` and `Queryverse.jl` packages to provide a set of fast and powerful interface functions for working with various kinds of table-like data structures through predictable access patters.

TensorFlow.jl acts as a wrapper around the popular *TensorFlow* machine learning framework from Google. It enables both input data parsing and post-processing of results to be done quickly via Julia's JIT compilation. It also provides the ability to specify models using native Julia looking code, and through Julia metaprogramming, simplifies graph construction, and reduces code repetition.

TensorOperations.jl is a package that enables fast tensor operations using a convenient Einstein index notation.

XGBoost.jl is a Julia interface of eXtreme Gradient Boosting, or XGBoost. It is an efficient and scalable implementation of gradient boosting framework. It includes efficient linear model solver and tree learning algorithms. The library is parallelized using OpenMP, and it can be more than 10 times faster than some existing gradient boosting packages. It supports various objective functions, including regression, classification, and ranking. The package is also made to be extensible, so that users are also allowed to define their own objectives easily. It is part of the Distributed (Deep) Machine Learning Community (dmlc).

Organizations

Much of the Julia package ecosystem on Github is grouped into organizations (or collections) of packages, often based on specific domains of knowledge. Currently there are over 35 different Julia organizations, and some of the more relevant ones for the statistician, data scientist, or machine learning practitioner are listed below.

JuliaCloud is a collection of Julia packages for working with cloud services.

JuliaDiff an informal organization which aims to unify and document packages written in Julia for evaluating derivatives. The technical features of Julia, namely, multiple dispatch, source code via reflection, JIT compilation, and first-class access to expression parsing make implementing and using techniques from automatic differentiation easier than ever before. Packages hosted under the JuliaDiff organization follow the same guidelines as for JuliaOpt; namely, they should be actively maintained, well documented, and have a basic testing suite.

JuliaData is a collection of Julia packages for data manipulation, storage, and I/O.

JuliaDiff is an informal organization for solving differential equations in Julia.

JuliaDiffEq is an organization for unifying the packages for solving differential equations in Julia, and includes packages such as DifferentialEquations.jl.

JuliaGeometry is a collection of packages that focus on computational geometry with Julia.

JuliaGPU contains a collection of Julia packages that support GPU computation.

JuliaGraphs is a collection of Julia packages for graph modeling and analysis.

JuliaImages is a collection of packages specifically focused on image processing and has many useful algorithms. Its main package is Images.jl.

JuliaInterop is a collection of packages that contains many different packages that enable interoperability between Julia and other various languages, such as C++, Matlab, and others.

JuliaMath contains a series of mathematics related packages.

JuliaML contains a series of Julia packages for Machine Learning.

JuliaOpt is a collection of optimization-related packages. Its purpose is to facilitate collaboration among developers of a tightly integrated set of packages for mathematical optimization.

JuliaParallel is a collection of packages containing various models for parallel programming in Julia.

JuliaPOMDP is a collection of POMDP packages for Julia.

JuliaPlots is a collection of data visualization plotting packages for Julia.

JuliaPy is a collection of packages that connect Julia and Python.

JuliaStats is the main collection of statistics and Machine Learning packages.

JuliaTeX is a collection of packages for TeX typesetting and rendering in Julia.

JuliaText is a JuliaLang Organization for Natural Language Processing, (textual) Information Retrieval, and Computational Linguistics

Junolab is the landing page for the Juno IDE (integrated desktop environment). Juno is a free environment for the Julia language, is built on the Atom editor, and is a powerful development tool. The `Juno.jl` package defines Juno's frontend API.

Bibliography

[AM10]. Albertos, P., Mareels, I.: Feedback and Control for Everyone. Springer Science & Business Media, Berlin (2010)

[AM07]. Antsaklis, P.J., Michel, A.N.: A Linear Systems Primer. Springer, Berlin (2007)

[BCP95]. Ball, M.O., Colbourn, C.J., Provan, J.S.: Network reliability. Handbooks in Operations Research and Management Science, vol. 7, pp. 673–762 (1995)

[BPRS18]. Baydin, A.G., Pearlmutter, B.A., Radul, A.A., Siskind, J.M.: Automatic differentiation in machine learning: a survey. J. Mach. Lear. Res. **18**(1), 5595–5637

[B11]. Bertsekas, D.P.: Dynamic Programming and Optimal Control, vol. II, 3rd edn. Athena Scientific, Belmont (2011)

[BGK10]. Botev, Z.I., Grotowski, J.F., Kroese, D.P.: Kernel density estimation via diffusion. Ann. Stat. **38**(5), 2916–2957 (2010)

[BL18]. Boyd, S., Vandenberghe, L.: Introduction to Applied Linear Algebra - Vectors, Matrices, and Least Squares. Cambridge University Press, Cambridge (2018)

[BAABLPP19]. Besançon, M., Anthoff, D., Arslan, A., Byrne, S., Lin, D., Papamarkou, T., Pearson, J.: Distributions.jl: Definition and Modeling of Probability Distributions in the JuliaStats Ecosystem (2019). arXiv:1907.08611

[B10]. Bulmer, M.: A portable introduction to data analysis. Publish on Demand Centre. University of Queensland Press (2010)

[CB01]. Casella, G., Berger, R.L.: Statistical Inference, 2nd edn. Cengage Learning, Boston (2001)

[CC11]. Chang, C.C., Lin, C.: LIBSVM: a library for support vector machines. ACM Trans. Intell. Syst. Technol. **2**(3) (2011)

[DG01]. Daley, D.J., Gani, J.: Epidemic Modelling: An Introduction. Cambridge University Press, Cambridge

[DS11]. DeGroot, M.H., Schervish, M.J.: Probability and Statistics, 4th edn. Pearson, London (2011)

[DL19]. Downey, A., Lauwens, B.: Think Julia - How to Think Like a Computer Scientist. O'Reilly Media, Newton (2019)

Y. Nazarathy and H. Klok, *Statistics with Julia*, Springer Series in the Data Sciences,
https://doi.org/10.1007/978-3-030-70901-3

[EH16]. Efron, B., Hastie, T.: Computer Age Statistical Inference. Cambridge University Press, Cambridge (2016)

[F68]. Feller, W.: An Introduction to Probability Theory and Its Applications, vol. 1, 3rd edn. Wiley, New York (1968)

[HTF01]. Friedman, J., Hastie, T., Tibshirani, R.: The Elements of Statistical Learning. Springer Series in Statistics. New York (2001)

[GBC16]. Goodfellow, I., Bengio, Y., Aaron, C.: Deep Learning. MIT Press, Cambridge (2016)

[G14]. Goodfellow, I., Jean, P., Mirza, M., Xu, B., Warde-Farley, D., Ozair, S., Courville, A., Bengio, Y.: Generative adversarial nets. In: Advances in Neural Information Processing Systems, pp 2672–2680 (2014)

[HB13]. Harchol-Balter, M.: Performance Modeling and Design of Computer Systems: Queueing Theory in Action. Cambridge University Press, Cambridge (2013)

[I18]. Innes, M.: Flux: Elegant machine learning with julia. J. Open Source Softw. (2018)

[I18b]. Innes, M.: Flux: Don't unroll adjoint: differentiating SSA-Form programs. arXiv:1810.07951

[KM1927]. Kermack, W.O., McKendrick, A.G.: A contribution to the mathematical theory of epidemics. In: Proceedings of the Royal Society of London. Series A, Containing Papers of a Mathematical and Physical Character (1927)

[KS18]. Kamiński, B., Szufel, P.: Julia 1.0 Programming Cookbook. Packt Publishing (2019)

[KB14]. Kingma, D.P., Jimmy, B.: Adam: a method for stochastic optimization (2014). arXiv:1412.6980

[K12]. Klebaner, F.C.: Introduction to Stochastic Calculus with Applications. World Scientific Publishing Company, Singapore (2012)

[K18]. Kwon, C.: Julia Programming for Operations Research, 2nd edn. CreateSpace Independent Publishing Platform, USA (2016), 3rd edn. (2018)

[KW19]. Kochenderfer, M., Wheeler, T.: Algorithms for Optimization. MIT Press, Cambridge (2019)

[KBTV19]. Kroese, D.P., Botev, Z.I., Taimre, T., Vaisman, R.: Data Science and Machine Learning: Mathematical and Statistical Methods. CRC Press, Boca Raton (2019)

[KTB11]. Kroese, D.P., Taimre, T., Botev, Z.I.: Handbook of Monte Carlo Methods. Wiley, New York (2011)

[LDL13]. Lafaye de Micheaux, P., Drouilhet, R., Liquet, B.: The R Software: Fundamentals of Programming and Statistical Analysis (Statistics and Computing). Springer, Berlin (2013)

[LG08]. Leon-Garcia, A.: Probability, Statistics and Random Processes for Electrical Engineering, 3rd edn. Pearson, London (2008)

[L09]. Littman, M.L.: A tutorial on partially observable Markov decision processes. J. Math. Psychol. (2009)

[L79]. Luenberger, D.G.: Introduction to Dynamic Systems; Theory, Models, and Applications. Addison-Wesley Reading, Boston (1979)

[L42]. Lukacs, E.: A characterization of the normal distribution. Ann. Math. Stat. **13**(1), 91–93 (1942)

[M07]. Mandjes, M.: Large Deviations for Gaussian Queues: Modelling Communication Networks. Wiley, New York (2007)

[MP18]. McNicholas, P.D., Tait, P.A.: Data Science with Julia. Chapman and Hall/CRC, Boca Raton (2019)

[M17]. Montgomery, D.C.: Design and Analysis of Experiments. Wiley, New York (2017)

[MR13]. Montgomery, D.C., Runger, G.C.: Applied Statistics and Probability for Engineers, 6th edn. Wiley, New York (2013)

[MG16]. Müller, A.C., Guido, S.: Introduction to Machine Learning with Python: A Guide for Data Scientists. O'Reilly Media, Inc., Newton (2016)

[N97]. Norris, J.R.: Markov Chains. Cambridge Series in Statistical and Probabilistic Mathematics. Cambridge University Press, Cambridge (1997)

[P14]. Puterman, M.: Markov Decision Processes: Discrete Stochastic Dynamic Programming. Wiley, New York (2014)

[R07]. Robert, C.: The Bayesian Choice: From Decision-theoretic Foundations to Computational Implementation. Springer Science & Business Media, Berlin (2007)

[R06]. Rosenthal, J.S.: A First Look at Rigorous Probability Theory. World Scientific Publishing Company, Singapore (2006)

[SBK75]. Selvin, S.: A problem in probability (letter to the editor). Amer. Stat. **29**(1), 67–71 (1975)

[SS17]. Shumway, R.H., Stoffer, D.S.: Time Series Analysis and Its Applications: With R Examples. Springer, Berlin (2017)

[S18]. Strang, G.: Linear Algebra and Learning from Data. Wellesley-Cambridge Press, Newton (2018)

[V12]. Van Buuren, S.: Flexible Imputation of Missing Data. Chapman and Hall/CRC, Boca Raton (2018)

[V16]. Voulgaris, Z.: Julia for Data Science. O'Reily, Newton (2018)

[V20]. Voulgaris, Z.: Julia for Machine Learning. Technics (2020)

List of Julia Code

Y. Nazarathy and H. Klok, *Statistics with Julia*, Springer Series in the Data Sciences,
https://doi.org/10.1007/978-3-030-70901-3

Index

Y. Nazarathy and H. Klok, *Statistics with Julia*, Springer Series in the Data Sciences,
https://doi.org/10.1007/978-3-030-70901-3

G

H

I

Printed in the United States
by Baker & Taylor Publisher Services